Materials and the Environment

Materials and the Environment
Eco-Informed Material Choice
Second Edition

Michael F. Ashby

ELSEVIER

AMSTERDAM • BOSTON • HEIDELBERG • LONDON
NEW YORK • OXFORD • PARIS • SAN DIEGO
SAN FRANCISCO • SINGAPORE • SYDNEY • TOKYO
Butterworth-Heinemann is an imprint of Elsevier

Butterworth-Heinemann is an imprint of Elsevier
225 Wyman Street, Waltham, MA 02451, USA
The Boulevard, Langford Lane, Kidlington, Oxford, OX5 1GB, UK

Notices
Knowledge and best practice in this field are constantly changing. As new research and experience broaden our understanding, changes in research methods, professional practices, or medical treatment may become necessary.

Practitioners and researchers must always rely on their own experience and knowledge in evaluating and using any information, methods, compounds, or experiments described herein. In using such information or methods they should be mindful of their own safety and the safety of others, including parties for whom they have a professional responsibility.

To the fullest extent of the law, neither the Publisher nor the authors, contributors, or editors, assume any liability for any injury and/or damage to persons or property as a matter of products liability, negligence or otherwise, or from any use or operation of any methods, products, instructions, or ideas contained in the material herein.

Library of Congress Cataloging-in-Publication Data
Ashby, M. F.
  Materials and the environment : eco-informed material choice / Michael F. Ashby. – 2nd ed.
    p. cm.
  ISBN 978-0-12-385971-6
  1. Materials–Environmental aspects. I. Title.

  TA403.6.A739 2012
  620.1'10286–dc23

  2011046988

British Library Cataloguing-in-Publication Data
A catalogue record for this book is available from the British Library.

For information on all Butterworth–Heinemann publications
visit our Web site at www.elsevierdirect.com

Printed in the United States of America
12 13 14 15 16 17   10 9 8 7 6 5 4 3 2 1

Contents

Preface and acknowledgments

The environment is a system. Human society, too, is a system. The systems co-exist and interact, weakly in some ways, strongly in others. When two already complex systems interact, the consequences can be hard to predict. One consequence has been the damaging impacts of industrial society on the environment and the ecosystem in which we live and on which we depend. Some of these impacts have been evident for more than a century, prompting remedial action that, in many cases, has been successful. Others are emerging only now, and among them, one of the most unexpected is our influence on global climate that, if allowed to continue, could become very damaging. These and many other eco-concerns derive from the changing ways in which we use energy and materials. If we are to manage both responsibly we must first understand the origins, the scale, and the consequences of the ways we use them now. And that needs *facts*.

The book. This text is a response. It aims to cut through some of the oversimplification and misinformation that is all too obvious in much discussion about the environment. It explains the ways in which we depend on and use materials and the consequences these have. It introduces methods for thinking about and designing with materials when one of the objectives is to minimize *environmental impact*, one that is often in conflict with others, particularly that of minimizing *cost*. It does not aim to provide ultimate solutions—that is a task for future scientists, engineers, designers, and politicians. Rather it is an attempt to provide perspective, background, methods, and data—a tool-box so to speak—to introduce one of the central issues of environmental concerns, that surrounding the use of materials. It provides tools and data that equip you to form your own judgments.

The text is written primarily for students of Engineering and Materials Science in any one of the four years of a typical undergraduate program. Chapters 1 to 14 develop the background and tools required for the materials scientist or engineer to analyze and respond to environmental imperatives. Chapter 15 is a collection of profiles of materials presenting the data needed for analysis. The two together allow case studies to be developed and provide resources on which students can draw to tackle the exercises at the end of each chapter (for which a solution manual is available) and to explore material-related eco-issues of their own finding.

To understand where we now are, it helps to look back over how we got here. Chapter 1 gives a history of our increasing dependence on materials and energy. Most materials are drawn from nonrenewable resources inherited from the formation of the planet or from geological and biological eras in its history. Like any inheritance, we have a responsibility to pass them on to further generations in a state that enables them to meet their aspirations as we now do ours. The volume of these resources is enormous, but so too is the rate at which we are using them. A proper perspective here needs both explanation and modeling. That is what Chapter 2 does.

Products, like plants and animals, have a life cycle, one with a number of phases starting with the extraction and synthesis of raw materials ("birth"), continuing with their manufacture into products, which are then transported, used ("maturity"), and at the end of life, sent to a landfill or to a recycling facility ("death"). Almost always, one phase of life consumes more resources and generates more emissions than all the others put together. The first job is to identify which one. Life-cycle assessment (LCA) seeks to do this, but there are problems: as currently practiced, life-cycle assessment is expensive, slow, and delivers outputs that are unhelpful for engineering design. One way to overcome them is to focus on the main culprits: one resource—*energy*—and one emission—*carbon dioxide, CO_2*. Materials have an *embodied energy* (the energy it takes to create them) and a *carbon footprint* (the CO_2 that creating them releases). The other phases of life and materials play a central role in these also. *Heating and cooling* and *transportation*, for instance, are among the most energy-gobbling and carbon-belching activities of an industrial society; the right choice of materials can minimize their appetite for both. This line of thinking is developed in Chapters 3 and 4, from which a strategy emerges that forms the structure of the rest of the book

Governments respond to environmental concerns in a number of ways applied through a combination of sticks and carrots, or, as they would put it, *command and control* methods and methods exploiting *market instruments*. This results in steadily growing volumes of legislation and regulation that, like it or not, require compliance. They are reviewed in Chapter 5.

As engineers and scientists, our first responsibility is to use our particular skills to guide design decisions that minimize or eliminate adverse eco-impact. Properly informed materials selection is a central aspect of this, and that needs data for the material attributes that bear most directly on environmental questions. Some, like *embodied energy* and *carbon footprint*, recycle *fraction* and *toxicity* have obvious eco-connections. But more often it is not these but mechanical, thermal, and electrical properties that have the greatest role in design to minimize eco-impact. The data sheets of Chapter 15 provide all of these. Data can be deadly dull. It can be brought to life (a little) by good visual presentations. Chapter 6 introduces the material attributes that are central to what follows and displays them in ways that give a visual overview.

Now to *design*. Designers have much on their minds; they can't wait for (or afford) a full LCA to decide between alternative concepts and ways of implementing them. What they need is an *eco-audit*—a fast assessment of product life phase-by-phase, and the ability to conduct rapid "what if?" studies to compare alternatives. Chapter 7 introduces audit methods illustrated by case studies in Chapter 8.

The audit points to the phase of life of most concern. What can be done about it? In particular, what material-related decisions can be made to minimize its eco-impact? Material selection methods are the subject of Chapter 9. They form a central part of the strategy that emerged from Chapter 3. It is important to see them in action. Chapter 10 presents case studies of progressive depth to illustrate ways to use them. The exercises suggest more.

Up to this point the book builds on established, well-tried methods of analysis and response, ones that form part of, or are easily accessible to, anyone with a background in engineering science. They provide essential background for an engineering-based approach to address environmental concerns, and they provide an essential underpinning for studies of broader issues. Among these are questions of *sustainability*, the subject of Chapter 11. Central to sustainability is reliable provision of *low-carbon power*, the subject of Chapter 12. Ultimately, sustainability requires that we maximize *material efficiency*, explored in Chapter 13. Finally we examine forces for change and responses to them under the heading *future options* (Chapter 14).

Chapter 15, forming the second part of the book, is a collection of 63 one-page data sheets for engineering metals, polymers, ceramics, composites, and natural materials. Each has a description and an image, a table of mechanical, thermal, and electrical properties, and a table of properties related to environmental issues. They provide a resource that is drawn upon in the main text, enabling its exercises and allowing the methods of the book to be applied elsewhere.

The CES software.[1] The audit and selection tools developed in the text are implemented in the CES Edu software, a powerful materials-information system that is widely used both for teaching and design. The book is self-contained—access to the software is not a prerequisite. The software is a useful adjunct to the text, enhancing the learning experience and providing access to data for a much wider range of materials. It allows realistic selection studies that properly combine multiple constraints and the construction of trade-off plots in the same format as those of the text.

What's new in the second edition? The basic structure of the book remains the same, but within this structure there are many changes, partly in response to feedback from users of the first edition, partly necessitated by the rapid evolution of the study of materials and the environment. Here is a summary.

- All chapters have been edited, expanded, and brought up to date.
- Worked in-text examples illustrate reasoning or the use of equations.
- The Exercises at the end of each chapter have been greatly expanded (a solution manual is available from the Publisher).

---

[1]Granta Design, 300 Rustat House, 62 Clifton Road, Cambridge CB1 7EG, UK.
www.grantadesign.com

- News-clips are incorporated into all the chapters. These are cuttings from the world press (almost all appearing in 2011) that help place materials issues into a broader context.
- A new chapter, "Case studies: eco-audits" (Chapter 8), illustrates the rapid audit method.
- A new chapter, "Materials for low-carbon power (Chapter 12)," is really an extended case study, examining the consequences on materials supply of a major shift from fossil-fuel–based power to power from renewables.
- A new chapter explores material efficiency (Chapter 13). This means designing and managing manufacturing to provide the services we need with the least production of materials.
- The datasheets of Chapter 15 have been updated and expanded to include natural and man-made fibers.
- "Further reading" sections at the end of each chapter have been brought up to date with 2009, 2010, and 2011 citations.

Feedback from readers has been a great help in guiding the development of the second edition. Criticisms and suggestions from readers of this second edition will be very welcome.

Acknowledgments

No book of this sort is possible without advice, constructive criticism, and ideas from others. Numerous colleagues have been generous with their time and thoughts. I would particularly like to recognize the suggestions and stimulus, directly or indirectly, made by Dr. Julian Allwood, Prof. David Cebon, Dr. Patrick Coulter, Dr. Jon Cullen, Prof. David MacKay, and Dr. Hugh Shercliff, all of Cambridge University; Professor Yves Bréchet of the University of Grenoble; Professor Ulrike Wegst of Dartmouth College; Professor John Abelson of the University of Michigan; Dr. Deborah Andrews of London South Bank University; and Julia Attwood, Fred Lord, and James Polyblank, at present research students at Cambridge University. Equally valuable has been the contribution of the team at Granta Design, Cambridge, responsible for the development of the CES software that has been used to make many of the charts that are a feature of this book. My special thanks are due to Heather Tighe of Elsevier for her patient, detailed editing of my imperfect text.

CHAPTER 1

Introduction: material dependence

CONTENTS

1.1 Introduction and synopsis

This book is about *materials and the environment*: the eco-aspects of their production, their use, and their disposal at end of life. It is also about ways to choose and design with them in ways that minimize the impact they have on the environment. Environmental harm caused by industrialization is not new. The manufacturing midlands of 18th century England acquired the name "Black Country" with good reason; and to evoke the atmosphere of 19th century London, Sherlock Holmes movies show scenes of fog—known as "pea-soupers"—swirling round the gas lamps of Baker Street. These were localized problems that have, today, largely been corrected. The

Renewable and non-renewable construction. Above: Indian village reconstruction. (Image courtesy of Kevin Hampton http://www.wm.edu/niahd/journals). Below: Tokyo at night. (Image courtesy of http://www.photoeverywhere.co.uk index).

change now is that some aspects of industrialization have begun to influence the environment on a global scale. Materials are implicated in this. As responsible materials engineers and scientists, we should try to understand the nature of the problem—it is not simple—and to explore what, constructively, can be done about it.

This chapter introduces the key role materials have played in advancing technology and the dependence—addiction might be a better word—that this has bred. Addictions demand to be fed, and this demand, coupled with the world's continued population growth, consumes resources at an ever-increasing rate. This has not, in the past, limited growth; the earth's resources are, after all, very great. But there is increasing awareness that the limits *do* exist, that we are approaching some of them, and that adapting to them will not be easy.

1.2 Materials: a brief history

Materials have enabled the advance of mankind from its earliest beginnings—indeed the ages of man are named after the dominant material of the day: the *Stone Age*, the *Copper Age*, the *Bronze Age*, the *Iron Age* (Figure 1.1). The tools and weapons of prehistory, 300,000 or more years ago, were bone and stone. Stones could be shaped into tools, particularly flint and quartz, which could be flaked to produce a cutting edge that was harder, sharper, and more durable than any other naturally occurring materials. Simple but remarkably durable structures could be built from the materials of nature: stone and mud bricks for walls; wood for beams; bark, rush, and animal skins for roofing.

Gold, silver, and copper, the only metals that occur in native form, must have been known about from the earliest time, but the realization that they were ductile, that is, that they could be beaten into a complex shape, and, once beaten, become hard, seems to have occurred around 5500 BC. By 4000 BC, there is evidence that technology to melt and cast these metals had developed, allowing for more intricate shapes. Native copper, however, is not abundant. Copper occurs in far greater quantities as the minerals azurite and malachite. By 3500 BC, kiln furnaces, developed for pottery, could reach the temperature and create the atmosphere needed to reduce these minerals, enabling the tools, weapons, and ornaments that we associate with the Copper Age to develop.

But even in the worked state, copper is not all that hard. Poor hardness means poor wear resistance; copper weapons and tools were easily blunted. Sometime around 3000 BC the probably accidental inclusion of a tin-based mineral, cassiterite, in the copper ores provided the next step in technology—the production of the copper-tin alloy *bronze*. Tin gives bronze a hardness that pure copper cannot match, allowing superior tools and weapons to be produced. This discovery of *alloying*—the hardening of one metal by adding another—stimulated such significant technological advances that it, too, became the name of an era: the Bronze Age.

"Obsolescence" sounds like 20th century vocabulary, but the phenomenon is as old as technology itself. The discovery, around 1450 BC, of ways to reduce ferrous

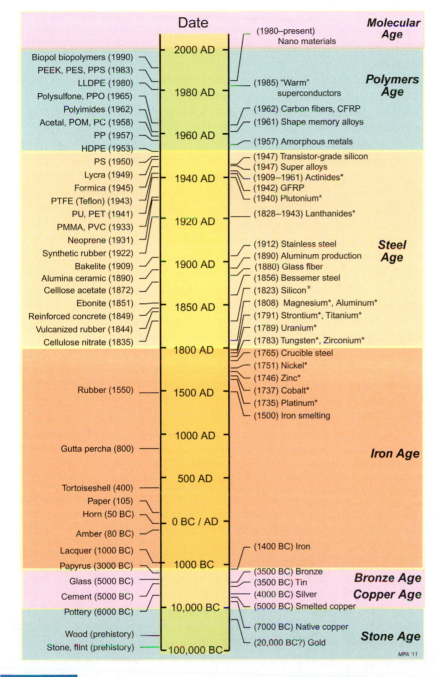

FIGURE 1.1 *The materials timeline. The scale is nonlinear, with big steps at the bottom, small ones at the top. A star (*) indicates the date at which an element was first identified. Unstarred labels give the date at which the material became of practical importance.*

oxides to make iron, a metal with greater stiffness, strength, and hardness than any other then available, rendered bronze obsolete. Metallic iron was not entirely new: tiny quantities existed as the cores of meteorites that had impacted the earth. The oxides of iron, by contrast, are widely available, particularly *hematite*, Fe_2O_3. Hematite is easily reduced by carbon, although it takes temperatures close to $1,100°C$ to do it. This temperature is insufficient to melt iron, so the material produced is a spongy mass of solid iron intermixed with slag; this mixture is then reheated and hammered to expel the slag, then forged to the desired shape. Iron revolutionized warfare and agriculture; indeed, it was so desirable that at one time it was worth more than gold. The casting of iron, however, presented a more difficult challenge, requiring temperatures around $1,600°C$. There is evidence that Chinese craftsmen were able to do this as early as 500 BC, but two millennia passed before, in 1500 AD, the blast furnace was developed, enabling the widespread use of cast iron. Cast iron allowed structures of a new type: the great bridges, railway terminals, and civil buildings of the early 19th century are testimony to it. But it was steel, made possible in industrial quantities by the Bessemer process of 1856, that gave iron the dominant role in structural design that it still holds today. For the next 150 years metals dominated manufacturing. It wasn't until the demands of the expanding aircraft industry in the 1950s that the emphasis shifted to the light alloys (those based on aluminium, magnesium, and titanium) and to materials that could withstand the extreme temperatures of the gas turbine combustion chamber (*super alloys*—heavily alloyed iron- and nickel-based materials). The range of application of metals expanded into other fields, particularly those of chemical, petroleum, and nuclear engineering.

The history of polymers is rather different. Wood, of course, is a polymeric composite, one used in construction from the earliest times. The beauty of amber—petrified resin—and of horn and tortoise shell—made up of the polymer keratin—attracted designers as early as 80 BC and continued to do so into the 19th century (in London, there is still a Horners' Guild, the trade association of those who work horn and shell). Rubber, which wasn't brought to Europe until 1550, was already known of and used in Mexico. Its use grew in importance in the 19th century, partly because of the wide spectrum of properties made possible by vulcanization—cross-linking by sulfur—to create materials as elastic as latex and others as rigid as ebonite.

The real polymer revolution, however, had its beginnings in the early 20th century with the development of Bakelite, a phenolic, in 1909, and of the synthetic butyl rubber in 1922. This was followed mid-century by a period of rapid development of polymer science, visible as the dense group at the upper left of Figure 1.1. Almost all the polymers we use so widely today were developed in a 20-year span from 1940 to 1960; among them were the bulk commodity polymers polypropylene (PP), polyethylene (PE), polyvinyl chloride (PVC), and polyurethane (PU), the combined annual tonnage of which now approaches that of steel. Designers seized on these new materials—they were cheap, brightly colored, and easily molded to complex shapes—to produce a spectrum of cheerfully ephemeral products. Design with polymers has since matured: they are now as important as metals in household products and automobile engineering.

The use of polymers in high-performance products requires a further step. "Pure" polymers do not have the stiffness and strength these applications demand; to provide it, they must be reinforced with ceramic or glass fillers and fibers, making them *composites*. Composite technology is not new. Straw-reinforced mud brick (adobe) is one of the earliest materials of architecture, one still used today in parts of Africa and Asia. Steel-reinforced concrete—the material of shopping centers, road bridges, and apartment blocks—appeared just before 1850. Reinforcing concrete with steel gave it tensile strength where previously it had none, thus revolutionizing architectural design; it is now used in greater volume than any other man-made material. Reinforcing metals, already strong, took much longer, and even today metal matrix composites are few.

The period in which we now live might have been named the Polymer Age had it not coincided with yet another technical revolution, that based on silicon. Silicon was first identified as an element in 1823, but found few uses until the realization, in 1947, that, when doped with tiny levels of impurity, it could act as a rectifier. This discovery created the fields of electronics and modern computer science, revolutionizing information storage, access and transmission, imaging, sensing and actuation, automation, and real-time process control.

The 20th century saw other striking developments in materials technology. Superconduction, discovered in mercury and lead when cooled to $4.2°K$ $(−269°C)$ in 1911, remained a scientific curiosity until, in the mid '80s, a complex oxide of barium, lanthanum, and copper was found to be superconducting at $30°K$. This triggered a search for superconductors with yet higher transition temperatures, leading, in 1987, to one that worked at the temperature of liquid nitrogen $(98°K)$, making applications practical, though they remain few.

During the early 1990s, scientists realized that material behavior depended on scale, and that the dependence was most evident when the scale was that of nanometers $(10^{-9}$ m$)$. Although the term *nanoscience* is new, technologies that use it are not. The ruby red color of medieval stained glasses and the diachromic behavior of the decorative glaze known as "lustre" derive from gold nanoparticles trapped in the glass matrix. The light alloys of aerospace derive their strength from nanodispersions of intermetallic compounds. Automobile tires have, for years, been reinforced with nanoscale carbon. Modern nanotechnology gained prominence with the discovery that carbon could form stranger structures: spherical C_{60} molecules and rod-like tubes with diameters of a few nanometers. Now, with the advance of analytical tools capable of resolving and manipulating matter at the atomic level, the potential exists to build materials the way that nature does it, atom by atom and molecule by molecule.

If we now step back and view the timeline of Figure 1.1 as a whole, clusters of activity are apparent—there is one in Roman times, one around the end of the 18th century, and one in the mid 20th century. What was it that triggered the clusters? Scientific advance, certainly. The late 18th and early 19th century was the time of the rapid development of inorganic chemistry, particularly electrochemistry, and it was this that allowed new elements to be isolated and identified. The mid 20th century saw the birth of polymer chemistry, spawning the polymers we use today and providing key concepts in unraveling the behavior of the materials of nature. But there may be more to it than that. Conflict stimulates science. The first of these two periods coincides with the Napoleonic Wars (1796–1815), one in which technology, particularly in France, developed rapidly. The second coincided with the Second World War (1939–1945), in which technology played a greater part than in any previous conflict. Defense budgets have, historically, been prime drivers for the development of new materials. One hopes that scientific progress and advances in materials are possible without conflict, and that the competitive drive of free markets can be an equally strong driver of technology. It is interesting to reflect that more than three quarters of all the materials scientists and engineers who have *ever* lived are alive today, and all of them are pursuing better materials and better ways to use them. Of one thing we can be certain: there are many more advances to come.

1.3 Learned dependency: the reliance on nonrenewable materials

Now back to the main point: the environmental aspects of the way we use materials. "Use" is too weak a word—it sounds as if we have a choice: use, or perhaps not use? We don't just "use" materials, we are totally dependent on them. Over time this dependence has progressively changed from a reliance on renewable materials—the way mankind existed for thousands of years—to one that relies on materials that consume resources that cannot be replaced.

As little as 300 years ago human activity subsisted almost entirely on renewables: stone, wood, leather, bone, and natural fibers. The few nonrenewables—iron, copper, tin, zinc—were used in such small quantities that the resources from which they were drawn were, for practical purposes, inexhaustible. Then, progressively, the nature of the dependence changed (Figure 1.2). Bit by bit, nonrenewables displaced renewables until, by the end of the 20th century, our dependence on them was, as already said, almost total.

Dependence is dangerous; it is a genie in bottle. Take away something on which you depend, meaning that you can't live without it, and life becomes difficult. Dependence exposes you to exploitation. While a resource is plentiful, market forces ensure that its price bears a relationship to the cost of its extraction. But the resources from which many materials are extracted, oil among them, are localized in just a few countries. While these countries compete for buyers, the price remains geared to the cost of production. But if demand exceeds supply or the producing nations reach arrangements to limit it, the genie is out of the bottle. Think, for instance, of the price of oil, which today bears no relationship to the cost of producing it.

Dependence, then, is a condition to be reckoned with. We will encounter its influence in subsequent chapters.

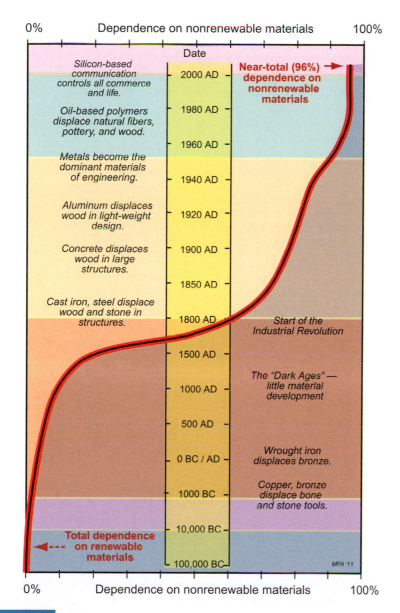

FIGURE 1.2 *The increasing dependence on nonrenewable materials over time, rising to 96% by weight today. This dependence is not of concern when resources are plentiful but is an emerging problem as they become scarce. (Data in part from USGS [2002].)*

> **News-clip: dangerous dependence**
>
> *Oil addiction puts us at the mercy of our enemies.*
> *Western nations rely on Saudi Arabia to pump more oil when prices rise to*
> *levels that threaten their prosperity. At the regular OPEC meeting this month*
> *the Saudi's proposal [to do so] was turned down. With rising domestic demand*
> *in producing countries, an output freeze would mean tight supplies and price*
> *rises.*
> *   There is no substitute fuel available in the foreseeable future to replace oil*
> *for transport. That means dependence on regimes in countries that are unsta-*
> *ble, or hostile, or both. . . .*
>
> **The Sunday Times**, June 19, 2011
>
> Much the same situation exists with a number of materials that are critical for
> modern manufacturing.

1.4 Materials and the environment

All human activity has some impact on the environment in which we live. The environment has some capacity to cope with this so that a certain level of impact can be absorbed without lasting damage, but it is clear that current human activities exceed this threshold with increasing frequency, diminishing the quality of the world in which we now live and threatening the wellbeing of future generations. At least part of this impact derives from the manufacture, use, and disposal of products, and products, without exception, are made from materials.

The materials consumption in the United States now exceeds 10 metric ton per person per year. The average level of global consumption is barely one eighth of this but is growing twice as fast. The materials (and the energy needed to make and shape them) are drawn from *natural resources*: ore bodies, mineral deposits, fossil hydrocarbons. The earth's resources are not infinite, but until recently, they have seemed so: the demands made on them by manufacturing throughout the 18th, 19th, and early 20th century seemed infinitesimal, the rate of new discoveries always outpacing the rate of consumption.

This perception has now changed. The realization that we may be approaching certain fundamental limits seems to have surfaced with surprising suddenness, but warnings that things can't go on forever are not new. Thomas Malthus, writing in 1798, foresaw the link between population growth and resource depletion, predicting gloomily that "the power of population is so superior to the power of the earth to produce subsistence for man that premature death must in some shape or other visit the human race." Almost 200 years later, in 1972, a group of scientists known as the Club of Rome reported their modeling of the interaction of population growth, resource depletion, and pollution, concluding that "if (current trends) continue unchanged ... humanity is destined to reach the natural limits of development within the next 100 years." The report generated both consternation and criticism, largely on the grounds that the modeling was over-simplified and did not allow for scientific and technological advance. But in the last decade, thinking about this broad issue has reawakened. There is a growing acceptance that, in the words of another distinguished report:

> ... *many aspects of developed societies are approaching ... saturation, in the sense that things cannot go on growing much longer without reaching fundamental limits. This does not mean that growth will stop in the next decade, but that a declining rate of growth is foreseeable in the lifetime of many people now alive. In a society accustomed ... to 300 years of growth, this is something quite new, and it will require considerable adjustment.* ***

The causes of these concerns are complex, but one stands out: population growth. Examine, for a moment, Figure 1.3. It is a plot of global population over the last 2,000 years. It looks like a simple exponential growth (something we examine in more depth in Chapter 2) but it is not. Exponential growth is bad enough—it is easy to be caught out by the way it surges upward. But this is far worse. Exponential growth has a constant doubling time—if it's exponential, a population doubles in size at fixed, equal time intervals. The doubling times for global population are marked on the figure. For the first 1,500 years, it is constant at about 750 years, but after that, starting with the Industrial Revolution, the doubling time halves, then halves again, then again. This behavior has been called "explosive growth"; it is harder to predict and results in a more sudden change. Malthus and the Club of Rome may have had the details wrong, but it seems they had the principle right.

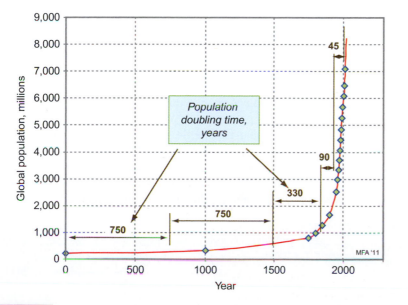

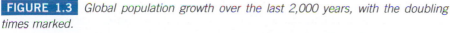

FIGURE 1.3 *Global population growth over the last 2,000 years, with the doubling times marked.*

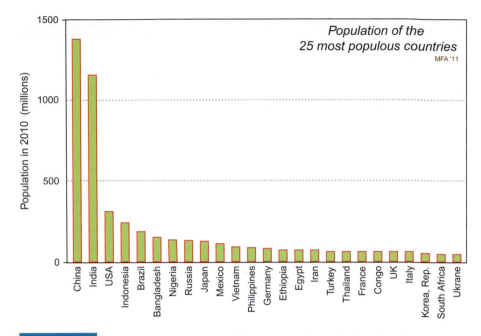

FIGURE 1.4 *The populations of the 25 most populous developed and developing countries in 2010.*

Global resource depletion scales with the population and with per-capita consumption. Per-capita consumption in developed countries is stabilizing, but in the emerging economies, as already said, it is growing more quickly. Figure 1.4 shows the distribution of population in the 25 most populous nations containing, between them, three quarters of the global total. The first two—China and India—account for 37% of the total, and it is in these two that material consumption is growing most rapidly.

Given all this, it make sense to explore the ways in which materials are used in design and how this might change as environmental prerogatives become increasingly pressing. The chapters that follow explore this.

News-clip: population, affluence, and consumption

Be a bull as China shops.
The world's largest population is enjoying rising wages and a growing disposable income. In short: 1.3 billion are rapidly becoming active consumers. . . .
                                                                    ***The Times**, May 21, 2011*

No further comment needed.

1.5 Summary and conclusions

Homo sapiens—that means us—differs from all other species in its competence in making things out of materials. We are not alone in the ability to *make*: termites build towers, birds build nests, beavers build dams; all creatures, in some way, make things. The difference lies in the *competence* demonstrated by man and in his extraordinary (there can be no other word) ability to expand and adapt that competence by research and development.

The timeline of Figure 1.1 illustrates this expansion. There is a tendency to think that progress of this sort started with the Industrial Revolution, but knowledge about and development of materials has a longer and more continuous history than that. The misconception arises because of the bursts of development in the 18th, 19th, and 20th centuries, and because the technological developments during the great eras of the Egyptian, Chinese, Greek, and Roman empires are forgotten. These empires did not just shape stone, clay, and wood, and forge and cast copper, tin, and lead, but they also found and mined the ores and imported them over great distances.

Importing tin from a remote outpost of the Roman empire (Cornwall, England, to Rome, Italy, 3,300 km by sea) to satisfy the demands of the Roman State hints at an emerging material dependence. The dependence has grown over time with the deployment of ever more man-made materials until today it is almost total. In reading this text, then, do so with the perspective that materials, our humble servants throughout history, have become, in another sense, our masters.

1.6 Further reading

Delmonte, J. (1985), *Origins of materials and processes*, Technomic Publishing Company, PA, USA. ISBN 87762-420-8. *(A compendium of information about materials in engineering, documenting the history)*

Flannery, T. (2010), *Here on earth*, The Text Publishing Company, Victoria, Australia. ISBN 978-1-92165-666-8. *(The latest of a series of books by Flannery documenting man's impact on the environment)*

Hamilton, C. (2010), *Requiem for a species: why we resist the truth about climate change*, Allen and Unwin, NSW, Australia. ISBN 978-1-74237-210-5. *(A profoundly pessimistic view of the future for humankind)*

Kent, R. (2009), "Plastics timeline," www.tangram.co.uk.TL-Polymer_Plastics_Timeline.html/. *(A web site devoted, like that of Material Designs, to the history of plastics)*

Lomberg, B., editor (2010), *Smart solutions to climate change: comparing costs and benefits*, Cambridge University Press, Cambridge, UK. ISBN 978-0-52113-856-7. *(A multiauthor text in the form of a debate ["The case for …," "The case against …"] covering climate engineering, carbon sequestration, methane mitigation, and market- and policy-driven adaptation)*

Lovelock, J. (2009), *The vanishing face of Gaia*, Penguin Books, Ltd., London, UK. ISBN 978-0-141-03925-1. *(James Lovelock reminds us that humans are just another species and that species have appeared and disappeared since the beginnings of life on earth.)*

Malthus, T.R. (1798), "An essay on the principle of population," London, Printed for Johnson, St. Paul's Church-yard. www.ac.wwu.edu/~stephan/malthus/malthus. *(The originator of the proposition that population growth must ultimately be limited by resource availability)*

Material Designs (2011), "A timeline of plastic," http://materialdesigns.wordpress.com/2009/08/06/a-timeline-of-plastics/. *(A web site devoted, like that of Kent, to the history of plastics)*

Meadows, D.H., Meadows, D.L., Randers, J., and Behrens, W.W. (1972), *The limits to growth*, Universe Books, New York, NY, USA. *(The "Club of Rome" report that triggered the first of a sequence of debates in the 20th century on the ultimate limits imposed by resource depletion)*

Meadows, D.H., Meadows, D.L., and Randers, J. (1992), *Beyond the limits*, Earthscan, London, UK. ISSN 0896-0615. *(The authors of The limits to growth use updated data and information to restate the case that continued population growth and consumption might outstrip the earth's natural capacities.)*

Nielsen, R. (2005), *The little green handbook*, Scribe Publications Pty Ltd., Carlton North, Victoria, Australia. ISBN 1-920769-30-7. *(A cold-blooded presentation and analysis of hard facts about population, land and water resources, energy, and social trends)*

Plimer, I. (2009), *Heaven and Earth—Global warming: the missing science*, Connor Publishing, Ballam, Victoria, Australia. ISBN 978-1-92142-114-3. *(Ian Plimer, Professor of Geology at the University of Adelaide, examines the history of climate change over a geological timescale, pointing out that everything that is happening now has happened many times in the past. A geo-historical perspective, very thoroughly documented.)*

Ricardo, D. (1817), "On the principles of political economy and taxation," John Murray, London, UK. www.econlib.org/library/Ricardo/ricP.html. *(Ricardo, like Malthus, foresaw the problems caused by exponential growth.)*

Schmidt-Bleek, F. (1997), *How much environment does the human being need—factor 10—the measure for an ecological economy*, Deutscher Taschenbuchverlag, Munich, Germany. ISBN 3-936279-00-4. *(Both Schmidt-Bleek and von Weizsäcker, referenced below, argue that sustainable development will require a drastic reduction in material consumption.)*

Singer, C., Holmyard, E.J., Hall, A.R., Williams, T.I., and Hollister-Short, G., editors (1954–2001), *A history of technology* (21 volumes), Oxford University Press, Oxford, UK. ISSN 0307-5451. *(A compilation of essays on aspects of technology, including materials)*

Tylecoate, R.F. (1992), *A history of metallurgy*, 2nd edition, The Institute of Materials, London, UK. ISBN 0-904357-066. *(A total-immersion course in the history of the extraction and use of metals from 6000 BC to 1976, told by an author with forensic talent and a love of detail)*

USGS (2002), Circular 2112, "Materials in the economy—material flows, scarcity and the environment," by L.W. Wagner, US Department of the Interior.

www.usgs.gov. *(A readable and perceptive summary of the operation of the material supply chain, the risks to which it is exposed, and the environmental consequences of material production)*

von Weizsäcker, E., Lovins, A.B., and Lovins, L.H. (1997), *Factor four: doubling wealth, halving resource use*, Earthscan, London, UK. ISBN 1-85383-406-8; ISBN-13: 978-1-85383406-6. *(Both von Weizsäcker and Schmidt-Bleek, referenced above, argue that sustainable development will require a drastic reduction in material consumption.)*

1.7 Exercises

E1.1. Use Google to research the history and uses of one of the following materials:

- Tin
- Glass
- Cement
- Bakelite
- Titanium
- Carbon fiber
- Cobalt
- Neodymium

Present the result as a short report of about 100–200 words (roughly half a page). Imagine that you are preparing it for school children. Who used the material first? Why? What is exciting or remarkable about it? What do we use it for now? Do we now depend on it or could we live comfortably without it?

E1.2. There is international agreement that it is desirable (essential, in the view of some) to reduce global energy consumption. Producing materials from ores and feedstock requires energy (its "embodied energy"). The table lists the energy per kg and the annual consumption of four materials of engineering. If consumption of each could be reduced by 10%, which material offers the greatest global energy saving? Which the least?

Material	Embodied energy MJ/kg	Annual global consumption (metric ton/yr)
Steel	26	2.3×10^9
Aluminium alloys	200	3.7×10^7
Polyethylene	80	4.5×10^7
Concrete	1.2	1.5×10^{10}
Device-grade silicon	3,000	5×10^3

E1.3. The ultimate limits of most resources are difficult to assess precisely, although estimates can be made. One resource, however, has a well-defined limit: that of usable land area. The surface area of the earth is 511 million square km, or 5.11×10^{10} hectares (a hectare is 0.01 sq. km). Only a fraction of this is land, and only part of that land is useful—the best estimate is that 1.1×10^{10} hectares of the earth's surface is biologically productive. Industrial countries require 6 hectares of biologically productive land per head of population to support current levels of consumption. The current (2011) global population is close to 6.7 billion (6.7×10^9). What conclusions can you draw from these facts?

Exercises using CES Level 2

E1.4. The CES Level 2 databases have a field called *"Date first used ("-" = BC)*. This field allows you to explore the history of development of materials. Plot four Yield strength–Density charts showing the materials available during the Stone Age (10,000 BC), the late Roman empire (300 AD), the Industrial Revolution (1800), and today. The way to do this is explained in the Science Note for the "Date first used" field that you will find if you double-click on the field name.

E1.5. Explore how the specific strength, σ_y/ρ has increased over time. [Here σ_y is the yield strength and ρ is the density]. To do so, make a chart with σ_y/ρ on the y-axis and Date first used on the x-axis, using linear scales for both.

CHAPTER 2

Resource consumption and its drivers

2.1 Introduction and synopsis

This chapter is about *orders of magnitude*: the round numbers describing the amounts of stuff we consume. You can't reach robust conclusions about man's influence on the environment without a feel for the numbers. As we saw in

The Bingham Canyon copper mine in Utah, now 1.2 km deep, 4 km across, and a Caterpiller truck that is part of the excavation equipment. (Images courtesy of the Kennecott Utah Copper.)

Chapter 1, manufacturing today is addictively dependent on continuous flows of materials and energy. How big are these? A static picture—that of the values today—is a starting point, and the quantities are enormous. And of course they are *not* actually static. Growth is the life-blood of today's consumer-driven economies. An economy that is not growing is "faltering," "stagnant," or "sick" (words use by Economics Correspondents of *The Times*). Business enterprises, too, seem to need to grow to survive. And all this growth causes the consumption of materials and energy to rise, or at least it has done so until now. Growth can be linear, increasing at a constant rate; it can be exponential, increasing at a rate proportional to its current size; or, as we saw in Chapter 1, it can sometimes increase even faster than that.

Exponential growth plays nasty tricks. Something cute and cuddly, growing exponentially, eventually evolves into an oppressive monster. Exponential growth is characterized by a doubling time: anything growing exponentially doubles its size at regular, equal intervals. Money invested in fixed-rate bonds has a doubling time, though it is usually a long one. The consumptions of natural resources—minerals, energy, water—grow in a roughly exponential way: they, too, have doubling times, and some of these are short. Certain resources are so abundant that there is no concern that we are using them faster and faster: the mineral resources from which iron and aluminium are drawn are examples. But others are not so abundant: their ores are localized and the amount that is economically accessible is limited. Then the doubling-up nature of exponential growth becomes a concern: consumption of these cannot continue to double forever. And extracting and processing any material, whether plentiful or scarce, uses energy—lots of energy—and it too is a resource under duress.

This sounds alarming, and many alarming statements have been made about it. But consider this: in 1930, the exhaustion time for the reserves of the ores of copper was estimated to be 30 years; today, 80 years later the exhaustion time of copper reserves is calculated as ... 30 years. There is obviously something going on here besides exponential growth. This chapter explores it.

First, however, a quick look at what the earth has to offer.

2.2 Where do materials come from?

We have 92 usable elements. Most of them are metals. If we want them for engineering purposes, we have to mine the naturally occurring minerals in which they are found, mostly as oxides, sulphides, or carbonates. They come mixed with other stuff, from which they have to be separated before the metal can be extracted. The mining, separating, and extracting processes all take energy. The more dilute the ore, the more stuff has to be dug, and the greater the effort of separation and the greater the energy demand. The metal content of the ore is called the *ore grade, G*, measured as the metric tons (mt) of metal per metric ton of ore as mined.

Figure 2.1 shows the distribution of elements in the earth's crust. They differ in abundance by a factor of over 10^{10}. The most abundant are the light elements, among them aluminium and magnesium. The eight elements that lie at the top of

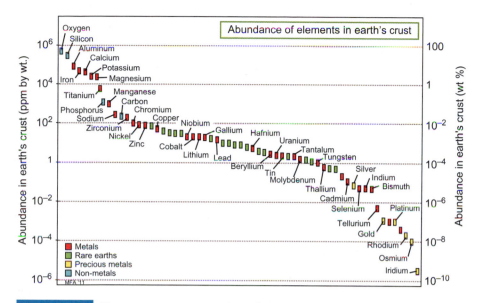

FIGURE 2.1 *The average concentration of the elements in the earth's crust, as parts-per-million (ppm) and as wt %. The top eight elements account for 98.5% of the total.*

the figure account for over 98% of the total—no others have a concentration above 0.1%. The "precious" metals (yellow bars on the figure), all lie below 0.00001%, the value for silver. The concentration of iridium is 10,000 times smaller than that. The earth's oceans, too, contain elements (Figure 2.2), but there, except for sodium and magnesium, the concentrations are even smaller.

Very approximately, the cost C_m ($/kg) of extracting metals scales inversely with the ore grade G (%) such that

$$C_m \approx \frac{10}{G} \tag{2.1}$$

If copper, as an example, had to be extracted from ore with a grade equal to its average concentration (50 parts per million, or ppm) in the crust, what might it cost? According to equation (2.1), it would cost $2,000 per kilogram (kg). In reality it costs less than one hundredth of that (as of this writing, copper costs $9.43/kg; a year ago it cost $6.50/kg). The difference, of course, is that copper isn't mined from the *average* crust but from localized, copper-rich deposits.

How do mineral rich deposits come about? When the earth was born things were pretty hot and mixed up. What, since then, has separated some elements out? At least four different processes, all very slow, are at work.

- Volcanism distills out the lower melting minerals and vaporizes the more volatile ones, condensing them again as they cool.

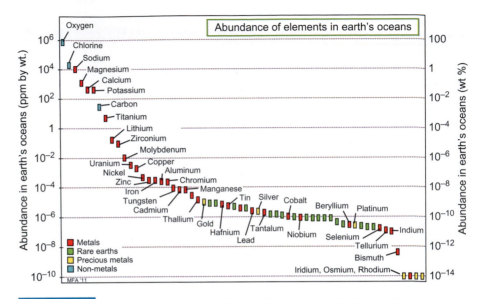

FIGURE 2.2 *The average concentration of the elements in the earth's oceans, as ppm and as %. Most are too dilute to allow for economical recovery.*

- Erosion by water and wind grind down the crust, sweeping the debris to calmer places where gravity sorts its components by density.
- Water seeks out minerals that are water-soluble in a selective way controlled by the pH of the water, dumping the dissolved minerals when they are concentrated by evaporation.
- Natural organisms are effective concentrators of minerals—the coal, oil, and gas deposits reflect the ability of plant life to concentrate carbon, and historically much of the world's phosphates have been mined from the guano of birds and bats.

Today we draw on the treasure-troves that these processes have created. All four took a long time to make them. We use them far faster than they can be re-created, which forces us to mine ever leaner, low G ores. Improved extraction technology allows leaner ores to be exploited, but, as equation (2.1) says, the leaner we go, the more it will cost.

With that background, we now turn to consumption.

2.3 Resource consumption

Materials. Speaking globally, we consume roughly 10 billion (10^{10}) metric tons of engineering materials per year, an average of 1.5 metric tons per person though it is not distributed evenly. Figure 2.3 gives a perspective: it is a bar chart of the primary

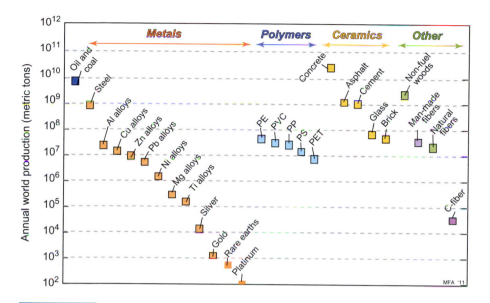

FIGURE 2.3 *The annual world production of 27 materials on which industrialized society depends. The scale is logarithmic. The log scale conceals the great differences; the production of steel, for instance, is one billion (10^9) times larger than that of platinum.*

production of the materials used in the greatest quantities. It has some interesting messages. On the extreme left, for calibration, are hydrocarbon fuels—oil and coal—of which we currently consume a colossal 9 billion (9×10^9) metric tons per year. Next, moving to the right, are metals. The scale is logarithmic, making it appear that the production of steel (the first metal) is only a little greater than that of aluminium (the next); in reality, the production of steel exceeds, by a factor of ten, that of all other metals combined. Steel may lack the high-tech image that attaches to materials like titanium, carbon-fiber reinforced composites, and (most recently) nanomaterials, but make no mistake, its versatility, strength, toughness, low cost, and wide availability are unmatched. At the other extreme are the platinum-group metals and the rareearths (15 elements near the bottom of the Periodic Table). Their quantities are small, but their importance is large. We return to them in later chapters.

Polymers come next. Fifty years ago their production was tiny. Today the production, in metric tons per year, of the five commodity polymers, polyethylene (PE), polyvinyl chloride (PVC), polypropylene (PP), polystyrene (PS), and polyethylene-terephthalate (PET), is comparable with that of aluminium; if measured in m^3 per year, they approach steel. The really big ones, though, are the materials of the construction industry. Steel is one of these, but the production of wood for construction purposes exceeds even that of steel when measured in metric tons per year (as in the diagram), and since it is a factor of ten lighter, wood totally eclipses

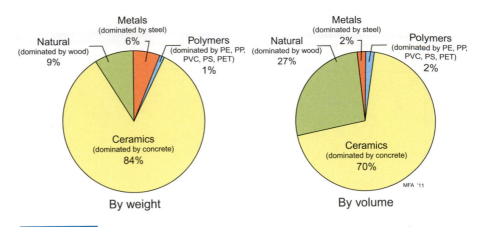

FIGURE 2.4 *A pie chart of materials usage by family, by weight on the left, by volume on the right. Ceramics dominate because of the enormous annual consumption of concrete.*

steel when measured in m³/year. Bigger still is the production of concrete, which exceeds that of all other materials combined. The other big ones are asphalt (roads), cement (most of which goes into concrete), brick, and glass. Fibers, too, are produced in very great quantities. Natural fibers have played a role in human life for tens of thousands of years, and continue to do so. Even larger, today, is the production of man-made fibers for both textiles and for industrial use. The final column illustrates things to come: it shows today's production of carbon fiber. Just 30 years ago this material would not have crept onto the bottom of this chart. Today its production is approaching that of titanium and is growing much more quickly.

Figure 2.4 presents some of this information in another way: as a pair of pie charts showing the tonnage-fraction and volume-fraction of each family of material: metals, polymers, ceramics, and natural. The logarithmic scale of Figure 2.3 can be deceptive: the true magnitude of the production of concrete and of wood only emerges when plotted in this second way. This is important when we come to consider the impact of materials on the environment, since impact scales with the quantity consumed.

Energy. Although this book is about materials, energy appears throughout; it is inseparable from the making of materials, their manufacture into products, their use, and their ultimate disposal. The SI unit of energy is the joule (J) but because a joule is very small, we generally use kJ (10^3 J), MJ (10^6 J), or GJ (10^9 J) as the unit. Power is joules/sec, or watts (W), but a watt, too, is small so we usually end up with kW, MW, or GW. The everyday unit of electrical energy is the kilowatt hour (kWh), one kW drawn for 3,600 seconds, so 1 kWh = 3.6 MJ.

Where does energy come from? There are ultimately just four sources:

■ The sun, which drives the winds, wave, hydro, photoelectric phenomena, and the photochemical processes that give biomass

- The moon, which drives the tides
- Nuclear decay of unstable elements inherited from the creation of the earth, providing geothermal heat and nuclear power
- Hydrocarbon fuels, the sun's energy in fossilized form

Harvesting these and the materials implications of doing so are the subjects of Chapter 12. For now we just note that while all four are ultimately finite, the time-scale for the exhaustion of the first three is so large that it is safe to regard them as infinite.

How much energy do we use in a year? When speaking of world consumption, the unit of convenience is the exajoule, symbol EJ, a billion billion (10^{18}) joules. The value today (2012) is about 500 EJ/year and, of course, it is rising. Figure 2.5 shows where it comes from. Fossil fuels dominate the picture, providing about 86% of the total (Figure 2.6a). Nuclear gives about 7%. Hydro, wind, wave, biomass, solar heat, and photovoltaics add up to just another 7%. These sun-driven energy pools are enormous, but unlike fossil and nuclear fuels, which are concentrated, they are distributed, making energy from them hard to capture.

Where does the energy go? Most of it goes into three big sectors: transportation, buildings (heating, cooling, lighting), and industry—including the production of materials (Figure 2.6b). Making materials consumes about 21% of global energy and is responsible for about the same fraction of carbon emitted to the atmosphere.

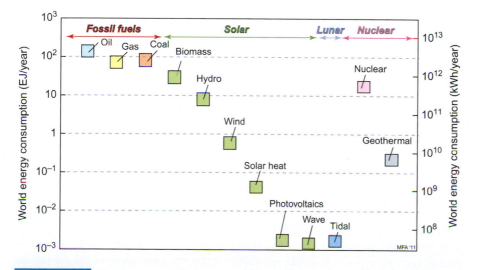

FIGURE 2.5 *The annual world consumption of energy by source. The units on the left are exajoules (10^{18} J); those on the right are the more familiar kWh. (Nielsen, 2005.)*

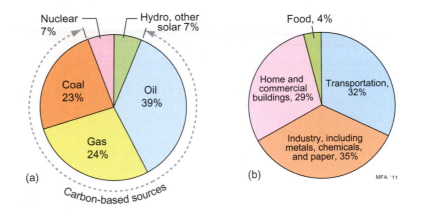

Nuclear 7%   Hydro, other solar 7%   Food, 4%

Coal 23%   Oil 39%   Home and commercial buildings, 29%   Transportation, 32%

Gas 24%   Industry, including metals, chemicals, and paper, 35%

(a)   Carbon-based sources   (b)   MFA '11

FIGURE 2.6   *World energy consumption (a) by source and (b) by use. The nonrenewable carbon-based fuels, oil, gas, and coal, account for 86% of the total.*

The energy demands of steel-making

Example: The current world production of steel is 2.3×10^9 metric tons per year. The embodied energy of steel is approximately 26 MJ/kg. The annual global consumption of energy is 500 EJ. What fraction of the world's annual energy is required to provide the steel we consume?

Answer: The annual energy embodied in steel is $(2.3 \times 10^9) \times 1{,}000 \times (26 \times 10^6) = 6 \times 10^{19}$ J = 60 EJ. This is 12% of global energy consumptiion.

Figures 2.5 and 2.6(a) focus on the *sources of energy*. For the energy to be useful you have to do something with it, and that, almost always, involves *energy conversion*. Figure 2.7 suggests the possible conversion paths. Almost always conversion carries energy "losses." The first law of thermodynamics tells us that energy is conserved, so it can't really be lost. In almost all energy conversion some energy is converted to heat. *High-grade heat* is heat at high temperature, as in the burning gas of a power station or the burning fuel in an internal combustion engine or gas turbine—it can be used to do work. *Low-grade heat* is heat at low temperature and it is not nearly so useful—indeed most of it is simply allowed to escape, and in this sense is "lost." Energy conversion generally has low-grade heat as a by-product with the result that the conversion efficiency, η, to *useful* energy is less than 100%. The refining of metals from their oxide, sulphide, or other ores, for instance, involves the conversion of thermal or electrical energy into chemical energy—the energy that could (in principle) be recovered by allowing the metal to re-oxidize or re-sulphidize. The recoverable energy, of course, is a lot less than the energy it took to do the refining because of

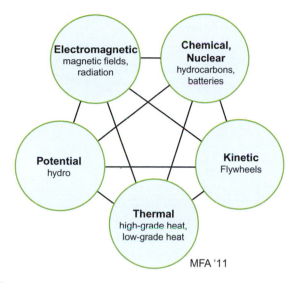

MFA '11

FIGURE 2.7 *Five types of energy. Each can be transformed into the others, as indicated by the linking lines, but at conversion efficiencies that differ greatly.*

heat losses. The conversion efficiency in material production is low, typically 5−35%.

The conversion efficiencies of the many paths in Figure 2.7 differ greatly even when "ideal," meaning that the conversion is as good as the laws of physics allow. Conversion of thermal to mechanical energy, for instance, is ultimately limited by the Carnot efficiency, η_c:

$$\eta_c = 1 - \frac{T_{out}}{T_{in}}$$

where T_{in} is the temperature of the steam or hot gas entering the heat engine and T_{out} is the temperature at which it exits; both temperatures are measured in Kelvin. Plotted, this looks like Figure 2.8. The maximum input temperature T_{in} is about 650°C for steam turbines or 1,400°C for gas turbines, limited by the materials of combustion chambers, blades, and disks. The exhaust temperature T_{out} is at least 150°C, giving a theoretical maximum efficiency of about 75%.

The real conversion efficiencies of heat engines are much less than this. Beam engines of the early 19th century had efficiencies of about 2%. The first electricity generating plant built in Holborn, London, in 1882 had an efficiency of just 6%. A modern steam-turbine plant, today, achieves 38% although the national average in a country like the US or the UK is less than this because many power stations are old. A plant driven by a gas turbine can reach 50%, a consequence of a higher T_{in} and lower T_{out}. Table 2.1 lists practical efficiencies of other energy conversion

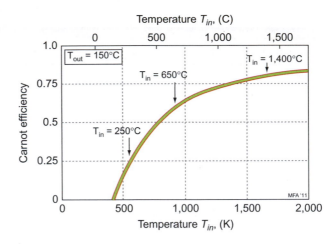

FIGURE 2.8 *Carnot efficiency η_c of heat engines as a function of temperature T_{in}, assuming $T_{out} = 150°C$*

processes. Many are low. Thus a more useful measure of efficiency η is simply the ratio

$$\eta = \frac{Power\ out}{Power\ in} \qquad (2.2)$$

where "Power" is energy—in any form—per unit time. This is the definition used in the rest of the book.

Many processes involve chains of energy-conversion steps. Figure 2.9 shows one. Oil (the primary energy) is drawn from a well. It is processed into a usable form with a loss of 3%. The fuel oil is converted to electrical energy with a conversion efficiency of 38%. The electrical energy is transmitted to the point of use with transmission losses of 10%, where a motor converts it to mechanical energy with an efficiency of 85%. The motor drives a hydraulic pump with losses of 10%, which in turn powers a stamping press that converts hydraulic pressure into kinetic energy with an efficiency, allowing for stand-by losses, of 35%. The overall conversion efficiency of primary oil to useful kinetic energy at the press is the product of each of these efficiencies.

Efficiencies

Example: What is the overall conversion efficiency of the chain of Figure 2.9?

Answer: The overall efficiency is $\eta_{tot} = 0.97 \times 0.38 \times 0.9 \times 0.85 \times 0.9 \times 0.35 = 0.1$. Ninety percent of the primary energy is lost as low-grade heat, performing no useful function.

Table 2.1	Some approximate efficiency factors for energy conversion		
Energy conversion path	**Efficiency, direct conversion (%)**	**Efficiency relative to oil equivalence (%)**	**Associated carbon (kg CO_2 per useful MJ)**
Gas to electric	37–40	37–40	0.18
Oil to electric	36–38	36–38	0.2
Coal to electric	33–35	33–35	0.22
Biomass to electric	23–26	23–26	0
Hydro to electric	75–85	75–85	0
Nuclear to electric	32–34	32–34	0
Fossil fuel to heat, enclosed system	100	100	0.07
Fossil fuel to heat, vented system	65–75	65–75	0.10
Biomass to heat, vented system	33–35	33–35	0
Oil to mechanical, diesel engine	20–22	19–21	0.17
Oil to mechanical, gasoline engine	13–15	12–14	0.15
Oil to mechanical, kerosene gas turbine	27–29	25–27	0.15
Electric to thermal	100	37–40	0.20
Electric to mechanical (electric motors)	80–93	30–32	0.23
Electric to chemical (lead-acid battery)	80–85	29–32	0.24
Electric to chemical (advanced battery)	85–90	31–33	0.23
Electric to em radiation (incandescent lamp)	12–15	4–6	1.17
Electric to em radiation (LED)	80–85	29–32	0.23
Light to electric (solar cell)	10–22	–	0

Using energy to extract and refine materials is an example of an energy-conversion process. The energy per unit mass used to make a material from its ores and feedstock is known as its *embodied energy*. The term is a little misleading: only part of this energy is really "embodied" in the sense that it is *in* the materials and that you could get it back. That bit is the free energy difference between the refined material and the ore from which is came. Take a metal—say, iron—as an example. Iron is made by the reduction of the oxide *hematite*, Fe_2O_3, in a blast furnace. Thermodynamics dictates that the minimum energy needed to do this is 6.1 MJ/kg—it is the free energy of oxidation of iron to this particular oxide. This energy could, in principle, be recovered by re-oxidizing the iron under controlled conditions (it is truly "embodied"). But the measured energy to make iron—the quantity that

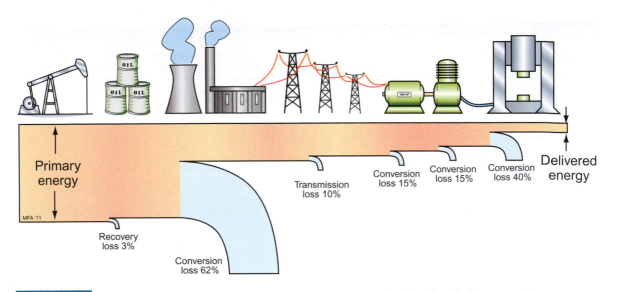

FIGURE 2.9 *A chain of energy-conversion and transmission steps, each with a "loss" of energy as low-grade heat*

is called its *embodied energy*—is three times larger, at about 18 MJ/kg, a conversion efficiency of 33%. Where has the rest gone? Mostly it has become heat carried away in the gases of the blast furnace. Similar losses reduce the efficiency of all material synthesis and refining.

Apparent production efficiency of copper

Example: Copper can be extracted from natural copper oxide, CuO. The free energy of formation of CuO is 127 kJ/mol. This is the amount of energy that *must* be provided to create one mole of copper from CuO. How does this compare with the reported embodied energy of copper, 71 MJ/kg? (The molecular weight of copper is 63.6 kg/kmol.)

Answer: The "ideal" energy per kg to reduce CuO to copper is

$$H = \frac{127}{63.6} \times 1{,}000 = 2{,}000 \text{ kJ/kg} = 2.0 \text{ MJ/kg}$$

This is less than 3% of the embodied energy, 71 MJ/kg. The difference arises because of the additional energy needed to mine and concentrate the CuO, and the energy lost as low-grade heat during the refining.

Apparent production efficiencies in general

Example: Is the low value of extraction efficiency of copper, calculated in the previous example, typical of metals in general?

Answer: If you use the same calculation as in the previous example, the ratios of energy of formation to embodied energies for other metals are as follows. The low apparent efficiencies have the same origin as that for copper.

Element	Free energy of formation		Embodied energy	Apparent efficiency
	kJ/mol	MJ/kg	MJ/kg	%
Al	796	30	203	15
Cu	134	2.1	68	3
Fe	374	6.7	29	23
Mg	626	26	360	7
Ni	233	4.0	127	3
Pb	233	1.1	53	2
Sn	552	4.7	40	12
Ti	907	19	600	3
Zn	339	5.2	65	8

Water. The third resource on which we are totally dependent is water. How much have we got? A great deal, but 97% of it is salt and two thirds of the rest is ice (Figure 2.10). Water is a renewable resource but only at the rate that the ecosystem allows. Increased demand now puts supply under growing pressure: the world-wide demand for water has tripled over the last 50 years. Forecasts suggest that water may soon become as important an issue as oil is today, with more than half the human race short of water by 2050. Agriculture is the largest consumer, taking about 65% of all fresh water (Figure 2.11). Industry consumes about 10% of the total, half of which is required for power generation. This suggests that material production is not at present a major driver of water use, but a shift to a greater use of bio-derived materials could make it so. That, however, is not a reason for ignoring it.

The water demands of materials and manufacturing are measured directly as factory inputs and outputs. The production of steel, for example, uses water in the extraction of the minerals (iron ore, limestone, and fossil fuels), for material conditioning (dust suppression), pollution control (scrubbers to clean up waste gases), and for cooling equipment and quenching ingots. Water consumption is measured as liters of water per kilogram of material produced, l/kg (or, equivalently, kg/kg since a liter of water weighs 1 kg). The range for engineering materials extends from 10 l/kg to over 1,000 l/kg.

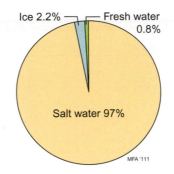

FIGURE 2.10 *The global distribution of water. Only a tiny fraction is accessible as fresh water.*

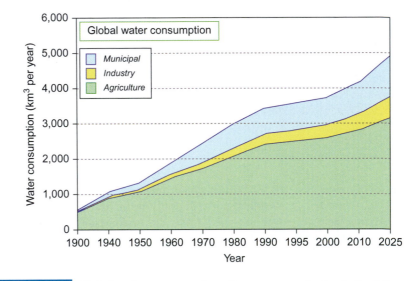

FIGURE 2.11 *Global water consumption. Energy accounts for half of the yellow "Industry" band. The other half includes material production.*

The water consumption for the growth of natural materials (biomass) requires a distinction between those that are irrigated and those that are not. Plant life that is the source of materials such as wood, bamboo, cork, and paper is not, as a general rule, irrigated, while plants used in the production of some bioplastics (cellulose polymers, polyhydroxyalkanoates, polylactides, starch-based thermoplastics) and for cattle feedstock (for leather) generally require irrigation. For this reason it is usual to split water usage into two parts: commercial water usage and total water usage. For most materials these two values are the same, but whereas the total water

Table 2.2	The water demands of energy
Energy source	**Liters water per MJ**
Grid electricity	24
Industrial electricity	11
Energy direct from coal	0.35
Energy direct from oil	0.3

usage in the growing of trees and plants includes nonirrigated water, the commercial water usage is just that which has been irrigated. The data in the data sheets in Chapter 15 are for commercial water usage.

The provision of energy, too, uses water for cooling cycles (with loss by evaporation) and for dust suppression and washing. Table 2.2 lists water usage per MJ of delivered energy for electricity, both produced and distributed via a public grid system, and electricity produced industrially (industrial electricity generation is more efficient as the hot gases produced can be used in other processes, whereas they are simply vented in electricity production for the grid).

2.4 Exponential growth and doubling times

Modern industrialized nations are heavily dependent on a steady supply of raw materials. Most materials are being produced at a rate that is growing exponentially with time, at least approximately, driven by increasing global population and standards of living. So we should look first at exponential growth and its consequences.

If the current rate of production of a material is P metric tons per year and this increases by a fixed fraction r % every year, then

$$\frac{dP}{dt} = \frac{r}{100}P \tag{2.3}$$

Integrating over time t gives

$$P = P_o \exp\left\{\frac{r(t-t_0)}{100}\right\} \tag{2.4}$$

where P_o is the production rate at time $t = t_o$. The upper part of Figure 2.12 shows how P grows at an accelerating rate with time. Taking logs of this equation gives

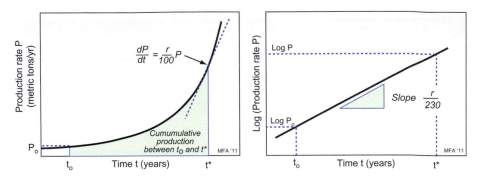

FIGURE 2.12 *Exponential growth. Production P doubles in a time* $t_d \approx 70/r$ *where r % per year is the annual growth rate.*

$$\log_e \left(\frac{P}{P_o} \right) = 2.3 \log_{10} \left(\frac{P}{P_o} \right) = \frac{r}{100} (t - t_o)$$

or

$$\log_e(P) = \log_e(P_o) + \frac{r}{230} (t - t_o) \tag{2.5}$$

so a plot of $\log_{10}(P)$ against time t, as in the lower part of Figure 2.12, is *linear* with a slope of $r/230$.

Calculating growth rates

Example: World production of silver in 1950 was 4,000 metric tons. By 2010 it had grown to 21,000 metric tons. Assuming exponential growth, what is the annual growth rate of production of silver?

Answer: Exponential growth of production P is described by equation (2.4). Setting $P = 21,000$ metric tons per year, $P_o = 4,000$ metric tons per year and $(t - t_o) = 60$ years, then solving for the growth rate r gives

$$r = \frac{100}{(t - t_o)} \ln \left(\frac{P}{P_o} \right) = 2.8\% \text{ per year}$$

Figure 2.13 shows the production of three metals and of carbon-fiber reinforced polymers (CFRP) over the past 100 years, plotted, above, on linear scales and, below, on semi-log scales, exactly as in the previous figure. The broken lines show the slopes corresponding to growth at $r = 2\%$, 5%, and 10% per year. Copper and zinc

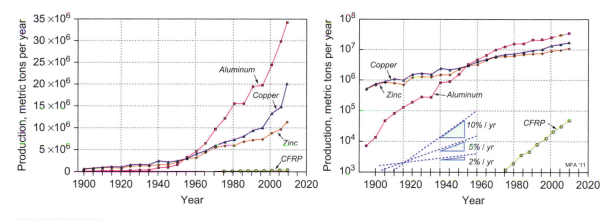

FIGURE 2.13   *The growth of production of three metals and of CFRP (carbon-fiber reinforced composites) over a 100-year interval, plotted on linear and semi-logarithmic scales*

production has grown at a consistent 3% per year over this period. Aluminium, initially, grew at nearly 7% per year but has now settled back to about 3.5%.

Exponential growth is characterized by a *doubling-time* t_D over which production doubles in size. Setting $P/P_o = 2$ in equation (2.5) gives

$$t_D = \frac{100}{r} \log_e(2) \approx \frac{70}{r} \qquad (2.6)$$

Predicting future growth rates

Example: The production of carbon fiber for structural use is growing exponentially at 10% per year. If the production volume in 2010 was 5.3×10^4 metric tons per year, what will it be in 2020?

Answer: Inserting the data into equation (2.4) of the text gives

$$P = P_o \exp \frac{r}{100}(t - t_o) = 5.3 \times 10^4 \exp (0.10 \times 10) = 1.4 \times 10^5$$

metric tons per year.

The cumulative production Q_t between times t_o and t^* is found by integrating equation (2.4) over time, giving

$$Q_t = \int_{t_o}^{t^*} P\, dt = \frac{100\, P_o}{r} \left(\exp \left\{ \frac{r(t^* - t_o)}{100} \right\} - 1 \right) \qquad (2.7)$$

This result illustrates the most striking feature of exponential growth: at a global growth rate of just 3% per year, we will mine, process, and dispose of more "stuff" in the next 25 years than in the entire 300 years since the start of the Industrial Revolution.[1] A sobering thought.

2.5 Reserves, the resource base, and resource life

The materials on which industry depends are drawn very largely from the earth's reserves of minerals. A *mineral reserve R* is defined as that part of a known mineral deposit that can be extracted legally and economically using today's technology. It is natural to assume that reserves describe the *total* quantity of mineral present in the ground that is accessible and, once used, is gone forever, but this is wrong. In reality reserves are an economic construct that grow and shrink under varying economic, technical, and legal conditions. Improved extraction technology can enlarge them, but environmental legislation or changing political climate may make them shrink. Demand stimulates prospecting, with the consequence that reserves tend to grow in line with consumption. The world reserves of lead, for instance, are three times larger today than they were in 1970. The annual production has increased by a similar factor.

The *resource base* (or just *resource*) of a mineral is the *real* total, and it is much larger. It includes not only the current reserves but also all other usable deposits that might be revealed by future prospecting and that, by various extrapolation techniques, can be estimated. It includes, too, known and unknown deposits that cannot be mined profitably now but that may become available in the future because of higher prices, better technology, or improved transportation. Although the resource base is much larger than the reserves, much of it is inaccessible using today's technology and its evaluation is subject to great uncertainty.

The distinction between reserves and resources is illustrated by Figure 2.14. It has axes showing the *degree of certainty* with which the mineral is known to exist, and the *ore grade*, a measure of the richness of the ore and, indirectly, of the ease and cost of extracting it. The largest rectangle represents the resource base. The smaller green-shaded rectangle is the reserves, of which a small part, unshaded, has been depleted by past exploitation. The reserves are extended to the right by prospecting and they are extended downward by improved mining technology or by an increase in price because this allows leaner ores to be mined profitably. A number of factors cause the reassignment of resources into and out of the *reserve* category. They include

■ Commodity price—As material prices rise, it becomes profitable to mine lower-grade ore.

---

[1]The proof of this statement can be found in the solution to exercise E 2.17, detailed in the Solution Manual for the exercises of this book.

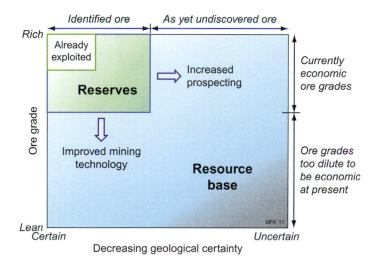

FIGURE 2.14   *The distinction between* reserves *and the* resource base. *The resource base is fixed. The reserves are the part of this that has been discovered and established as economically viable for extraction. The more dubious corners of the reserves and the resource base are shaded in grey.*

- Improved technology—New extractive methods can increase the economically workable ore grade.
- Production costs—Rising fuel or labor costs can make deposits uneconomic.
- Legislation—Tightening or loosening environmental laws can increase or decrease production costs or enable or deprive access to exploitable deposits.
- Depletion—Mining consumes reserves; prospecting enlarges them. A rate of production that exceeds that of discovery causes the reserve to shrink, a sign of problems ahead.
- Export restrictions—Limiting the export of a critical material by a producing country for economic or political reasons can make previously uneconomic reserves in other countries profitable.

Thus the reserves from which materials derive are elastic. It is nonetheless important to have a figure for their current value so that mining companies can assess their assets and governments can ensure availability of materials critical to the economy. There are procedures for estimating the current size of reserves—the US Geological Survey, for example, does so annually. The data sheets in Chapter 15 list the most recent values available at the time of writing.

If you know the size of a material's *reserve* and its *annual world production* (also listed in the data sheets) it might seem that you could estimate *resource criticality* by dividing one by the other. That, too, is wrong. Here is why.

Resource criticality: time to exhaustion. The availability of any commodity depends on the balance between supply and demand. The material supply chain,

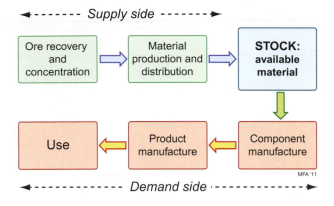

FIGURE 2.15   *The supply chain for a material. If the market works efficiently, the supply side and the demand side remain in balance. Scarcity arises when demand exceeds supply.*

sketched in Figure 2.15, has a supply side from which material flows into stock. The demand side draws material out, depleting the stock. In a free market, market forces keep the two in long-term balance, though, in the short term, there can be an imbalance while the supply side adjusts to increased demand. Material scarcity appears when the supply chain fails to respond in this way. The failure can have many origins. The most obvious is that the resource becomes so depleted that it can no longer be exploited economically.

This has led to attempts to predict resource life. If you have *D* dollars in the bank and you spend it at the rate *S* dollars per year without replenishing it, you can expect a letter from your bank manager in about *D/S* years from now. The equivalent when speaking of reserves is the exhaustion time, measured by the *static index of exhaustion*, $t_{ex,s}$

$$t_{ex,s} = \frac{R}{P} \tag{2.8}$$

where *R* (metric tons) is the reserve and *P* (metric tons per year) is the production rate. But this ignores the growth. As we have seen, the production rate *P* is not constant, but generally increases with time at a rate *r* per year. Allowing for this means that the reserves will be consumed in a shorter time:

$$t_{ex,d} = \frac{100}{r} ln\left\{ \frac{rR}{100P} + 1 \right\} \tag{2.9}$$

known as the *dynamic index*.

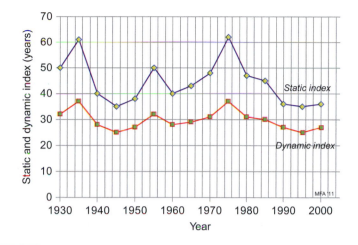

FIGURE 2.16 *The static and dynamic index for copper over the last 70 years*

Indices of exhaustion

Example: The current reserves of copper stand at 550 million metric tons. The annual world production of copper is 15 million metric tons, growing at 3% per year. What are the values of the static and dynamic indices of exhaustion for copper?

Answer: Inserting these data into equations (2.8) and (2.9) gives $t_{ex,s} = 37$ years and $t_{ex,d} = 32$ years.

Figure 2.16 shows the static and dynamic index for copper over the last 70 years. The static index has hovered around 40 years for the whole of that time; the dynamic index has done the same at 30. As a prediction of resource exhaustion, neither one inspires confidence. The message they convey is *not* that we will run out of copper in 30 years time. It is that a comfort zone exists for the value of the index, and that it is around 30 years. Only when it falls below this value is there sufficient incentive to prospect for more. We need to approach the problem of resources criticality in another way.

Scarcity: market balance and breakdown. Scarcity arises when supply fails to meet demand. Thus far we have assumed (without saying so) that the market works effectively, keeping the supply side and the demand side in balance. When it does, increased demand is met by increased prospecting, improved technology, and increased production. The price of a material then reflects the true cost of its extraction. But what happens when market forces don't work? That is a thought that troubles both industry and government.

Some resources are widely distributed. Others are not; for these the ore bodies that are rich enough to be worked economically are located in just one or a few countries—it is called *supply chain concentration*. Then political unrest, economic upheaval, rebellion, or forced change of government in one of these countries or its neighbors (through which the ore must be transported) disrupts supply in ways to which the market cannot immediately respond. And there is the potential, too, for the few suppliers to reach agreement to limit supply, thereby driving up prices, a process known as *cartel action*. The market cannot immediately respond because it takes time to set up new extraction facilities and establish new supply chains.

If the supplier that stocks your favorite wine closes, you seek another supplier. If they, already under pressure, have only limited stock, you could try to buy it all and store it, creating a stockpile. If they refuse, you have to consider reducing your consumption or—horrors—finding a substitute, keeping your favorite wine for special occasions. That, too, is the reaction of the demand side of the market: seek other suppliers, stockpile, explore substitutes for all but the most demanding applications, and (something that does not work with wine) increase recycling.

News-clips: Price volatility

Walkout lifts copper price.
Copper prices are set to jump after a strike in Chile spread yesterday.
The Sunday Times, July 31, 2011

Silver at its highest price for 31 years.
In six months the silver market has leapt by more than 100% [to $47 per ounce]. The industrial uses of sliver (particularly in mobile communication and solar panels) represent half the annual consumption of this metal.
The Figaro, April 25, 2011

Gold, safe-haven in bad times.
The trading price of gold has multiplied by a factor of seven in the past 10 years [to $1520 per ounce].
The Figaro, April 30, 2011

Price volatility of this sort, due either to scarcity or to hedging against unstable currencies, makes life very difficult for manufacturers of products that require these metals.

Can scarcity be predicted? Scarcity caused by failure of the supply chain becomes less likely when the chain is diverse, meaning that the ore deposits are widely distributed and there are many producers and distributors. When the reverse is true, manipulation of price by limiting supply becomes more likely.

With this background, let us return to the question of criticality.

More realistic indicators of criticality. The argument here gets a little more complex. Figure 2.17 helps set it out. The resource-base—the biggest rectangle of Figure 2.14—is, of course, finite. The green hill of Figure 2.17 shows schematically how the reserve—the exploitable part of the resource base—at first grows as prospecting and improved extraction technology allow access to more material. Its exploitation (the yellow hill) begins to eat it away, but initially the rate of discovery reassuringly exceeds that of exploitation and no alarm bells ring. But there comes a point at which the finding of new deposits and further improved technology become more difficult. Then reserves, while still growing, grow at a rate that falls behind that of production. This decline in discovery rate is followed, with a characteristic time lag, by a decline in production.

This progression is reflected in the price of minerals. When first discovered and exploited, a mineral is expensive (blue line on Figure 2.17). As extractive technology improves and prospecting unveils richer deposits, the price falls. While reserves remain large, the price, when corrected for inflation, settles to a stable plateau. But then, with about the same time lag as that for the fall in production, depletion makes itself felt and the price climbs. The crossing point is significant: once past this point the reserves start to shrink—they are being used faster than they are being replenished. There are indicators of criticality here that can be taken seriously:

- The rate of discovery falls below the rate of production.
- The production rate curve peaks and starts to decline.
- Materials must be extracted from leaner ore grade.
- Price fluctuations cease to be intermittent and prices rise permanently.

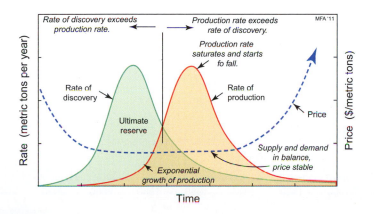

FIGURE 2.17   *The evolution of the rates of discovery of a resource (broken curve) and its exploitation, the rate of production (red curve), and the price (blue curve)*

In reality the curves are not smooth. Figure 2.17 is a schematic. Reality looks more like Figure 2.18. It shows oil discoveries, production, and the price corrected to the value of the dollar in 2008, over time since 1900. Here some of the fluctuations are included. Major discoveries do not occur every year, which explains the lumpy profile of the green blocks. Anticipated discovery between now and 2040 are the grey blocks. Another big lump or two is certainly possible—the competing claims by Russia, Norway, Canada, and the US for territorial rights in the Arctic suggest that there may be large deposits there. Ignoring this for the moment, a smoothed curve—the green hill—has been sketched in. Growth in production—the yellow hill—is much smoother; the demand has grown steadily over the last 100 years. There have been some fluctuations (not shown) but these are small. That of price—the blue curve on the right-hand scale—is much more erratic. Oil price does not reflect the actual cost of extraction, but rather the degree to which the rate of production is controlled to maintain price. As national reserves become depleted, those of the greatest consumers—the US and Western Europe—cannot meet demand, creating dependence on imports and vulnerability to production quotas. The spike in the 1970s and that of 2008 are examples. The fluctuations of the real data are distracting, but if these are smoothed out, say, by using a 10-year moving average of the data, the evolution begins to look like that of the schematic. We are past the crossover point of the discovery and production-rate curves. It seems likely that this rise in price is not a temporary fluctuation but is here to stay. Price increases as large as this, however, open new doors. Oil sands and offshore Arctic oil, for example, have, until now, been largely inaccessible because of the cost of exploitation. They now become economically attractive, extending the tail of the estimated production curve in Figure 2.18.

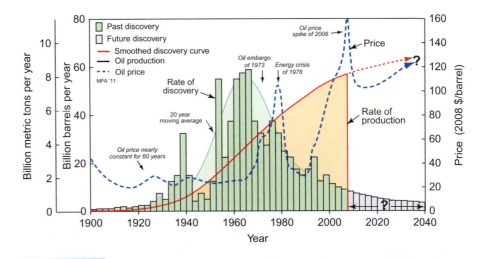

FIGURE 2.18 *The evolution of the rates of discovery, the rate of production, and the price of oil*

The crossover points of the discovery and production curves for most minerals from which engineering materials are drawn have not yet been reached. For many it is still far off. But at least we now have measures that mean something.

Strategic (critical) materials. The mineral resource bases from which many materials are drawn are so large and widely distributed that the health of the supply chain is not a concern. The resource bases supporting the steel and aluminium industry are examples—both are vital to the economy; it is the resource of energy rather than that of material that could limit their production. But there are others that are a cause of concern. They are the materials for which the known reserves are limited in size or are localized in countries from which supply cannot be guaranteed, or both. Governments classify materials as "strategic" or "critical" if their supply is concentrated in one country or could be restricted by few corporate interests, and because they are used in products that are important economically or for national security.

Table 2.3 gives some examples of materials classified as strategic. It is interesting to ask, "Why?" The answer lies partly in their specialized uses and partly in the global location of the main ore bodies from which they are drawn. The members of

Table 2.3	Elements that, by reason of critical use or extreme localization, are deemed "strategic"	
Element	**Critical applications**	**Principal global sources***
General engineering		
Copper	Electrical conduction in all electro-mechanical things	Canada, Chile, Mexico
Manganese	Essential alloying element in steel	South Africa, Russia, Australia
Columbium (Niobium)	Micro-alloyed steels; super alloys; superconductors	Brazil, Canada, Russia
Tantalum	Ultra compact capacitors for mobile electronic equipment; alloying element in steel	Australia, China, Thailand
Vanadium	High-speed tool steel, micro-alloyed steel	South Africa, China
Cobalt	Cobalt-based super alloys; alloying element in steel	Zambia, Canada, Norway
Titanium	Light, high-strength, corrosion resistant alloys	China, Russia, Japan
Rhenium	High-performance turbines	Chile
Electronics		
Lithium	Lithium-ion batteries; Al-Li alloys for aircraft	Russia, Kazakhstan, Canada
Gallium	Gallium arsenide PV devices; semiconductors	Canada, Russia, China
Indium	Transparent conductors; InSb semiconductors; LEDs	Canada, Russia, China
Germanium	Solar cells	China

(Continued)

Table 2.3	(Continued)	
Element	**Critical applications**	**Principal global sources***
Platinum group		
Platinum	Catalyst in chemical engineering and auto exhausts	South Africa, Russia
Palladium	Catalyst in chemical engineering and auto exhausts	South Africa, Russia
Rhodium	Catalyst in chemical engineering and auto exhausts	South Africa
Rare earths		
Lanthanum	High refractive index glass; hydrogen storage; battery-electrodes, particularly for hybrid cars	China, Japan, France
Cerium	Catalysis; alloying element in aluminum alloys	China, Japan, France
Praseodymium	Rare-earth magnets; materials for lasers	China, Japan, France
Neodymium	Rare-earth permanent magnets; lasers	China, Japan, France
Promethium	Nuclear batteries (beta-emissions to electric power)	China, Japan, France
Samarium	Rare-earth magnets; lasers; neutron capture; masers	China, Japan, France
Europium	Red and blue phosphors; lasers; mercury-vapor lamp	China, Japan, France
Gadolinium	Rare-earth magnets; high-refractive index glass or garnets; laser; x-ray tube; computer memory; neutron capture	China, Japan, France
Terbium	Green phosphors; laser; fluorescent lamp	China, Japan, France
Dysprosium	Rare-earth magnets lasers	China, Japan, France
Holmium	Lasers	China, Japan, France
Erbium	Lasers; vanadium steel	China, Japan, France
Ytterbium	Infrared laser; high-temperature superconductors (YBCO)	China, Japan, France
Lutetium	Catalysts in petroleum industry	China, Japan, France
Lanthanum	High-refractive index glass; hydrogen storage; battery-electrode; camera lens	China, Japan, France

Sources: USGS (2002); US Department of Energy (2010); Jaffee and Price (2010)

the first group have applications in general engineering. The high electrical conductivity of copper gives it exceptional economic importance—replacing it by any affordable substitute results in increased resistive losses, lower electrical efficiencies, and higher energy costs. Manganese has no satisfactory substitute as an alloying addition in carbon and stainless steel. Columbium (niobium), tantalum, and vanadium, through their ability to form very stable carbides, have become essential ingredients of high-strength steel. The alloys of titanium have a unique position as light, high-strength, corrosion resistant alloys.

The elements in the other three groups have more specialized applications. Electronic engineering, with its dependence on semiconducting devices, has created demand for a much wider spectrum of elements (it is said that a mobile phone contains most of the Periodic Table). Chemical and petroleum engineering rely heavily on catalysts drawn from the platinum group. Laser, magnet, and battery technologies increasingly rely on rare earths for their performance. Many of these elements are drawn from small, highly localized resource bases: 80% of the world production of platinum group elements comes from South Africa; 90% of all rare earth supplies come from China. For these materials it is the unique properties, the rarity, and the extreme localization of the sources that makes them "strategic."

News-clip: Supply-side breakdown

US digs deep to secure the technology of the future.
It's a deep pit in the Mojave Desert but it could hold the key to the US challenging China's technological domination of the 21st century. The mine (which closed eight years ago) is the largest known deposit of rare-earth elements outside China. Now the owners have approval to restart operations. As far as the US Government is concerned, the mine cannot open soon enough. A Department of Energy report (December, 2010) warned that the US risks losing control over the production of a host of technologies, from smart phones to smart bombs, electric car batteries to wind turbines, because China has a virtual monopoly on the rare-earth metals essential to their production.

The Guardian, January 1, 2011

China controls 97% of global rare-earth metals production. The consequences of such total domination of a strategic resource became impossible to ignore when, in October 2010, China cut exports of the metals by 70% compared with the previous year, disrupting manufacturing in Japan, Europe, and the US.

News-clip: Demand-side breakdown

Japan crisis hurts uranium miners.
Uranium miners are suffering at the hands of investors amid fears that Japan's nuclear crisis (the melt-down at the Fukushima plant) could restrict the growth of atomic energy.

The Times, March 18, 2011

In March 2011 uranium share prices fell 35%, takeovers were cancelled, China suspended approval permits for new reactors, and Germany indicated it would phase out its reactors altogether. The problems are not always on the supply side.

News-clip: Resources, a blessing or a curse?

Miner pays Aborigines $2bn to gain land rights.
Rio Tinto has struck a $2 billion, 40 year deal with five Aboriginal groups to secure access to their land in resource-rich Western Australia.

The Times, June 4, 2011

Will this deal lift living standards for these Aborigines? Since 1990 alone, the petroleum industry has invested more than $20 billion in exploration and production activity in Africa, but the wealth has destroyed traditional lifestyles and generated armed conflicts. Diamonds and gold have brought similar societal breakdown. It is a paradoxical fact that countries and regions with an abundance of natural resources, particularly minerals and oil, tend to have less economic growth and worse development outcomes than countries with less—an effect known as the *resource curse.*

2.6 The materials-energy-carbon triangle

The national press, periodically, carries reports that we will soon run out of one material or another. The reports derive from a misinterpretation of the time-to-exhaustion estimates discussed in Section 2.3. We will not run out of materials. Market forces will see to that. But they will become more expensive.

As resources become depleted, it takes more energy to extract them—they become more *energy-intensive*. Given boundless cheap, carbon-free energy, materials can be extracted from ever leaner ores, from the ocean, from landfill sites, and from products at the end of life even when the quantities they contain are miniscule. The size of these combined resources is enormous. With cheap, carbon-free energy the material supply can continue (almost) forever. But most of the energy we use at present is not carbon-free. Nor is it boundless or cheap: North Sea oil and gas are increasingly depleted, and prospecting for more now involves drilling through kilometers of rock beneath a kilometer of water.

The global response has been to seek energy sources that are "carbon-free" (none are completely that), replacing energy from fossil fuels with energy harvested, directly and indirectly, from the sun (photovoltaics, wind, hydro, biomass), from the moon (tidal power), or from radioactive decay (nuclear, geothermal). But the *energy densities* of all but nuclear power are low. That means that the harvesting structures have to cover large areas to capture useful quantities of energy and it takes a lot of material to make them.

Thus "carbon-free" energy is *material-intensive*. Materials, as we saw in Section 2.3, are *energy-intensive*. Energy, the way we make it now, is *carbon-intensive* (Figure 2.19). Any move to reduce one has an impact on the others; it is a classic example of conflicting objectives. To confront this quandary more rationally, we must examine the materials-energy-carbon triangle in more depth, but we will defer that to Chapter 12.

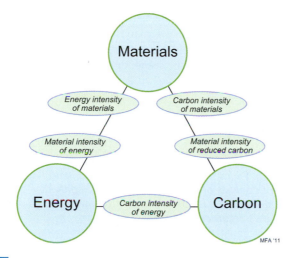

FIGURE 2.19 *The materials-energy-carbon triangle. Each depends on the other two such that reducing the drain on one often increases the drain on another. Using resources efficiently requires a balance between the three.*

2.7 Summary and conclusions

The growth of global population and affluence increases the demand for energy and materials. The growth in demand is approximately exponential, meaning that consumption grows at a rate that is proportional to its current value—for most materials it is between 3 and 6% per year. Exponential growth has a number of consequences. One is that consumption doubles every $70/r$ years, where r is the growth rate in % per year. It also means that the total amount consumed (the integral of the consumption over time) also doubles in the same time interval.

Most materials are drawn from the minerals of the earth's land masses and oceans. The *resource base* from which they are drawn is large, but it is not infinite. Its magnitude is not easy to estimate so that at any point in time, only a fraction of it, the *reserve*, is established as accessible and economically viable. Market forces, when they operate properly, ensure that reserves remain adequate, keep supply in line with demand, and ensure that prices remain stable and fair.

The balance between supply and demand can, however, be disrupted. Depletion of the resource base causes scarcity, driving up prices. Reserves that are localized can become vulnerable to cartel action or cut off by local political unrest. Market disruption can be economically damaging, so foreseeing and anticipating it is necessary if stability is to be maintained. Economic forecasting of this sort is based on the tracking of discovery and production rates of material resources and the identification of their sources, flagging those for which a single source dominates world markets as vulnerable.

Behind all this is a bigger issue: the tight coupling between materials, energy, and carbon. Energy is needed to make materials. Materials are needed to make usable energy. Emissions are unavoidable if materials are made and energy is used. The final figure of the chapter sets the scene for a fuller exploration of this coupling in Chapter 12.

2.8 Further reading

Allwood, J.M., Ashby, M.F., Gutowski, T.G., and Worrell, E. (2011), "Material efficiency, a White Paper," Resources, Conservation, and Recycling, 55, pp 362−381 (*An analysis of the need for material efficiency, possible ways of achieving it, and the obstacles to implementing them. Much of the reasoning of this chapter follows arguments developed in this paper.*)

Alonso, E., Gregory, J., Field, F., and Kirchain, R. (2007), "Material availability and the supply chain: risks, effects and responses," *Environmental Science and Technology*, 41, pp 6649−6656. (*An informative analysis of the causes of instability in material price and availability*)

Chapman, P.F. and Roberts, F. (1983), *Metal resources and energy*, Butterworths Monographs in Materials, Butterworth and Co., Thetford, UK. ISBN 0-408-10801-0. (*A monograph that analyzes resource issues, with particular focus on energy and metals*)

Cullen, J.M. (2010), "Engineering fundamentals of energy efficiency," PhD Thesis, Engineering Department, Cambridge University, Cambridge, UK. (*Cullen analyzes the big picture—the efficiency of the global use energy and the scope for improving this.*)

Cullen, J.M. and Allwood, J.M. (2010), "The efficient use of energy: tracing the global flow of energy form fuel to service," *Energy Policy*, 38, pp 75−81. (*A study of the "losses" associated with the energy-conversion and energy-transmission steps in energy-using processes*)

Harvey, L.D.D. (2010), *Energy and the new reality 1: energy efficiency and the demand for energy services*, Earthscan Ltd, London, UK. ISBN 978-1-84971-072-5. (*An analysis of energy use in buildings, transport, industry, agriculture, and services, backed up by comprehensive data*)

Jaffe, R. and Price, J. (2010), "Critical Elements for New Energy Technologies," American Physical Society Panel on Public Affairs (POPA) study, American Physical Soc. USA.

McKelvey, V.E. (1974), *Technology Review*, MIT Press, (March/April) p 13. (*The original presentation of the McKelvey diagram*)

MRS (2010), "Energy critical materials—securing materials for emerging technologies," A report by the APS Panel on Public Affairs & the Materials Research Society, American Physical Society, Washington, DC, USA. (*An assessment of the metals that are critical for future energy production and use, and the security—or otherwise—of the supply chain. See also, US Department of Energy, 2010, below.*)

Shell Petroleum (2007), *How the energy industry works*, Silverstone Communications Ltd., Towchester, UK. ISBN978-0-9555409-0-5. *(Useful background on energy sources and efficiency)*

Shiklomanov, I.A. (2010), "World water resources and their use," UNESCO International Hydrological Programme, www.webworld.unesco.org, accessed July 2010. *(A detailed analysis of world water consumption and emerging problems with supply)*

US Department of Energy (2010a), "Critical materials strategy," Office of Policy and International Affairs, materialstrategy@hq.doe.gov, www.energy.gov. *(A broader study than MRS 2010, above, but addressing many of the same issues of material critical to the energy, communication, and defense industries, and the priorities for securing adequate supply)*

US Department of Energy (2010b), "Critical materials strategy for clean technology," www.energy.gov/news/documents/criticalmaterialsstrategy.pdf. *(A report on the role of rare-earth elements and other materials in low-carbon energy technology)*

USGS Circular 2112 (2002), "Materials in the economy—material flows, scarcity and the environment," by L.W. Wagner, US Department of the Interior (www.usgs.gov). *(A readable and perceptive summary of the operation of the material supply chain, the risks to which it is exposed, and the environmental consequences of material production)*

USGS Mineral Information (2007), "Mineral yearbook" and "Mineral commodity summary," http://minerals.usgs.gov/minerals/pubs/commodity/. *(The gold standard information source for global and regional material production, updated annually)*

Wolfe, J.A. (1984), *Mineral resources—a world review*, Chapman & Hall, London. ISBN 0-4122-5190-6. *(A survey of the mineral wealth of the world, both for metals and nonmetals, describing the extraction and the economic importance of each)*

2.9 Exercises

E2.1. Explain the distinction between reserves and the resource base.

E2.2. Use the Internet to research rare-earth elements. What are they? Why are they important? Why is there concern about their availability?

E2.3. The average concentration of iridium in the earth's crust is $3 \times 10^{-10}\%$ by weight. The price of iridium is about \$14,000/kg. Use equation (2.1) to estimate the grade of the ore from which iridium is drawn. How does this compare with the average concentration in the earth's crust?

E2.4. The global production of platinum in 2011 stands at about 178 metric tons per year, most from South Africa. The catalytic converter of a car requires about 1 gram of precious metal catalyst, most commonly, platinum. Car manufacture in 2011 was approximately 52 million vehicles. If all have platinum catalysts, what fraction of the world production is absorbed by the auto industry? If the production rate of cars is growing at 4% per year and that of platinum is constant, how long will it be before the demand exceeds the total global supply?

E2.5. The world consumption rate of CFRP is rising at 10% per year. How long does it take to double?

E2.6. Derive the dynamic index

$$t_{ex,d} = \frac{100}{r} \log_e \left(\frac{rR}{100\,P_o} + 1 \right)$$

starting with equation (2.3) of the text.

E2.7. A total of 16 million cars were sold in China in 2010; in 2008 the sale was 6.6 million. What is the annual growth rate of car sales, expressed as % per year? If there were 16 million cars already on Chinese roads by the end of 2010 and this growth rate continues, how many cars will there be in 2020, assuming that the number that are removed from the roads in this time interval can be neglected?

E2.8. Understanding reserves: copper. The table lists the world production and reported reserves of copper over the 13 years between 1995 and 2007.

Year	Price (US $/kg)	World production (million metric tons/yr)	Reserves (million metric tons)	Reserves/world production (yrs)
1995	2.93	9.8	310	
1996	2.25	10.7	310	
1997	2.27	11.3	320	
1998	1.65	12.2	340	
1999	1.56	12.6	340	
2000	1.81	13.2	340	
2001	1.67	13.7	340	
2002	1.59	13.4	440	
2003	1.78	13.9	470	
2004	2.86	14.6	470	
2005	3.7	14.9	470	
2006	6.81	15.3	480	
2007	7.7	16	480	

- Examine trends (plot price, production, and reserves against time)—what do you conclude?
- Tabulate the Reserves/world Production to give the static index of exhaustion. What does the result suggest about reserves?

E2.9. The table shows the annual production and the reserves of five metals over a period of 10 years. What has been the growth rate of production? What is that of the reserves? What can you conclude about the criticality of each material?

Metal	Year	Annual production, metric tons/ year	Reserves, metric tons	Growth rate of production in% per year	Growth rate of reserves in % per year
Platinum	2005	217	71×10^3		
	1995	145	56×10^3		
Nickel	2005	1.49×10^6	64×10^6		
	1995	1.04×10^6	47×10^6		
Lead	2005	3.27×10^6	67×10^6		
	1995	2.71×10^6	55×10^6		
Copper	2005	15.0×10^6	480×10^6		
	1995	10.0×10^6	310×10^6		
Cobalt	2005	57.5×10^3	7×10^6		
	1995	22.1×10^3	4×10^6		

E2.10. The table lists the production of titanium, in metric tons per year, over time. Plot the data onto a copy of the lower part of Figure 2.12 of the text, plotting log (consumption) against time.

What, approximately, is the average growth rate of production of titanium between 1960 and 2007? How long will it be before the production of CFRP exceeds that of titanium?

E2.11. Tabulate the annual world production (metric tons/year) and the densities (kg/m^3) of carbon steel, PE, softwood, and concrete. You will find the data for these materials in Chapter 15. Use an average of the ranges given in the data sheets. Calculate, for each, the annual world production measured in m^3/year. How does the ranking change?

E2.12. The prices of cobalt, copper, and nickel have fluctuated wildly in the past decade. Those of aluminum, magnesium, and iron have remained much more stable. Why? Research this by examining uses (which metal are used in high value-added products?) and the localization of the producing mines. The USGS web site listed under Further reading is a good starting point.

E2.13. The production of zinc over the period 1992–2006 increased more or less steadily at a rate of 3.1% per year. The reserves, over the same period, increased by 3.5%. What conclusions can you draw from this about the criticality of the zinc supply? What is your reasoning?

E2.14. The production of platinum, vital for catalysts and catalytic converters, has risen from 145 to 217 metric tons per year over the last 10 years. The ores are highly localized in South Africa, Russia, and Canada. The reserves have risen from 56,000 to 71,000 metric tons in the same time interval. Would you classify platinum as a critical material? Base your judgment on these facts, making use of the relative growth rates of production and reserves and on the dynamic index (equation (2.9) of the text) calculated using 2005 data.

E2.15. Global water consumption has tripled in the last 50 years. What is the growth rate, $r\%$, in consumption, C, assuming exponential growth? By what factor will water consumption increase between now (2012) and 2050?

E2.16. Plot the *Annual world production* of metals against their *Price*, using mean values from the data sheets at the end of this book. What trend is visible?

E2.17. Prove the statement made in the text that, "at a global growth rate of just 3% per year we will mine, process, and dispose of more 'stuff' in the next 25 years than in the entire 300 years since the start of the industrial revolution."

Exercises using CES Eco Level 2

E2.18. Use CES to plot the *Annual world production* of materials against their *Price*. What trend is visible?

E2.19. Make a plot of the static index of exhaustion for metals (the reserves in metric tons divided by the annual production rate in metric tons per year), using the "Advanced" facility in CES to plot the ratio. Which metals have the longest apparent resource life? Why is this index not a reliable measure of the true resource life of the metal?

CHAPTER 3

The material life cycle

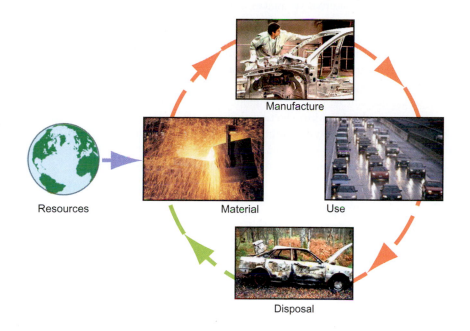

Resources · Material · Manufacture · Use · Disposal

3.1 Introduction and synopsis

The materials of engineering have a life cycle. Materials are created from ores and feedstock. These are manufactured into products that are distributed and used. Products, like us, have finite life, at the end of which they become scrap. The materials they contain, however, are still there; some, unlike us, can be resurrected and enter a second life as recycled content in a new product.

Life-cycle assessment (LCA) traces this progression, documenting the resources consumed and the emissions excreted during each phase of life. The output is a sort of biography, documenting where the materials have been, what they have done, and the consequences of this for their surroundings. It can take more than one form. It can be a full LCA that scrutinizes every aspect of life (arduous and expensive in time and money); or it can be a brief character-sketch painting, an approximate (but still useful) portrait; or it can be something in between.

Image of casting courtesy of Skillspace; image of car making courtesy of US Department of Energy EERE program, image of cars courtesy of Reuters.com)

Responsible design, today, aims to provide safe, affordable services while minimizing the drain on resources and the release of unwanted emissions. To do this, the designer needs an ongoing eco-audit of the design (or redesign) as it progresses. To be useful the eco-audit must be fast, allowing quick "what if?" exploration of the consequences of alternative choices of material, use pattern, and end-of-life scenarios. A full LCA is not well adapted for this task—it is slow and expensive. *Streamlined LCA* and the *eco-audit methods* have evolved to fill the gap. They are approximate but still have sufficient resolution to guide decision making.

This chapter is about the life cycle of materials and its assessment: how an LCA works, its precision (or lack of it), the difficulties of implementing it, and ways these difficulties can be bypassed to guide material choice in product design. The chapter starts with a brief introduction to the design process itself—we need that to see how the assessment and auditing methods mesh with design. It ends by introducing a strategy that is developed in the chapters that follow. There is also an appendix describing currently available LCA software.

3.2 The design process

The starting point of a design is a *market need* or a *new idea*; the end point is the full *specification* of a product that fills the need or embodies the idea. It is essential to define the market need precisely, that is, to formulate a *need statement*, often in the form: "a device is required to perform task X," expressed as a set of *design*

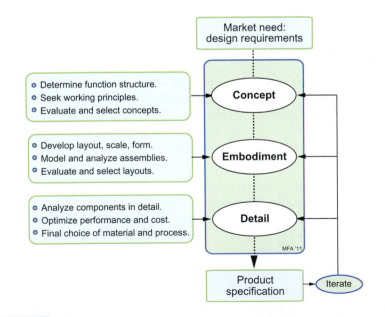

FIGURE 3.1   *The design process: requirements, concept, embodiment, detail, production*

requirements. Between the need statement and the product specification lie the set of stages shown in Figure 3.1: the stages of *concept*, *embodiment*, and *detailed design*.

The design proceeds by developing concepts to perform the functions in the design requirements, each based on a *working principle*. At the concept stage of design all options are open: the designer considers alternative concepts and the ways in which these might be separated or combined. The next stage, embodiment, takes the promising concepts and seeks to analyze their operation at an approximate level. This involves sizing the components and selecting materials that will perform properly in the ranges of stress, temperature, and environment suggested by the design requirements, examining the implications for performance and cost. The embodiment stage ends with a feasible layout, which is then passed to the detailed design stage. Here specifications for each component are drawn up. Critical components may be subjected to precise mechanical or thermal analysis. Optimization methods are applied to components and groups of components to maximize performance. A final choice of geometry and materials is made and the methods of production are analyzed and priced. The output of the detailed stage is a detailed production specification.

The environmental impact that a product has over its subsequent life is largely determined by decisions taken during the design process. The concept, the embodiment, the detail, and the choice of materials and manufacturing process all play a role. A complete assessment of this impact requires a scrutiny of the entire life cycle.

3.3 The materials life cycle

The idea of a *life cycle* has its roots in the biological sciences. Living organisms are born; they develop, mature, grow old and, ultimately, die. The progression is built-in—all organisms follow broadly the same path—but the way they develop on the way, and their behavior, lifespan, and influence depend on their interaction with their *environment*—the surroundings in which they live. Life sciences track the development of organisms and the ways in which they interact with their environment.

The life cycle idea has since been adapted and applied in other fields. In the social sciences it is the study of the interaction of individuals with their social environment. In the management of technology it is the study of the birth, maturity, and decline of an innovation in the business environment. In product design it is the interaction of products with the natural, social, and business environments. Concern about resource depletion (the Club of Rome report, already described), the oil crisis of the early 1970s, followed by the first evidence of carbon-induced global warming, focused attention on yet another field: the life cycle of manufactured products and their interaction, above all, with the natural environment. Products are made of materials—materials are their flesh and bones, so to speak—and these are central to the interaction. The study of a product and its associated material life cycle involves assessing the environmental impacts associated with its life, from the extraction of raw materials to their return to the ecosphere as "waste"—from birth to death, or (if you prefer) cradle to grave. That means tracking materials through life. So let us explore that.

Figure 3.2 is a sketch of the materials life cycle. Ore, feedstock, and energy are drawn from the planet's natural resources and processed to produce materials. These are further processed to create the materials that are subsequently manufactured into products, which are distributed, sold, and used. Products have a useful life at the end of which they are discarded; a fraction of the materials they contain might enter a recycling loop, the rest is committed to incineration or landfill.

Energy and materials are consumed at each point in the life cycle of Figure 3.2, depleting natural resources. There is an associated penalty of carbon dioxide, CO_2, oxides of sulfur, SO_x, and of nitrogen, NO_x, and other emissions in the form of gaseous, liquid, and solid waste and low-grade heat. In low concentrations most of these are harmless, but as their concentrations build, they become damaging. The problem, simply put, is that the sum of these unwanted by-products now often exceeds the capacity of the environment to absorb them. For some, the damage is local and the originator of the emissions accepts the responsibility and cost of containing and fixing it (the environmental cost is said to be *internalized*). For others the damage is global and the creator of the emissions is not held directly

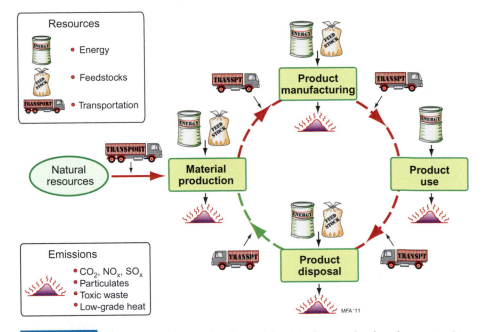

FIGURE 3.2   *The material life cycle. Ore and feedstock are mined and processed to yield a material. This is manufactured into a product that is used and, at the end of its life, discarded, recycled, or, less commonly, refurbished and reused. Energy and materials are consumed in each phase, generating waste heat and solid, liquid, and gaseous emissions.*

responsible, so the environmental cost becomes a burden on society as a whole (it is *externalized*). The study of resource consumption, emissions, and their impacts is called *life-cycle assessment* (LCA).

News-clip: Externalized costs

Nitrogen pollution.
A study evaluates the cost of nitrogen run-off as 150–740 euros per year per EU inhabitant.

Le Monde, **April 14, 2011**

Nitrates increase the productivity of land and help meet the increasing demand for food that accompanies population growth. The damage caused by nitrate run-off to the ecology of rivers and coastal waters has long been known, but its cost has not been factored into the economics of agriculture. We know the gain in agricultural output that nitrates provide, but up until now, no figure has been placed on the damage they cause. This study sets a very broad range on this figure and allows a first estimate of the net gain or loss associated with the use of nitrates. But until this cost is attached to the price of agricultural produce from nitrated soil, it remains externalized, a hidden cost falling on all EU inhabitants.

3.4 Life-cycle assessment: details and difficulties

Formal methods for life-cycle assessment first emerged in a series of meetings organized by the Society for Environmental Toxicology and Chemistry (SETAC) of which the most significant were held in 1991 and 1993. This led, from 1997 on, to a set of standards for conducting an LCA, issued by the International Standards Organization (ISO 14040 and its subsections 14041, 14042, and 14043). These prescribe procedures for (and here I quote)

> *defining goal and scope of the assessment, compiling an inventory of relevant inputs and outputs of a product system; evaluating the potential impacts associated with those inputs and outputs; interpreting the results of the inventory analysis and impact assessment phases in relation to the objectives of the study.*

The study must (according to the ISO standards) examine energy and material flows in raw material acquisition, processing and manufacturing, distribution and storage (transport, refrigeration, and so forth), use, maintenance and repair, recycling options, and waste management.

There is a lot here and there is more to come. A summary in plainer English might help.

- *Goals and scope:* Why do the assessment? What is the subject and which bit(s) of its life are assessed?

- *Inventory compilation:* What resources are consumed? What emissions are excreted?
- *Impact assessment:* What do the resource consumption and emissions do to the environment—particularly, what bad things?
- *Interpretation:* What do the results mean? If they are bad, what can be done about it?

We look now at what each involves.

Goals and scope. Why do the study? Here are some possible answers.

- To guide the design of more environmentally friendly products
- To demonstrate that you are an environmentally responsible manufacturer
- To allow the public to form their own judgment about your products
- To demonstrate that your products are greener than those of your competitor
- To be able to claim conformity to standards such as ISO 14040 and PAS 2050 (described later)
- Because the enterprise to which you are a supplier or subcontractor requires that you do so so that they can claim conformity to standards.

There is a wide spread of motives here—it would be surprising if one assessment method fit the needs of all.

And there is the question of scope: where should the LCA start and finish? Figure 3.3 shows the four phases of life, each seen as a self-contained unit, with notional "gates" through which inputs pass and outputs emerge. If you were the manager of the manufacturing unit, for example, your purpose might be to assess your plant, ignoring the other three phases of life because everything outside your gates is beyond your control. This is known as a "gate-to-gate" study; its scope is limited to the activity inside the box labelled System Boundary A. There is a tendency for the individual life phases to seek to minimize energy use, material waste, and internalized emission costs spontaneously because it saves money to do so. But this action by one phase may have the result of raising resource consumption and emissions of the others. An example: if minimizing the manufacturing energy and material costs for a car results in a heavier vehicle and one harder to disassemble at end of life, then the gains made in one phase have caused losses in two of the others. Put briefly: the individual life phases tend to be self-optimizing; the system as a whole does not. We return to this in Chapters 9 and 10 where the necessary trade-off methods are developed.

If the broader goal is to assess the resource consumption and emissions of the product over its entire life, the boundary must enclose all four phases (System Boundary B). The scope becomes that of product birth to product death, including, at birth, the ores and feedstock and, at death, the consequences of disposal.

Some LCA proponents see a still more ambitious goal and grander scope (System Boundary C). If ores and feedstock are included (as they are within System Boundary B), why not the energy and material flows required to make the

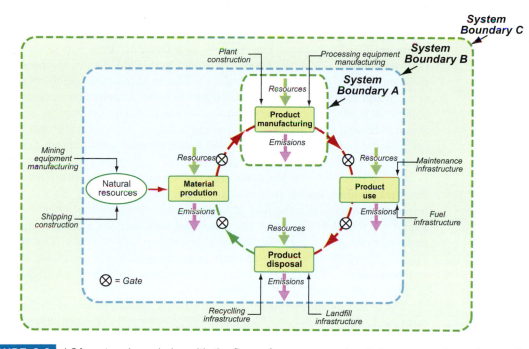

FIGURE 3.3 *LCA system boundaries with the flows of resources and emissions across them. System Boundary A encloses a single phase of the lifecycle. System Boundary B encloses the direct inputs and emissions of the entire life. It does not make sense to place the system boundary at C, which has no well-defined edge.*

equipment used to mine them? And what about the resource and emission flows to make the equipment that made *them*? It is the "infinite recession" problem. Here an injection of common sense is needed. Setting the boundaries at infinity gets us nowhere. Equipment-making facilities make equipment for other purposes, too, and this produces a dilution effect: the remoter they are, the smaller the fraction of their resources and emissions that is directly linked to the product being assessed. The standards are vague on how to deal with this point, merely instructing that the system boundary "shall be determined," leaving the scope of the assessment as a subjective decision. Input-output analysis gives a formal structure for dealing with these more remote contributions, but we shall leave that for later. For now, the practical way forward is to include only the primary flows directly required for the materials, manufacturing, use, and disposal of the product, excluding the secondary ones needed to make the primary ones possible.

Inventory compilation. Setting the boundaries is the first step. The second is data collection: amassing an inventory of the resource flows passing into the system and the emissions passing out. But how should it be measured? Per kilogram (kg) of final product? Yes, if the product is sold and used by weight. Per cubic meter (m^3)

of final product? Yes, if it is sold by volume. But few products are sold and used in this way. More usually it is neither of these but per *unit of function*, a point we will return to in later chapters. The function of a container for a soft drink (a Coke bottle, a plastic water bottle, a beer can) is to contain fluid. The bottle maker might measure resource flows per bottle, but if the idea is to compare containers of different size and material, then the logical measure is the resources consumed *per unit volume of fluid contained*. Refrigerators provide a cooled environment and maintain it over time. The maker might measure resource flows per fridge, but the logical measure from a life-cycle standpoint is the resource consumption *per unit of cooled volume per unit time* (cold space/m^3/year).

We will find that the functional units of resources entering one phase are not the same as those leaving it. There is nothing subtle about this, it's just to make accounting easier. Thus the flow of materials leaving Phase 1 of life and entering Phase 2 *is* traded by weight, so the functional unit here is "per unit weight": the embodied energy of copper, for instance, is listed as 68–74 MJ/kg. The output of Phase 2 is a product; here "per product" might be used. In the use phase, the function performed by the product is of central importance and here the logical measure is "per unit of function."

The inventory analysis, then, assesses resource consumption and emissions per functional unit. It is also necessary to decide on the level of detail—the granularity—of the assessment. It doesn't make sense to include every nut, bolt, and rivet. But where should the cut-off come? One proposal is to include the components that make up 95% of the weight of the product, but this is risky: electronics, for instance, don't weigh much, but the resources and emissions associated with their manufacture can be large, a point we return to in Chapter 6.

Figure 3.4 is a schematic of the start of an inventory analysis—the identification of the main resources and emissions for a washing machine. Most of the parts are made of steel, copper, plastics, and rubber. Both materials production and product manufacturing require carbon-based energy with associated emissions of CO_2, NO_x, SO_x, and low-grade heat. The use phase consumes water as well as energy, with contaminated water as an emission. Disposal of the washing machine creates burdens typical of any large appliance.

Impact assessment. The inventory, once assembled, lists resource consumption and emissions but they are not all equally malignant—some are of more concern than others. Impact categories include *resource depletion, global warming potential, ozone depletion, acidification, eutrophication*,[1] *human toxicity*, and more. Each impact is calculated by multiplying the quantity of each inventory item by an *impact assessment factor*[2]—a measure of how profoundly a given inventory type

---

[1]The over-enrichment of a body of water with nutrients—phosphates, nitrates—resulting in excessive growth of organisms and depletion of oxygen concentration.

[2]Normalization and impact assessment factors can be found in PAS 2050 (2008) or Saling et al. (2002).

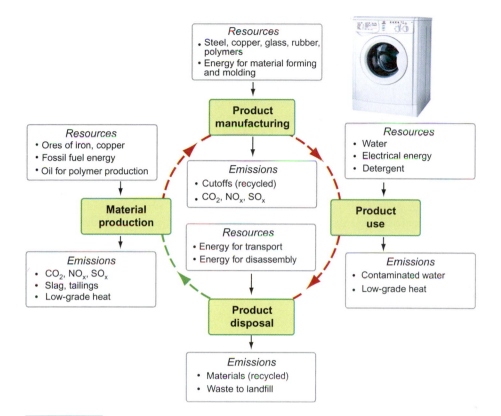

FIGURE 3.4 *The principal resource emissions associated with the lifecycle of a washing machine*

Table 3.1	Example of global warming potential impact assessment factors
Gas	**Impact assessment factor**
Carbon dioxide, CO_2	1
Carbon monoxide, CO	1.6
Methane, CH_4	21
Di-nitrous monoxide, N_2O	256

contributes to each impact category. Table 3.1 lists some examples of that for assessing global warming potential. The overall impact contribution of a product to each category is found by multiplying the quantity of each emission by the appropriate impact assessment factor, and summing the contributions of all the components of the product for all four phases of life.

Interpretation. What do these inventory and impact values mean? What should be done to reduce their damaging qualities? The ISO standard requires answers to these questions but gives little guidance about how to reach them beyond suggesting that it is a matter for specialists.

All this makes a full LCA a time-consuming matter requiring experts. Expert time is expensive. A full LCA is not something to embark on lightly. And while it is very detailed, it is not necessarily very precise.

The output and its precision. Figure 3.5 is part of the output of an LCA—one for the production of aluminum cans (it stops at the exit gate of the manufacturing plant, so this is a "cradle to gate" and not a "cradle to grave" study). The functional unit is "per 1,000 cans." There are three blocks of data: the first is an inventory of resources of ores, feedstock, and energy; the second is a catalog of emissions of gases and particulates; the third is an assessment of impacts—only some of them are shown in the figure.

Despite the formalism that attaches to LCA methods, the results are subject to considerable uncertainty. *Resource* and *energy* inputs can be monitored in a straightforward and reasonably precise way. The *emissions* rely more heavily on sophisticated monitoring equipment—few are known to better than ±10%. Assessments of *impacts* depend on values for the marginal effect of each emission on each impact category; many of these have much greater uncertainties.

And there are two further difficulties, both troublesome. First, what is a designer supposed to do with these numbers? The designer, seeking to cope with the many interdependent decisions that any design involves inevitably finds it hard to know how best to use data like those of Figure 3.5. How are energy, or CO_2 and SO_x, emissions to be balanced against resource depletion, energy consumption,

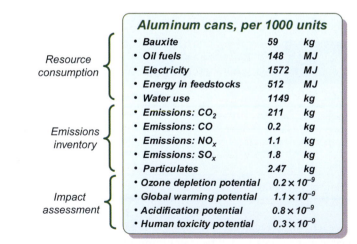

FIGURE 3.5 *Typical LCA output showing three categories: resource consumption, emission inventory, and impact assessment (data in part from Boustead, 2007)*

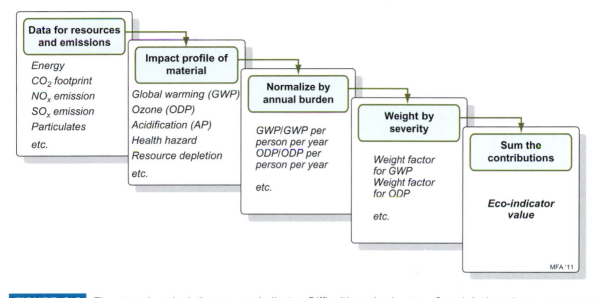

FIGURE 3.6 *The steps in calculating an eco-indicator. Difficulties arise in steps 3 and 4: there is no agreement on how to choose the weight factors.*

global warming potential, or human toxicity? They are not measured in the same units and they differ, in the example of Figure 3.5, by six orders of magnitude. And second, how is the assessment to be paid for? A full LCA takes days or weeks. Does the result justify this considerable expense? LCA has value as a *product assessment tool*, but it is not a *design tool.*

Aggregated measures: eco-indicators. The first of these difficulties has led to efforts to condense the LCA output into a single measure called an *eco-indicator*. To do this, four steps are necessary, shown in Figure 3.6. The first is that of *classification* of the data listed in Figure 3.5 according to the impact each causes (global warming, ozone depletion, acidification, etc.). The second step is that of *normalization* to remove the units and reduce the data to a common scale (0–100, for instance). The third step is that of *weighting* to reflect the perceived seriousness of each impact. Thus global warming might be seen as more serious than resource depletion, and therefore, it is given a larger weight. In the final step, the weighted, normalized measures are *summed* to give the indicator.[3] Eco-indicators are most used in condensing eco-information for the first phase of life, that of material production. Values for materials, when available, are included in the data sheets of Chapter 15.

The use of eco-indicators is criticized by some. The grounds for criticism are that there is no agreement on normalization or weighting factors, that the method is

---

[3]Details can be found in EPS (1993), Idemat (1997), EDIP (1998), and Wenzel et al. (1997).

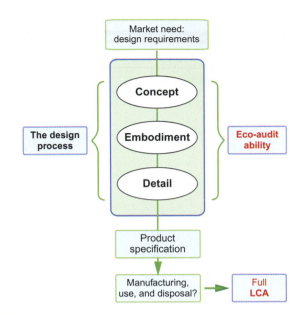

FIGURE 3.7 *An LCA is an end-of-life assessment tool. A streamlined LCA and an eco-audit are design tools.*

opaque since the indicator value has no simple physical significance, and that defending design decisions based on a measurable quantity like *energy consumption* or *carbon release to atmosphere* carries more conviction than doing so with an indicator.

In summary, a full LCA offers the most complete and exhaustive analysis of the environmental impact of products, but it is an expensive, time-consuming tool that requires great detail, much of it unavailable until the product has been manufactured and used. To guide design decisions, particularly the choice of materials, we need tools of a different sort, ideally with the ability to carry out rapid "what if?" audits that allow the designer to explore alternative options (Figure 3.7).

3.5 Streamlined LCA and eco-auditing

Emerging legislation imposes ever increasing demands on manufacturers for eco-accountability. The EU Directive 2005/32/EC on energy-using products (EuPs), for example, requires that manufacturers of EuPs must demonstrate "that they have considered the use of energy in their products as it relates to materials, manufacture, packaging, transport, use and end of life." This sounds horribly like it requires that the manufacturers conduct a full LCA on each one of their products. Many manufacturers make hundreds of different products. The expense both in money and time would be prohibitive.

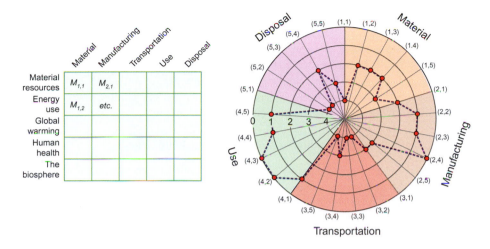

FIGURE 3.8 *An example of a streamlined LCA matrix and a target plot displaying the rankings in each element of the matrix. In this example the use phase gets poor ratings.*

As already explained, the complexity of a full LCA makes it unworkable as a design tool. This perception has stimulated two lines of development: simplified or "streamlined" methods of assessment that focus on the most significant inputs, neglecting those perceived to be secondary; and software-based tools that ease the task of a conducting an LCA. Software solutions are documented in the appendix to this chapter. Let's turn now to streamlining.

The matrix method. The detail required for a full LCA precludes its use as a design tool; by the time the necessary detail is known, the design is too far advanced to allow radical change. *Streamlined LCA* attempts to overcome this by basing the study on a reduced inventory of resources, accepting a degree of approximation. One approach is to simplify while still attempting a *quantitative* analysis—one using numbers—described in Chapters 7 and 8 of this book. The other—one developed by Graedel[4] and others, and used in various forms by a number of industries—is *qualitative.* The matrix on the left of Figure 3.8 shows the idea. The life phases appear as the column headers; the impacts as the row headers. An integer between 0 (highest impact) and 4 (least impact) is assigned to each matrix element M_{ij}, based on experience guided by checklists, surveys, or protocols.[5] The overall *Environmentally Responsible Product Rating, R_{erp},* is the sum of the matrix elements.

$$R_{erp} = \sum_{i}\sum_{j} M_{ij} \qquad (3.1)$$

---

[4]Graedel (1998); Todd and Curran (1999)—see Further reading at the end of the chapter.

[5]Graedel (1998) provides an extensive protocol.

Alternative designs are ranked by this rating.

The information in the matrix is displayed in a more visual way as a target plot, shown on the right of Figure 3.8. It has five concentric circles corresponding to the ranking 0 (highest impact) to 4 (least impact); the elements of the matrix are plotted as dots on radial lines, one line for each element. For an "ideal" product, all the dots lie on the innermost ring, scoring a "bull's-eye." A product with its dots near the outermost circle has much room for improvement.

An eco-comparison of 1950s and 1990s cars[6]

Example: *The task.* Table 3.2 is a low-resolution bill of materials and fuel consumption for typical cars of the 1950s and the 1990s. The 1950s car is heavier, made of relatively few materials, none of them of recycled origin, has poor fuel efficiency, and was dumped at end of life. The more modern car is lighter, made of a more complex

Table 3.2	Estimated material content of generic automobiles*	
Material	**1950s auto (kg)**	**1990s auto (kg)**
Iron	220	207
Steel	1290	793
Aluminum	0	68
Copper	25	22
Lead	23	15
Zinc	25	10
Plastics	0	101
Rubber	85	61
Glass	54	38
Platinum	0	0.001
Fluids	96	81
Other	83	38
Total weight (kg)	**1901**	**1434**
Fuel consumption	**15 mpg**	**27 mpg**

*From Graedel (1998)

[6]Data and basic methods from Graedel (1998).

mix of materials, some derived from recycling, has better fuel efficiency, and will be 80% recycled at end of life. Compare the eco-profiles of the two vehicles.

Answer: The assessor chooses energy efficiency, carbon efficiency, and material efficiency as three eco-criteria to use in the assessment ("efficient" means that the function, private transport, is provided with the minimum use of material and energy resources and of carbon emissions). The assessment is to be cover life. Using this background information and considerable experience, the assessor assigns the rankings of 0 to 4 to each element of the matrices shown in the upper part of Figure 3.9. The 1950s car scores an R_{erp} value of 18. The 1990s car scores 39. The lower part of the figure shows the corresponding target plots. Unsurprisingly, the eco-character of the 1990s car in this example is rather better than that of the 1950s, particularly in its use and disposal phases. All very instructive, but how did the assessor arrive at the rankings? The answer is buried in the store of experience the assessor brings to bear on the task. And do the absolute values of the numbers have any significance? Clearly not. The energy used to propel a car over its life greatly exceeds that required to manufacture it or to create the materials of which it is made. The matrix and target plot capture the issues, but not their relative importance. For that, we need numbers.

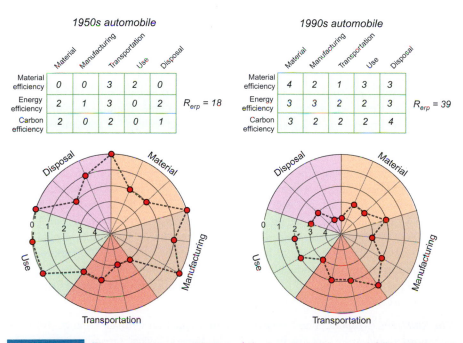

FIGURE 3.9 *The assessment matrices and the target plots for cars of the 1950s and of 1990s. The more modern car has a higher value of R_{etp} and a smaller enclosed area on the target plot.*

FIGURE 3.10 *It is now standard practice to report official fuel economy figures for cars (e.g., Combined: 42–46 mpg [5.9–6.4 liter/100 km]), CO_2 emissions: 143–154 g/km) and energy ratings for appliances (e.g., 330 kWh/year, efficiency rating: A).*

There are many variants of the matrix approach, differing in the impact categories of the rows and the life (or other) categories of the columns. The method's benefits include that it is flexible, easily adapted to a variety of products, carries a low overhead in time and effort, and—in the hands of practitioners of great experience—can take the subtleties of emissions and their impacts into account. It has the drawback that it relies heavily on experience and judgment. It is not a tool to put in the hands of a novice. Is there an alternative?

One resource, one emission. There is, as yet, no consensus on a metric for the eco-impact of product life that is both workable and able to guide design. On one point, however, there is a degree of international agreement[7]: a commitment to a progressive reduction in carbon emissions, generally interpreted as meaning carbon dioxide (CO_2) or carbon dioxide equivalent ($CO_{2,eq}$), a value corrected for the global warming potential of the other gaseous emissions. At the national level the focus is more on reducing energy consumption, but since this and CO_2 production are closely related, reducing one generally reduces the other. Thus there is a certain logic in basing design decisions on one resource—energy—and one emission—CO_2. They carry more conviction than the use of a more obscure indicator, as evidenced by the now-standard reporting of both energy efficiency and the CO_2 emissions of cars, and the energy rating and efficiency ranking of appliances (Figure 3.10) dealing with the use-phase of life. To justify this further, we digress briefly to glance at the IPCC report of 2007.

The 2007 IPCC report. The Intergovernmental Panel on Climate Change (IPCC)—an international study set up by the World Meteorological Organization and the

---

[7]The Kyoto Protocol of 1997 and subsequent Treaties and Protocols, detailed in Chapter 5.

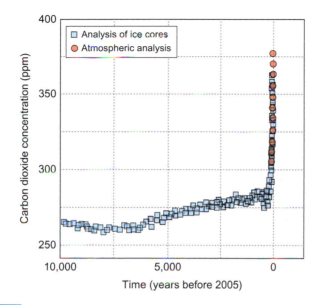

FIGURE 3.11 *Atmospheric concentration of CO_2 over the past 10,000 years measured from ice cores and atmospheric samples. Redrawn from the IPCC report of 2007.*

United Nations Environmental Panel—publishes a series of reports on the effect of industrial activity on the biosphere and the human environment. The most recent of these (IPCC, 2007) is of such significance that familiarity with it is a prerequisite for thinking about sustainability and the environment. Briefly, the conclusions it reaches are these:

- The average air, ocean, and land surface temperatures of the planet are rising. The increase is causing widespread melting of snow and ice cover, rising sea levels, and changes of climate.
- Climate change, measured, for instance, by the annual averages of the air, ocean, and land temperatures, affects natural ecosystems, agriculture, animal husbandry, and human environments. An increase in average global temperature of just 1°C can have a significant effect on all of them. A rise of 5° would create great difficulties.
- The global atmospheric concentration[8] of CO_2 has increased at an accelerating rate since the start of the industrial revolution (around 1750) and is now at its highest level for the past 600,000 years. Most of the increase has been between 1950 and the present day (Figure 3.11).

---

[8]Throughout this book carbon release to the atmosphere is measured in kg of CO_2. One kg of elemental carbon is equivalent to 3.6 kg of CO_2. For a wide range of materials the value of $CO_{2,eq}$ can be equated to $1.06 \times CO_2$, both measured in kg/kg.

- Increasingly accurate geophysical measurements allow the history of temperature and atmospheric carbon to be tracked, and increasingly precise meteorological models allow scenario exploration and prediction of future trends in both. Both suggest that climate-temperature rise is caused by greenhouse gases, and that anthropomorphic (man-made) CO_2 is the probable cause.

The point is that, of the many emissions associated with industrial activity, it is CO_2 that is of greatest current concern. It is global in its impact, causing harm both to the nations that generate most of it and those that do not. It is closely related to the consumption of fossil fuels, themselves a diminishing resource and one that is a source of international tension. If the IPCC report is to be taken seriously, the urgency to cut carbon emissions is great. At this stage in structuring our thinking about materials and the environment, taking energy consumption and the release of atmospheric CO_2 (or $CO_{2,eq}$) as metrics is a logical simplification.

3.6 The strategy

The need is for an assessment strategy that addresses current concerns and combines acceptable cost burden with sufficient precision to guide decision making. It should be flexible enough to accommodate future refinement and simple enough to allow rapid "what if?" exploration of alternatives. To achieve this, it is necessary to strip off much of the detail, multiple targeting, and complexity that make standard LCA methods so cumbersome.

The approach developed here has three components:

1. *Adopt simple metrics of environmental stress.* As already discussed, energy consumption and CO_2 emissions are logical choices as simple metrics for environmental stress. The two are related and are understood by the public at large. Energy has the merit that it is the easiest to monitor, can be measured with relative precision, and, with appropriate precautions, can if necessary be used as a proxy for CO_2.

2. *Distinguish the phases of life.* Figure 3.12 suggests the breakdown—assigning a fraction of the total life-energy demands of a product to material creation, product manufacturing, transport, product use, and disposal. Product disposal can take many forms, some carrying an energy penalty, some allowing energy recycling or recovery. Because of this ambiguity, disposal has a chapter (Chapter 4) to itself.

   When this distinction is made, it is frequently found that one of the phases of life dominates the picture. Figure 3.13 presents the evidence. The upper row shows an approximate energy breakdown for three classes of energy-using products: a civil aircraft, a family car, and an appliance. For all three the use-phase consumes more energy than the sum of all the others.

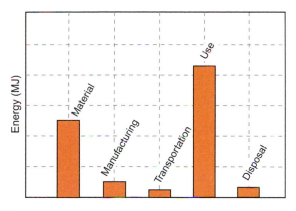

FIGURE 3.12 *Breakdown of energy into that associated with each life phase*

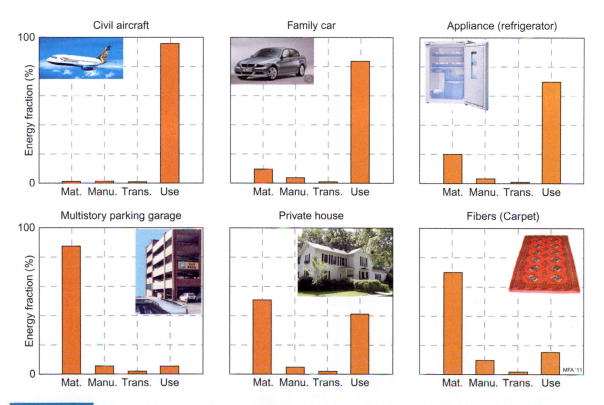

FIGURE 3.13 *Approximate values for the energy consumed at each phase of Figure 3.2 for a range of products. The disposal phase is not shown because there are many alternatives for each product.*

The lower row shows products that still require energy during the use-phase of life, but not as much as those of the upper row. For these, the embodied energies of the materials of which they are made are frequently the largest contribution.

Two conclusions can be drawn. The first: when one phase of life dominates, it is this dominant phase that becomes the first target for redesign since it is here that a given fractional reduction makes the biggest contribution. The second: when differences are as great as those in Figure 3.13, great precision is not essential because it is the ranking that matters. Modest changes to the input data leave the ranking unchanged. It is the nature of people who measure things to wish to do so with precision, and precision must be the ultimate goal. But it is possible to move forward without it: precise judgments can be drawn from imprecise data.

3. ***Base the subsequent action on the energy or carbon breakdown.***
Figure 3.14 suggests how the strategy can be implemented. If material production is the dominant phase, then the logical way forward is to choose materials with low embodied energy and to minimize the amount of it that is used. If manufacturing is an important energy-using phase of life, reducing processing energies becomes the prime target. If transportation

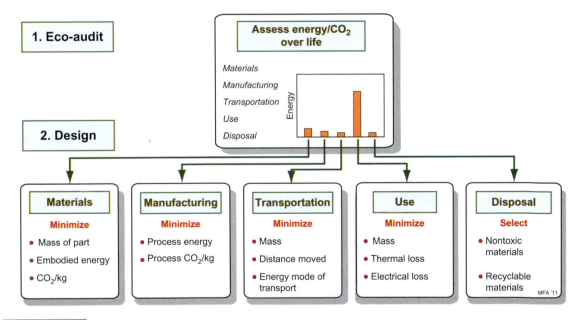

FIGURE 3.14 *Rational approaches to the ecodesign of products start with an analysis of the phase of life to be targeted. Its results guide redesign and materials selection to minimize environmental impact. The disposal phase, shown here as part of the overall strategy, is not included in the current version of the tool.*

makes a large contribution, then seeking a more efficient transportation mode or reducing transportation distance becomes the first priority. When the use-phase dominates, the strategy is that of

- minimizing mass and rolling resistance if the product is part of a system that moves,
- increasing thermal efficiency if the product is a thermal or thermo-mechanical system, or
- reducing electrical losses if the product is an electromechanical system.

In general, the best material choice to minimize one phase will not be the one that minimizes the others, requiring trade-off methods (Chapter 9) to reach an appropriate compromise.

Implementation requires tools. Two tools are needed, one to perform the eco-audit sketched in the upper part of Figure 3.14, the other to enable the analysis and selection of the lower part. The first, the eco-audit tool, is described in Chapter 7 and 8. The second, that of optimized selection, is the subject of Chapters 9 and 10. Tools require data. Data sheets for materials, documenting their engineering and eco-properties,[9] appear in Chapter 15.

3.7 Summary and conclusions

Products, like organisms, have a life, during the course of which they interact with their environment. Their environment is also ours; if the interaction is a damaging one, it diminishes the quality of life of all who share it.

Life-cycle assessment (LCA) is the study and analysis of this interaction, quantifying the resources consumed and the waste emitted. It is holistic, spanning the entire life from the creation of the materials, through the manufacture of the product, its use, and its subsequent disposal. Although standards (the ISO 14040 series) now prescribe procedures for conducting an LCA, they remain vague, allowing a degree of subjectivity. Implementing them requires skill and access to much detail, making a full LCA expensive in both money and time, and one that delivers outputs that are not well adapted to the needs of designers.

No surprise. The technique of LCA is relatively new and is still evolving. The way forward is to adopt a less rigorous but much simpler approach, streamlining the assessment by restricting it to the key eco-aspects of most immediate concern. The matrix method, of which there are many variants, assigns a ranking for each impact category in each phase of life, summing the rankings to get an Environmentally Responsible Product Rating. Another approach, better adapted to guiding material choice, is to limit the impact categories to one resource—energy—and one emission—CO_2—auditing designs or products for their demands on both.

---

[9]The data sheets are a subset of those contained in the CES (2011) software, which also implements both the tools described here.

Providing that the resolution of the audit is sufficient to draw meaningful conclusions, the results can guide design decisions without imposing an unacceptable burden of analysis.

3.8 Further reading

Aggregain (2007), The Waste and Resources Action Program (WRAP), www.wrap.org.UK. ISBN 1-84405-268-0. *(Data and an Excel-based tool to calculate energy and the carbon footprint of recycled road-bed materials)*

Allwood, J.M., Laursen, S.E., de Rodriguez, C.M., and Bocken, N.M.P. (2006), *Well dressed? The present and future sustainability of clothing and textiles in the United Kingdom*, University of Cambridge, Cambridge, UK. ISBN 1-902546-52-0. *(An analysis of the energy and environmental impact associated with the clothing industry)*

"Boustead Model 5" (2007), Boustead Consulting, West Sussex, UK, www.boustead-consulting.co.uk. Accessed December 2011. *(An established life-cycle assessment tool)*

"Eco-indicator "(1999), PRé Consultants, Amersfoort, Netherlands, www.pre.nl/eco-indicator99/eco-indicator_99.htm. Accessed December 2011. (An explanation of the Eco-indicator method.)

EPS (1992), "The EPS enviro-accounting method: an application of environmental accounting principles for evaluation and valuation in product design," *Report B1080*, IVL Swedish Environmental Research Institute, by Steen, B., and Ryding, S.O. Gothenburg, Sweden. (The EPS method is an alternative to the Eco-indicator approach.)

EU Directive on Energy Using Products (2005), "Directive 2005/32/EC of the European Parliament and of the Council of 6 July 2005 establishing a framework for the setting of ecodesign requirements for energy-using products and amending Council Directive 92/42/EEC and Directives 96/57/EC and 2000/55/EC of the European Parliament and of the Council." Strasburg, France *(One of several EU Directives relating to the role of materials in product design)*

GaBi (2008), PE International, Leinfelden-Echterdingen, Germany. www.gabi-software.com/. Accessed December 2011. *(GaBi is a software tool for product assessment to comply with European legislation.)*

Goedkoop, M., Effting, S., and Collignon, M. (2000), "The Eco-indicator 99: A damage oriented method for Life-cycle Impact Assessment, Manual for Designers," www.pre.nl. Accessed December 2011. *(An introduction to eco-indicators, a technique for rolling all the damaging aspects of material production into a single number)*

Graedel, T.E., and Allenby, B.R. (2003), *Industrial ecology*, 2nd edition, Prentice Hall, NJ, USA. ISBN 978-0131252387. *(An established treatise on industrial ecology)*

Graedel, T.E. (1998), *Streamlined Life-cycle Assessment*, Prentice Hall, NJ, USA. ISBN 0-13-607425-1. *(Graedel is the father of streamlined LCA methods. The first half of this book introduces LCA methods and their difficulties. The second half develops his streamlined method with case studies and exercises. The appendix details protocols for informing assessment decision matrices.)*

GREET (2007), Argonne National Laboratory and the US Department of Transportation, www.transportation.anl.gov/. Accessed December 2011. *(Software for analyzing vehicle energy use and emissions)*

Guidice, F. La Rosa, G., and Risitano, A. (2006), *Product design for the environment*, CRC/Taylor and Francis, London, UK. ISBN 0-8493-2722-9. *(A well-balanced review of current thinking on ecodesign)*

Heijungs, R. (editor) (1992), "Environmental life-cycle assessment of products: background and guide," Netherlands Agency for Energy and Environment, Amsterdam, the Netherlands.

"Idemat Software" version 1.0.1 (1998), Faculty of Industrial Design Engineering, Delft University of Technology, Delft, The Netherlands. *(An LCA tool developed by the University of Delft)*

ISO 14040 (1998), Environmental management—Life-cycle assessment—Principles and framework.

ISO 14041 (1998), Goal and scope definition and inventory analysis.

ISO 14042 (2000), Life-cycle impact assessment.

ISO 14043 (2000), Life-cycle interpretation, International Organization for Standardization, Geneva, Switzerland. *(The set of standards defining procedures for life-cycle assessment and its interpretation)*

Kyoto Protocol (1997), United Nations, Framework Convention on Climate Change. Document FCCC/CP1997/7/ADD.1, http://unfccc.int/resource/docs/convkp/kpeng.pdf. Accessed December 2011. *(An international treaty to reduce the emissions of gases that, through the greenhouse effect, cause climate change)*

MEEUP Methodology Report, final (2005), VHK, Delft, Netherlands, www.pre.nl/EUP/. Accessed December 2011. *(A report by the Dutch consultancy VHK commissioned by the European Union, detailing their implementation of an LCA tool designed to meet the EU Energy-using Products directive)*

MIPS (2008), The Wuppertal Institute for Climate, Environment and Energy, www.wupperinst.org/en/projects/topics_online/mips/index.html. Accessed December 2011. *(MIPS software uses an elementary measure to estimate the environmental impacts caused by a product or service.)*

National Academy of Engineering and National Academy of Sciences (1997), *The Industrial Green Game: Implications for Environmental Design and Management*, National Academy Press, Washington, DC, USA. ISBN 978-0309-0529-48. *(A monograph describing best practices that are being used by a variety of industries in several countries to integrate environmental considerations in decision making)*

PAS 2050 (2008), *Specification for the assignment of the life-cycle greenhouse gas emissions of goods and services*, ICS code 13.020.40, British Standards Institution, London, UK. ISBN 978-0-580-50978-0. *(A proposed European Publicly Available Specification (PAS) for assessing the carbon footprint of products)*

Saling, P, Kicherer, A., Dittrich, B.Wittlinger, R., Zombik, W., Schmidt, I., Schrott, W., and Schmidt, S. (2002), "Eco-efficiency analysis by BASF: the method," *Int. J. Life-cycle Assess*, 7, pp 203–218. (A description of the eco-efficiency metric devised for use by BASF.)

SETAC (1991), "A technical framework for life-cycle assessment," Fava, J.A., Denison, R., Jones, B., Curran, M.A., Vignon, B., Selke, S., and Barnum, J., (Eds.), Society of Environmental Toxicology and Chemistry, Washington, DC, USA. *(The meeting at which the term Life-cycle Assessment was first coined)*

SETAC (1993), "Guidelines for life-cycle assessment—a code of practice," Consoli, F., Fava J.A., Denison, R., Dickson, K., Kohin, T., and Vigon, B. (Eds.), Society of Environmental Toxicology and Chemistry, Washington, DC, USA. *(The first formal definition of procedures for conducting an LCA)*

Todd, J.A., and Curran, M.A. (1999), "Streamlined life-cycle assessment: a final report from the SETAC North America streamlined LCA workshop," Society of Environmental Toxicology and Chemistry, Washington, DC, USA. *(One of the early moves toward streamlined LCA)*

3.9 Appendix: software for LCA

The most common uses of life-cycle assessment are for product improvement ("how can I make my products greener?"), support of strategic choices ("is this or that the greener development path?"), benchmarking ("how do our products compare?"), and for communication ("our products are the greenest."). Most of the software tools designed to help with this use ISO 14040 to 14043 as a prescription. In doing so, they commit themselves to a process of considerable complexity.[10] There is no compulsion to follow this route and some do not. Some of these are aimed at specific product sectors (vehicle design, building materials, paper making). Others are aimed at the early stages of product design and these, of necessity, are simpler in their structure. Two, at least, have education as its target. So there is quite a spectrum, 11 of which are listed in Table 3.3. Some of these tools are free, some can be bought, and others are available only through the services of a consultant— an understandable precaution, given their complexity.

SimaPro (2008). SimaPro 7.1 is a widely used tool to collect, analyze, and monitor the environmental performance of products and services developed by Pré Consultants in the Netherlands. Life cycles can be analyzed in a systematic way, following the ISO 14040 series recommendations. There is an educational version. A free demo is available from the Pré web site.

Boustead Model 5 (2007). The Boustead Model is a tool for life-cycle inventory calculations broadly following the ISO 14040 series recommendations. Ian Boustead, the author of the software, has many years of experience in cycle assessment working with European polymer suppliers.

TEAM (2008). TEAM is Ecobilan's Life-cycle Assessment software. It allows the user to build and use a large database and to model systems associated with products and processes following the ISO 14040 series of standards.

---

[10]Pré Consultants estimate that the time needed to perform a "screening" LCA is about 8 days, that for a full LCA is about 22 days.

Table 3.3	LCA and LCA-related software
Tool name	**Provider**
SimaPro	Pré Consultants (www.pre.nl)
Boustead model 5	Boustead Consultants (www.boustead-consulting.co.uk)
TEAM (EcoBilan)	PricewaterhouseCoopers (www.ecobalance.com/)
GaBi	PE International (www.gabi-software.com/)
MEEUP method	VHK, Delft, Netherlands (www.pre.nl/EUP/)
GREET	US Department of Transport (www.transportation.anl.gov/)
MIPS	Wuppertal Institute (www.wupperinst.org/)
CES Eco '12	Granta Design, Cambridge, UK (www.grantadesign.com)
Aggregain	WRAP (www.aggregain.org.uk/)
KCL-ECO 3.0	KCL Finland (www.kcl.fi)
Eiloca	Carnegie Mellon Green Design Institute, USA (www.eiolca.net/)
Okala Ecodesign guide	Industrial Design Society of America (www.idsa.org/okala-ecodesign-guide)
LCA Calculator	IDC, London, UK(www.lcacalculator.com/)

GaBi (2008). GaBi 4, developed by PE International, is a sophisticated tool for product assessment to comply with European legislation. It has facilities for analyzing cost, environment, social, and technical criteria and optimization of processes. A demo is available.

MEEUP method (2005). The Dutch Methodology for Ecodesign of Energy-using Products (MEEUP) is a response to the EU directive on energy-using products (the EuP directive) described in Chapter 5. It is a tool for the analysis of products—mostly appliances—that use energy, following the ISO 14040 series of guidelines.

GREET (2007). The Greenhouse Gasses, Regulated Emissions and Energy Use in Transportation Model (GREET) is a free spreadsheet running in Microsoft Excel developed by Argonne National Laboratory for the US Department of Transportation. There are two versions, one for fuel-cycle analysis and one for vehicle-cycle analysis. They deal with specific emissions, not with impacts and weighted combinations. For a given vehicle and fuel system, the model calculates energy consumption, emissions of CO_2-equivalent greenhouse gases—primarily carbon dioxide (CO_2), methane (CH_4), and nitrous oxide (N_2O)—and six criteria pollutants: volatile organic compounds (VOCs), carbon monoxide (CO), nitrogen oxide (NO$_x$), particulate matter with size smaller than 10 micron (PM10), particulate matter with size smaller than 2.5 micron (PM2.5), and sulfur oxides (SO$_x$).

MIPS (2008). MIPS stands for Material Input per Service Unit. MIPS is an elementary measure to estimate the environmental impacts caused by a product or service. The full life cycle from cradle to grave (extraction, production, use, waste/recycling) is considered. It allows the environmental implications of products, processes, and services to be assessed and compared. It enables material intensity analysis both at the micro-level (focusing on specific products and services) and at the macro-level (focusing on national economies).

CES Edu (2012). Granta Design specializes in materials information-management software. One of their products, CES Edu, is a widely used tool for teaching the selection and use of materials and processes. It includes modules that implement the eco-audit methods described in Chapter 7 and the eco-selection procedures of Chapter 9.

Aggregain (2008). Aggregain, developed and distributed by WRAP, is a free analysis tool that runs in Microsoft Excel and is used for promoting the supply and use of recycled and secondary aggregates (including recycled concrete from construction, demolition waste material, and railway ballast) for the construction and road-building industries.

KCL-ECO 3.0. KCL represents the paper-making industry. KCL-Eco is an LCA tool designed specifically for this industry.

Eio-lca (2008). Economic input-output LCA (Eio-lca) of Carnegie Mellon University calculates sector emissions based on input-output data for the sectors of the North American Industry Classification Scheme (NAICS). It is not designed for the assessment of products. Demo available.

Okala Ecodesign Guide (2010). Okala provides an introduction to ecological and sustainable design for practicing and beginning designers; it was developed with the support from Eastman Chemical, Whirlpool, and the Industrial Design Society of America (ISDA).

LCA Calculator (2011). This is a quick and intuitive way for designers and engineers to understand, analyze, and compare environmental impacts of products and particular design decisions.

3.10 Exercises

E3.1. (a) Which phase of life would you expect to be the most energy intensive (in the sense of consuming fossil fuel) for the following products?

- A toaster
- A two-car garage
- A bicycle
- A motorcycle

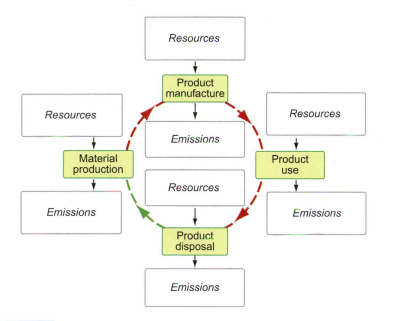

FIGURE 3.15 *A template for listing the principal resources and emissions associated with the life of a product*

- A refrigerator
- A coffee maker
- An LPG-fired patio heater

(b) Pick one of these and list the resources and emissions associated with each phase of its life along the lines of Figure 3.4 (template provided in Figure 3.15).

E3.2. Functional units. Think of the basic need filled by the products listed here. List what you would choose as the functional unit for an LCA.

- Washing machines
- Refrigerators
- Home heating systems
- Air conditioners
- Lighting
- Home coffee maker
- Public transport
- Hand-held hair dryers

E3.3. (a) What is meant by "externalized" costs and costs that are "internalized" in an environmental context?

(b) Now a moment of introspection. List three externalized costs associated with your lifestyle. If your life is so pure that you have less than three, then list some of other people you know.

E3.4. What, in the context of life-cycle assessment, is meant by "system boundaries"? How are they set?

E3.5. Describe briefly the steps prescribed by the ISO to guide life-cycle assessment of products.

E3.6. What are the difficulties with a full LCA? Why would a simpler, if approximate, technique be helpful?

E3.7. Pick two of the products listed in Exercise E.3.1 and, using your judgment, attempt to fill out the simplified streamlined LCA matrix in Figure 3.16 to give an Environmentally Responsible Product Rating, R_{epr}.

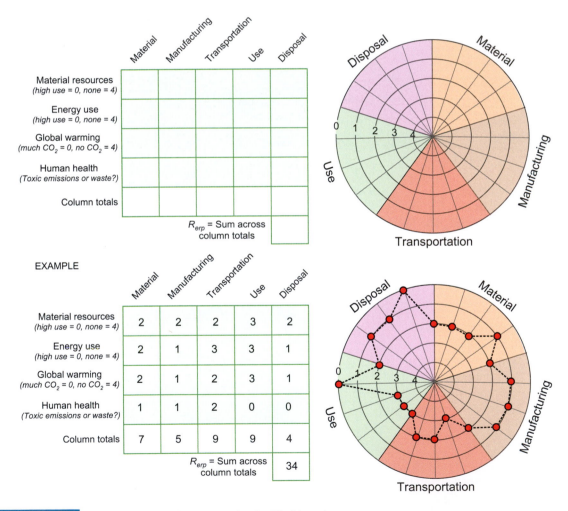

EXAMPLE

	Material	Manufacturing	Transportation	Use	Disposal
Material resources (high use = 0, none = 4)	2	2	2	3	2
Energy use (high use = 0, none = 4)	2	1	3	3	1
Global warming (much CO_2 = 0, no CO_2 = 4)	2	1	2	3	1
Human health (Toxic emissions or waste?)	1	1	2	0	0
Column totals	7	5	9	9	4
R_{erp} = Sum across column totals					34

FIGURE 3.16 *A blank target and an example of a filled target.*

Make your own assumptions (and report them) about where the product was made, how far it has been transported thus far, and whether it will be recycled. Assign an integer between 0 (highest impact) and 4 (least impact) to each box and then add them to give an environmental rating, providing a comparison. Try the following protocol:

- Material: Is it energy-intensive? Does it create excessive emissions? Is it difficult or impossible to recycle? Is the material toxic? If the answer to these questions is *yes*, score 4. If *no*, score 0. Use the intermediate integers for other combinations.
- Manufacturing: Is the process one that uses much energy? Is it wasteful (meaning cut-offs and rejects are high)? Does it produce toxic or hazardous waste? Does make use of volatile organic solvents? If yes, score 4. If no, score 0, etc.
- Transportation: Is the product manufactured far from its ultimate market? Is it shipped by air freight? If yes to both, score 4. If no to both, score 0.
- Use: Does the product use energy during its life? Is the energy derived from fossil fuels? Are any emissions toxic? Is it possible to provide the use-function in a less energy-intensive way? Scoring as above.
- Disposal: Will the product be sent to a landfill at end of life? Does disposal involve toxic or long-lived residues? Scoring as above.

What difficulties did you have? Do you feel confident that the results are meaningful?

CHAPTER 4

End of first life: a problem or a resource?

4.1 Introduction and synopsis

When stuff is useful, we show it respect and call it *material*. When the same stuff ceases to be useful, we cease to respect it and we call it *waste*. Waste is deplorable, and it is much deplored, that from packaging particularly so. Is it inevitable? The short answer is *yes*—it is a consequence of one of the inescapable laws of physics: that entropy can only increase. A fuller answer is *yes, but*. The "but" has a number of aspects. That is what this chapter is about.

First, a calibration. We (the global we) are consuming materials at an ever faster rate (Chapter 2). The first owner of a product, at end of life, rejects it as waste. So waste, too, is generated at an ever growing rate. What happens to it? In five words: *landfill, combustion, recycling, re-engineering*, or *reuse*. That sounds comprehensive—it must be feasible to find a home for cast-off products in one of these. Ah, but. The capacity of a channel for dealing with products at end of first life, to be

Is this waste or is it a resource?. (Image courtesy Envirowise - Sustainable Practices, Sustainable Profits, a UK Government programme managed by AEA Technology Plc.)

effective, must match the rate of rejection. Only one of the five has any real hope of achieving this. And then there are the economics. End of life is not simple.

To start at the beginning: why do we throw things away?

4.2 What determines product life?

The rapid turnover of products we see today is a comparatively recent phenomenon. In earlier times, furniture was bought with the idea that it would fill the needs not just of one generation but of several—treatment that, today, is reserved for works of art. A wristwatch, a gold pen, these were things you used for a lifetime and then passed on to your children. No more. Behind all this is the question of whether the *value* of a product increases or decreases with age.

A product reaches the end of its life when it's no longer valued. The cause of death is, frequently, not the obvious one—that the product just stopped working. The life expectancy is the least of[1]

- The *physical life*, meaning the time in which the product breaks down beyond economic repair;
- The *functional life*, meaning the time when the need for it ceases to exist;
- The *technical life*, meaning the time at which advances in technology have made the product unacceptably obsolete;
- The *economical life*, meaning the time at which advances in design and technology offer the same functionality at significantly lower operating cost;
- The *legal life*, meaning the time at which new standards, directives, legislation, or restrictions make the use of the product illegal;
- And finally the *desirability life*, meaning the time at which changes in taste, fashion, or aesthetic preference render the product unattractive.

One obvious way to reduce resource consumption is to extend product life, making it more durable. But durability has more than one meaning: we've just listed six. Materials play a role in them all—something that we return to later. Accept, for the moment, that a product *has*, for one reason or another, reached the end of its first life. What are the options?

4.3 End-of-first-life options

Figure 4.1 introduces the choices: landfill, combustion for heat recovery, recycling, re-engineering, and re-use.

Landfill. Much of what we now reject is committed to landfill. Already there is a problem—the land available to "fill" in this way is already, in some European countries, almost full. Recall one of the results of Chapter 2: if the consumption of

---

[1]This list is a slightly extended version of one presented by Woodward (1997), "Life-cycle costing," *Int. J. Project Management*, 15, pp 335–344.

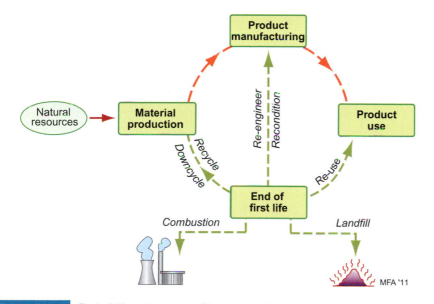

FIGURE 4.1 *End-of-life options: landfill, combustion, re-engineering, reconditioning, and reuse*

materials grows by 3% per year, we will use and—if we discard it—throw away as much stuff in the next 25 years as in the entire history of industrialization. Landfill is not going to absorb that. Administrations react by charging a landfill tax—currently somewhere near €50 per metric ton and rising, seeking to divert waste into the other channels of Figure 4.1. These must be capable of absorbing the increase. None, at present, can.

Combustion for heat recovery. Materials, we know, contain energy. Rather than throwing them away it would seem better to retrieve and reuse some of their energy by controlled combustion, capturing the heat. But this is not as easy as it sounds. First, combustibles must be separated from non-combustible material (Figure 4.2). Then the combustion must be carried out under controlled conditions that do not generate toxic fumes or residues, requiring high temperatures, sophisticated control, and expensive equipment. The energy recovery is imperfect partly because it is incomplete and partly because the incoming waste carries a moisture content that has to be boiled off. The efficiency of heat recovery from the combustion process is at best 50%, and if the recovered heat is used to generate electricity, it falls to 35%. And communities don't like an incinerator at their back door. Thus useful energy *can* be recovered by the combustion of waste, but the efficiency is low, the economics are unattractive, and the neighbors can be difficult.

Despite all this, combustion for heat recovery is, in some circumstances, practical and attractive. The most striking example is the cement industry, one with an enormous energy budget and CO_2 burden because of the inescapable step of

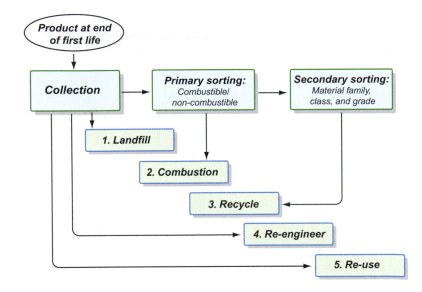

FIGURE 4.2   *End-of-life scenarios: landfill, combustion for heat recovery, recycling, refurbishment, and reuse. Different levels of sorting and cleaning are required for each.*

calcining in its production. Increasingly, combustion of vehicle tires and industrial and agricultural wastes are used as a heat source, reducing the demand on primary fuels, but not of course the attendant release of CO_2.

Recycling. Waste is only waste if nothing can be done to make it useful. It can also be a resource. Recycling is the reprocessing of recovered materials at the end of product life, returning them into the use-stream. It is the end-of-life scenario that is best adapted to extracting value from the waste-stream. We return to this in Section 4.5 for a closer look.

Re-engineering or refurbishment. There is the story of the axe—an excellent axe—that, over time, had two new heads and three new handles. But it was still the same axe. Refurbishment, for some products, is cost-effective and, compared with total replacement, energy efficient. Aircraft, for instance, don't wear out; instead, replacement of critical parts at regular inspection periods keeps the plane, like the axe, functioning just as it did when it was new. The Douglas DC3, a 70-year-old design, is still flying, though not, of course, in the hands of its original owner. Premium airlines fly premium aircraft, so older models are sold on to operators with smaller budgets.

Re-engineering is the refurbishment or upgrading of the product or of its recoverable components. Certain criteria must be met to make it practical. One is that the design of the product is fixed—as it is with aircraft once an airworthiness certificate is issued—or that the technology on which it is based is evolving so slowly that there remains a market for the restored product. Here are some examples: housing, office space, road and rail infrastructure, all are sectors with enormous appetites for materials. Some more examples: office equipment, particularly printing equipment and copying machines and communication systems. These are services; the product providing them is unimportant to those who need the service as long as it works well. It makes more sense to lease a service (as we all do with telephone lines, mobile phones, Internet service, municipal fresh water, and much else) because it is in the leasers' interests to maximize the life of the equipment.

And there is another obstacle to re-engineering: that fashion, style, and perceptions change, making a reconditioned product unacceptable even though it works perfectly well. Personal image, satisfaction, and status are powerful drivers of conspicuous consumption.

Reuse. The cathedrals of Europe, almost all of them, are built on the foundations of earlier structures, often from the 10th or 11th century, built, in turn, on still earlier 5th or 6th century beginnings. If the structure is in a region that was once part of the Roman Empire, then columns, friezes, fragments of the forum, and other structural elements of yet greater antiquity have found their way into the structure too. Reuse is not a new idea, but it is a good one.

Put more formally: reuse is the redistribution of the product to a consumer sector that is willing to accept it in its used state, perhaps to reuse it for its original purpose (a second-hand car, for instance), perhaps to adapt to another use (converting the car to a hot-rod or a bus into a mobile home).[2] The key issue here is that of communication. Housing estate (realtor) listings and used car and boat magazines exist precisely to provide channels of communication for used products. Charity shops pass on clothing, objects, and junk,[3] acquiring them from those for whom they had become waste and selling them to others who perceive them to have value. The most effective tool ever devised to promote product reuse is probably eBay, successful in this precisely because it provides a global channel of communication.

4.4 The problem of packaging

Few applications of materials attract as much criticism as their use of packaging. The functional life of packaging ends as soon as the package is opened. It is ephemeral, it is trite, it generates mountains of waste, and most of the time, it is unnecessary.

Or is it? Think for a moment about the most highly developed form that packaging takes: the way we package ourselves. Clothes provide protection from heat

---

[2] What most of us see as waste can become the material of invention to the artist. For remarkable examples of this, visit the Museé International des Arts Modestes, 23 quai du Maréchal de Lattre de Tassigny, 34200 Sète, France (www.miam.org).

[3] I remember a sign above a store in an English town: "We buy junk. We sell antiques."

and cold, from sun and rain. Clothes convey information about gender and ethnic and religious background. Uniforms identify membership and status, most obviously in the military and the church, but also in other hierarchical organizations: airlines, hotels, department stores, and even utility companies. And at a personal level, clothes do much more: they are an essential part of the way we present ourselves. While some people make the same clothes last for years, others wear them only once or twice before—for them—they become "waste" and are given to a charity shop.

Fine, you may say, we need packaging of that sort; but products are inanimate. What's the point of packaging for them? The brief answer: products are packaged for precisely the same reasons that we need clothes: protection, information, affiliation, status, and presentation.

So let us start with some facts. Packaging makes up about 18% of household waste, but only 3% of landfill. Its carbon footprint is 0.2% of the global total. Roughly 60% of packaging in Europe, a bit less in the United States, is recovered and used for energy recovery or recycling. Packaging makes possible the lifestyle we now enjoy. Without it, supermarkets would not exist. By protecting foodstuffs and controlling the atmosphere that surrounds them, packaging extends product life, allows access to fresh products all year round, and reduces food waste in the supply chain to about 3%; without packaging, the waste is far higher. Tamper-proof packaging protects the consumer. Pack information identifies the product, its sell-by date (if it has one), and gives instructions for use. Brands are defined by their packaging—the Coca-Cola bottle, the Campbell's soup can, Kellogg's products—essential for product presentation and recognition.

The packaging industry[4] is well aware of the negative image that packaging holds and it strives to minimize its weight and volume. There are, of course, exceptions but there has been progress in optimizing packaging—providing all its functionality with the minimum use of materials. The most used of these—paper, cardboard, plastic, glass, aluminium, and steel—have established recycling markets (see below). Much packaging ends up in household waste, the most difficult type to sort. The answer to this is better waste-stream management in which the sorting is done by the consumer via marked containers. The protective function of much packaging requires material multilayers that cannot be recycled but that, if sorted, are still a source of energy.

So, the bottom line. Legislating packaging out of existence would require major adjustments to lifestyles, greatly increase the waste stream, and deprive consumers of convenience, product protection, and hygienic handling. The challenge is that of returning as much of it as possible into the materials economy at end of life.

The role of industrial design. What have you discarded lately that still worked or, if it didn't, could have been fixed? Changing trends, promoted by seductive advertizing, reinforce the desire for the new and urge the replacement of the old.

---

[4]See, for instance, the Packaging Federation, www.packagingfedn.co.uk, or the Flexible Packaging Association, www.flexpack.org.

Industrial design carries a heavy responsibility here—it has, at certain periods, been directed toward creative obsolescence: designing products that are desirable only if new, and urging the consumer to buy the latest models, using marketing techniques that imply that acquiring them is a social and psychological necessity.

But that is only half the picture. A well-designed product can acquire value with age, and—far from becoming unwanted—can outlive its design life many times over. The auction houses and antique dealers of New York, London, and Paris thrive on the sale of products that, often, were designed for practical purposes but are now valued more highly for their aesthetics, associations, and perceived qualities. People do not throw away things for which they feel emotional attachment. So there you have it: industrial design both as villain and as hero. Where can it provide a lead?

When your house no longer suits you, you have two choices: you can buy a new house or you can adapt the one you have, and in adapting it, you make it more personally yours. Houses allow this. Most other products do not. An old product (unlike an old house) is often perceived to be incapable of change and to have such low value that it is simply discarded. That highlights a design challenge: to create products that can be adapted and personalized so that they acquire, like a house, a character of their own and transmit the message "keep me, I'm part of your life." This suggests a union of technical and industrial design to create products that can accommodate evolving technology, but at the same time, are made with a quality of material, design, and adaptability that give them a lasting and individual character, something to pass on to your children.[5]

4.5 Recycling—resurrecting materials

Of the five end-of-life options shown in Figure 4.1, only one meets the essential criteria:

- that it can return waste materials into the supply chain and
- that it can do so at a rate that, potentially, is comparable with that at which the waste is generated.

Landfill and combustion fail to meet the first criterion, and re-engineering and reuse fail the second. That leaves *recycling* (Figure 4.2).

Quantification of the process of material recycling is difficult. Recycling costs energy, and this energy carries its burden of emissions. But the *recycle energy* is generally small compared to the initial embodied energy, making recycling—when it is possible at all—an energy-efficient proposition. It may not, however, be one that is cost-efficient; that depends on the degree to which the material has become dispersed. In-house scrap, generated at the point of production or manufacturing, is

---

[5]For an organization with such an ideal, see www.eternally-yours.nl.

localized and is already recycled efficiently (near 100% recovery). Widely distributed "scrap"—material contained in discarded products—is more expensive to collect, separate, and clean. Many materials cannot be recycled, although they may still be reused in a lower-grade activity; continuous-fiber composites, for instance, cannot yet be re-separated economically into fiber and polymer in order to reuse them, though they can be chopped and used as fillers. Most other materials require an input of virgin material to avoid build-up of uncontrollable impurities. Thus the fraction of a product that can ultimately re-enter the cycle of Figure 4.1 depends both on the material itself and on the product into which it has been incorporated.

Metals. The recycling of metals in the waste stream is highly developed. Metals differ greatly in their density, in their magnetic and electrical properties, and even in their color, making separation comparatively easy. The value of metals, per kilogram, is greater than that of other materials. All this helps make metal recycling economically attractive. There are many limitations on how recycled metals are used, but there are enough good uses that the contribution of recycling to today's consumption is large.[6]

Polymers. The same cannot be said of polymers. Commodity polymers are used in large quantities, many in products with short lives, and they present major problems in waste management, all of which, you would think, would encourage effective recycling. But polymers all have nearly the same density, have no significant magnetic or electrical signature, and can take on any color that the manufacturer likes to give them. They can be identified by X-ray fluorescence or infrared spectroscopy, but these are not infallible and they are expensive. Add to this that many are blends and contain fillers or fibers. Add further that the recycling process itself involves a large number of energy-consuming steps. Unavoidable contamination can prevent the use of recycled polymers in the product from which they were derived, condemning them to more limited use. For this reason the value of recycled polymers—the price they command—is typically about 60% of that of virgin material.

A consequence of this is that the recycling of many commodity polymers is low. Increasing the recycle fraction is a question of identification, and here there is progress. Figure 4.3 shows, in the top row, the standard recycle marks, ineffective because they do not tell the whole story. The lower row shows the emerging identification system. Here polymer, filler, and weight fraction are all identified. The string, built up from the abbreviations listed in the Appendix of this chapter, gives enough information for effective recycling.

---

[6]A recent study reveals that the concentration of gold in the circuit boards and microprocessors of mobile phones and personal computers averages 0.2% by weight. The ores from which gold is currently extracted contain only 0.002% of the metal.

> PP-T-20 <   > PC-ABS-GF <

FIGURE 4.3   *Above: Recycle marks for the most commonly used commodity polymers. Below: More explicit recycle marks detailing blending, fillers, and reinforcement. The first is polypropylene with 20% talc powder. The second is a polycarbonate-ABS blend with glass fiber. The coding is explained in the Appendix to this chapter.*

Identifying polymers

Example: A product contains components with the recycle marks *TPA* and *PMMI-CF-30*. What are they?

Answer: Refer to the Appendix at the end of this chapter. It identifies the first as polyamide thermoplastic elastomer and the second as polymethylmethacrylimide with 30% carbon fiber.

The economics of recycling. Although recycling has far-reaching environmental and social benefits, it is market forces that—until recently—have determined whether or not it happens. Municipalities collect recyclable waste, selling it through brokers to secondary processors who reprocess the materials and sell them, at a profit, to manufacturers. The recycling market is like any other, with prices that fluctuate according to the balance of supply and demand. In a free market the materials that are recycled are those from which a profit can be made. These include almost all metals but few polymers (Table 4.1).

Scrap arises in more than one way. *New* or *primary scrap* is the cutoffs from billets, risers from castings, and turnings from machining that are a by-product of the manufacturing of products; it can be recycled immediately, often in-house. *Old* or *secondary scrap* appears when the products themselves reach the end of their useful life (Figure 4.4). The value of recyclable waste depends on its origin. New scrap carries the highest value because it is uncontaminated and easy to collect and reprocess. Old scrap from commercial sources, such as offices and restaurants, is more valuable than that from households because it is more homogeneous and needs less sorting.

Producers of secondary materials must, of course, compete with those producing virgin materials. It is this that couples the price of the first to that of the

Table 4.1	Recycling markets	
Material family	**Developed end-uses for recycled materials**	**Secondary uses exist but are not developed as a market**
Metals	Steel and cast iron Aluminum Copper Lead Titanium All precious metals*	Paper—metal foil packaging
Polymers and elastomers	Polyethylene terephthalate (PET) High density polyethylene (HDPE) Polypropylene (PP) Polyvinylchloride (PVC)	All other polymers and elastomers, notably tires
Ceramics and glasses	Bottle glass Brick Concrete and asphalt	Non-bottle glass
Other materials	Cardboard, paper, newsprint	Wood Textiles: cotton, wool, and other fibers

Typical concentration of gold in ores from which it is extracted: 5 grams/metric ton. Typical concentration in a mobile phone: 150 grams/metric ton.

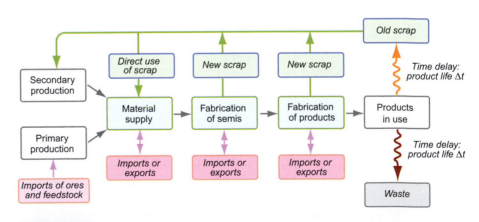

FIGURE 4.4 *The material flows in the economy, showing recycling paths. New scrap arises during manufacturing and is reprocessed almost immediately. Old scrap derives from products at end of life; it re-enters production only after a delay of Δt, the product life.*

second. Virgin materials are more expensive than those that have been recycled because their quality, both in engineering terms and that of perception, is greater. Manufacturers using recycled materials require assurance that this drop in quality will not compromise their products.

The profitability of a market can be changed by economic intervention: subsidies, for instance, or penalties. Legislation setting a required level of recycling for vehicles and for electronic products at the end of their lives is now in force in Europe; other nations have similar programs and plans for more. Municipalities, too, have recycling laws requiring the reprocessing of waste that, under free market conditions, would have zero value. In a free market these would end up in the landfill, but the law prohibits it. When this is so municipalities sell the waste for a negative price—that is, they pay processing firms to take it. The negative price, too, fluctuates according to market forces and may, if technology improves or demand increases, turn positive, removing the need for the subsidy.

Where laws requiring recycling do not exist, recycling must compete also with landfill. Landfill, too, carries a cost. What is recycled and what is dumped then change as market conditions—the level of a landfill tax, for instance—change and businesses seek to minimize the cost of managing waste.

News-clip: The fluctuating value of waste

Back at junk value, recyclables are piling up.
The economic downturn has decimated the market for recycled materials, leaving more headed for landfills.

The New York Times, December 8, 2008

The value of recycled materials can go down as well as up, at least in the short term. As landfill charges rise, the economics of recycling become more attractive. But a recycling plant requires investment, and investors do not like unstable markets.

News-clip

Japan recycles rare earth minerals from used electronics.
Recent problems with Chinese supplies of rare earths have sent Japanese traders and companies in search of alternative sources. [The new source] is not underground, but in what Japan refers to as urban mining—recycling metals and minerals from the country's huge stockpiles of used electronics. "We've literally discovered gold in cellphones," said Tetsuzo Fuyushiba, a former land minister.

The New York Times, October 5, 2010

... and the value of recycled materials can go up as well as down. Japan's National Institute for Materials Science (NIMS), estimates that used electronics in

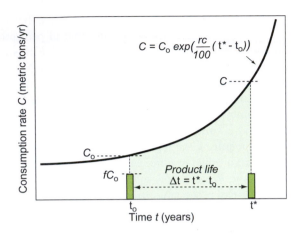

FIGURE 4.5 *Material dispersed as products does not appear as scrap for recycling until the product comes to the end of life. If consumption grows, long-lived products contribute less than those with short lives.*

Japan hold an estimated 300,000 tons of rare earths. They have a vital role in battery, electric motor, and laser technology. Restricted supply of virgin materials makes their recovery from waste products increasingly attractive.

All this may give the impression that waste management is a local issue, driven by local or national market forces. But the insatiable appetite of the fast-developing nations, particularly China and India, turn the "waste" of Europe and the United States into what, for them, is a resource. Low labor costs, sometimes less restrictive environmental regulation, and different manufacturing quality standards drive a world market in both waste and recycled materials.

The contribution of recycling to current supply. Suppose that a fraction f of the material of a product with a life of Δt years becomes available as old scrap. Its contribution to today's supply is the fraction f of the consumption Δt years ago. Material consumption, generally, grows with time, so this delay between consumption and availability as scrap reduces the contribution it makes to the supply of today. Figure 4.5 illustrates this. Suppose, for the moment, that a material exists that is used for one purpose only in a product with a life span Δt and that, at end of life, a fraction f (about 0.6 in the figure) is recycled into supply for current consumption, which has been growing at a rate $r_c\%$ per year. If the consumption rate when the product was made at time t_0 was C_0 metric tons per year, then the consumption today, at time t^*, is

$$C = C_0 \exp\left(\frac{r_c}{100}(t^* - t_0)\right) = C_0 \exp\left(\frac{r_c}{100}\Delta t\right) \tag{4.1}$$

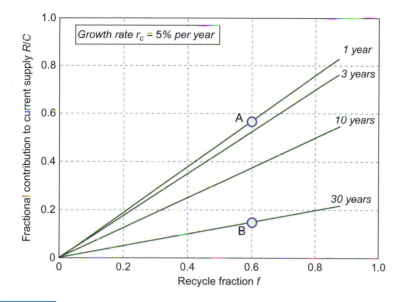

FIGURE 4.6 *Plot of equation (4.3) for recycling effectiveness when the growth rate in consumption is 5% per year, for various product lives and recycle fractions.*

where $\Delta t = t^{\star} - t_0$. The recovered fraction R is

$$R = fC_0 \qquad (4.2)$$

shown as a green bar on the figure. Its fractional contribution to supply today is

$$\frac{R}{C} = \frac{f}{\exp\left(\frac{r_c}{100}\,\Delta t\right)} \qquad (4.3)$$

Figure 4.6 shows what this looks like for a product with a growth rate of 5% per year. If the fraction is $f = 0.6$, and the product life Δt is 1 year, the contribution is large, about 0.58 of current supply (Point A on Figure 4.6). But if the product life is 30 years, the contribution falls to 0.17 (Point B). The recycle contribution increases with f, of course. But it decreases quickly if the product has a long life or a fast growth rate.

In reality most materials are used in many products, each with its own lifespan Δt_i, recycle fraction f_i. Consider one of these—product i—that accounts for a fraction s_i of the total consumption of the material. Its fractional contribution to supply today is

$$\frac{R_i}{C} = \frac{s_i f_i}{\exp\left(\frac{r_c}{100}\,\Delta t_i\right)} \tag{4.4}$$

where r_c is the overall growth rate of consumption of the material. The total contribution of recycling is the sum of terms like this for the material in all the products that use it.

Recycle contribution of gizmos and widgets

Example: A material is used to make both gizmos and widgets; each accounts for 25% of the total consumption of the material. Gizmos last for 20 years; their sales have grown steadily at 10% per year. At the end of life gizmos are dismantled and all the material is recovered ($f_{gizmo} = 1$). Widgets, on the other hand, have an average life of 4 weeks and are difficult to collect; their recycle fraction is only $f_{widget} = 0.5$. Their sales have grown slowly at 1% per year for the recent past. Which contributes most to today's consumption?

Answer: If we insert these data into equation 4.3, we find an unexpected result: gizmos, all of which are recycled, contribute the tiny fraction of 0.03 to current supply whereas widgets, only half of which are recycled, contribute a much larger fraction of 0.125. The main effect here is the product lifetime: products with short lives make larger contributions to recycling than those that last for a long time. This reveals one of the many unexpected aspects of the materials economy: making products that last longer can reduce demand, but it also reduces the scrap available for recycling. If, in making them last longer they have to be made more robust, meaning that they need more material, the effect can actually increase the demand for primary materials.

4.6 Summary and conclusions

The greater the number of us that consume and the greater the rate at which we do so, the greater is the volume of materials that our industrial system ingests and then ejects as waste. Real waste is a problem—a loss of resources that cannot be replaced—and there has to be somewhere to put it and that, too, is a diminishing resource.

But waste can be seen differently: as a resource. It contains energy and it contains materials, and, since most products still work when they reach the end of their first life, waste contains components or products that can still be useful. There are a number of options for treating a product at the end of its first life: extract the energy via combustion, extract the materials and reprocess them, replace the bits that are worn and sell it again, or, simplest of all, put it on eBay or other trading system and sell it as-is.

All have merit. But only one—recycling—can begin to cope with the volume of waste that we generate and transform it into a useful resource.

4.7 Further reading

Chapman, P.F., and Roberts, F. (1983), *Metal resources and energy*, Butterworth's Monographs in Materials, Butterworth and Co., Thetford, UK. ISBN 0-408-10801-0. *(A monograph that analyzes resource issues, with particular focus on energy and metals)*

Chen, R.W., Navin-Chandra, D., Prinz, F.B. (1993), "Product design for recyclability: a cost benefit analysis," Proceedings of the IEEE International Symposium on Electronics and the Environment, Vol. 10–12, pp. 178–183. ISEE.1993.302813. *(Recycling lends itself to mathematical modeling. Examples can be found in the book by Chapman and Roberts (above) and in this paper, which takes a cost-benefit approach.)*

Guidice, F., La Rosa, G., and Risitano, A. (2006), *Product design for the environment*, CRC/Taylor and Francis, London, UK. ISBN 0-8493-2722-9. *(A well-balanced review of current thinking on eco-design)*

Hammond, G., and Jones, C. (2010), *Inventory of carbon and energy (ICE)*, Annex A: methodologies for recycling, The University of Bath, Bath, UK. *(An analysis of alternative ways of assigning recycling credits between first and second lives)*

Henstock, M.E. (1988), *Design for recyclability*, Institute of Metals, London, UK. *(A useful source of background reading on recycling)*

Imhoff, D. (2005), *Paper or plastic: searching for solutions to an overpackaged world*, University of California Press. ISBN-13: 978-1578051175. *(What it says: a study of packaging taking a critical stance)*

PAS 2050 (2008), *Specification for the assignment of the life-cycle greenhouse gas emissions of goods and services*, ICS code 13.020.40, British Standards Institution, London, UK. ISBN 978-0-580-50978-0. *(This Publicly Available Specification (PAS) deals with carbon-equivalent emissions over product life, with a prescription for the way to assess end of life.)*

4.8 Appendix: designations used in recycle marks

(a) Base polymers

E/P	ethylene-propylene plastic
EVAC	ethylene-vinyl acetate plastic
MBS	methacrylate-butadiene-styrene plastic
ABS	acrylonitrile-butadiene-styrene plastic
ASA	acrylonitrile-styrene-acrylate plastic
C	cellulose polymers
COC	cycloolefin copolymer
EP	epoxide; epoxy resin or plastic
Imod	impact modifier
LCP	liquid-crystal polymer
MABS	methacrylate-acrylonitrile-butadiene-styrene plastic

MF	melamine-formaldehyde resin
MPF	melamine-phenolic resin
PA11	homopolyamide (Nylon) based on 11-aminoundecanoic acid
PA12	homopolyamide (Nylon) based on ω-aminododecanoic acid or on laurolactam
PA12/MACMI	copolyamide (Nylon) based on PA12, 3, 3-dimethyl-4, 4-diaminodicyclo-hexylmethane and isophthalic acid
PA46	homopolyamide (Nylon) based on tetramethylenediamine and adipic acid
PA6	homopolyamide (Nylon) based on ε-caprolactam
PA610	homopolyamide (Nylon) based on hexamethylenediamine and sebacic acid
PA612	homopolyamide (Nylon) based on hexamethylenediamine and dodecane-diacid (1,10-decandicarboxylic acid)
PA66	homopolyamide (Nylon) based on hexamethylenediamine and adipic acid
PA66/6T	copolyamide based on hexamethylenediamine, adipic acid, and terephthalic acid
PA666	copolyamide based on hexamethylenediamine, adipic acid, and ε-caprolactam
PA6I/6T	copolyamide based on isophthalic acid, adipic acid, and terephthalic acid
PA6T/66	copoylamide based on adipic acid, terephthalic acid, and hexamethylenediamine
PA6T/6I	copolyamide based on hexamethylenediamine, terephthalic acid, adipic acid, and isophthalic acid
PA6T/XT	copolyamide based on hexamethylenediamine, 2-methyl-penta-methylene-diamine, and terephthalic acid
PAEK	polyaryletherketon
PAIND/INDT	copolyamide based on 1, 6-diamino-2,2,4-trimethylhexane, 1, 6-diamino-2, 4, 4-trimethylhexane, and terephthalic acid
PAMACM12	homopolyamide based on 3,3'-dimethyl-4,4'-diaminodicyclohexyl-methane and dodecandioic acid
PAMXD6	homopolyamide based on m-xylylenediamine and adipic acid
PBT	poly(butylene terephthalate)
PC	polycarbonate
PCCE	poly(cyclohexane dicarboxylate)
PCTA	poly(cyclohexylene dimethylene terephthalate), acid
PCTG	poly(cyclohexylene dimethylene terephthalate), glycol
PE	polyethylene
PEI	polyetherimide
PEN	polyethylene naphthalate
PES	polyethersulfone
PET	polyethylene terephthalate
PETG	polyethylene terephthalate, glycol

PF	phenol-formaldehyde resin
PI	polyimide
PK	polyketone
PMMA	poly(methyl methacrylate)
PMMI	polymethylmethylacrylimide
POM	polyoxymethylene, polyacetate, polyformaldehyde
PP	polypropylene
PPE	poly(phenylene ether)
PPS	poly(phenylene sulfide)
PPSU	poly(phenylene sulfone)
PS	polystyrene
PS-SY	polystyrene, syndiotactic
PSU	polysulfone
PTFE	polytetrafluoroethylene
PUR	polyurethane
PVC	polyvinyl chloride)
PVDF	poly(vinylidene fluoride)
SAN	styrene-acrylonitrile plastic
SB	styrene-butadiene plastic
SMAH	styrene-malefic anhydride plastic
TEEE	thermoplastic ester- and ether-elastomers
TPA	polyamide thermoplastic elastomer
TPC	copolyester thermoplastic elastomer
TPO	olefinic thermoplastic elastomer
TPS	styrenic thermoplastic elastomer
TPU	urethane thermoplastic elastomer
TPV	thermoplastic rubber vulcanisate
TPZ	unclassified thermoplastic elastomer
UP	unsaturated polyester

(b) Fillers

CF	carbon fiber
CD	carbon fines, powder
GF	glass fiber
GB	glass beads, spheres, balls
GD	glass fines, powder
GX	glass not specified

K	calcium carbonate
MeF	metal fiber
MeD	metal fines, powder
MiF	mineral fiber
MiD	mineral fines, powder
NF	natural organic fiber
P	mica
Q	silica
RF	aramid fiber
T	talcum
X	not specified
Z	others not included in this list

4.9 Exercises

E4.1. Many products are thrown away and enter the waste stream even though they still work. What are the reasons for this?

E4.2. Do you think manufacturing without waste is possible? "Waste," here, includes low-grade heat, emissions, and solid and liquid residues that cannot be put to a useful purpose. If waste is inevitable, why?

E4.3. What options are available for coping with the waste-stream generated by modern industrial society?

E4.4. Recycling has the attraction of returning materials into the use-stream. What are the obstacles to recycling?

E4.5. Car tires create a major waste problem. Use the Internet to research ways in which the materials contained in car tires can be used, either in the form of the tire or in some decomposition of it.

E4.6. List five functions of packaging.

E4.7. As a member of a brain storming group, you are asked to devise ways of reusing polystyrene foam packaging—the sort that encases TV sets, computers, appliances, and many other things when they are transported. Use free thinking: no suggestion is too ridiculous.

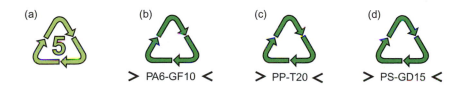

(a) (b) PA6-GF10 (c) PP-T20 (d) PS-GD15

E4.8. You are employed to recycle German washing machines, separating the materials for recycling. You encounter components with the recycle marks seen in Figure 4.7:

How do you interpret them?

Answer. (a) Polypropylene. (b) Polyamide 6 (Nylon 6) with 10% glass fiber. (c) Polypropylene with 20% talc. (d) Polystyrene with 15% glass fines (powdered glass).

E4.9. The consumption of lead is growing at 4% per year. It has a number of uses, principally as electrodes in storage batteries, as roofing and pipe-work on buildings, and as pigment for paints. The first two of these allow recycling, the third does not. Batteries consume 38% of all lead, have an average life of 4 years, a growth rate of 4% per year, and are recycled with an efficiency of 80%. Architectural lead accounts for 16% of total consumption. The lead on buildings has an average life of 70 years after which 95% of it is recycled. What is the fractional contribution of recycled lead from each source to current supply?

E4.10. A material M is imported into a country principally to manufacture one family of products with an average life of Δt years and a growth rate of r_c% per year. The material is not currently recycled at end of life, but it could be. The government is concerned that imports should not grow.

    (a) What recycle fraction, f_{crit}, is necessary to make this possible?
    (b) The longer the average product life, Δt, the smaller is its fractional contribution to future supply. What is the limiting life beyond which recycling cannot keep up with increased demand?

E4.11. The rare-earth metal neodymium is an ingredient of high-field permanent magnets. Its consumption is growing at $r_c = 5\%$ per year. If 50% of all neodymium entering service is used for the electric motors of hybrid and electric cars with an average life of 15 years, what fraction will recycling 95% of the neodymium in these cars contribute to future consumption? Computer hard disk drives, too, contain neodymium magnets. If 15% of current neodymium consumption is that in hard disk drives with an average life of 3 years, what fraction will they contribute to future consumption if they, too, have a 95% recycle rate?

CHAPTER 5

The long reach of legislation

5.1 Introduction and synopsis

The prophet Moses, seeking to set standards for the ways in which his people behaved, created or received (according to your viewpoint) ten admirably concise commandments. Most start with the words "Thou shalt not...," with simple, easily understood incentives (heaven, hell) to comply. Today, as far as materials and design are concerned, it is environmental protection agencies and European commissions that issue commandments, or, in their language, *Directives*. The consequences of not following them are not as Old Testament in their severity as those of the original ten, but if you want to grow your business, compliance becomes a priority.

This involves some obvious steps:

- Being aware of directives or other binding controls that touch on the materials or processes you use.
- Understanding what is required to comply with them.
- Having (or developing) tools to make compliance as painless as possible.

Warning signs that relate to materials. Clockwise from the top left: Dangerous, Highly flammable, Explosive, Very corrosive, Environmentally hazardous, Poisonous.

- Exploring ways to make compliance profitable rather than a burden—exploiting compliance as a marketing tool, for example.

This chapter is about controls and economic instruments that impinge on the use of engineering materials. It reviews current legislation and describes an example of tools to help with compliance.

5.2 Growing awareness and legislative response

Table 5.1 lists nine documents that have had profound influence on current thinking about the effects of human activity on the environment. The publications span

Table 5.1	Required reading: landmark publications
Date, author, and title	**Subject**
1962, Rachel Carson, *Silent Spring*	Meticulous examination of the consequences of the use of the pesticide DDT and of the impact of technology on the environment.
1972, Club of Rome, *Limits to Growth*	The report that triggered the first of a sequence of debates in the 20th century on the ultimate limits imposed by resource depletion.
1972, The Stockholm Earth Summit	The first conference convened by the United Nations to discuss the impact of technology on the environment.
1987, The UN World Commission on Environment and Development (WCED) report, "Our Common Future"	Known as the Brundtland Report, it defined the principle of sustainability as "Development that meets the needs of today without compromising the ability of future generations to meet their own needs."
1987, Montreal Protocol	The international protocol that called for the phase-out of chemicals that deplete ozone in the stratosphere.
1992, Rio Declaration	An international statement of the principles of sustainability, building on those of the 1972 Stockholm Earth Summit.
1998, Kyoto Protocol	An international treaty to reduce the emissions of gases that, through the greenhouse effect, cause climate change.
2001, Stockholm Convention	The first of an ongoing series of meetings to agree upon an agenda for the control and phase-out of persistent organic pollutants (POPs).
2007, IPCC Fourth Assessment Report. "Climate Change 2007: The Physical Basis"	This report of the Intergovernmental Panel on Climate Change (IPCC) establishes beyond any reasonable doubt the correlation between carbon in the atmosphere and climate change.

a little less than 50 years. Over this period the response of industry to a pollution or environmental problem has evolved through a number of phases,[1] best summarized in the following way.

- *Ignore it*: pretend it isn't there.
- *Dilute it*: make the smoke stack taller, or pump it farther out to sea.
- *Fix it with as little disruption of production as possible*: the "end-of-pipe" approach.
- *Prevent it* **in the first place**: the first appearance of design for the environment.
- *Aim for sustainability*: seek ways to establish equilibrium with the environment—the phase we are in now.

Today national legislation and international protocols and agreements set environmental standards. The international agreements tend to be broad statements of intent. The national legislation, by contrast, tends to be specific and detailed.

Historically, environmental legislation has targeted isolated problems as they arise—dumping of toxic waste, sewage in water, lead in gasoline, ozone depletion in the atmosphere—taking a *command and control* approach: basically, this is "thou shalt not" cast in modern terms. However, there is a growing recognition that this can lead to perverse effects where action to fix one isolated problem just shifts the burden elsewhere and may even increase it. For this reason, there has been a shift from command and control legislation toward the use of *economic instruments*— green taxes, subsidies, trading schemes—that seek to use market forces to encourage the efficient use of materials and energy. We have already seen that some activities create environmental burdens that have costs that are not paid for by the provider or user. These are called external costs or *externalities*. A more effective approach is to transfer the cost back to the activity creating it, thereby *internalizing* it. But that is not always easy, as we shall see.

5.3 International treaties, protocols, and conventions

It is exceedingly difficult to negotiate enforceable treaties that bind all the nations of the planet to a single course of action; the differences of culture, national priorities, economic development, and wealth are too great. The best the international community can achieve is an agreement, declaration of intent, or protocol[2] that a

---

[1] Details can be found in books on industrial ecology such as Ayres and Ayres (2002)—see Further reading.

[2] A *protocol* is a memorandum of resolutions arrived at in negotiation, signed by the negotiators, as a basis for a final convention or treaty. In fact, the Kyoto Protocol is more than that, because it is a binding treaty to meet certain agreed objectives. The distinction between the protocol and the convention, in current usage, is that while the convention encourages countries to stabilize emissions, the protocol commits them to do so.

subset of nations feels able to sign. Such agreements directly influence policy in the nations that sign them. By defining the high ground, they exert moral pressure both on signatories and non-signatories alike. Two have been particularly significant in their influence on government policy on materials.

> **The Montreal Protocol** (1989) is a treaty aimed at reducing the use of substances that deplete the ozone layer of the stratosphere. Ozone depletion allows more UV radiation to reach the surface of the earth, damaging living organisms. The culprits are typified by CFCs—chlorofluorocarbons—that were widely used as refrigerants and as blowing agents for polymer foams, particularly those used for house insulation. All have now been replaced by less harmful substitutes. This protocol has largely achieved its aims.
>
> **The Kyoto Protocol** (1997) is an international treaty to reduce the emissions of gases that, through the greenhouse effect, cause climate change. It sets binding targets for the 44 industrialized countries that have signed it, committing them to reduce greenhouse gas emissions over the 5-year period 2008–2012.

International directives and protocols are usually based on *principles*—statements of what are seen as fundamental rights—rather than on laws that cannot be agreed or enforced. They exert moral rather than legal pressure. Here are examples of some that have emerged from the protocols and conventions listed in Table 5.1.

- *Principle 1* (Stockholm Convention): the right to exploit one's own environment.
- *Principle 2* (Rio Declaration): the right to industrial development provided it does not damage others.
- *Precautionary principle* (WCED report): where there are possibilities of large irreversible impacts, the lack of scientific certainty should not prevent preventative action from being taken.
- *Polluter pays principle*: transfers the responsibility and cost of pollution to the polluter ("internalizing externalities" in eco-speak).
- *Sustainable development principle*: protection of the environment, equity of burden.

The idea is that these should provide a framework within which strategies and actions are developed.

5.4 National legislation: standards and directives

"The Council of the European Union, having regard to A, B and C, acting in accordance with procedures P, Q and R of activities X, Y and Z, and whereas...(there follows a list of 27 further "whereases") HAS ADOPTED THIS DIRECTIVE...." That was a paraphrase of the start of an EU directive. Environmental legislation makes for heavy reading. It is cast in legal language of such gothic formality and baroque intricacy that organizations spring up with the sole purpose of interpreting it. But since much of it impinges, directly or indirectly, on the use of materials, it is important to get the central message.

National legislation, as typified by the US Environmental Agency Acts or the European Union Environmental Directives, takes four broad forms.

- Setting standards
- Negotiating voluntary agreements with industry
- Imposing binding legislation, with penalties if its terms are not met
- Devising economic instruments that harness market forces to induce change: taxes, subsidies, and trading schemes (discussed in Section 5.5)

Standards. ISO 14000 of the International Standards Organization defines a family of standards for environmental management systems.[3] It contains the set ISO 14040, 14041, 14042, and 14043 published between 1997 and 2000, which prescribe broad but vague procedures for the four steps described in Chapter 3, Section 3.4 (*goals and scope, inventory compilation, impact assessment,* and *interpretation*). The standard is an attempt to bring uniform practice and objectivity into life-cycle assessment (LCA) and its interpretation, but it is not binding in any way.

ISO 14025 is a standard that guides the reporting of LCA data as an Environmental Product Declaration (EPD) or a Climate Declaration (CD).[4] The goal is to communicate information about the environmental performance of products as a "declaration" in a standard, easily understood format (there is an example in Chapter 8). The data used for the declaration must follow the ISO 14040 family of procedures and must be independently validated by a third party. The EPD describes the output of a full LCA, or (if declared so) of part of one. To make it a little easier, the CD is limited to emissions that contribute to global warming: CO_2, CO, CH_4, and N_2O. But it is still a big job.

Publicly Available Specification (PAS) 2050 (2008) is an attempt to reduce the burden further. It focuses on greenhouse gas (GHG) emissions, ignoring all others. It seeks to provide a consistent way to assess the life-cycle GHG emissions of goods and services by defining a common basis for comparison and for communicating the results. It is a consultative document, not, at present, a standard, but it is increasingly accepted by the manufacturing industry in Europe as a product-assessment tool that is less burdensome to implement than the ISO standard.

In practice LCA procedures are used primarily for in-house product development, benchmarking, and to promote the environmental benefits of one product over another. They are rarely used as the basis for regulation because of the difficulties, described in Chapter 3, of setting system boundaries, of double counting, and of limited coverage across products.

---

[3]See www.iso-14001.org.uk/iso-14040 for a summary.

[4]For the more intimate details, see www.environdec.com.

Voluntary agreements and binding legislation. Current legislation is aimed at internalizing costs and conserving materials by increasing manufacturers' responsibilities, placing on them the burden of cost for disposal. Here are some examples.

The US Resource Conservation and Recovery Act (RCRA), enacted in 1976, is a federal law of the United States. The Environmental Protection Agency (EPA) monitors compliance. RCRA's goals are
- to protect the public from harm caused by waste disposal
- to encourage reuse, reduction, and recycling
- to clean up spilled or improperly stored wastes

The US EPA 35/50 Program, enacted in 1988, identified 17 priority chemicals, italicized in Table 5.2, with the aim of reducing industrial toxicity by voluntary action by industry over a 10-year period.

The CAFE Standard. The Arab oil embargo of 1973/4 created a spike in oil prices and the realization that dependence on imported oil carried risks (see the spike in Figure 2.18). It stimulated the US Congress to pass the Energy Policy Conservation Act of 1975, establishing the Corporate Average Fuel Economy (CAFE) standards, penalties, and credits. The motive was to raise the fuel efficiency of new cars sold in the United States from an average of around 15 mpg (miles per US gallon[5]) to 27.5 mpg by 1985. The Energy Independence and Security Act, passed 32 years later (2007), raised the bar, aiming for a progressive increase to 35 mpg by 2020.

The CAFE standards set targets for the average fleet fuel consumption, measured in mpg, for each car maker. *Fleet* means all the cars, of all sizes, sold in a given model year. Failure to meet the target incurs a penalty per car of $5.50 per tenth of a mpg below the target. Exceeding the target creates a credit, calculated in the same way, which can be set against penalties in adjacent years. The sums of money involved are large: between 1983 and 2010 manufacturers have paid more than $500 million. Most European car makers regularly pay CAFE penalties of up to $20 million per year. Electric cars and cars using biofuel are penalized less heavily or not at all.

Figure 5.1 shows how the CAFE standard has risen since 1975, quickly at first, then in 1990 stabilizing at 27.5 mpg where it remained for the next 20 years. The proposed future increase appears as a broken line. Fuel price, the other major driver for efficiency, is shown as the green line with the scale on the right. The US national average fuel economy is shown as the blue line. It has slightly exceeded the standard, largely because of imports of fuel-efficient cars from Japan, until the steep rise in fuel price in 2005 that appears to have pushed it up.

---

[5]1 mpg$_{US}$ = 1.2 mpg$_{Imperial}$ = 0.245 Liters/100 km.

Table 5.2	Priority chemicals and materials

Volatile organic compounds (VOCs)	Applications
Benzene	Intermediate in production of styrene, thus many polymers
Carbon tetrachloride	Solvent for metal degreasing, lacquers
Chloroform	Solvent
Methyl ethyl keytone	Solvent for metal degreasing, lacquers
Tetrachoroethylene	Solvent for metal degreasing
Toluene	Solvent
Trichlorethylene	Solvent, base of adhesives
Xylenes	Lacquers, rubber adhesives
Toxic metals or salts of metals	
Asbestos	Fiberboard reinforcement, thermal and electrical insulation
Antimony	Bearings, pigments in glasses
Beryllium + compounds	Space structures, copper-beryllium alloys
Cadmium + its compounds	Electrodes, plating, pigment in glasses and ceramics
Chromium compounds	Electroplating, pigments in glasses and ceramics
Cobalt + compounds	Super alloys, pigments in glasses and glazes
Lead + compounds	Storage batteries, bearing alloys, solders
Mercury + compounds	Control equipment, liquid electrode in chemical production
Nickel + compounds	Nickel carbonyl as intermediate in nickel production
Radioactive materials	Materials science and medicine
Toxic chemicals	
Cyanides	Electroplating, extraction of gold and silver

Why would automakers choose to pay millions of dollars in penalties when, by reducing weight and engine size, they could avoid it? It is because it is more profitable for premium companies like Mercedes and BMW to give the customer what they want (size, power, luxury) and pay the penalty than it is to damage brand image by failing to meet expectations. If, as an

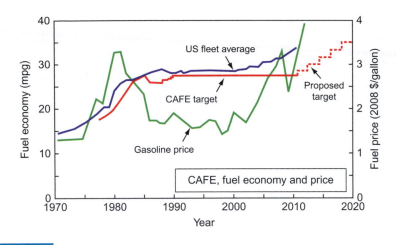

FIGURE 5.1 *The average fuel economy of US cars (blue line) has tracked the CAFE standard (red line) until recently, when the steep rise in fuel price (green line) appears to have pushed it up.*

example, the fleet economy lies a full 2 mpg below the CAFE target, the penalty per car is a mere $110, which can be passed on to the purchaser without the manufacturer feeling pain. By contrast, that same 2 mpg difference in fuel economy translates, over a vehicle life of 150,000 miles, into an additional 430 gallons of fuel, which, at $4 per gallon, costs $1,700. Simple economics thus suggests that fuel price should be a more powerful driver for fuel economy than the CAFE standard, but because fuel cost is spread over life, it may not appear so.

If the proposed increase in the CAFE standard to 2020 is implemented, an automaker who does not respond by increasing fuel efficiency could face penalties approaching $1 billion per year. This creates a more pressing incentive to build fuel-efficient cars, driving research on alternative materials to reduce weight. We return to this topic in Chapter 10.

News-clip: Eco-nudges

New fuel economy labels for cars and trucks are unveiled.

In the most extensive overhaul of the decals in 30 years, the U.S. Environmental Protection Agency and the Department of Transportation said Wednesday that 2013 model year cars and trucks will have more comprehensive labels detailing projected fuel costs and emissions. In addition to the miles per gallon, the labels will show, on a scale of 1 to 10, how a vehicle stacks up against competitors for smog, tailpipe emissions and fuel economy. Electric vehicles and plug-in hybrids will have a mpg equivalent number as well as

details about driving range and charging times. The new government labels supplement the rules enacted last year requiring the average fuel economy for new cars and trucks to reach at least 34.1 mpg by 2016.
Los Angeles Times, May 25, 2011

The US EPA Code of Federal Regulation (CFR). Protection of the environment (CFR Part 302) deals with protection of the environment and human health, imposing restrictions on chemicals released into the environment during manufacture, life, and disposal. Like REACH (discussed in a minute), it requires manufacturers to register the use of a long list of chemicals and materials (Table 302.2 of the Regulation) if the quantity used exceeds a threshold.

Product Liability Directive (1985). The European Directive 85/374/EEC imposes liability responsibility for damage caused by defective products.

Landfill Directive (1999). The European Directive 1999/31/EC lays down binding procedures for dealing with hazardous and nonhazardous wastes. It bans the landfilling of many products, among them tires, for which alternatives must be found.

Volatile Organic Compounds (VOCs, 1999). The European Directive EC 1999/13 aimed to limit the emissions of VOCs caused by the use of organic solvents and carriers like those in organic-based paints and industrial cleaning fluids. Compliance became mandatory in 2007.

End-of-Life Vehicles (ELV, 2000). The European Community Directive, EC2000/53, establishes norms for recovering materials from dead cars. The initial target, a rate of reuse and recycling of 80% by weight of the vehicle and the safe disposal of hazardous materials, was established in 2006. By 2015 the target is a limit of 5% by weight to landfill and a recycling target of 85%. The motive is to encourage manufacturers to redesign their products to avoid using hazardous materials and to maximize ease of recovery and reuse.

Hazardous Substances Directive (RoHS, 2002). The curious acronym stands for "the restriction of the use of certain hazardous substances in electrical and electronic equipment." This Directive bans the placing on the EU market of new electrical and electronic equipment containing more than agreed levels of six materials: lead, cadmium, mercury, hexavalent chromium used in pigments, paints, and electro plate and various flame and retardants. It is closely linked with the even more curiously named WEEE Directive.

Waste Electrical and Electronic Equipment (WEEE, 2002). The WEEE Directive (EC 2002/96 and 2003/108) sets collection, recycling, and recovery targets for electrical goods and is part of a legislative initiative to solve the problem of toxic waste arising from electronic products. It requires that producers finance the collection, recovery, and safe disposal of their products and meet certain recycling targets. Products failing to meet the requirement must be marked with a "naughty" mark (a crossed-out wheeled box).

Energy Using Products (EuP, 2003). The EU Directive EC 2003/0172 establishes a framework of eco design requirements for products that use energy—appliances, electronic equipment, pumps, motors, and the like. It requires that manufacturers of any product that uses energy "shall demonstrate that they have considered the use of energy in their product as it relates to materials, manufacture, packaging transport and distribution, use, and end of life. For each of these the consumption of energy must be assessed and steps to minimise it identified."

Energy-related Products (ErP, 2009). As of 2009 the EuP Directive has been replaced by the ErP Directive EC 2009/125, which has a similar intent. The EuP Directive only covered products that actually used energy, such as microwaves, washing machines, or televisions. The new ErP Directive covers not only energy-using products, but also products that are energy-related even thought they don't use energy directly, such as double-glazed windows, sink taps, and showerheads.

Registration, Evaluation, Authorization and Restriction of Chemical Substances (REACH, 2006). The Directive EC 1907/2006 came into force in June 2007, to be phased in over the following 11 years. The directive places more responsibility on manufacturers to manage risks from chemicals and to find substitutes for those that are most dangerous. The list is a long one—it has some 30,000 chemicals on it. "Chemical" includes metals and alloys in quantities greater than 1 metric ton per year. And there is more. Manufacturers in Europe and importers into Europe must register the restricted substances they use by providing detailed technical dossiers for each, listing their properties, by providing an assessment of their impacts on the environment and human health, and by stating the risk-reduction measures they have adopted. It is illegal for manufacturers and importers to place substances on the market without first registering them.

Here are just a few examples of restrictions that matter for engineering materials. The restrictions are an amalgam of the Directives and Regulations. Some are draconian—a total ban. Others are mild—keep it below 0.1% in things for children.

- Asbestos (prohibited)
- Flame retardant compounds (limit 0.1%)
- Arsenic compounds (prohibited in wood products)
- Cadmium compounds (0.25% in galvanization, 0.01% in electronics)
- Chromium, hexavalent compounds (limited to 0.1% in electronics)
- Lead compounds (limit 0.1% in electronics)
- Mercury compounds (limit 0.1% in electronics)
- Nickel compounds (limit in articles in prolonged contact with skin)
- Organostannic (tin) compounds (limit 0.1% in products)

- Ozone-depleting BCMs, CFCs, halons, solvents (prohibited)
- Phthalate plasticizers (toys and childcare limit of 0.1%)
- Hydrocarbon solvents (limit 0.1% in products)

The trouble, if you make things, is that you have to know them all to avoid getting into trouble. Many polymers contain flame retardants or plasticizers. Chromium plating involves hexavalent chromium. Ordinary solder contains lead. Rechargeable batteries contain cadmium and long-life, low-drain batteries contain mercury. Nickel is a component of resistors, heating elements, and magnets. Many manufacturing processes involve the use of organic solvents.

News-clip: Banned materials

To universal surprise, the National Assembly (of France) bans phthalates.
The Assembly yesterday passed into law a proposal to forbid the manufacture, importation, sale or distribution of products containing phthalates....
Le Figaro, May 4, 2011

Phthalates are esters of phthalic acid ($C_6H_6(COOH)_2$), derived from benzene. They are used to plasticize polymers, particularly PVC, increasing flexibility, transparency, and durability. They are used in man-made fibers and fabrics, plastic and elastomeric toys, furniture, fittings (notably car interiors), and medical equipment such as blood bags. Phthalates decompose slowly so they accumulate in water, soil, and the atmosphere of enclosed spaces such as offices. In sufficient concentration they are harmful.

The suddenness of the legislation has caught manufacturers in France by surprise. There are alternatives to phthalates but they are poorly researched and may also be toxic. The law, if ratified without leaving a period for adjustment, would seriously disadvantage the French fabrics industry.

Battery Directive (2006). The Directive 2006/66/EC prohibits the sale of batteries containing more than 0.0005% of mercury or 0.002% of cadmium, prohibits the dumping or incineration of batteries, and seeks to implement collection and recycling schemes to recover the materials they contain.

Waste Framework Directive (WFD, 2008). The Directive 2008/98/EC sets a hierarchy for waste management: prevention, reuse and recycling, and energy recovery. It requires that those generating waste must keep records of the waste type, quantity, origin and destination, frequency, mode of transport, and treatment method. The full cost of waste disposal must be borne by the holder of the waste (the "Polluter pays" principle).

Reusability, Recyclability and Recoverability Directives (RRR, 2008). Directive 2005/64/EC aims to force vehicle makers to develop recycling strategies, ensure sustainable product development, and make use of energy and natural resources. It requires that 85% of the vehicle mass must be reusable or recyclable and that 95% be reusable or recoverable by 2015. Makers of passenger cars and light commercial vehicles that do not meet the Directive will be denied access to EU markets.

5.5 Economic instruments: taxes, subsidies, and trading schemes

"Economic instruments manipulate market forces to influence the behaviour of consumers and manufacturers in ways that are more subtle and effective than conventional controls, and they generally do so at lower cost."[6] Well, that's the idea. Taxation—one "economic instrument"—may not strike you as subtle,[7] but it does seem to work better than "command and control" methods. Here are some examples.

> *Green taxes.* Many countries now operated a *landfill tax*, currently standing at around €50 ($70) per metric ton in Europe, as a tool to reduce waste and foster better waste management. An *aggregate tax* on gravel and sand, about €2 ($2.8) per metric ton, recognizes that extracting aggregates has environmental costs; it is designed to reduce the use of virgin aggregates and stimulate the use of waste from construction and demolition. Most nations impose a *fuel duty*—a tax on gasoline and diesel fuel—to encourage a shift to fuel-efficient vehicles and to increase the use of biofuels by taxing them at a lower rate. Increasingly governments impose a *carbon tax* (presently €20 [$28] per metric ton of carbon), often based on energy consumption (using energy as a proxy for carbon), and *NO_x and SO_x emission taxes*. Roughly half of the US states charge a deposit—a returnable tax—on bottles and cans, a scheme that has proved effective in returning these into the recycling loop.
>
> But imposing a tax—a carbon tax, for instance—has two difficulties. First it does not guarantee the environmental outcome of reducing CO_2—industries that can afford it will simply pay it. The second difficulty is one of public acceptance. Taxes carry high administration costs, and people don't trust governments to spend the tax on the environment; fuel taxes, for example, don't get spent on roads or pollution-free vehicles. Of those two certainties of life—death and taxes—it is taxes that people try hardest to avoid.

---

[6]UK Department for Environment, Food and Rural Affairs (DEFRA), www.defra.gov.uk/environment/index.html.

[7]"Nothing is certain but death and taxes." Benjamin Franklin, writing 250 years ago.

News-clip: Externalized costs

The controversial idea of a carbon tax in Europe resurfaces.
A mandate now before the European Commission proposes a tax of at least 20 euros/tonne on carbon emissions from homes, transport and agriculture but not on the industrial sectors already covered by carbon-trading legislation.

Le Monde, April 8, 2011

A carbon tax is more effective than an energy tax because it disadvantages only energy from fossil fuels. At this date only a few countries have a carbon tax (Sweden, Denmark, inland, Ireland, and the UK). The EC meeting was set up to make the carbon tax Europe-wide, but the meeting did not result in agreement. France saw the mandate as disadvantaging its agriculture. Germany saw it as impeding the expansion of its automobile sector. There is still no Europe-wide carbon tax.

Subsidies. If people don't do what you want, you can try paying them to change their minds. At present, most developed nations subsidize the building of wind and solar farms to generate low-carbon power, motivated by commitments they have made to reduce greenhouse gas emissions. Without the subsidy, these alternative energy systems are currently not economical. The longer-term idea is to kick-start these industries, relying on technical and economic developments that will make them viable without state support in the future. The essential ingredient is consistency: offering a subsidy one year and cancelling it the next undermine investor confidence in a nascent manufacturing base before it has had time to become established.

News-clip: Slippery subsidies

Subsidy review dims solar hopes.
The future viability of Britain's biggest solar power schemes has been thrown into doubt by government proposals to cut their public subsidy.

The Financial Times, March 19, 2011

Wind power suffered a similar fate.

Trading schemes. Another way of putting a value on something is to create a market for it. The stock market is an example: a company issues shares, the total number of which represents its "value." The shares are traded— sold or purchased for real money—and they therefore float in value, rising if they are seen as undervalued, falling if they are seen as overvalued. At any moment in time, the share price sets a value on the company. A notion that emerged from the Kyoto Meeting of 1987 was to adapt this "market principle" to establish a value for emissions. To see how it works and the difficulties associated with it, we need to digress to explore emissions trading.

Emissions trading is a market-based scheme that allows participants to buy and sell permits for emissions or credits for reduction in emissions (a different thing) in certain pollutants. Taking carbon (meaning CO_2) as an example, the regulator first decides on a total acceptable emissions level and divides this into tradable units called *permits*. These are allocated to the participants, based on their actual carbon emissions at a chosen point in time. The actual carbon emissions of any one participant change with time, falling if they develop more efficient production technology or rising if they increase capacity. A company that emits more than its allocated permits must purchase allowances from the market, while a company that emits less than its allocations can sell its surplus. Unlike regulation that imposes emission limits, emissions trading gives companies the flexibility to develop their own strategy to meet emission targets. The environmental outcome is reduced over concerns grow. The buyer is paying a charge for polluting while the seller is rewarded for having reduced emissions. Thus those who can easily reduce emissions most cheaply will do so, achieving pollution reduction at the lowest cost to society.

Emissions trading has another dimension—that of offsetting carbon release by buying credits in activities that absorb or sequester carbon or that replace the use of fossil fuels by energy sources that are carbon-free: tree planting, and solar, wind, or wave power, for example. By purchasing sufficient credits the generator of CO_2 can claim to be "carbon neutral."

Carbon off setting has its critics. Three of the more telling criticisms are that

- It provides an excuse for enterprises to continue to pollute as before by buying credits and passing the cost on to the consumer.
- The scheme only achieves its aim if the mitigating project runs for its planned life, and this is often very long. Trees, for instance, have to grow for 50 to 80 years to capture the carbon with which they are credited— fell them sooner for quick profit and the offset has not been achieved. Wind turbines and wave power, similarly, achieved their claimed offset only at the end of their design life, typically 25 years.
- It is hard to verify that the credit payments actually reach the mitigating projects for which they were sold—the tree-planters or wind turbine builders. Too much of the payment gets absorbed in administrative costs.

The market has its dark side—insider trading, market fixing—to which the carbon market is as vulnerable as any other.

News-clips: Permit pirates

Carbon market stays closed after cyber thieves make off with €28 million.
The Times, January 30, 2011

Paris carbon stock-market arms itself against pirates.
The Figaro, May 3, 2011

A permit to emit one metric ton of carbon currently trades at around €15. Inadequate security has allowed the theft and subsequent resale of large numbers of carbon certificates—the *Times* reports a 500% increase in cyber raids in the past two years. New security measures seek to make this kind of cyber-theft more difficult.

5.6 The consequences

The burden this legislation places on the materials and manufacturing industries is considerable. The requirements are far-reaching. To recapitulate:

- They must document the use of any one of 30,000 listed chemicals.
- They must analyze energy and material use in all their energy-using products.
- They must find substitutes for VOCs and other restricted substances.
- They must take back, disassemble, and acceptably dispose of an increasingly large range of products.

Figure 5.2 summarizes these interventions, suggesting where they influence the flows of the life cycle. The intent of the legislation—that of reducing resource consumption and damaging emissions, and particularly of internalizing the costs these generate—makes sense. The difficulty is that implementing it generates administrative, reporting, and other costs in addition to the direct costs of clean-up, thereby adding to the burden on industry.

These additional costs are minimized by the use of well-designed software and other tools. Most of the LCA packages described in the Appendix of Chapter 3 were created to help companies analyze their products, using standards that meet the requirements of the legislation. A Web search on any one of the acts and directives listed in Section 5.4 reveals more tools designed to help implement them. The development of tools that integrate with existing Product Data Management (PDM) systems can make compliance semiautomatic, flagging the use of any material that is, in any sense, restricted, and automatically generating the reports that the legislation requires.

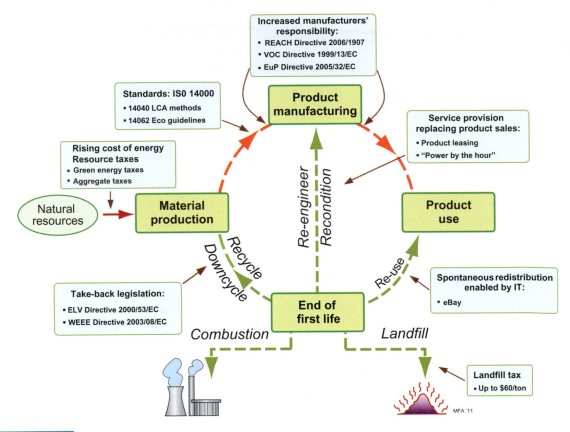

FIGURE 5.2 *Interventions and other mechanisms that influence the flows in the material life cycle.*

5.7 Summary and conclusions

Governments intervene when they wish to change the way people and organizations behave. Many now accept that the way they behave at present is damaging the environment in ways that could be irreversible. Some of this damage is local and can be tackled at a national level by making the polluter pay or rewarding those that do not pollute. National and multinational regulations, controls, and directives impose reporting requirements, set tax levels, and establish trading schemes to create incentives for change, with the ultimate aim of making design for the environment a priority.

Some impacts, however, are on a global scale. The externalized costs fall both on the nations responsible for the impact and on those that are not. Solutions here require international agreements. Binding, universal, and enforceable agreements here are out of reach—the diversity of wealth, national priorities, and political systems are too great. Nonetheless protocols and statements can be negotiated that many nations feel able to sign. It is in furthering these international agreements that we must place our hopes for the future.

But meanwhile, the volume of legislation grows and grows. One only wishes that environmental agencies could aspire to be as concise as Moses.

5.8 Further reading

Ayres, R.U., and Ayres, L.W., editors (2002), *A handbook of industrial ecology*, Edward Elgar, Cheltenham, UK. ISBN 1 84064 506 4. *(Industrial ecology is industrial because it deals with product design, manufacture and use, and ecology because it focuses on the interaction of this with the environment in which we live. This handbook brings together current thinking on the topic.)*

Brundtland, G.H., Chairman (1987), *Our common future*, Report of the World Commission on Environment and Development, Oxford University Press, Oxford, UK. ISBN 0-19-282080-X. *(Known as the Brundtland Report, it defined the principle of sustainability as "Development that meets the needs of today without compromising the ability of future generations to meet their own needs.")*

Carson, R. (1962), *Silent Spring*, Houghton Mifflin, republished by Mariner Books (2002), New York, NY. ISBN 0-618-24906-0. *(Meticulously examines the consequences of using the pesticide DDT and the impact of technology on the environment)*

ELV (2000), "The Directive EC 2000/53 Directive on End-of-life vehicles (ELV)," *Journal of the European Communities*, L269, 21/10/2000, pp 34–42. *(European Union Directive requiring take-back and recycling of vehicles at end of life)*

Hardin, G. (1968), "The tragedy of the commons," *Science*, 162, pp 1243–1248. *(An elegantly argued exposition of the tendency to exploit a common good such as a shared resource (the atmosphere) or pollution sink (the oceans) until the resource becomes depleted or over-polluted)*

IPCC (2007), "Fourth Assessment Report of the IPCC," The Intergovernmental Panel on Climate Change, UNEP, www.ipcc.ch. *(The report that establishes beyond any reasonable doubt the correlation between carbon in the atmosphere and climate change)*

ISO 14040 (1998), Environmental management—Life-cycle assessment—Principles and framework.

ISO 14041 (1998), Goal and scope definition and inventory analysis.

ISO 14042 (2000), Life-cycle impact assessment.

ISO 14043 (2000), Life-cycle interpretation, International Organization for Standardization, Geneva, Switzerland. *(The set of standards defining procedures for life-cycle assessment and its interpretation)*

Meadows D.H., Meadows D.L., Randers, J., and Behrens, W.W. (1972), *The limits to growth*, Universe Books, New York, USA. *(The "Club of Rome" report that triggered the first of a sequence of debates in the 20th century on the ultimate limits imposed by resource depletion)*

National Highway Traffic Safety Administration (NHTSA) (2011), "CAFE Overview," www.nhtsa.gov/cars/rules/cafe/overview.

PAS 2050 (2008), "Specification for the assignment of the life-cycle greenhouse gas emissions of goods and services," *ICS code 13.020.40*, British Standards Institution, London, UK. ISBN 978-0-580-50978-0. *(This Publicly Available Specification (PAS) deals with carbon-equivalent emissions over product life.)*

RoHS (2002), The Directive EC 2002/95/EC on the Restriction of the Use of Certain Hazardous Substances in Electrical and Electronic Equipment, European Commission, Strasbourg, France. *(This Directive, commonly referred to as the Restriction of Hazardous Substances Directive, or RoHS, was adopted by the European Union in February 2003 and came into force on 1 July 2006.)*

WEEE (2002), The Directive EC 2002/96 on Waste electrical and electronic equipment (WEEE), *Journal of the European Communities 37*, 13/02/2003, pp 24–38.

5.9 Exercises

E5.1. What is a protocol? What do the Montreal Protocol and the Kyoto Protocol commit the signatories to do?

E5.2. What is meant by "internalized" and "externalized" environmental costs? If a company is required to "internalize its previous externalities" what does it mean?

E5.3. What is the difference between *command and control* methods and the use of *economic instruments* to protect the environment?

E5.4. How does emissions trading work?

E5.5. Carbon trading sounds like the perfect control mechanism to enable emissions reduction. But nothing in this world is perfect. Use the Internet to research the imperfections in the system and report your findings.

E5.6. What are the merits and difficulties associated with (a) taxation and (b) trading schemes as economic instruments to control pollution?

E5.7. The fleet average fuel economy of one car maker fails to meet the present CAFE target (27.5 mpg) by 3 mpg. As we know, this will incur a penalty of $165 per car. Assuming that the first owner of a typical car bought from this car maker drives it for 100,000 miles before selling it, how much more gasoline will it burn than one that meets the standard? If the carbon footprint of gasoline is 2.9 kg/liter and 1 US gallon = 3.79 liters how much extra carbon does the car emit? If gasoline costs $4 per US gallon, what is the cost penalty of the deficit of 3 mpg?

E5.8. The CAFE target is raised to 35 mpg. A car maker with US sales of 500,000 vehicles per year elects to continue to market a range with a fleet market of 25 mpg. What CAFE penalty will the car maker incur? What is this expressed as a percentage of sales if the car-maker receives 70% of the average show-room price of $40,000 per car?

E5.9. Your neighbor with a large 4 × 4 proudly tells you that his car, despite its great size, is carbon-neutral. What does he mean (or think he means)?

E5.10. In December 2007, Saab posted advertisements urging consumers to "switch to carbon-neutral motoring," claiming that "every Saab is green." In a press release, the company said they planned to plant 17 native trees for each car bought. The company claimed that its purchase of offsets for each car sold made Saab the first car brand to make its entire range carbon free. What is misleading about this statement? (The company has since withdrawn it).

E5.11. What tools are available to help companies meet the VOC regulations? Carry out a Web search to find out and report your findings.

E5.12. What tools are available to help companies meet the EuP regulations? Carry out a Web search to find out and report your findings.

E5.13. What is REACH? What help can I get to implement REACH in my company? Use the Internet to find out.

CHAPTER 6

Eco-data: values, sources, precision

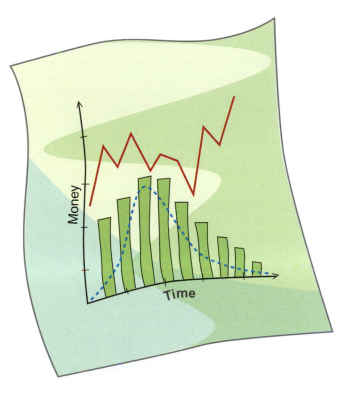

6.1 Introduction and synopsis

Decisions need data—that means numbers. You can speculate without numbers, but if you want your conclusions to rest on a solid foundation, you need real numbers. This chapter sets out the generic data—the environmental numbers that crop up all the time when designing with materials. Chapter 15 provides details for specific materials. There are an awful lot of tables and—well—numbers in both. The important thing is to know where to find them when you need them. They provide the input to the eco-audit and selection methods developed in Chapters 7, 8, 9, and 10. Eco-property charts give overviews of what the numbers look like. You will find six in this chapter and more in later chapters.

119

The precision of much eco-data is low. The values of some are known to within 10%, others with much less certainty. So it is worth first asking the general question: how much precision do you need to deal with a given problem? The short answer: just enough to distinguish the viable alternatives. As we shall see, precise judgments can be based on imprecise data. That is a good place to start.

6.2 Data precision: recalibrating expectations

The engineering properties of materials—their mechanical, thermal, and electrical attributes—are well characterized. They are measured with sophisticated equipment following internationally accepted standards and are reported in widely accessible handbooks and databases. They are not *exact*, but their precision—when it matters—is reported; many are known to three-figure accuracy, some to more. A pedigree like this gives confidence. They are data that can be trusted.

Additional properties are needed to incorporate eco-objectives into the design process. They include measures of the energy committed and carbon released into the atmosphere when a material is extracted or synthesized—its *embodied energy* and *carbon footprint*—and similar data for processing of the material to create a shaped part. There are more, introduced in the following pages. But before the introductions it helps to know what to expect.

Take embodied energy as an example. It is the energy needed to produce unit mass (usually, one kilogram) of a material from—well—whatever it is made from. It is a key input to any eco-tool. Unlike the engineering properties, many with a provenance stretching back 200 years, embodied energy is an upstart with a brief and not very creditable history. There are no sophisticated test machines to measure it. International standards, detailed in ISO 14040 and discussed in Chapter 3, lay out procedures, but these are vague and not easily applied. So just how far can values for this and other eco-properties be trusted? An analysis, documented in Section 6.3, suggests a standard deviation of at *leaste* ± 10%.

Bad news? Not necessarily. It depends on how you plan to use the data. Methods for selecting materials based on environmental criteria must be fit for a purpose. The distinctions they reveal and the decisions drawn from them must be *significant*, meaning that they stand despite the imprecision of the data on which they are based.

The data sheets of Chapter 15 deal with this by listing all properties as ranges: *Aluminum: embodied energy* 200–220 MJ/kg, for example. The ranges allow "best case" and "worst case" scenarios to be explored. When point (single-valued) data are needed, as they are in the examples of this and other chapters, take the mean of the range.

6.3 The eco-attributes of materials

Table 6.1 lists eco-data for materials in a similar format to the data sheets of Chapter 15. Here we step though the blocks, explaining what the names mean.

Table 6.1   Eco-attributes of a material

Aluminum alloys

Eco properties: material	
Global production, main component	37×10^6 metric ton/yr
Reserves	2.0×10^9 metric ton
Embodied energy, primary production	200−220 MJ/kg
CO_2 footprint, primary production	11−13 kg/kg
Water usage	495−1490 l/kg
Eco-indicator	710 millipoints/kg
Eco properties: processing	
Casting energy	11−12.2 MJ/kg
Casting CO_2 footprint	0.82−0.91 kg/kg
Deformation processing energy	3.3−6.8 MJ/kg
Deformation processing CO_2 footprint	0.19−0.23 kg/kg
End of life	
Embodied energy, recycling	22−39 MJ/kg
CO_2 footprint, recycling	1.9−2.3 kg/kg
Recycle fraction in current supply	41−45%

The data themselves are drawn from many sources. They are listed under Further reading at the end of this chapter.

Eco-properties: materials. The first block of data in Table 6.1 contains information about the resource base from which the material is drawn and the rate at which it is being exploited. The *annual world production* is simply the mass of the material extracted annually from ores or feedstock, expressed in terms of the metric tons of metal (or other engineering material) it yields. The *reserve*, as explained in Chapter 2, is the currently reported sizes of the economically recoverable ores or feedstock from which the material is extracted or created. The value is that of the metric tons of metal that it contains.

The *embodied energy* H_m of a material is the energy that must be committed to create 1 kg of usable material—1 kg of steel stock, or of PET pellets, or of cement powder, for example—measured in MJ_{oe}/kg (MJ_{oe} means "megajoules, oil equivalent"). Some materials are made using fossil fuels as the primary source of energy—the reduction of iron ore using coke in a blast furnace is an example. Others are made using electrical energy, some part of which is usually made from fossil fuels, so to allow embodied energies to be comparable, the electrical energy has to be corrected back to a fossil-fuel equivalent, for which oil is the standard—that is what

MJ_{oe} means. We won't write MJ_{oe} all the time but will adopt the convention that chemical energies (like that of oil) are given in MJ and electrical energies in kilowatt hours (kWh).[1]

The *CO_2 footprint* of a material is the mass of CO_2 released into the atmosphere per unit mass of material, units kg/kg.[2] CO_2 is a concern because of its global warming potential (GWP) caused by its ability to absorb and trap infrared radiation from the sun. Other emissions, such as carbon monoxide, CO, and methane, CH_4, also have global warming potential. It is common practice to report what is called the "carbon-equivalent," symbol $CO_{2.eq}$ (units still kg/kg), which is the mass of CO_2 with the same GWP as that of the real emissions. As a rule of thumb, 1 kg of CO_2 = 1.06 kg $CO_{2.eq}$, though the exact relationship depends on the material and the way it is made. The data for carbon footprint in Chapter 15 are for $CO_{2.eq}$. This distinction is important in conducting a rigorous LCA, but the 6% difference lies in the data-noise of case studies we examine in this book, so we shall ignore it and just speak of carbon footprint or CO_2.

It is tempting to try to estimate embodied energy via the thermodynamics of the processes involved. Extracting aluminum from its oxide, for instance, requires the provision of the free energy of oxidation of aluminum to liberate it. This much energy must be provided, it is true, but it is only the beginning. The thermodynamic efficiencies of processes are low, seldom reaching 50%. Only part of the output is usable—the scrap fraction ranges from a few % to more than 20%. The feedstocks used in the extraction or production themselves carry embodied energy. Transportation is involved. The production plant itself has to be lit, heated, and serviced. And if it is a dedicated plant, one that is built for the sole purpose of making the material or product, there is an *energy mortgage*—the energy consumed in building the plant in the first place. Embodied energies are more properly assessed by *resource flow analysis*. For a material such as ingot iron, cement powder, or PET granules, the embodied energy/kg is found by monitoring over a fixed period of time the total energy input to the production plant (including that smuggled in, so to speak, as embodied energy of feedstock) and dividing this by the quantity of usable material shipped out of the plant.

The meaning of these quantities is best illustrated by taking a tour through the life cycle of a single product, a PET bottle, for example. Figure 6.1 starts the tour. It shows, much simplified, the inputs to a PET production facility: oil derivatives such as naptha and other feedstock, direct power (which, if electric, is generated in part from fossil fuels with a conversion efficiency of about 38%), and the energy of transporting the materials to the facility.

---

[1] 1 kWh = 3.6 $MJ_{electrical}$ = $3.6/\eta$ MJ_{oe} where η is the conversion efficiency of fossil fuel to electrical energy.

[2] The atomic weight of carbon is 12, that of oxygen is 16, so one kilogram of carbon, when burnt, produces 12 + (2 × 16) = 3.6 kg of CO_2.

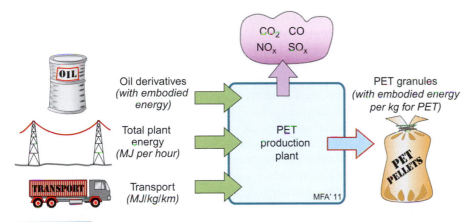

FIGURE 6.1 *The idea of embodied energy. Energy, in various forms, enters or is required by the plant. Its output is a material. The energy per kg of usable material is the embodied energy of the material.*

The material inputs to any processing operation are referred to as *feedstock*. Some are inorganic, some organic. Accounting for inorganic feedstock is straightforward because, during processing, they appear either in the final product or in the waste output. Accounting for hydrocarbon feedstock is more complex because it can be used either as a material input or as a fuel input. When used as a fuel, the hydrocarbon is burned to provide energy, and once burned, it is gone forever. When hydrocarbons are used as a feedstock instead (for plastics, for example), the energy content is not lost, but is rolled up and incorporated in the product. In describing the embodied energy of materials or products, it is important to include both the fuel and the feedstock energies in the total because they both represent a demand for a resource. By contrast, hydrocabons used as fuels create an immediate release of carbon to the atmosphere, but when they are used as feedstock, they do not.

The plant of Figure 6.1 has an hourly output of usable PET granules. The embodied energy of the PET, $(H_m)_{PET}$, with usual units of MJ/kg, is then given by

$$(H_m)_{PET} = \frac{\sum Energies\ entering\ plant\ per\ hour}{Mass\ of\ useable\ PET\ granules\ produced\ per\ hour}$$

The CO_2 footprint of a material is assessed in a similar way. The carbon emission that is the result of creating unit mass of material includes that associated with transport, the generation of the electric power used by the plant, and that of feedstocks and hydrocarbon fuels. The CO_2 footprint, with the usual units of kg of CO_2/kg, is then the the sum of all the contributions to CO_2 release, per unit mass of usable material exiting the plant.

Digression: dealing with natural materials. Trees grow by absorbing CO_2 from the atmosphere and H_2O from the earth to build hydrocarbons, notably cellulose and lignin, the building blocks of wood. In this way wood "sequesters" carbon, removing it from the atmosphere and storing it. This has led to the statement that the carbon footprint of wood is negative; wood absorbs carbon rather than releasing it.

Many eco-statements need close scrutiny and this is one. Coal is a hydrocarbon, derived, like wood, from plant life. The carbon in coal was once in the atmosphere; plant life of the carboniferous era captured it and sequestered it as coal. But we do not credit coal with a negative carbon footprint because, when we use it, we do not replace it. The carbon it sequestered millennia ago is returned to the atmosphere and not recaptured. A carbon credit for wood is only real if the cycle is closed, meaning that the wood is grown as fast as we use it, or that it is used for long-lived structural purposes, not burnt as fuel.

Some plant life, useful in engineering, can be grown "sustainably." Hemp, used in rope and fabrics, building construction and, increasingly, to reinforce polymer composites, is one. Hemp can be grown without fertilizers and it grows fast; it can be grown as fast as it is used. The carbon it sequesters is retained in the fabric, building, or composite made from it, so its use genuinely removes carbon from the atmosphere, returning it when it is burnt or decomposes at end of life. A carbon-neutral cycle is possible here.

So the question: is wood like coal or is it like hemp? At an international level, the world's forests are being cut down for construction, fuel, pulping, and mere clearing at a far greater rate than they are replanted. Trees take 80 years to grow to maturity; planting trees to offset the carbon footprint of driving a car (average life, 14 years) does not add up. Until stocks are replaced as fast as they are used, wood has to be viewed as a resource that is more like coal than hemp, with a corresponding positive carbon footprint.

Today some woods *are* derived from sustainable, managed, forests. When woods are known to come from such sources, it is legitimate to credit them with sequestering carbon,[3] giving them a negative carbon footprint—a point to which we return in Chapter 11.

Wood-based products—plywood, chipboard, fiberboard, and the like—have the merit that they use more of the trunk than solid wood beams or panelling does, and they can be made from lower-grade timber. But all such products involve a significant component of a polymer adhesive, driving up the energy content and carbon footprint yet higher.

Data precision. Figure 6.2 plots some of the reported values of the embodied energy of aluminum over time, starting with the earliest measurements around 1970. The mean value of the data plotted here is 204 MJ/kg. The standard deviation is 58 MJ/kg, and that is 25% of the mean. A closer examination of the data

---

[3]Allow for sequestered energy and carbon in woods by subtracting 25 MJ/kg and 2.8 kg CO_2 per kg from the values given in the data sheet.

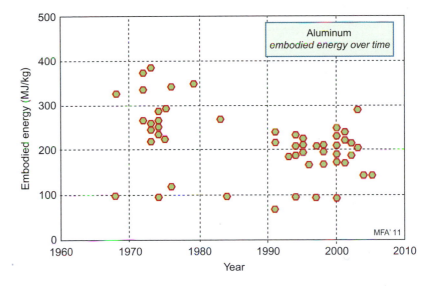

FIGURE 6.2 *Data for the embodied energy of aluminum. The mean is 204 MJ/kg, with a standard deviation of 58 MJ/kg. Using the best-characterized data only gives a mean of 210 MJ/kg with a standard deviation of 20 MJ/kg.*

sources shows some to be more rigorous than others, so a more selective plot shifts the mean to 210 MJ/kg and reduces the standard deviation to about 10%. Embodied energies for other materials, when plotted in this way, show a similar spread.

Where does this variability come from? The differences in the process routes by which materials are made in different production facilities; the differences in energy-mix in electrical power in different countries; the difficulty in setting system boundaries; and the procedural problems in assessing energy, CO_2, and the other eco-attributes all contribute to the imprecision. So—the bottom line—when dealing with embodied energies such as the 210 MJ/kg quoted earlier, read them thus: the first figure (the 2) can be trusted; the second (the 1) is debatable; the third is meaningless.

Using data for embodied energy

Example: A cooking pot has a cast virgin-aluminum body and a molded phenolic handle. The body weighs 0.8 kg and the handle weighs 0.1 kg. What is the total embodied energy of the materials of the pot? If the aluminum body were replaced by one made of virgin cast iron weighing three times more, would the total embodied energy of the pot be less?

Answer: The data sheets of Chapter 15 have data for embodied energy:

Material	Embodied energy, virgin material (MJ/kg)
Aluminum alloys	210
Cast iron	29
Phenolic	79

The total embodied energy of the aluminum pot is

$$0.8 \times 210 + 0.1 \times 79 = 176 \text{ MJ}$$

The total embodied energy of the cast iron pot is

$$2.4 \times 29 + 0.1 \times 79 = 78 \text{ MJ}$$

The cast iron pot has an embodied energy that is less by 98 MJ than that of the aluminum one, despite its greater mass. In reality, the aluminum and the cast iron used in an application like this would contain considerable recycled content. This is illustrated in the next example.

Using materials with recycled content

Example: The cooking pot of the last example is, in fact, made from aluminum with a typical (43%) recycled content. The cast iron pot, too, contains the typical (67%) recycled content. How much difference does this make to the embodied energies of the pots?

Answer: The data sheet for aluminum alloys of Chapter 15 has data for typical recycled content in current supply and the embodied energy of recycled aluminum and cast iron:

Material	Embodied energy, virgin material (MJ/kg)	Typical recycle content (%)	Embodied energy, recycling (MJ/kg)
Aluminum alloys	210	43	25
Cast iron	29	67	8
Phenolic	79	0	—

The embodied energy of the materials of the aluminum pot now total

$$0.8 \times (0.57 \times 210 + 0.43 \times 25) + 0.1 \times 79 = 112 \text{ MJ}$$

Using aluminum with recycled content reduces the energy by 36%. The embodied energy of the materials of the cast iron pot now total

$$2.4 \times (0.33 \times 29 + 0.67 \times 8) + 0.1 \times 79 = 44 \text{ MJ}$$

a reduction of 44%.

Special case (1): precious metals. Precious metals (platinum, gold, silver) are used in small quantities but they carry a heavy burden of embodied energy, carbon, and, of course, cost. Many have unique properties: exceptional electrical conductivity, exceptional resistance to corrosion, and exceptional functional properties as electrodes, sensors, and catalysts. A rule of convenience in energy accounting, used later, is to ignore (or approximate) the contribution of parts of a product that, collectively, contribute less than 5% to the weight, but we can't do that with precious metals—even a small quantity adds significantly to the energy and carbon totals. Table 6.2 lists the data needed to include them.

Precious metals in engineering structures

Example: A catalytic converter has a steel container weighing 1 kg and an extruded alumina honeycomb core weighing 0.5 kg, which is coated with 1 gram of a platinum catalyst. Which of the three components contributes most to the embodied energy of the materials of the converter?

Answer: From Table 6.2 and the data sheets of Chapter 15.

Material	Embodied energy (MJ/kg)
Low-carbon steel (Ch.15)	26.5
Alumina (Ch. 15)	52.5
Platinum (Table 6.5)	270,000

Table 6.2	Precious metals (used as catalysts, electrodes, contacts)		
Metal	**Embodied energy (MJ/kg)**	**Carbon footprint (kg/kg)**	**Unique applications**
Silver	$1.4 \times 10^3 - 1.55 \times 10^3$	95−105	Photosensitive compounds, dentistry
Gold	$240 \times 10^3 - 265 \times 10^3$	14,000−15,900	Corrosion-free contacts, dentistry
Palladium	$5.1 \times 10^3 - 5.9 \times 10^3$	404−447	Catalysis, hydrogen purification
Rhodium	$13.5 \times 10^3 - 14.9 \times 10^3$	1,000−1,200	Catalysis
Platinum	$257 \times 10^3 - 284 \times 10^3$	14,000−15,500	Catalysis, electrodes, contacts
Iridium	$2 \times 10^3 - 2.2 \times 10^3$	157−173	Catalysis, electrodes, high-temperature igniters

The contribution of 1 kg of steel to the embodied energy of the converter is 26.5 MJ, that of 0.5 kg of alumina ceramic is 26 MJ, and that of 1 gram of platinum is 270 MJ. The platinum contributes much more than the sum of the others, despite the tiny quantity.

Special case (2): electronics. Electronic components—integrated circuits, surface-mount devices, displays, batteries, and the like—are now an integral part of almost all appliances. They don't weigh much, but they are energy-intensive to make. Electronics are made of materials, but by process routes that are so complex that calculating the embodied energy from that of the materials themselves is not possible. Instead we assemble data derived from life-cycle assessments of electronic systems, subsystems, and components. They are not easy to find. Table 6.3 lists sufficient data to include them at an approximate level. What does it say? Principally that electronics have embodied energies and carbon footprints that are larger, by a factor of 10 to 50, than those we associate with conventional engineering materials.

Energy and carbon footprint of electronics

Example: A fire alarm has two AA Ni-Cd batteries, small-device electronics weighing 30 grams, and a phenolic casing weighing 400 grams. Which of these makes the largest contribution to the embodied energy and carbon footprint?

Answer: From Table 6.3 and the data sheet for phenolic in Chapter 15 we have

Component and material	Embodied energy	Carbon footprint
Ni-Cd AA battery	3 MJ/unit	0.2 kg/unit
System, small electronic device	3,000 MJ/kg	300 kg/kg
Phenolic	79 MJ/kg	2.65 kg/kg

Thus the embodied energy of the batteries is $2 \times 3 = 6$ MJ, that of the electronics is $0.03 \times 3,000 = 90$ MJ, and that of the casing is $0.4 \times 79 = 32$ MJ. The electronics accounts for more energy than the other components combined. The same is true for the carbon footprint.

Special case (3): Materials of architecture and construction. Domestic and commercial buildings consume materials on an enormous scale. The structure uses the most; the envelope enclosing and insulating the structure comes second, the services third, and the interior finishing and decoration requires the least, but all are large. The materials of construction are a little more specialized than those of general engineering. For that reason, they are assembled with their embodied energies and carbon footprints in Table 6.4 under the subheadings Primary structure, Enclosure, Services, and Interior. The table makes it easy to find data when exploring case studies for building projects.

Table 6.3	Approximate embodied energies and CO_2 footprints for electronic components	
Product	**Energy, OE (MJ)***	**Carbon, CO_2 equivalent (kg)**
Systems, unit as specified		
Desktop computer, without screen	4,783	274
Laptop computer	3,013	250
Small electronic devices, depending on scale, (per kg)	2,000–4,000	200–400
LCD displays, per m^2	2,950–3,750	295–375
Printer, laser jet	907	68
Subsystems, per unit		
Printed wiring board, desktop PC motherboard, per kg	2,830	162
Printed wiring board, laptop PC motherboard, per kg	4,670	267
Hard disk drive	65–216	4–12
CD-ROM/DVD-ROM drive	100–300	5–15
Power supply	574	30
Fan	254	12
Keyboard, per unit	468	27
Mouse device, optical, with cable, per unit	93	5
Toner module, laser jet	215–220	10–11
Components, unit as specified		
Integrated circuit, IC, logic or memory type, per kg	9,700–16,000	500–1013
Transformers, per kg	70–105	4–6
Transistors, per kg	2,700–3,000	140–147
Diodes and LEDs, per kg	4,600–4,700	230–235
Capacitors, per kg	950–1,150	48–60
Inductors, per kg	850–1,500	45–84
Resistors, per kg	700–2,000	35–120
Cable, per meter	5–10	0.2–0.5
Plugs, inlet and outlet, per unit	8–13	0.4–1

(*Continued*)

Table 6.3 (Continued)		
Product	**Energy, OE (MJ)***	**Carbon, CO_2 equivalent (kg)**
Power sources, unit as specified		
Li-Ion batteries for laptops, per kg	900–935	74–100
Ni-MH batteries for laptops, per kg	933	57
Lead-acid batteries for cars (39 Wh per kg), per kg	19	1
Li-Ion battery, scooters, cars (99 Wh per kg), per kg	324	8
Alkaline AA cell battery, per unit	1	0.1
Li-Ion AA cell battery, per unit	3	0.2
Ni-Cd AA battery, per unit	3	0.2
Ni-Cd C-cell battery, per unit	5	0.4
Materials, unit as specified		
Single crystalline silicon, electronics, per kg	4,966	251
Single crystalline silicon, photovoltaics, per kg	2,239	103
Mono-crystal-Si wafer, electronics, per m^2	6,017	305
Mono-crystal-Si wafer, photovoltaics, per m^2	2,804	129
Poly-silicon wafer, per m^2	2,000	90
Solder, Sn/Ag4/Cu 0.5	234	20

*Data from IDEMAT (2009), Knapp and Jester (2000), MEEUP report (2005), and Hammond and Jones (2009)

Embodied energies of structure and enclosure

Example: The structure of a steel-framed building requires, per square meter of floor area, 625 kg of standard concrete (foundation) and 86 kg of steel (100% recycled). The enclosure, per m^2 of floor, requires 0.2 m^3 of fiberglass insulation and 6.6 m^2 of ¾-inch (19 mm) plywood. What, approximately, is the embodied energy of structure plus enclosure?

Answer: Drawing data from Table 6.4, we find the embodied energy of the structure is

$$625 \times 1.14 + 86 \times 7.3 = 1,340 \text{ MJ} = 1.34 \text{ GJ}$$

That for ¾-inch plywood is

$$0.019 \times 5,720 = 109 \text{ MJ/m}^2$$

So the embodied energy of the enclosure is

$$0.2 \times 970 + 6.6 \times 109 = 913 \text{ MJ} = 0.91 \text{ GJ}$$

The total embodied energy is thus 3.02 GJ, of which the structure forms 60% and the enclosure 40%.

Table 6.4 Approximate embodied energies and carbon footprints of building materials

	Density (kg/m^3)	Embodied energy		Carbon footprint	
		MJ/kg	MJ/m^3	kg/kg	kg/m^3
Primary structure					
Aggregate	1,500	0.10	150	0.006	9
Brick, common	2,100	2.8	5,880	0.22	462
Concrete 30 MPa	2,450	1.3	3,180	0.095	233
Concrete block	2,500	0.94	2,350	0.061	151
Concrete, high-volume fly-ash	2,010	1.14	2,290	0.068	137
Concrete, precast	1,390	2.00	2,780	0.12	167
Concrete, reinforced (8 wt% steel)	2,910	2.49	7,250	0.21	611
Concrete, standard	2,390	1.14	2,700	0.1	239
Soil cement (rammed earth)	1,950	0.42	819	0.03	55
Steel, virgin	7,850	27	212,000	1.8	14,100
Steel, 100% recycled	7,850	7.3	57,300	0.57	4,470
Steel, typical 42% recycled content	7,850	18.7	147,000	1.23	9,660
Straw bale	125	0.22	27	−099	−12
Timber, structural	550	7.3	1,380	0.42	230
Enclosure					
Aluminum, virgin	2,700	210	567,000	12	324,000
Aluminum, 43% recycled content	2,700	131	354,000	7.7	220,800
Duck (cellulose-based fabric)	980	72	7,060	4.5	4.230
GFRP (sheet molding compound)	1,900	115	220,000	8.1	1,540

(Continued)

Table 6.4	(Continued)					
	Density (kg/m³)	Embodied energy			Carbon footprint	
		MJ/kg	MJ/m³		kg/kg	kg/m³
Glass	2,400	15.9	37,550		0.76	1,820
Gypsum wall board	1,000	6.0	6,000		0.33	3,300
Insulation, cellulose	40	3.3	112		0.1	4.0
Insulation, fiberglass	32	30.3	970		2.1	67
Insulation, mineral wool	9.5	14.6	139		1.0	9.5
Insulation, polystyrene	32	117	3,700		4.2	40
Insulation, vermiculite	101	25	2,530		1.37	138
Particle board	550	8.0	4,400		0.6	330
Plywood	600	10.4	5,720		0.8	480
Sealants: silicone, neoprene, epoxy	1,100	120	132,000		7.5	8,250
Shingles, asphalt	540	9.0	4,930		0.6	329
Services						
Brass, fittings	8,100	54	437,000		3.5	28,400
Copper: piping, cladding	8,940	60	536,000		3.7	33,100
Lead: cladding, piping	11,300	27	305,000		1.9	21,500
PVC: piping	1,370	68	93,200		3.1	2,600
Steel, galvanized	7,845	35	275,000		2.1	16,500
Zinc, cladding	7,140	46	328,000		3.3	23,600
Interior						
Carpet, synthetic	570	148	84,900		8.8	
Hardwood	980	8.0			0.47	
Leather	922	107			4.29	
Linoleum	1,300	116	150,930		5.8	
Paint	1,260	93	117,500		4.6	
Tiles, terracotta	2,290	3.0			0.22	
Particle board, chipboard	550	8.0	4,400		0.6	
Softwood	550	7.3	1,380		0.42	230

Sources: Canadian Architect (2011), EcoInvent (2010), CES (2011)

Water usage. The data sheets of Chapter 15 list the approximate quantity of commercial water, in liters, used in the production of 1 kg of material. The data are discussed further in Section 6.5.

Eco-indicators. As explained in Chapter 3, attempts have been made to roll up the energy, water, and gaseous, liquid, and solid emissions associated with material production into a single number or eco-indicator. The records of Chapter 15 include the value when it is available.

Material processing: energy and CO_2 footprint. Product manufacturing requires that materials be shaped. The processing energy H_p associated with a material is the energy, in MJ/kg, used to do this. Metals, typically, are cast, rolled, or forged. Polymers are molded or extruded. Ceramics are shaped by powder methods, composites are created by molding or lay-up methods (Table 6.5(a)). The energy consumed by a casting furnace or an injection molding machine can be measured directly, but the production plant as a whole uses more energy than this through the provision of transport, heating, lighting, management, and maintenance. A more realistic

Table 6.5(a)	Primary shaping processes		
Material	**Shaping process**	**Typical range of energies (MJ/kg)**	**Carbon, $CO_{2,eq}$ (kg/kg)**
Metals	Casting	8–12	0.4–0.6
	Rough rolling, forging	3–5	0.15–0.25
	Extrusion, foil rolling	10–20	0.5–1.0
	Wire drawing	20–40	1.0–2.0
	Metal powder forming	20–30	1–1.5
	Vapor phase methods	40–60	2–3
Polymers	Extrusion	3.1–5.4	0.16–0.27
	Molding	11–27	0.55–1.4
Ceramics	Ceramic powder form	20–30	1–1.5
Glasses	Glass molding	2–4	0.1–0.2
Hybrids	Compression molding	11–16	1.6–0.5
	Spray-/Lay up	14–18	0.7–0.9
	Filament winding	2.7–4.0	0.14–0.2
	Autoclave molding	100–300	5–15

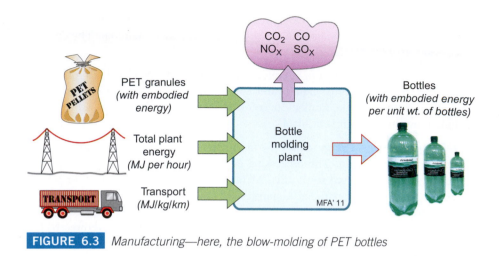

FIGURE 6.3 *Manufacturing—here, the blow-molding of PET bottles*

measure of processing energy is the total energy entering the manufacturing plant (but excluding that embodied in the materials themselves since that is attributed to the material production phase) divided by the weight of usable shaped parts that it delivers, giving a value in MJ/kg. The carbon footprint is calculated in a similar way.

Now back to the tour with the PET bottle. The granules now become the input (after transportation) to a facility that blow-molds PET bottles, as shown in Figure 6.3. The molding operation consumes energy and feedstock, and emissions and waste are excreted. The products—PET blow-molded bottles—emerge. The output of the analysis is the energy committed per unit weight of bottles produced. The CO_2 footprint is evaluated in a similar way or by calculating it from the energy use and fuel type (see Table 6.7 later in this section).

Secondary processes take a shaped part and add features, join, and finish it. There are many such processes. The energy demands of a basic subset are assembled, as typical ranges, in Table 6.5(b) (machining and grinding), Table 6.5(c) (joining by welding, fasteners, and adhesives), and Table 6.5(d) (painting and plating). They have different units: MJ and kg carbon per kg of material removed for machining; MJ and kg carbon per m^2 for painting, plating, and adhesives; and MJ and kg carbon per m for welding and riveting.

What does this table tell us? That the energy and carbon footprint to process a material is a lot less than that to make it in the first place (an example of the fact that chemical energies are generally bigger than mechanical ones—explosives more destructive than battering rams). But that is not the end of the story. The material is made just once. Its subsequent processing often involves a chain of shaping, joining, and finishing steps. The sum of these steps, in terms of energy, carbon, and cost, can be comparable with the embodied energy of the material itself.

Table 6.5(b)	Secondary processes: machining		
Process type	**Variants**	**Range of energies (MJ/kg removed)**	**Carbon, $CO_{2,eq}$ (kg/kg removed)**
Machining	Heavy	0.8–2.5	0.06–0.17
	Finishing (light)	6–10	0.4–0.7
	Grinding	25–35	1.8–2.5
	Water jet, EDM, Laser	500–5000	35–350

Table 6.5(c)	Secondary processes: joining		
Process type	**Variants**	**Typical range of energies**	**Typical range of carbon, $CO_{2,eq}$**
Welding	Gas welding	1–2.8 MJ/m	0.055–0.15 kg/m
	Electric welding	1.7–3.5 MJ/m	0.12–0.25 kg/m
Fasteners	Fasteners, small	0.02–0.04 MJ/fastener	0.0015–0.003/fastener
	Fasteners, large	0.05–0.1 MJ/fastener	0.0037–0.0074/fastener
Adhesives	Adhesives, cold	7–14 MJ/m^2	1.3–2.8 kg/m^2
	Adhesives, heat-curing	18–40 MJ.m^2	3.2–7.0 kg/m^2

Table 6.5(d)	Secondary processes: finishing		
Process type	**Variants**	**Typical range of energies (MJ/m^2)**	**Typical range of $CO_{2,eq}$ (kg/m^2)**
Painting	Painting	50–60	0.63–0.095
	Baked coatings	60–70	0.9–1.3
	Powder coating	67–86	3.7–4.6
Plating	Electroplating	80–100	4.4–5.3

Notes: 1. Data sources are listed under the table headings in Section 6.7, "Further reading."

2. The primary shaping and machining data listed here are plotted in Figure 6.13.

Using process energies

Example: A large connecting rod is cast from virgin aluminum. The bores are finished by light machining, removing 0.05 kg of metal. The finished connecting rod weighs 3.2 kg. It is suggested that machining the connecting rod from a solid, cold-rolled aluminum blank would give a higher-quality product. The starting blank weighs 5 kg.

Heavy (coarse) machining is used for all but the finishing of the bores. Assuming virgin aluminum is used and that the machinings are not recycled, what is the energy penalty of taking the machine-from-solid option?

Answer: Drawing data from Table 6.5(b) and from the data sheets of Chapter 15 we find:

Material and process step: aluminum alloy connecting rod	Energy	Units
Embodied energy (Ch.15)	210	MJ/kg
Casting energy (Ch.15)	11.6	MJ/kg
Deformation processing energy (Ch.15)	5.1	MJ/kg
Heavy machining energy (Table 6.2)	1.7	MJ/kg removed
Light machining energy (Table 6.2)	8.0	MJ/kg removed

Summing the embodied energy of the aluminum, here are the primary process energy and the machining energies we find for the cast connecting rod

$$3.2 \times (210 + 11.6) + 0.05 \times 8.0 = 710 \text{ MJ}$$

and for the con-rod machined from solid

$$5 \times 210 + (5 - 3.2) \times 1.7 + 0.05 \times 8.0 = 1{,}053 \text{ MJ}$$

The machine-from-solid route requires 48% more energy than the casting route. But this ignores the energy recovery possible by promptly recycling the machinings. We return to this when we discuss recycling.

The relative magnitudes of energies of material and processing

Example: Flat, rolled mild steel panels, $0.5 \text{ m} \times 0.5 \text{ m} \times 1 \text{ mm}$ are required for cladding. The density of steel is $7{,}900 \text{ kg/m}^3$. The panels have a baked-on coating on one surface only. Compare the energies associated with material production, primary shaping, rolling, and the coating for the panel using data from Table 6.5(d) and the record for mild steel in Chapter 15.

Answer: One panel has a coated surface area of 0.25 m^2 and weighs 2 kg. Drawing data from the data sheet for low-carbon steel of Chapter 15 and from Table 6.5(d) we find

Material and process step: mild steel panel	Energy	Units
Embodied energy (Ch.15)	26.5	MJ/kg
Deformation processing energy (Ch.15)	4.5	MJ/kg
Energy for a baked coating (Table 6.5(d))	65	MJ/m^2

The embodied energy of the material of the panel is

$$2 \times 26.5 = 53 \text{ MJ}$$

The primary deformation processing of the panel is

$$2 \times 4.5 = 9.0 \text{ MJ}$$

The energy required to bake-coat the panel is

$$0.25 \times 65 = 16 \text{ MJ}$$

The largest contribution is that of the embodied energy of the steel. The total processing energy is less than half that of the embodied energy of the material.

Recycling and end of life. On with the tour of life as a PET bottle, which is now transported to the point of sale, passes to the consumer, is used, and ultimately reaches the end of its useful life (Figure 6.4). Underlying data for transport and use are explored in Section 6.4, in a moment. Here we jump to the last block of attributes listed in Table 6.1: those relating to *recycling* (Figure 6.5). The existence of

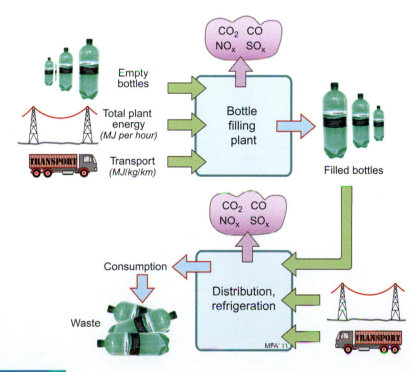

FIGURE 6.4   *The use-phase of PET water bottles: filling, distribution, and refrigeration. Energy is consumed in transportation and refrigeration.*

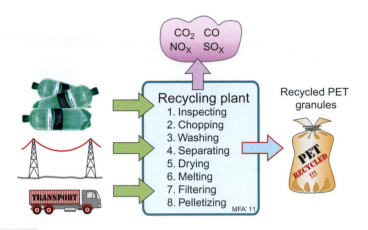

FIGURE 6.5 *Recycling. Many steps are involved, all of which consume some energy, but the embodied energy of the material is conserved.*

embodied energy has another consequence: the energy needed to recycle a material is sometimes much less than that required for its first production because the embodied energy is retained. The recycled fraction in current supply (Table 6.5) is a useful indicator of the viability of recycling. Typical values for metals lie in the range of 30−60%. Glass (22%) and paper (71%), too, are recycled extensively, but polymers are not: only PET is recycled to any large degree. This is because metals are easy to identify and separate, and because the energy needed to recycle them, typically, is about one fifth of that required to make them in the first place. Polymers are more difficult to identify automatically, and the energy saving offered by recycling is smaller (Table 6.6).

Many products are shredded before sorting and separation at end of life. The energy for shredding is approximately 0.1 MJ/kg.

Recycling of prompt scrap

Example: An earlier example compared a cast aluminum connecting rod with one machined from a solid blank. The cast rod required 710 MJ for materials and manufacturing; the one machined from a solid blank took 1,053 MJ, assuming that the 1.8 kg of metal machined-off was not recycled. In practice 100% of it would be recycled. The difference between the embodied energy of virgin and recycled aluminum is a measure of the energy recovered by doing so. How much does this change the energy total for the connecting rod machined from a solid blank?

Answer: Drawing data from the data sheet for aluminum alloys of Chapter 15 we find

Table 6.6	Typical recycling data			
Material	**Recycled fraction in current supply[1] (%)**	**Embodied energy, virgin material (MJ/kg)**	**Embodied energy, recycled material (MJ/kg)**	**Ratio of recycled to virgin energies (%)**
Aluminum	36	210	26	12
Steel	42	26.5	7.3	27
Copper	42	58	13.5	23
Lead	72	27	7.4	27
PET	21	85	39	46
PP	5	74	50	67
Glass	24	10.5	8.2	78
Paper	72	45	20	44

Notes. *Data for recycled fraction are from USGS Circular 1221 (2002); USGS (2007). Other data sources are listed under the table headings in Section 6.7, "Further reading." See the data sheets of Chapter 15 for more data.*

Material and process step: aluminum alloy connecting rod	Energy	Units
Embodied energy (Ch.15)	210	MJ/kg
Embodied energy, recycling (Ch. 15)	25	MJ/kg

The estimated energy recovered by recycling is

$$210 - 25 = 185 \text{ MJ/kg}$$

Thus the energy "credit" from recycling the machined metal is

$$1.8 \times 184 = 331 \text{ MJ}$$

This reduces the total energy demand of the machined-from-solid connecting rod to 722 MJ, barely more than that of the cast connecting rod.

6.4 Energy and CO_2 footprints of energy, transport, and use

Energy is used to transport the materials and products from where they are made to where they are used. The products themselves use energy during their life—some use a great deal. This energy is provided predominantly by fossil fuels (oil, gas, coal)

Table 6.7	The energy intensity of fuels and their carbon footprints					
Fuel type	**kg OE***	**MJ/liter**	**MJ/kg**	**CO_2 (kg/liter)**	**CO_2 (kg/kg)**	**CO_2 (kg/MJ)**
Coal, lignite	0.45	—	18–22	—	1.6	0.080
Coal, anthracite	0.72	—	30–34	—	2.9	0.088
Crude oil	1.0	38	44	3.1	3.0	0.070
Diesel	1.0	38	44	3.1	3.2	0.071
Gasoline	1.05	35	45	2.9	2.89	0.065
Kerosene	1.0	35	46	3.0	3.0	0.068
Ethanol	0.71	23	31	2.8	2.6	0.083
LNG	1.2	25	55	3.03	3.03	0.055
Hydrogen	2.7	8.5	120	0	0	0

*Kilograms oil equivalent (the kg of oil with the same energy content)

and by electric power, much of which is also generated from fossil fuels. These energy sources differ in their energy intensity and carbon release.

Energy intensities of fossil fuels and their carbon footprints are listed in Table 6.7. Reading across, there is the fuel type, the oil equivalent (OE—the kg of crude oil with the same energy content as 1 kg of the fuel), the energy content per unit volume and per unit weight, and the CO_2 release per unit liter, per kg, and per MJ. The units (kg, MJ, etc.) used here are those standard in the SI system. Conversion factors to other systems (lbs, BTU, etc.) can be found at the end of the book.

The oil and carbon equivalence of electric power. Electricity is the most convenient form of energy. Today most is generated by burning fossil fuels, but the pressure on the fossil fuel supply and the problems caused by the emissions they release are motivating governments to switch to nuclear and renewable sources, most of which generate electric power. The *energy mix* in the electricity supply of a country is the proportional contribution of each source to the total. The first four columns of Table 6.8 give examples of this mix for countries that span the extremes. Australia relies on fossil fuels for almost all of its electricity. China and India, with the world's largest populations, rely heavily on them. In France the source is predominantly nuclear. Norway relies almost entirely on hydroelectric power.

The relevant numbers from an environmental point of view are those in the last three columns: the efficiency of electricity generation from fossil fuels and the oil equivalence of electrical energy and the associated release of CO_2. They differ

Table 6.8	Electricity generation, energy mix, MJ oil per kWh, and CO_2 per kWh*					
Country	Fossil fuel %	Nuclear %	Renewables %	Efficiency[a] %	MJ_{oe}[b] per kWh[d]	CO_2,[c] kg per kWh[d]
Australia	92	0	8	33	10.0	0.71
China	83	2	15	32	9.3	0.66
France	10	78	12	40	0.9	0.06
India	81	2.5	16.5	27	10.8	0.77
Japan	61	27	12	41	5.4	0.38
Norway	1	0	99	—	0	0
UK	75	19	6	40	6.6	0.47
USA	71	19	10	36	7.1	0.50
OEDC (Europe)	62	22	16	39	5.7	0.41
World average	**67**	**14**	**19**	**36**	**6.7**	**0.48**

*Data from IEA (2008)

(a) Conversion efficiency of fossil fuel to electricity

(b) MJ_{oe} of fossil fuel (oil equivalent) used in energy mix per kWh of delivered electricity from all sources

(c) CO_2 release per kWh of delivered electricity from all sources

(d) 1 kWhr is 3.6 $MJ_{electric}$

greatly, mainly because of the differing energy mixes, to a lesser extent because of the differing conversion efficiencies. We need these numbers in later chapters. To keep things simple, we will use values for a "typical" developed country with an energy mix of 75% fossil fuel and a conversion efficiency of 38%, giving an oil equivalence of 7 MJ_{oe} per kWh and a carbon footprint of 0.5 kg CO_2 per kWh.

Transportation. Manufacturing is now globalized. Products are made where it is cheapest to make them and then they are transported, frequently over large distances, to the point of sale. Transportation is an energy-conversion process: primary energy (oil, gas, coal) is converted into mechanical power, and this is used to provide motion, sometimes with an intermediate conversion to electrical power. As in any energy-conversion process, there are losses, here most conveniently expressed as the energy consumed per metric ton per kilometer (MJ/metric ton · km), carrying with it an associated CO_2 footprint (kg/metric · ton.km).

Reported data for transport energy and carbon footprint vary considerably. This is partly because the efficiency of transport systems varies from country to country, and partly because of a lack of agreement on what should be included. Do you base the assessment simply on the fuel consumption per metric ton per km, or do you include the embodied energy of the roads, rail, or other infrastructure on which the system depends? The second estimate gives values that are roughly twice as large as the first. Including infrastructure creates a system boundary problem—do you also include the energy to make the equipment that built the roads? Here we adopt the simpler, lower, estimate based on fuel (and other direct consumables) alone.

Table 6.9 lists data for transportation energy and carbon emissions. Transportation energy and carbon footprint are calculated by multiplying the weight of the product by the distance traveled and the fuel-vehicle coefficients listed

Table 6.9 The approximate energy and carbon footprint of transportation*		
Transportation type and fuel	Energy (MJ/ metric ton $\cdot$ km[+])	Carbon footprint (kg CO_2/metric ton $\cdot$ km[+])
Ocean shipping—Diesel	0.16	0.015
Coastal shipping—Diesel	0.27	0.019
Barge—Diesel	0.36	0.028
Rail—Diesel	0.25	0.019
Articulated HGV (up to 55 metric tons)—Diesel	0.71	0.05
40 metric ton truck—Diesel	0.82	0.06
32 metric ton truck—Diesel	0.94	0.067
14 metric ton truck—Diesel	1.5	0.11
Light goods vehicle—Diesel	2.5	0.18
Family car—Diesel	1.4–2.0	0.1–0.14
Family car—Gasoline	2.2–3.0	0.14–0.19
Family car—LPG	3.9	0.18
Family car—Hybrid gasoline-electric	1.55	0.10
Super sports car and SUV—Gasoline	4.8	0.31
Long haul aircraft—Kerosene	6.5	0.45
Short haul aircraft—Kerosene	11–15	0.76
Helicopter (Eurocopter AS 350)—Kerosene	55	3.30

*Data sources are listed under Further reading.

[+]1 ton $\cdot$ mile = 1.46 metric ton $\cdot$ km

in the table. Passenger transportation is usually assessed per seat per km. A 5-seat family car weighing 1,400 kg requires about 0.6 MJ/seat·km. Intercity trains consume in the range of 0.09–0.23 MJ/seat·km, with high-speed trains (300 km/hr or more) at the upper end of this range. Aircraft consume between 1.8 and 4.5 MJ/seat·km.

Using transport data

Example: Cars manufactured in China are shipped 19,000 km to Europe where they are transported 500 km by a 32-metric ton truck to the point of sale. If the car weighs 1,400 kg, how much energy is used in this transport cycle?

Answer: From Table 6.9 the energy required for ocean shipping is 0.16 MJ/metric ton · km and that for a 32-metric ton truck is 0.46 MJ/metric ton · km. Thus the energy of the transport cycle is

$$1.4(19,000 \times 0.16 + 500 \times 0.46) = 4,578 \text{ MJ}$$

Use-energy and carbon footprint. Many products consume energy, or energy is consumed on their behalf, during the use phase of life. As we shall see in Chapter 7, this use-phase energy is often larger than that of any other phase. Most of it derives from fossil fuels; some is consumed in that state. Much is first converted to other forms of energy before it is used. The most obvious is electricity.

When fossil fuels are used directly (as in the use of gasoline to power cars) the primary energy and CO_2 can be read directly from Table 6.7. When instead it is used as electricity, the relevant fossil fuel energy and CO_2 depend on the energy mix and generation efficiency, and these differ from country to country. It is then necessary to convert the electrical energy, usually given in kWh, to the MJ and CO_2 oil equivalent by multiplying by the conversion factors in the last two columns of Table 6.8.

Use-energy and carbon

Example: A commercial clothes dryer is rated at 10 kW. It has an anticipated duty cycle averaging 30 hours per week over its design life of 5 years. It is installed in a state with an electric power energy mix 80% of which is derived from oil-fired power stations with a conversion efficiency of 36%. What is the expected life use-energy and carbon footprint of the dryer?

Answer: The energy used over life is

$$(10 \times 30 \times 52 \times 5) = 78,000 \text{ kWh of electric power}$$

A fraction, 0.8, of this power, 62,400 kWh is derived from oil. Multiplying this by 3.6 to convert to $MJ_{electric}$ and then dividing by the conversion efficiency 0.36 gives the oil-equivalent energy:

$$62,400 \times 3.6/0.36 = 624,000 \text{ MJ}_{oe}$$

Multiplying this by the carbon per MJ of crude oil, 0.07/MJ, from Table 6.7 gives the life carbon footprint of the dryer:

$$624,000 \times 0.07 = 43,680 \text{ kg of } CO_2$$

In summary, the dryer consumes 104 GWh of energy and is responsible for *nearly* 44 metric tons of carbon.

Special topic: energy, carbon, and cars. The fuel consumption and CO_2 emission of cars increase with their weight. Figures 6.6 and 6.7 show the evidence. The first is a plot of energy against mass on log scales, allowing a power-law fit for the energy consumption, H_{km}, in MJ/km as a function of the vehicle mass, m, in kg, listed in the second column of Table 6.9. The second, Figure 6.7, shows the carbon rating in g/km as a function of energy per km (MJ/km) on linear scales. The two are proportional, with the constants of proportionality marked on the figure. The CO_2 rating (g/km) as a function of mass (kg) in the third column of the table is found by multiplying this by the energy/km in column 2.

From these we calculate the energy penalty associated with one kilogram of increased weight, evaluated here for a car of weight 1,000 kg, by differentiating the expressions for H_m in Table 6.10. The results are listed in the last column of the table. We now have the inputs we need for modeling and selection.

Saving energy by reducing weight

Example: A small gasoline-powered car weighs 1,000 kg. The rear seating weighs 25 kg. How much energy and carbon would be saved over a duty-life of 150,000 km by taking out the rear seating?

Answer: The use-energy per unit mass of a gasoline-powered 1,000 kg car, from Table 6.10, is 0.0021 MJ/km·kg. Thus reducing the mass by 25 kg will, over 150,000 km, reduce the energy the car requires by

$$0.0021 \times 25 \times 150,000 = 7,875 \text{ MJ}$$

Table 6.7 gives the carbon per unit energy of gasoline as 0.065 kg/MJ. Thus the reduced mass will reduce carbon emission over life by

$$0.065 \times 7,875 = 512 \text{ kg}$$

In summary, removing the seating can save 7.8 GJ of energy and just over half a metric ton of carbon.

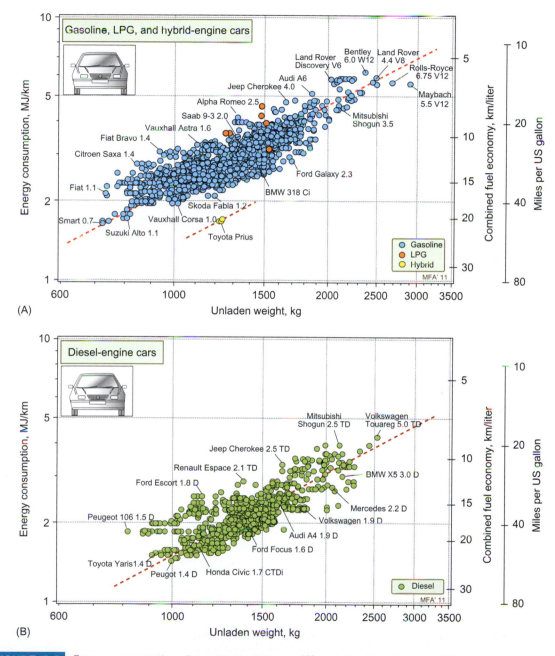

FIGURE 6.6 *Energy consumption of gasoline engine cars (A) and diesel-engine cars (B)*

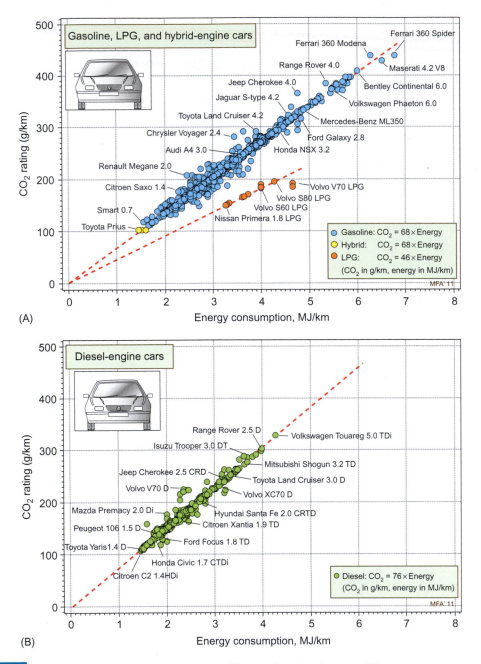

FIGURE 6.7 *CO₂ emission of gasoline engine cars (A) and diesel-engine cars (B).*

Table 6.10	The energy and CO_2 rating of cars as a function of their mass		
Fuel type	**Energy per km · mass** (H_{km} in MJ/km, m in kg)	**CO_2 per km · mass** ($CO_{2/km}$ in g/km, m in kg)	**dH_{km}/dm MJ/km · kg** ($m = 1000$ kg)
Gasoline power	$H_{km} \approx 3.7 \times 10^{-3} \, m^{0.93}$	$CO_2/km \approx 0.25 \, m^{0.93}$	2.1×10^{-3}
Diesel power	$H_{km} \approx 2.8 \times 10^{-3} \, m^{0.93}$	$CO_2/km \approx 0.21 \, m^{0.93}$	1.6×10^{-3}
LPG power	$H_{km} \approx 3.7 \times 10^{-3} \, m^{0.93}$	$CO_2/km \approx 0.17 \, m^{0.93}$	2.2×10^{-3}
Hybrid power	$H_{km} \approx 2.3 \times 10^{-3} \, m^{0.93}$	$CO_2/km \approx 0.16 \, m^{0.93}$	1.3×10^{-3}

Data life. If an aircraft, designed and built 25 years ago, requires replacement parts, they have to be made of the same alloy that was used when the plane was first built—a change can invalidate the airworthiness certificate. When a nuclear reactor is decommissioned, the chemical composition of its materials determines the life and intensity of fission products that have to be contained. These are just two examples of the importance of ensuring data life. It is not a short-term problem. Boeing keeps records of all materials used in their planes for 60 years. Nuclear safety inspectors, when asked about data life, say "infinite." Emerging standards ISO 14721 (2003), and ISO 10303 part 45 and 235 relate to the safe long-term storage of material property data.

6.5 Exploring the data: property charts

If you want to select a material to meet eco-criteria, you need to be able to compare it with all the alternatives. Data sheets like those of Chapter 15 of this book tabulate the properties of individual materials but they don't make it easy to compare one with another. The way to do that is to plot *material property charts*.

Material property charts. Property charts are of two types: *bar charts* and *bubble charts*. A bar chart is simply a plot of the value-ranges of one property. Figure 6.8 shows an example: it is a bar chart for Young's modulus, E, the mechanical property that measures stiffness. The largest value is more than ten million times greater than the smallest—many other properties have similarly large ranges—so it makes sense to plot them on logarithmic[4] scales (as here), not linear ones. The length of each bar shows the range of the property for each of the materials, here segregated by family. The differences between the families now become apparent. Metals and ceramics have high moduli. Those of polymers are smaller by a factor

---

[4]Logarithmic means that the scale goes up in constant multiples, usually often. We live in a logarithmic world—our senses, for instance, all respond in that way.

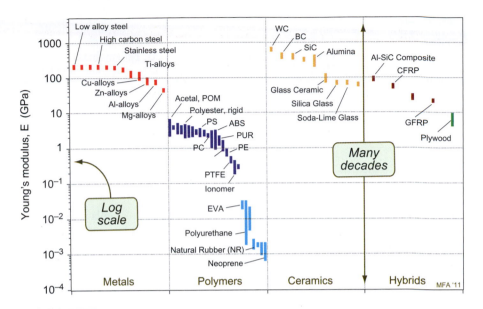

FIGURE 6.8 *A bar chart of the modulus. It reveals the difference in stiffness between the families.*

of about 50 than those of metals. Those of elastomers are some 500 times smaller still.

More information is packed into the picture if two properties are plotted to give a *bubble chart*, as in Figure 6.9, here showing modulus E and density ρ. As before, the scales are logarithmic. Now families are more distinctly separated. Ceramics lie in the yellow envelope at the very top: they have moduli as high as 1,000 GPa. Metals lie in the reddish zone near the top right; they, too, have high moduli but they are heavy. Polymers lie in the dark blue envelope in the center, elastomers in the lighter blue envelope below, with moduli as low as 0.0001 GPa. Materials with a lower density than polymers are porous: man-made foams and natural cellular structures like wood and cork. Each family occupies a distinct, characteristic field. Yet more information can be displayed by using functions of properties (like E/ρ) for the axes of the charts. Examples of these appear later.

Material property charts are a core tool.[5]

■ They give an overview of the physical, mechanical, functional, and environmental properties of materials, presenting the information about them in a compact way.

---

[5]Further descriptions and a wide range of charts can be found in the sources listed under Material property charts in Section 6.7.

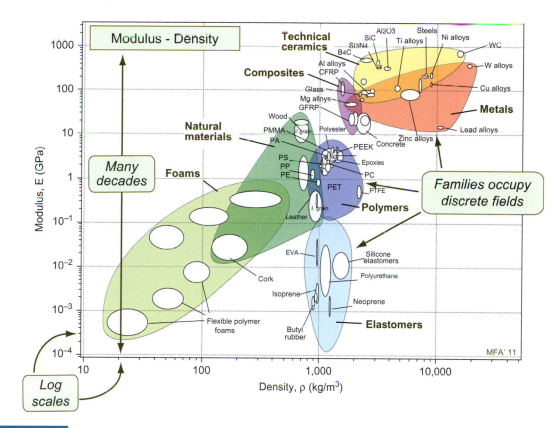

FIGURE 6.9 *A bubble chart of modulus and density. Families occupy discrete areas of the chart.*

- They reveal correlations between properties and are helpful in understanding the underlying science and in estimating properties when no direct measurements are available.
- They provide a tool for optimized selection of materials to meet given design requirements, and they help us understand the use of materials in existing products.
- They allow the properties of new materials and hybrids to be displayed and compared with those of conventional materials, bringing out their novel characteristics and suggesting possible applications.

Property charts appear in later chapters. Right now we use them to explore the eco-data.

Embodied energies per unit mass of materials are compared in the bar chart of Figure 6.10. The light alloys based on aluminum, magnesium, and titanium have high values, approaching 800 MJ/kg for titanium. Precious metals lie much higher still (Table 6.2). Polymers all cluster around 100 MJ/kg, less than the light alloys

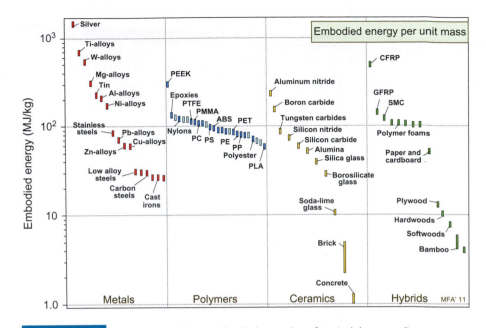

FIGURE 6.10 *A bar chart of the embodied energies of materials per unit mass*

but considerably more than steels and cast irons, which have energies between 20 and 40 MJ/kg. Technical ceramics such as aluminum nitride have high energies; those for glass, cement, brick, and concrete are much lower. Composites, too, have a wide spread. High-performance composites—here we think of CFRP (carbon-fiber reinforced polymers)—lie at the top, well above most metals. At the other extreme, paper, plywood, and timber are comparable with the other materials of the construction industry.

But is embodied energy *per unit mass* the proper basis of comparison? Suppose, instead, the comparison is made *per unit volume* (Figure 6.11). The picture changes. Now metals as a family lie above the others. Polymers cluster around a value that is lower than most metals—by this measure, they are not the energy-hungry materials they are sometimes made out to be. The nonmetallic materials of construction—concrete, brick, wood—lie far below all of them. CFRP is now only a little greater than aluminum.

This raises an obvious question: if we are to choose materials with the objective of minimizing their embodied energy, what basis of comparison should we use? A mistaken choice invalidates the comparison, as we have just seen. The right answer is to compare embodied energy *per unit of function*. We return to this, in depth, in Chapters 9 and 10.

Carbon footprint. Material production pumps enormous quantities of CO_2 into the atmosphere—some 20% of the global total arises in this way. So it is interesting to

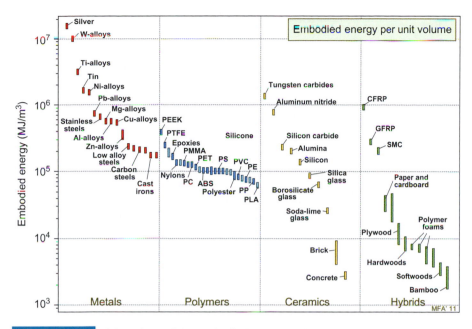

FIGURE 6.11 *A bar chart of the embodied energies of materials per unit volume*

ask: which materials contribute the most? That depends on the carbon footprint per kg and on the number of kg per year that are produced. The data sheets have the information to explore this. Figure 6.12 answers the question and illustrates how the data can be used. It was made by multiplying the annual world production by the carbon footprint for material production to give the tonnage of CO_2 per material per year. The big four are iron and steel, aluminum, concrete (cement), and paper and cardboard. Between them, they account for more than all the rest put together.

News clip: Low carbon concrete

Concrete is remixed with environment in mind.

Portland cement has been around since the early 1800s. . . . Aesthetic considerations aside, concrete is environmentally ugly. The manufacturing of Portland cement is responsible for about 5 percent of human-caused emissions of the greenhouse gas carbon dioxide. "The new twist over the last 10 years has been to try to avoid materials that generate CO_2" said Kevin A. MacDonald, vice president for engineering services of the Cemstone Products Company. In his mixes, Dr. MacDonald replaced much of the Portland cement with two industrial waste products—fly ash, left over from burning coal in power plants, and blast-furnace slag. Both are what are called pozzolanas, reactive materials that help make the concrete stronger. Because the CO_2

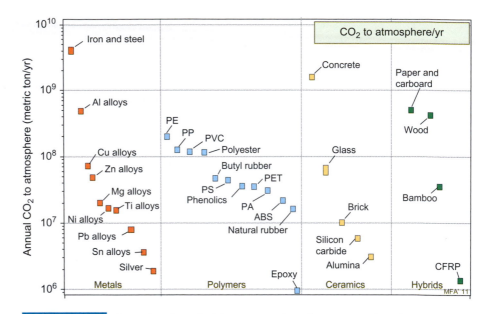

FIGURE 6.12 *Annual carbon dioxide to atmosphere from material production*

> *emissions associated with them are accounted for in electricity generation and steel making, they also help reduce the concrete's carbon footprint.*
> **The New York Times**, March 31, 2009
>
> Reducing the carbon footprint of concrete by 20%, possible by using pozzalanas, reduces global carbon emissions by 1% in one go.

Water usage is compared in Figure 6.13. Materials with high embodied energy tend to have high water usage—not surprising, given the water demands of energy listed in Table 2.2. There is not much else to be said except that the water consumptions plotted here are small compared with those required, per kg, for water-intensive agricultural crops like rice and cotton, or for materials derived from animal husbandry like meat, leather, and wool.

Shaping-process energies (MJ/kg of shaped output) are plotted on the four columns on the left of Figure 6.14. The efficiencies of these processes are low, meaning that they use many times more energy than the ideal minimum. The "ideal" energy for casting is to raise the metal to its melting point plus that required to melt it. In practice casting energies are five or more times greater than this. The metal has to be held in something and it, too, has a heat capacity. There are heat losses by conduction, convection, and radiation. If the heat source is fossil-fuel-generated electricity, there is another multiplier of 3 to allow for conversion efficiency. There is

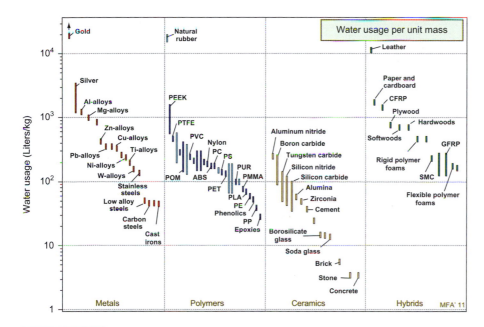

FIGURE 6.13 *Water usage bar chart. The demands of water for material are small compared with those for agriculture.*

the energy to make and bake the molds, to trim the castings, and to account for metal "losses" as cut-offs and imperfect castings. Other shaping processes have similar low efficiencies.

Deformation processing of metals is less energy-intensive than casting, making it a more attractive process from an environmental point of view, though it, too, is far from ideal. Vapor processing and powder methods stand out as particularly energy-intensive. The extrusion and molding of polymers, composites, and glass, too, consume much more energy than that simply required to heat the material to the melting temperature.

Machining-process energies (MJ/kg of material removed) appear on the right of Figure 6.14. The energy is usually given as energy per unit volume removed, less commonly as energy per unit weight removed. The energy increases with the strength of the material. Coarse ("heavy") machining is more efficient (uses less energy per unit volume removed) than fine ("finishing" or "light") machining, which can require up to five times more. Grinding, a special kind of machining, takes more than ten times the energy of coarse machining, although the quantity of material removed is usually very small.

Recycling. Figure 6.15 presents the data for the recycled fraction in current supply. As discussed in Chapter 4, the recycling of metals is highly developed and its

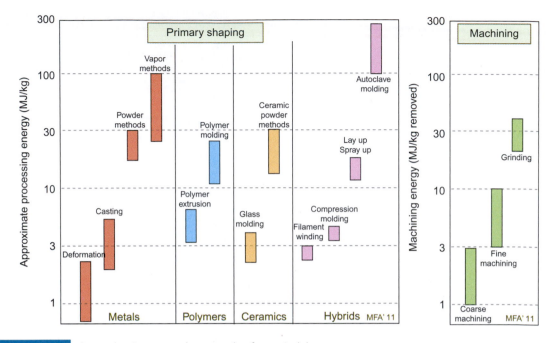

FIGURE 6.14 *Approximate processing energies for materials*

contribution to current supply is large. The same cannot be said of polymers. The commodity polymers are used in large quantities, many in products with short lives, and they present major problems in waste management, all of which, you would think, would encourage effective recycling. But the economics of polymer recycling are unattractive, with the result that the contribution to current supply is small.

Special topic: adjusting for country-specific energy mixes. The carbon footprint of materials, processes, transportation, and use depends on the energy mix in the energy they consume, and that varies from country to country. Ideally we would like directly measured values for eco-data that are country-specific, but such data just don't exist. It does not feel right to ignore the country-specific aspect of material production and product use just because we lack data for it. The extensive use of biofuels in Brazil, of geothermal electricity generation in Iceland, and of hydroelectricity in Norway significantly changes the carbon footprints of materials made and used there. Is there not some sort of fix—an approximate correction—that could be applied to the data we *do* have to accommodate this?

We will regard the energy required to make and process materials as country-independent, and seek adjustment factors for the carbon footprint to accommodate country-specific energy mixes used to do it. We take oil as the norm for hydrocarbons and electricity supply from a gas-fired power station with 38% efficiency as the norm for electrical power. Table 6.11 lists the correction factor to be applied to

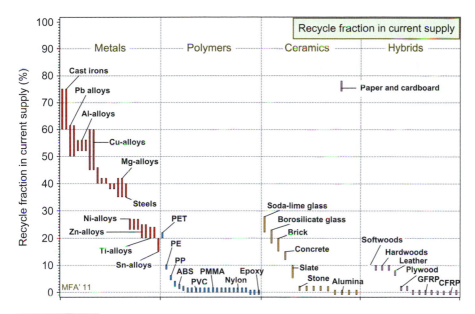

FIGURE 6.15   *Recycle fraction bar chart. Metals are extensively recycled. Most other materials are not.*

Table 6.11	Adjustment factors for carbon footprint for alternative fuel and electricity energy mixes		
Hydrocarbon fuel source	Correction factor for carbon footprint*	Electricity energy source	Correction factor for carbon footprint[+]
100% oil (diesel)	1	75% oil/gas (EU average)	1
100% kerosene	0.96	100% coal	2
100% gasoline	0.92	100% nuclear	0.06
100% natural gas	0.77	100% solar PV	0.2
100% biodiesel	0	100% wind	0.08
100% bio-ethanol	0	100% hydro	0.06
100% hydrogen	0–0.1	100% geothermal	0.05

*The correction factors for hydrocarbon-based power are the ratios of CO_2/MJ of fuel to that of oil, using data from Table 6.7.

[+] The correction factors for electrical power are based on the carbon footprints of power sources analyzed in Chapter 12.

carbon footprint from fuels and electrical energy use for 100% conversion to the listed source. The correction factor for actual energy mixes can then be found by linear interpolation. The two examples that follow show how these data can be used.

Carbon footprint of products using biofuel

Example: Transportation within Brazil makes extensive use of biofuel. If the typical fuel mix is 85% biodiesel and 15% conventional diesel, what is the carbon footprint of transport by a 44-metric ton truck in Brazil?

Answer: Table 6.9 gives the carbon emissions for a 44-metric ton truck using diesel oil as fuel at 0.06 kg CO_2, per metric ton.km. Interpolation gives the correction factor for 85% biodiesel, using values from Table 6.11, gives

$$0.85 \times 0 + 0.15 \times 1 = 0.15$$

Thus the carbon footprint of transport by a 44-metric ton truck is reduced to $0.06 \times 0.15 = 0.009$ kg CO_2 per metric ton.km.

Hydro-aluminum

Example: Global warming has increased the potential for generating hydroelectric power from Greenland's melting glaciers. Aluminum production using electric power from fossil fuel carries a large carbon footprint: 12.6 kg/kg, according to the data sheet in Chapter 15. It is proposed to establish an aluminum production facility in the most promising of Greenland's fjords. Doing so will create the need for additional shipping to transport bauxite to Greenland and aluminum back to where it will be used—the estimate is 20 metric ton · km of sea freight per kg of delivered aluminum. What do these aspects of production do to the carbon footprint of the aluminum?

Answer: Table 6.11 gives the carbon footprint of hydropower relative to that of conventional fossil-fuel sourced electricity as 0.06, reducing the production footprint to

$$12.6 \times 0.06 = 0.76 \text{ kg/kg}$$

Table 6.9 gives the carbon footprint of sea freight as 0.015 kg/metric ton · km. Thus the additional 20 metric ton · km of sea transport creates an additional carbon footprint of

$$20 \times 0.015 = 0.3 \text{ kg/kg}$$

giving a total of 1.06 kg/kg as the final footprint of the aluminum. On this basis the Greenland scheme looks attractive. But water flow from glacial melt fluctuates with the seasons. If the plant is to meet delivery obligations, it may need back-up power from more conventional sources, with an inevitable increase in emissions.

6.6 Summary and conclusions

You can't answer technical questions without numbers. "Choice X is better than Choice Y..." is a statement that is on solid ground only if you have data to demonstrate that it is indeed so. Concern for the environment, today, sometimes leads to statements based more on emotion than reason, clouding issues and breeding deception. So, boring though numbers can be, they are essential for what follows.

To use numbers, you have to know where to find them, what they mean, and how accurate (or inaccurate) they are. This chapter introduced the ones used later in this book, presenting them as bar charts that display relationships and correlations. The thing to remember about them is that their precision is low. If you are going to base decisions on their values, make sure that the decision still stands if the numbers are wrong by ±10%. Such uncertainty does not prevent decision making provided its presence is recognized and allowed for. The charts help here. When the bars or bubbles overlap, the differences between the materials they represent are not significant. When they don't overlap, there is a significant difference.

The poor coverage of eco-property data generates two further challenges. The first is that some property values for some materials are simply not there—no one has measured them yet. It then becomes necessary to estimate them. Useful correlations exist between embodied energy, ore grade, and material price, and between carbon footprint and embodied energy. These provide sanity checks for existing data and ways of making approximate estimates when no real measurements are available.

The second difficulty arises from the differences in energy mix of hydrocarbon fuel and of electrical power from country to country. These differences mean that carbon footprints of materials made or processed in one country differ from those of another. The carbon footprints of transportation, too, are country-specific. We lack direct measurements that allow comparisons between countries, so we have to fall back on adjusting the data we do have by applying approximate correction factors, as explained in this chapter.

6.7 Further reading

The references are segregated by data type.

General engineering properties of materials

Definitions of material properties can be found in numerous general texts on engineering materials, among them those listed here.

Ashby, M.F., Shercliff, H.R. and Cebon, D. (2009), *Materials: engineering, science, processing and design*, 2nd edition, Butterworth Heinemann, Oxford, UK. ISBN-13: 978-1-85617-895-2. *(An elementary text introducing materials through material property charts, and developing the selection methods through case studies)*

Budinski, K.G. and Budinski, M.K. (2010), *Engineering materials, properties, and selection*, 9th edition, Prentice Hall, NY, USA. ISBN 978-0-13-712842-6. *(An established materials text that deals with both material properties and processes)*

Callister, W.D. (2006), *Materials science and engineering, an introduction*, 7th edition, John Wiley, New York, USA. ISBN 9-780-471-73696-7. *(A well-respected materials text, now in its 7th edition, widely used for materials teaching in North America)*

Charles, J.A., Crane, F.A.A. and Furness, J.A.G. (1997), *Selection and use of engineering materials*, 3rd edition, Butterworth Heinemann, Oxford, UK. ISBN 0-7506-3277-1. *(A materials-science approach to the selection of materials)*

Dieter, G.E. (1999), *Engineering design, a materials and processing approach*, 3rd edition, McGraw-Hill, New York, USA. ISBN: 9-780-073-66136-0. *(A well-balanced and respected text focusing on the place of materials and processing in technical design)*

Farag, M.M. (2008), *Materials and process selection for engineering design*, 2nd edition, CRC Press, Taylor and Francis, London, UK. ISBN 9-781-420-06308-0. *(A materials-science approach to the selection of materials)*

Kalpakjian, S. and Schmid, S.R. (2003), *Manufacturing processes for engineering materials*, 4th edition, Prentice Hall, Pearson Education, Inc., New Jersey, USA. ISBN 0-13-040871-9. *(A comprehensive and widely used text on material processing)*

Shackelford, J.F. (2009), *Introduction to materials science for engineers*, 7th edition, Prentice Hall, NJ, USA. ISBN 978-0-13-601260-3. *(A well-established materials text with a design slant)*

Material property charts

Ashby, M.F. (2011), *Materials selection in mechanical design*, 4nd edition, Butterworth Heinemann, Oxford, UK. ISBN 0-7506-6168-2. *(An advanced text developing material selection methods in detail)*

Geo-economic data

Chemsystems (2006), www.chemsystems.com. Accessed December 2011 *(Data for polymers)*

Cheresources (2007), www.cheresources.com/polystyz5.shtml. Accessed December 2011 *(Data for polymers)*

Geokem (2007), www.geokem.com/global-element-dist1.html. Accessed December 2011 *(Data for metals and minerals)*

International Rubber Study Group (IRSG) (2007), Vol 61 No 4/Vol 61 No 5, January/February. *(Data for rubber)*

Pilkington Group Ltd. (2007), Prescot Road, St. Helens, Merseyside UK. *(Data for glass)*

US Geological Survey (2007), http://minerals.usgs.gov/minerals/pubs/commodity/. Accessed December 2011 *(Data for metals and minerals)*

Material production: embodied energy and CO_2, engineering materials

Aggregain (2007), "The Waste and Resources Action Program (WRAP)," www
.wrap.org.UK. ISBN 1-84405-268-0. *(Data and an Excel-based tool to calculate energy and carbon footprint of recycled road-bed materials)*

AMC (2006). Australian Magnesium Corporation, www.aph.gov.au/house/committee/environ/greenhse/gasrpt/Sub65-dk.pdf.

APME (1997, 1998, 1999, 2000), "Eco-profiles of the European plastics industry," Association of plastics manufacturers in Europe, Brussels, Belgium. www.lca.apme.org.

BCA (2007), "A carbon strategy for the cement industry," British Cement Association, www.cement-industry.co.uk. Accessed December 2011.

Boustead Model 5 (2007), Boustead Consulting, West Sussex, UK. (www.boustead-consulting.co.uk *(An established life-cycle assessment tool)*

Building Research Establishment (2006), BRE Environmental Profiles database, BRE Environment Division, BREEM Center, Watford UK. (A respected source of data for building materials)

Chapman, P.F. and Roberts, F. (1993), *Metals resources and energy*, Butterworths, London, UK. ISBN 0-408-10802-9. (An early analysis of resource availability)

Chemlink Australasia (1997), www.chemlink.com.au/mag&oxide.htm. Accessed December 2011.

Ecoinvent Version 2.2 (2011), Competence Centre of the Swiss Federal Institute of Technology, Zürich, Switzerland. www.ecoinvent.org. Accessed December 2011. *(The ecoinvent database is a compilation of Life Cycle Inventory (LCI) data for materials and products).*

ELCD (2008), lca.jrc.ec.europa.eu. Accessed December 2011. *(A high-quality Life Cycle Inventory (LCI). Core data sets of this first version of the Commission's "European Reference Life Cycle Data System (ELCD)")*

Energy Information Association (2008), www.eia.doe.gov. *(Official energy statistics from the US Government)*

European Aluminium Association (2000), www.eaa.net.

European Reference Life Cycle Database (ELCD) of the Sustainability Unit of the Joint Research Centre of the European Commission (2010), Petten, the Netherlands.

GREET (2007), Argonne National Laboratory and the US Department of Transportation, www.transportation.anl.gov/. *(Software for analyzing vehicle energy use and emissions)*

Hammond, G. and Jones, C. (2011), "Inventory of Carbon and Energy (ICE)," Department of Mechanical Engineering, University of Bath, Bath, UK.

International Aluminum Institute (2000), "Life cyclse inventory of the worldwide aluminum industry," Part 1—automotive, www.world-aluminum.org/. Accessed December 2011.

Kemna, R., van Elburg, M., Li, W., and van Holsteijn, R. (2005), "Methodology study eco-design of energy-using products," Van Holsteijn en Kemna BV (VHK), Delf, Netherlands.

Kennedy, J. (1997), "Energy minimisation in road construction and maintenance," a Best Practice Report for the Department of the Environment, UK.

Lafarge Cement, Lafarge Cement Oxon, UK. Lawrence Berkeley National Laboratory report (2005), "Energy use and carbon dioxide emissions from steel production in China," LBNL-47205, April.

Lawson, B. (1996), "Building materials, energy and the environment: Towards ecologically sustainable development," RAIA, Canberra www.greenhouse.gov .au/yourhome/technical/fs31.htm.

Lime Technology (2007), www.limetechnology.co.uk/whylime.

MEEUP Methodology Report, final (2005), VHK, Delft, Netherlands. www.pre.nl/ EUP/. *(A report by the Dutch consultancy VHK commissioned by the European Union, detailing their implementation of an LCA tool designed to meet the EU Energy-using Products directive)*

Ohio Department of Natural Resources, Division of Recycling (2005), www.dnr .state.oh.us/recycling/awareness/facts/tires/rubberrecycling.htm.

Pilz, H., Schweighofer, J. and Kletzer, E. (2005), "The contribution of plastic products to resource efficiency," Gesellschaft fur umfassende analysen (GUA), Vienna, Austria.

Schlesinger, M.E. (2007), *Aluminum recycling*, CRC Press, Taylor and Francis, London UK. ISBN 0-8493-9662-X.

Stiller, H. (1999), *Material intensity of advanced composite materials*, Wuppertal Institut fur Klima, Umvelt, Energie. ISSN 0949-5266.

Sustainable Concrete (2008), www.sustainableconcrete.org.uk/. *(A web site representing the UK concrete producers carrying useful information about carbon footprint)*

Szargut, J., Morris, D.R., Steward, F.R. (1988), *Energy Analysis of Thermal Chemical and Metallurgical Processes*, Hemisphere, New York, USA.

Szokolay, S.V. (1980), *Environmental science handbook: for architects and builders*, Lancaster: Construction Press, , Broklyn, Victoria, Australia.

The Nickel Institute North America (2007), Nickel Institute, Toronto, Canada, www.nickelinstitute.org/. Accessed December 2011.

Material production: embodied energy and CO$_2$, precious metals

GREET (2007), Argonne National Laboratory and the US Department of Transportation, www.transportation.anl.gov/. Accessed December 2011. *(Software for analyzing vehicle energy use and emissions)*

London Platinum and Palladium Market (2006), www.lppm.org.uk. Accessed December 2011.

Lonmin Plc. (2005), 2005 Corporate Accountability Report, www.lonmin.com/. Accessed December 2011.

TEAM (2008), Tool for Environmental Analysis and Management, www .ecobalance.com/. Accessed December 2011. *(TEAM is Ecobilan's Life-Cycle Assessment software. It allows the user to build and use a large database and to model systems associated with products and processes following the ISO 14040 series of standards)*

Electronic components

EcoInvent (2010), EcoInvent Certre, Swiss Centre for Life Cycle Inventories, www
.ecoinvent.org. Accessed December 2011. *(A massive compilation of
environmental data for materials hosted by the University of Delft)*

IDEMAT (2009), The University of Delft, Delft, Netherlands.

Knapp, K.E. and Jester, T.L. (2005), "An empirical perspective on the energy payback
time for photovoltaic modules," Solar 2000 Conference, Madison, Wisconsin.

Kuehr, R. and Williams, E., editors (2003), *Computers and the environment:
understanding and managing their impacts*, Kluwer Academic Publishers and
United Nations University, Aarhus, Denmark. ISBN 1-4020-1679-4. *(A multiauthor monograph on the environmental aspects of electronic devices, with
emphasis on the WEEE regulations relating to end of life. Chapter 3 deals with
the environmental impacts of the production of personal computers.)*

MEEUP (2006), Kemna, R., van Elburg, M., Li W. and van Holsteijn, R.,
"Methodology study eco-design of energy-using products," Final report, VHK,
Delft, Netherlands and the European Commission, Brussels, Belgium. *(A study
commissioned by the European Union into the development of software to
meet the Energy-using Product Directive)*

Architecture and the built environment

Cole, R.J. and Kernan, P.C. (1996), "Life-cycle energy use in office buildings,"
Building and Environment, 31, 4, pp 307−317. *(An in-depth analysis of energy
use per unit area of steel, concrete, and wood-framed construction)*

Canadian Architect (2011), www.canadianarchitect.com/asf (accessed July 2011).
*(Canadian Architect is a journal for architects. Their web site provides a helpful
information source.)*

EcoInvent (2010), EcoInvent Certre, Swiss Centre for Life Cycle Inventories, www
.ecoinvent.org. *(A massive compilation of environmental data for materials
hosted by the University of Delft)*

Hammond, G. and Jones, C. (2011), "Inventory of Carbon and Energy (ICE),"
Department of Mechanical Engineering, University of Bath, Bath, UK. *(A major
compilation of embodied energy data for the principal materials of construction)*

Water

AZoM (2008), "A to Z of Materials," *Journal of Materials Online*, www.azom.com.
Accessed December 2011.

Chiang, S.H. and Moeslein, D. (1978−1980), "Analysis of water use in nine
industries," Parts 1−9, Department of Chemical and Petroleum Engineering,
University of Pittsburgh, Pittsburg, PA, USA.

Davis, J.R. (1995), in *ASM Specialty Handbook*, ASM International, Metals Park,
Ohio, USA.

Implicit Price Deflators in National Currency and US Dollars (2006), United
Nations statistics division, http://unstats.un.org/unsd/snaama/dnllist.asp.
Accessed December 2011.

Lenzena, M. (2001), "An input-output analysis of Australian water usage," *Water Policy* 3, pp. 321–340.

Leontief, W. (1970), "Environmental repercussions and the economical structure: An input-output approach," *The Review of Economics and Statistics*, 52, 3 pp 262–271.

Pearce, F. (2006), "Earth: The parched planet," *New Scientist*, 2540, February.

Proops, J.L.R. (1997), "Input-output analysis and energy intensities: a comparison of some methodologies," *Appl. Math. Modelling*, 1, March.

Shiklomanov, I.A. (2010), "World water resources and their use," UNESCO International Hydrological Programme, www.webworld.unesco.org (accessed July 2010. *(A detailed analysis of world water consumption and emerging problems with supply)*

UNESCO (2006), "World water assessment programme," United Nations educational scientific and cultural organization, www.unesco.org/water. Accessed December 2011.

Vela'zquez, T.E. (2006), "An input-output model of water consumption: analyzing intersectoral water relationships in Andalusia," *Ecological Economics*, 56, pp 226–240.

Aggregated measures: eco-indicators

EPS (1993), "Life cycle analysis in product engineering," *Environmental Report 49*, Volvo Car Corp. Gothenburg, Sweden.

Goedkoop, M., Effting, S., and Collignon, M. (2000), "The Eco-indicator 99: A damage oriented method for Life Cycle Impact Assessment," *Manual for Designers* (14th April 2000) www.pre.nl. Accessed December 2011.

Idemat Software version 1.0.1 (1998), Faculty of Industrial Design Engineering, Delft University of Technology, Delft, The Netherlands.

Material shaping processing: energy and CO_2

General

Allen, D.K. and Alting, L. (1986), "Manufacturing processes," Brigham Young University, Provo, Utah, USA.

Boustead Model 4 (1999), Boustead Consulting, West Sussex, UK, www.boustead-consulting.co.uk. Accessed December 2011.

Ecoinvent Version 2.2 (2011), Competence Centre of the Swiss Federal Institute of Technology, Zürich, Switzerland. www.ecoinvent.org. Accessed December 2011. *(The ecoinvent database is a compilation of Life Cycle Inventory (LCI) data for materials and products).*

European Reference Life Cycle Database (ELCD) of the Sustainability Unit of the Joint Research Centre of the European Commission (2010), Petten, the Netherlands.

MEEUP (2006), "Methodology study eco-design of energy-using products," Final report, VHK, Delft, Netherlands and the European Commission, Brussels, Belgium www.pre.nl/EUP/. *(A study commissioned by the European Union into the development of software to meet the Energy Using Product Directive)*

Deformation processing of metals

Abdul Samad, M. and Rio, R.S. (2001), *J. Manufacturing Science and Engineering*, Minimization of energy in the multipass rolling process, Vol. 123, pp 135–141.

Allen D.K. and Alting, L. (1986), "Manufacturing processes," student manual, Brigham Young University, Provo, Utah, USA.

Boustead Model 4 (1999), Boustead Consulting, West Sussex, UK, www .boustead-consulting.co.uk. Accessed December 2011.

Cast Metal Coalition (2005), http://cmc.aticorp.org/examples.html. Accessed December 2011. *(Based on US National figures)*

Eco-Invent database (2007).

IDEMAT (2009), The University of Delft, Delft, Netherlands.—*(Most of this data is from EcoInvent 07)*

LBNL (2005), "Energy use and carbon dioxide emissions from steel production in China," Lawrence Berkeley National Laboratory report, LBNL-47205. April 2005.

MEEUP Methodology Report, Final (2005), VHK for European Commission, R. Kemna, M.van Elburg, W. Li, and R. van Holsteijn, Delftech Park, Delft, Netherlands.

US Department of Energy (1997), "Supporting industries energy and environmental profile," Energy Efficiency and Renewable Energy, Washington DC.

Casting of metals

Allen D.K. and Alting, L. (1986), "Manufacturing processes," student manual, Brigham Young University, Provo, Utah, USA.

Boustead Model 4 (1999), Boustead Consulting, West Sussex, UK, www.boustead-consulting.co.uk. Accessed December 2011.

Cast Metal Coalition (2005), http://cmc.aticorp.org/examples.html. Accessed December 2011. *(Based on US National figures)*

Cast Metal Coalition (2010), http://cmc.aticorp.org/datafactors.html. Accessed December 2011. *(Energy data based on US national figures)*

Dalquist, S. and Gutowski, T. (2004), "Life cycle analysis of conventional manufacturing techniques: die casting," MIT report LMP-MIT-03-12-09-2004.

Dalquist, S. and Gutowski, T. (2004), "Life cycle analysis of conventional manufacturing techniques: sand casting," Proc. 2004 ASME IMECE meeting, Anaheim, CA.

Energetics Inc. (1999), "Energy and Environmental Profile of the US Metal Casting Industry," http://www1.eere.energy.gov/industry/metalcasting/pdfs/profile.pdf. Accessed December 2011.

Eurecipe (2005), Reduced Energy Consumption in Plastics Engineering, from the 2005 European Benchmarking Survey of Energy Consumption, www.eurecipe.com.

LBNL (2005), "Energy use and carbon dioxide emissions from steel production in China," Lawrence Berkeley National Laboratory report, LBNL-47205. April 2005.

Vapor forming

Branham, M. (2008), "Energy and materials use in the integrated circuit industry," M.S. Thesis, Department of Mechanical Engineering, MIT.

Ecoinvent Version 2.2 (2011), Competence Centre of the Swiss Federal Institute of Technology, Zürich, Switzerland. www.ecoinvent.org. Accessed December 2011. *(The ecoinvent database is a compilation of Life Cycle Inventory (LCI) data for materials and products)*

Gutowski, T., Dahmus, J., Branham, M., and Jones, A.A. (2007), "A thermodynamic characterization of manufacturing processes," IEEE International Symposium on Electronics and the Environment, Orlando, FL, USA. May 2007.

Krishnan, N., Boyd, S., Somani, A., Raous, S., Clark, D. and Dornfeld, D. (2008), "A hybrid life cycle inventory of nano-scale semiconductor manufacturing," *Environmental Science and Technology*, 42, pp 3069–3075.

Murphy, C.F., Kenig, G.A., Allen, D., Laurent, J-P. and Dyer, D.E. (2003), "Development of parametric material, energy and emissions inventories for wafer fabrication in the semiconductor industry," *Environ. Sci. Technol.* 37, pp 5373–5382.

Cabuk (2010), www.cabuk1.co.uk.

Polymer molding

Boustead Model 4 (1999), Boustead Consulting, West Sussex, UK, www.boustead-consulting.co.uk. Accessed December 2011.

Ecoinvent Version 2.2 (2011), Competence Centre of the Swiss Federal Institute of Technology, Zürich, Switzerland. www.ecoinvent.org . Accessed December 2011. *(The ecoinvent database is a compilation of Life Cycle Inventory (LCI) data for materials and products)*

Eurecipe (2005), "Reduced Energy Consumption in Plastics Engineering," from the 2005 European Benchmarking Survey of Energy Consumption, www.eurecipe.com. Accessed December 2011.

Gutowski, T.S., Branham, M., Dahmus, J.B., Jones, A.J. and Thiriez, A. (2009), "Thermodynamic analysis of resources used in manufacturing processes," *Environmental Science and Technology*, 43, pp 1584–1590.

Kemna R., van Elburg M., Li W. and van Holsteijn R. (2005) VHK, MEEUP Methodology Report, Final (2005), VHK for European Commission, Delftech Park, Delft, Netherlands, 28 Nov. 2005.

VHK (2005), MEEUP Methodology Report, Final (2005) VHK for European Commission, R. Kemna, M.van Elburg, W. Li, and R. van Holsteijn, Delftech Park, Delft, Netherlands, 28 Nov. 2005.

Kent, R. (2008), "Energy management in plastics and processing: Strategies, targets, techniques and tools." Plastics Information Direct, Bristol, England.

Suzuki, T. and Takahashi, J. (2005), The Ninth Japan International SAMPE symposium Nov. 29–Dec. 2. *(A detailed energy breakdown of energy of materials for cars)*

Thiriez, A. (2006), "An environmental analysis of injection molding," Master's thesis, Massachusetts Institute of Technology, Cambridge, MA, USA.

Thiriez, A. and Gutowski, T. (2006), "An Environmental Analysis of Injection Molding," IEEE International Symposium on Electronics and the Environment, San Francisco, California, USA, May 8–11.

US Department of Energy (2008), "Supporting industries energy and environmental profile," www1.eere.energy.gov/industry/energy_systems/pdfs/si_profile.pdf.

US Environmental Protection Agency (1995), Washington, DC. www.epa.gov. Accessed December 2011.

Glass molding

Worrell, E., Galitsky, C., Masanet, E., Graus, W. (2008), "Energy efficiency improvements and cost saving opportunities for the glass industry: An energy star guide for energy and plant managers," Environmental Energy Technologies Division of Ernest Orlando Lawrence Berkeley National Laboratory.

Composite shaping

Suzuki, T. and Takahashi, J. (2005), The Ninth Japan International SAMPE symposium Nov. 29–Dec. 2. *(A detailed energy breakdown of energy of materials for cars)*

Material machining and grinding processing: energy and CO_2

Draganescu, F., Gheorghe, M. and Doicin, C.V. (2003), "Models of machine tool efficiency and specific consumed energy," *Jnl. of Materials Processing Technology*, 141, pp 9–15.

Ghosh Chattopadhyay, S. and Paul S. (2008), "Modelling of specific energy requirement during high-efficiency deep grinding," *International Journal of Machine Tools and Manufacture*, 48: 11, pp 1242–1253.

Groover, M.P. (1999), *Fundamentals of modern manufacturing*, John Wiley & Sons, Inc, New York, NY, USA. ISBN 0-471-36680-3.

Gutowski, T.S., Branham, M.S., Dahmus, J.B., Jones, A.J. and Thiriez, A. (2009), "Thermodynamic analysis of resources used in manufacturing processes," *Environ. Sc. Technol.* 43, pp 1584–1590.

Gutowski, T. S., Dahmus, J.B., Branham, M.S. and ,Jones, A.J. (2009), "A thermodynamic characterization of manufacturing processes," IEEE Symposium on Electronics and the Environment, Orlando, FL, USA.

Kurd, M. (2004), "The material and energy flow through abrasive water machining," BSc Thesis, Massachusetts Institute of Technology, cited by (Gu 09).

McGeough, J.A. (1988), *Advanced methods of machining*, Chapman and Hall, New York, NY, USA.

Todd, R.H., Allen, D.K. and Alting, L. (1994), "Manufacturing Processes Reference Guide," Welding. ISBN 0-8247-9914-3.

Groover, M.P. (1999), *Fundamentals of Modern Manufacturing*, John Wiley & Sons, Inc, New York, NY, USA. ISBN 0-471-36680-3.

Idemat (2005), "LCA software tool," Delft University of Technology. www.tudelft.nl.

MEEUP report (2005), Kemna, R., van Elburg, M., Li, W. and van Holsteijn, R., Delft Tech Park, Netherlands. *(A study commissioned by the European Union into the development of software to meet the Energy Using Product Directive)*

Misha, R.S. and Mahoney, M.W. (2007), *Friction stir welding and processing*, ASM International, Metals Park, Ohio, USA.

US Department of Energy (1997), Washington DC. www.doe.gov. Accessed December 2011.

Fasteners

Bookshar, D. (2001), *Energy consumption of pneumatic and DC electric assembly tools*, Stanley Assembly Technologies, Cleveland, OH, USA. www.stanley assembly.com. Accessed December 2011.

Adhesives

Bradley, R. Griffiths, A., and Levitt, M. (1995), "Paints and coatings, adhesives and sealants," Construction Industry Research and Information Association (CIRIA), Vol. F, ISBN 8 6017 8161.

Hammond, G., and Jones, C. (2009), "Inventory of Carbon and Energy (ICE)," Department of Mechanical Engineering, University of Bath, Bath, UK.

Material finishing processes—painting, polymer coating, and plating energies

BBC (2010), Paint calculator, www.bbc.co.uk/homes/diy/paintcalculator.shtml.

Bradley, R., Griffiths, A., and Levitt, M. (1995), "Paints and coatings, adhesives and sealants," Construction Industry Research and Information Association (CIRIA), Vol. F, ISBN 8 6017 8161.

Centre for Building Performance Research (2003), "EE & CO2 coefficients for New Zealand building materials," and "Table of embodied energy coefficients," at www.victoria.ac.nz/cbpr/documents/pdfs/ee-co2_report_2003.pdf and www.victoria.ac.nz/cbpr/documents/pdfs/ee-coefficients.pdf (viewed 19 October 2010).

Franklin Associates (2005), VRP (Vehicle Recycling Partnership) data for automotive painting provided by James Littlefield, private communication.

Geiger, O. (2010), "Embodied energy in strawbale houses," www.grisb.org/publications/pub33.htm. Accessed December 2011.

Groover, M.P. (1999), *Fundamentals of Modern Manufacturing*, John Wiley & Sons, Inc., New York, USA. ISBN 0-471-36680-3.

Hammond, G. and Jones, C. (2009), "Inventory of Carbon and Energy (ICE)," Department of Mechanical Engineering, University of Bath, Bath, UK.

Idemat Software version 1.0.1 (1998), Faculty of Industrial Design Engineering, Delft University of Technology, Delft, The Netherlands.

MEEUP report (2005), Kemna, R., van Elburg, M., Li, W. and van Holsteijn, R. Delft Tech Park, Delft, The Netherlands. *(A study commissioned by the European Union into the development of software to meet the Energy Using Product Directive)*

Misha, R.S. and Mahoney, M.W. (2007), *Friction stir welding and processing*, ASM International, Metals Park, Ohio, USA.

US National Renewable Energy Laboratory (2010), www.nrel.gov.lci. Accessed December 2011.

Recycling and end-of-life

Aggregain (2007), The Waste and Resources Action Program (WRAP), www.wrap.org.UK. ISBN 1-84405-268-0.

AMC (2006), Australian Magnesium Corporation, www.aph.gov.au/. Accessed December 2011.

Chemlink Australasia (1997), www.chemlink.com.au/. Accessed December 2011.

Geokem (2007), www.geokem.com/global-element-dist1.html. Accessed December 2011.

Hammond, G. and Jones, C. (2006), "Inventory of carbon and energy (ICE)," Dept. of Mechanical Engineering, University of Bath, Bath, UK.

International Aluminum Institute (2000), "Life cycle inventory of the worldwide aluminum industry," Part 1—automotive, www.world-aluminum.org/. Accessed December 2011.

Kemna, R., van Elburg, M., Li, W., and van Holsteijn, R. (2005), "Methodology study eco-design of energy-using products," Van Holsteijn en Kemna BV (VHK), Delft, The Netherlands.

Lafarge Cement (2007), Lafarge Cement UK, Chilton, Oxon, UK.

Lawson, B. (1996), "Building materials, energy and the environment," RAIA, Canberra www.greenhouse.gov.au/. Accessed December 2011.

Ohio Department of Natural Resources (2005), Division of Recycling, www.dnr.state.oh.us/. Accessed December 2011.

Pilz, H., Schweighofer, J., and Kletzer, E. (2005), "The contribution of plastic products to resource efficiency," Gesellschaft fur umfassende analysen (GUA), Vienna, Austria.

Schlesinger, M.E. (2007), *Aluminum recycling*, CRC Press, New York, USA. ISBN 0-8493-9662-X.

Sustainable Concrete (2008), www.sustainableconcrete.org.uk/.

The Nickel Institute North America (2007), Nickel Institute, Toronto, Canada www.nickelinstitute.org/. Accessed December 2011.

US Environmental Agency (2007), www.eia.doe.gov. Accessed December 2011.

US Geological Survey (2007), http://minerals.usgs.gov/. Accessed December 2011.

Waste on line (2007), www.wasteonline.org.uk/.

Transport and use energies

General

Abare (2009), http://abare.gov.au/interactive/09_ResearchReports/EnergyIntensity/htm/chapter_5.htm.*(Australian Government statistics on energy intensities of transport)*

AggRegain (2006), CO_2 emissions estimator tool, published by WRAP (Waste & Resources Action Programme), The Old Academy Banbury, UK, www.aggregain .org.uk.

Carbon Trust (2007), "Carbon footprint in the supply chain," www.carbontrust. co.uk/. Accessed December 2011.

Harvey, L.D.D. (2010), *Energy and the new reality 1: energy efficiency and the demand for energy services*, Earthscan Ltd, London, UK. ISBN978-1-84971-072-5. *(An analysis of energy use in buildings, transport, industry, agriculture, and services, backed up by comprehensive data)*

IEA Scoreboard 2009 (2009), www.scribd.com/doc/53697399/34/Energy-effciency-in-freight-transport. *(A comprehensive survey of energy trends)*

NREL (2010), National Renewable Energy Laboratory, www.nrel.gov/lci. Harvey, L.D.D. (2010), Energy and the new reality.

Weber, C.L. and Matthews, H.S. (2008), "Food miles and the relative impacts of food choices in the United States," *Environ. Sci. Technol.* 42, pp 3508–3513.

Aircraft

Green, J.E., Cottington, R.V., Davies, M., Dawes, W.N., Fielding, J.P., Hume, C.J., Lee, D.J., McClarty, J., Mans, K.D.R., Mitchell, K., and Newton, P.J. (2003), "Air travel—greener by design: the technology challenge," Report of the Technology Subgroup, www.greeenerbydesigh.org.uk. Accessed December 2011.

Cars

See data in Chapter 9, Figures 9.11 and 9.12.

Trucks

Manicore (2008), www.manicore.com/. *(Useful discussion of the definition of oil-equivalence of energy sources)*

Transport Watch UK (2007), www.Transwatch.co.uk/. Accessed December 2011.

TRL UPR (1995), "Energy consumption in road construction and use," Transport Research Laboratory, UK.

Rail freight

Network Rail (2007), www.networkrail.co.uk/. *(The web site of the UK rail track provider)*

Shell Petroleum (2007), *How the energy industry works*, Silverstone Communications Ltd., Towchester, UK. ISBN978-0-9555409-0-5.

Bureau of Transport Statistics (2011), www.bts.gov/publications/ national_transportation_statistics/html/table_04_25_m.html. Accessed December 2011.

Shipping

Congressional Budget Office (1982), "Energy use in freight transportation," US Congress, Washington, DC, USA.

Henningsen, R.F. (2000), "Study of greenhouse gas emissions from ships," Norwegian Mariine Technology Research Institute (MARINTEK), Trondheim, and the International Maritime Organsation (IMO) London, UK.

International Chamber of Shipping (2005), International Shipping Federation, Annual review 2005, US Department of Transport, Maritime Administration, Pittsburgh, PA. www.marisec.org/. Accessed December 2011.

Shipping Efficiency (2010), www.shippingefficiency.org/. Accessed December 2011. *(An initiative to grade energy and CO_2 of ships, shipping being responsible for about 3% of global CO_2)*

US Department of Transportation Maritime Administration (1994), "Environmental advantages of Inland Barge transportation." www.port .pittsburgh.pa.us/docs/eaibt.pdf. Accessed December 2011.

Fuel mix in electrical energy

Boustead Model 5 (2007), Boustead Consulting, West Sussex, UK. www.boustead-consulting.co.uk. *(An established life-cycle assessment tool)*

ELCD (2008), http://ca.jrc.ec.europa.eu/. *(A high-quality Life Cycle Inventory (LCI). Core data sets of this first version of the Commission's "European Reference Life Cycle Data System (ELCD)")*

International Energy Agency (IEA) (2008), *Electricity Information*, IEA publications. ISBN 978-9264-04252-0. *(An authoritative source of statistical data for the electricity sector. This is one of a series of IEA statistical publications about energy resources)*

Food and drink

Carbon neutral (2011), www.goeco.com.au. Accessed December 2011.

6.8 Exercises

E6.1. What is meant by *embodied energy per kilogram* of a metal? Why does it differ from the free energy of formation of the oxide, carbonate, or sulfide from which it was extracted?

E6.2. *Embodied energies.* Window frames are made from extruded aluminum. It is argued that making them instead from extruded PVC would be more environmentally friendly (meaning that less embodied energy is involved). If the section shape and thickness of the aluminum and the PVC windows are the same, and both are made from virgin material, is the claim justified? You will find embodied energies and densities for the two materials in the data sheets of Chapter 15. Use the mcan values of the ranges given there.

E6.3. *Recycle energies.* The aluminum window frame of Exercise E6.2 is, in reality, made not of virgin aluminum but of 100% recycled aluminum. Recycled PVC is not available, so the PVC window continues to use virgin material. Which frame now has the lower embodied energy?

E6.4. *Recycle energies.* It is found that the quality of the window frame of Exercise E6.3, made from 100% recycled aluminum, is poor because of the impurities that were picked up. It is decided to use aluminum with a "typical" recycled content of 44% instead. The PVC window is still made from virgin material. Which frame now has the lower embodied energy?

E6.5. *Precious metals.* A chemical engineering reactor consists of a stainless steel chamber and associated pipe work weighing 3.5 metric tons, supported on a mild steel frame weighing 800 kg. The chamber contains 20 kg of loosely packed alumina spheres coated with 200 grams of palladium, the catalyst for the reaction. Compare the embodied energies of the components of the reactor, using data from the data sheets of Chapter 15 and from Table 6.2.

E6.6. What is meant by the process *energy per kilogram* for casting a metal? Why does it differ from the latent heat from melting the metal?

E6.7. *Carbon footprint.* Make a bar chart of the CO_2 footprint divided by embodied energy, using data from the data sheets of Chapter 15, for

- Cement
- Low-carbon steel
- Copper
- Aluminum alloys
- Softwood

Which material has the highest ratio? Why?

E6.8. *Sequestered carbon.* The embodied energies and CO_2 footprints for woods, plywood, and paper do not include a credit for the energy and carbon stored in the wood itself, for the reasons explained in the text. Recalculate these, crediting them with sequestering energy and carbon by subtracting out the stored contributions (take them to be 25 MJ/kg and 2.8 kg CO_2 per kg). Is there a net saving?

E6.9. *Embodied energy.* Rank the three common commodity materials, low-carbon steel, aluminum alloy, and polyethylene, by embodied energy/kg, H_m, and embodied energy/m³, $H_m\rho$, where ρ is the density, using data drawn from the data sheets of Chapter 15 (use the means of the ranges given in the databases). Finally rank them by embodied energy per unit stiffness (measured by $H_m\rho/E$ where E is Young's modulus).

E6.10. *Ideal and real embodied energy.* Iron is made by the reduction of iron oxide, Fe_2O_3, with carbon; aluminum is made by the electrochemical reduction of bauxite, basically Al_2O_3. The enthalpy of oxidation of iron is 5.5 MJ/kg; that of aluminum to its oxide is 20.5 MJ/kg. Compare these with the embodied energies of cast iron and of aluminum retrieved from the data sheets of Chapter 15 (use means of the ranges given there). What conclusions do you draw?

E6.11. *Ideal and real process energies.* The melting point of aluminum is 645°C. Its specific heat is $C_p = 810$ J/kg $\cdot$ K, and its latent heat of melting is $L = 390$ kJ/kg. Estimate the theoretical minimum energy needed to melt aluminum and compare this with the casting energy listed under "Eco-properties: processing" in the data sheet for aluminum in Chapter 15. What do you deduce from the comparison?

E6.12. *Eco-indicators.* The data sheets of Chapter 15 list eco-indicator values where these are available. As explained in the text, the eco-indicator value is a normalized, weighted sum involving resource consumption, emissions, and estimates of impact factors. Plot eco-indicator values against embodied energy (a much simpler measure of impact) for cast iron, carbon steel, low-alloy steel, and stainless steel. Is there a correlation?

E6.13. *Process energies.* A bicycle maker manufactures frames from drawn low-carbon steel tubing by gas welding, followed by the application of a baked-on paint coating. One frame weighs 11 kg, the length of weld is 0.4 m, and the surface area of the frame is 0.6 m². Calculate the energy contributions of material and processing (use the data for "Deformation processing" for tube drawing) from the data sheet for mild steel in Chapter 15, and for welding and baked coating from Tables 6.5(c) and (d), and sum them to give a final energy total. Which contribution is the largest?

E6.14. *Transportation energies.* Cast iron scrap is collected in Europe and shipped 19,000 km to China where it is recycled. The energy to recycle cast iron is 5.2 MJ/kg. How much does the transportation stage add to the total energy for recycling by this route? Is it a significant increase?

E6.15. *Transportation energies.* Bicycles, weighing 15 kg, are manufactured in South Korea and shipped to the West Coast of the United States, a distance of 9,000 km. On unloading, they are transported by a 32 metric ton truck to the point of sale, Chicago, a distance of 2,900 km. What is the transportation energy per bicycle?

To meet Christmas demand, a batch of the bicycles is air-freighted from South Korea directly to Chicago, a distance by air of 10,500 km. What is the transportation energy then?

The bikes are made almost entirely out of aluminum. How do these transport energies compare with the total embodied energy of the aluminum of which the bike is made?

E6.16. *Embodied energy.* A range of office furniture includes a chunky hardwood table weighing 25 kg and a much lighter table with a 2.0 kg virgin aluminum frame and a 3.0 kg glass top. Which of the two tables has the lower embodied energy? Use data from the data sheets of Chapter 15 to find out.

E6.17. *Recycle content.* The aluminum-glass table of the previous question is, in fact, made from aluminum with a typical recycled content. How much difference does this make to its embodied energy?

E6.18. *Process energies.* A cast-iron cistern cover is cast to its initial shape and then rough-machined to its final shape, removing 5% of its mass. If the initial casting weighed 16 kg, what, approximately, is the carbon footprint associated with the material and the processing of the cover? Which step contributes the most? Table 6.5(b) and the data sheet for cast iron of Chapter 15 have the necessary data.

E6.19. *Material versus processing energies.* A low-carbon steel car door skin (the outer panel) of area 0.8 m^2 and a thickness 1.2 mm is shaped by deformation processing. It is then given a baked coating on its outer face and assembled using fasteners; it requires 14 of them. Rank the approximate carbon footprint of the material and the process steps. Tables 6.5(c) and (d) and the record for low-carbon steel of Chapter 15 have the necessary data.

E6.20. *Prompt scrap.* The chassis and casing for a high-end portable computer is milled from a solid high-strength aluminum alloy block initially weighing 2 kg. In doing so, 80% of the block is removed by rough machining and a further 5% by finish machining. How does the machining energy compare with the initial embodied energy of the block if made from virgin aluminum? If instead 100% recycled aluminum is used for the block, does the ranking change?

E6.21. *Transportation energies.* You are an Assistant Professor of history at a German university and an active member of the Green Party. You have just been appointed to a Chair of History at the University of Sydney, Australia, 24,000 km away. You have a large library of books, which you cherish. The total weight of the books is 1,200 kg. You could send them by sea freight, but that would take months and there would be a risk of damage. Federal Express could air-freight them and they would get there before you, safe and sound. Weigh up the environmental consequences of these two options in terms of carbon release to the atmosphere. Table 6.9 has the required information.

E6.22. *Precious metals.* You have developed an environmental conscience. You have also met the girl of your dreams whom you are about to marry. The wedding ring you would like to buy for her is 24 carat (100%) gold and weighs 10 grams. She prefers one made of a 50−50 platinum-rhodium alloy weighing 15 grams. How different are the embodied energies of the two rings? (Table 6.2 has data for precious metals.)

Since this is a silly question already, we'll add the light bulb test. With a conversion efficiency of 0.38 for primary energy to electrical energy, for how many extra hours could you run a 100-watt bulb if your girlfriend is willing to settle for the gold ring? (Remember 1 kWh = 3.6 MJ.)

E6.23. *Electronics*. A product line of portable radios has 4 AA Ni-Cd batteries, small-device electronics weighing 100 grams, two speakers with Alnico magnets weighing 700 grams, a transformer weighing 500 grams, and ABS casing weighing 400 grams. Alnico has an embodied energy of 89 MJ/kg. Data for the other components can be found in Table 6.3 and the record for ABS is in Chapter 15. Which of these makes the largest contribution to the embodied energy?

Exercises using the CES software

E6.24. Make a bar chart of the CO_2 footprint divided by the embodied energy using the "Advanced" facility in the CES Level 2 package. Which material has the highest ratio? Why?

E6.25. Figures 6.10 and 6.11 of the text are plots of the embodied energy of materials per kg and per m^3. Use CES to make similar plots for the carbon footprint. Use the "Advanced" facility in the axis selection window to make the one for kg CO_2/m^3 by multiplying kg CO_2/kg by the density in kg/m^3.

E6.26. Plot a bar chart for the embodied energies of metals and compare it with one for polymers, on a "per unit yield strength" basis, using CES. Create the function

$$\frac{H_m\,\rho}{\sigma_y},$$

where H_m is the embodied energy per kg, ρ is the density, and σ_y is the yield strength. Do this by using the "Advanced" facility in the Axis selection box to form

(Embodied energy $*$ Density)/Yield strength (elastic limit)

Which materials are attractive by this measure?

E6.27. Compare the eco-indicator values of materials with their embodied energy. To do so, make a chart with (Embodied energy $\times$ Density) on the x-axis and Eco-indicator value on the y-axis. (Ignore the data for foams since these have an artificially inflated volume.) Is there a correlation between the two? Is it linear? Given that the precision of both could be in error by 10% are they significantly different measures? Does this give a way of estimating, approximately, eco-indicator values where none is available?

E6.28. Use CES to plot material price against annual production in metric tons per year—is there a correlation?

E6.29. Recycling materials at end of life carries a carbon footprint. If they are combustible, like paper or most plastics, there is an alternative—combustion with energy recovery—though it, too, carries a carbon footprint. Create a chart with these two footprints as axes. Select linear scales and plot a line along which the two are equal. Hence identify three materials for which the recycling CO_2 emission is greater than that of combustion.

E6.30. Recycling saves energy—an estimate of the saving is the difference between the embodied energy for primary production and that for recycling. Explore whether more energy is saved in this way than could be recovered if the material were combusted for energy recovery. Create a chart with these two energies as axes. Select linear scales and plot a line along which the two are equal. Hence identify any material for which more energy could be recovered by combustion than by recycling.

CHAPTER 7

Eco-audits and eco-audit tools

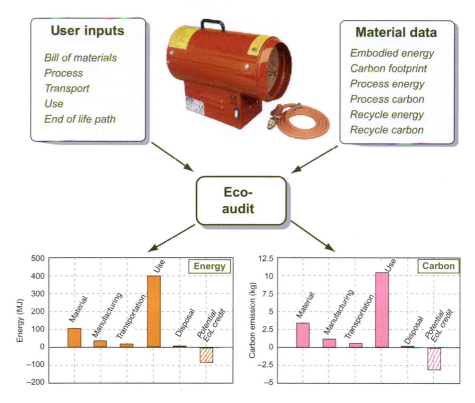

User inputs

Bill of materials
Process
Transport
Use
End of life path

Material data

Embodied energy
Carbon footprint
Process energy
Process carbon
Recycle energy
Recycle carbon

Eco-audit

7.1 Introduction and synopsis

An *eco-audit* is a fast initial assessment of the energy demands of carbon emissions of the life of a product. It identifies the phase of life—material, manufacturing, transportation, use, disposal—that carries the highest demand for energy or creates the greatest burden of CO_2. It points the finger, so to speak, identifying where the problems lie. Often, one phase of life is, in eco-terms, dominant, accounting for 80% or more of the energy and carbon totals. This difference is so large that the imprecision in the data and the ambiguities in the modeling, discussed in Chapter 3, are not an issue; the dominance remains even when the most extreme data values are used. It then makes sense to focus first on this dominant phase, since it is here that the potential gains through innovative material choice are

175

greatest. As we shall see later, material substitution has more complex aspects—there are trade-offs to be considered (Chapters 9 and 10)—but for now we focus on the simple audit.

The main purpose of an eco-audit is one of *comparison*, allowing alternative design choices to be explored rapidly. To do this it is unnecessary to include the last nut and bolt—indeed, with the exception of electronics and precious metals, it is usually enough to account for the few components that make up 95% of the mass of the product, assigning a "proxy" energy and CO_2 to those that are not included directly. The output, of course, is approximate, but if the comparison reveals differences that are large, robust conclusions can be drawn.

This chapter introduces the eco-audit method and the data needed to implement it. The next illustrates its use with a set of case studies. Software packages now exist that make the job easier. One is introduced in the Appendix.

7.2 Eco-audits

Figure 7.1 shows the procedure for the eco-audit of a product. The inputs are of two types. The first step is to assemble a *bill of materials*, a *process choice*, *transportation requirements*, *duty cycle* (the details of the energy and intensity of use), and *disposal route*, shown at the top left. Second, data for embodied energies, process energies, recycle energies, and carbon intensities are drawn from a database of material properties like that in Chapter 15, and those for the energy and carbon intensity of transportation and the use-energy are drawn from look-up tables like those in Chapter 6 (top right of the figure). The *outputs* are the energy or carbon footprint of each phase of life, presented as bar charts and in tabular form. The procedure is best illustrated by a case study of extreme simplicity—that of a PET drink bottle—since this allows the inputs and outputs to be shown in detail.

One brand of bottled water—we will call it Alpure—is sold in 1-liter PET bottles with polypropylene caps (Figure 7.2). One bottle weighs 40 grams; its cap weighs 1 gram. The bottles and caps are molded, filled with water at a source of sparkling purity located in the French Alps, and transported 550 km to London, England, by a 14-metric-ton truck. Once there, they are refrigerated, on average, for 2 days before appearing on the tables of the restaurant where they are consumed, adding significantly to the bill. The restaurant has an environmental policy: all plastic and glass bottles are sent for recycling. We use these data for the case study, taking 100 bottles as the unit of study, requiring approximately 1 m^3 of refrigerated space.

The eco-audit procedure has five steps, described here for energy. An audit for carbon emission follows the same steps.

> ***Step 1, materials.*** A bill of materials is drawn up, listing the mass of each
> component used in the product and the material of which it is made, as
> on the left of Table 7.1. Data for the embodied energy (MJ/kg) and $CO_{2,eq}$
> (kg/kg) per unit mass for each material is retrieved from the database—here,

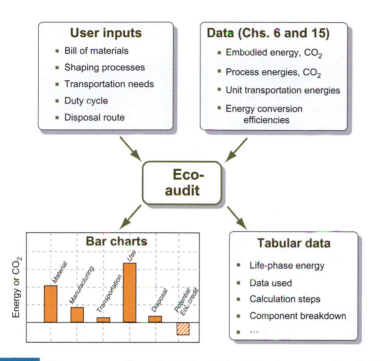

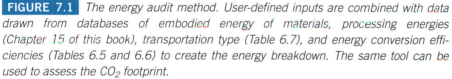

FIGURE 7.1 *The energy audit method. User-defined inputs are combined with data drawn from databases of embodied energy of materials, processing energies (Chapter 15 of this book), transportation type (Table 6.7), and energy conversion efficiencies (Tables 6.5 and 6.6) to create the energy breakdown. The same tool can be used to assess the CO_2 footprint.*

the data sheets of Chapter 15 of this book, using the averages of the ranges listed there (Table 7.2). Multiplying the mass of each component by its embodied energy and summing give the total material energy—the first bar of the bar chart and the first line of Table 7.3. The $CO_{2,eq}$ emission is calculated in a similar way.

Step 2, manufacturing. The audit focuses on primary shaping processes since they are generally the most energy-intensive steps of manufacturing. These are listed against each material, as in Table 7.1. The process energies per unit mass are retrieved from the data sheets of Chapter 15 (Table 7.2). Multiplying the mass of each component by its primary shaping energy and summing give an estimate of the total processing energy—the second bar of the bar chart of Figure 7.2 and the second row of Table 7.3. A similar calculation gives the CO_2 emissions.

On a first appraisal of the product, it is frequently sufficient to enter data for the components with the greatest mass, accounting for perhaps 95% of the total. The residue is included by adding an entry for "residual components"

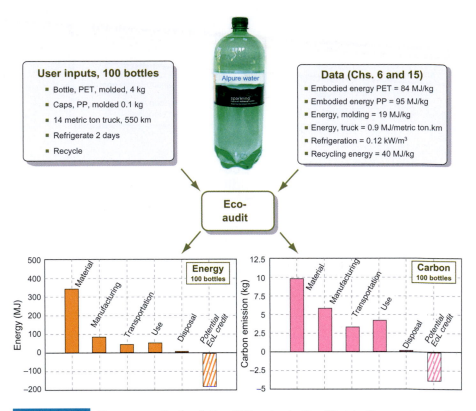

FIGURE 7.2 *The eco-audit of a 1-liter PET water bottle, filled in France, trucked to England, and refrigerated for 2 days*

Table 7.1	The bill of materials and processes for 100 bottles		
Component	**Material**	**Process**	**Mass, *m* (kg)**
Bottle	PET	Polymer molding	4.0
Cap	PP	Polymer molding	0.1
Dead weight (1 liter of water)	Water		100
		Total mass	104.1

giving it the mass required to bring the total to 100% and selecting a proxy material and process: "polycarbonate" and "molding" are good choices because their energies and CO_2 lie in the mid range of those for commodity materials.

Table 7.2	Data from the data sheets of Chapter 15 (mean values of ranges)			
	Material		**Process**	
Material	Energy, H_m (MJ/kg)*	CO_2 (kg/kg)*	Energy, H_p (MJ/kg)*	CO_2 (kg/kg)*
PET	85	2.35	19.5	1.48
PP	74	3.1	21	1.6

*From the data sheets of Chapter 15

Table 7.3	PET bottle, energy, and CO_2 summary, 100 units			
Life phase	Energy (MJ)	Energy (%)	CO_2 (kg)	CO_2 (%)
Material	348	66	9.6	43
Manufacture	80	15	6.1	27
Transport	51	9	3.5	16
Use	62	10	3.0	14
Disposal	0.8	0	0.1	0
Totals, first life	524	100	22	100
Potential EoL credit	−188		−0.2	

Step 3, transportation. This step estimates the energy for transportation of the product from the manufacturing site to point of sale. For the water bottle, this is dominated by the transportation of the filled bottles from the French Alps to London, a distance of 550 km. The energy demands of transportation modes were described in Chapter 6 (Table 6.8); that for the 14-metric-ton truck used in this example is 0.9 MJ/metric ton · km. Multiplying this by the mass of the product and the distance travelled provides the estimate. It is not just the bottles that travel 550 km but also the water they contain. This is included in the bill of materials shown in Table 7.1 to ensure that its mass is included in the auditing of the transportation phase. The result is a transportation energy of 0.49 MJ per bottle, shown as the third bar of the bar chart and row of Table 7.3.

Step 4, the use phase. The use phase requires a little explanation. There are two different classes of contributions.

■ Some products are (normally) static but require energy to perform their function: electrically powered products like hairdryers, electric kettles, refrigerators, power tools, and space heaters are examples. Even apparently nonpowered products like household furnishings or unheated buildings still

consume some energy in cleaning, lighting, and maintenance. The first
class of contribution, then, relates to the power consumed by, or on behalf
of, the product itself.

- The second class is associated with transportation. Products that form
  part of, or are carried by, a transportation system add to its mass and
  thereby augment its energy consumption and CO_2 burden. The
  transportation table, Table 6.8, lists the energy and CO_2 penalty per unit
  weight and distance. Multiplying this by the product weight and the
  distance over which it is carried gives an estimate of the associated use-
  phase energy and CO_2.

All energies are related back to primary energy, meaning oil, via oil-equivalent
factors for energy conversion discussed earlier (Tables 6.6 and 6.7).
Retrieving these and multiplying by the power and the duty cycle give an
estimate of the oil-equivalent energy of use.

The PET bottle is a static product. Energy is consumed on its behalf via
refrigeration for 2 days. The power requirements for refrigeration, based on
A-rated appliances, are 0.12 kW/m^3 for refrigeration at 4°C and 0.15 kW/m^3
for freezing at −5°C, using electrical power in both cases. Thus refrigerating
1 m^3 for 2 days takes 5.76 kWh. Converting this to MJ/m^3 by multiplying
by 3.6, and to an oil-equivalent energy by dividing by an energy-conversion
efficiency of 33%, gives 62 MJ/m^3. One bottle requires about 0.01 m^3 of
refrigerator space, so the energy to refrigerate it for 2 days is a surprising
0.6 MJ per bottle—more than the energy to truck it from France.

Step 5, disposal. There are five options for disposal at the end of life: *landfill,
combustion for energy recovery, recycling, re-engineering,* and *reuse* (look
back at Figure 4.2). A product at end of first life has the ability to return
part or all of its embodied energy. This at first sounds wrong—much of the
"embodied" energy was not *embodied* at all but was lost as low-grade heat
via the inefficiencies of the processing plant, and even when it is still there,
it is, for metals and ceramics, inaccessible since the only easy way to
recover energy directly is by combustion, not an option for steel, concrete,
or brick. But think of it another way. If the materials of the product are
recycled or the product itself is re-engineered or reused, a need is filled
without drawing on virgin material, thereby saving energy. Carbon release
works in the same way, with one little twist: one end-of-life option,
combustion, recovers some energy but in doing so it releases CO_2.

Table 7.4 lists the path for each option and first-order estimates for the
associated energies and carbon emission. These allow the approximate
evaluation of end-of-life choice in the case studies and exercises that
follow. The energy cost of transportation to landfill is small compared
with the other energies associated with life. Recycling recovers the
difference between the original embodied energy and the energy of recycling.
Re-engineering and reuse retain almost all the original embodied energy.

Table 7.4	Disposal route and associated energy and carbon emission*	
Disposal route	**First-order estimate for energy (MJ/kg)**	**First-order estimate for carbon (kg/kg)**
1. **Landfill.** Collect and transport to landfill site.	EoL debit $H_d \approx 0.1$	EoL debit $C_d \approx 0.01$
2. **Combust for heat recovery.** Collect, combust, recover heat.	EoL credit $\eta_c H_c$ (η_c = combustion efficiency = 0.25 H_c = heat of combustion)	EoL debit $C_c = \alpha H_c$ (C_c = combustion carbon $\alpha = 0.07$ kg CO_2 /MJ)
3. **Recycle.** Collect, sort by material family and class, recycle.	$\tilde{H} + (1-r)H_d$ EoL credit $r(\tilde{H} - H_{rc})$	$\tilde{C} + (1-r)C_d$ EoL credit $r(\tilde{C} - C_{rc})$
4. **Re-engineer.** Collect, dismantle, replace or upgrade components, re-assemble.	Recover most of embodied energy H_m. Use $0.9\tilde{H}$ as EoL credit.	Recover most of carbon footprint C_m. Use $0.9\tilde{C}$ as EoL credit.
5. **Reuse.** Market as "pre-owned" product via trading outlets, websites etc.	Recover mean embodied energy $\tilde{H}$ as EoL credit	Recover mean carbon footprint $\tilde{C}$ as EoL credit

*We define the effective embodied energy $\tilde{H}$ and carbon footprint $\tilde{C}$ as

$$\tilde{H} = RH_{rc} + (1-R)H_m \text{ MJ/kg} \quad \text{and} \quad \tilde{C} = RC_{rc} + (1-R)C_m \text{ kg/kg}$$

where H_m is the embodied energy and C_m the carbon footprint of virgin material (the values in the data sheets of Chapter 15), H_{rc} and C_{rc} are the corresponding values for recycled material and the recycle content of the material at start of life is R.

Table 7.5	Recycle energy and CO_2 and heats of combustion for PET and PP			
Material	**Embodied energy H_m MJ/kg***	**Material $CO_{2,eq}$ C_m kg/kg***	**Recycle energy H_{rc} MJ/kg***	**Recycle $CO_{2,eq}$ C_{rc} kg/kg***
PET	85	2.35	39	2.3
PP	74	3.1	50	2.1

*From the data sheets of Chapter 15.

The data sheets of Chapter 15 provide estimates for recycled energy H_{rc} and carbon release C_{rc} for recyclable materials, and the heat of combustion H_c and associated carbon C_c for those that can be burned. Recovered energy and carbon credits appear as negatives on the eco-audit bar charts.

The bottle was originally made of virgin PET and PP. Both are recycled at end of life. The potential end-of-life (EoL) energy-credit is then $r(H_m - H_{rc})$, where r is the fraction recycled at EoL, which we take to be 100%. Drawing data from the data sheets (Table 7.5) then gives the EoL credits shown in the bottom row of Table 7.3.

But is it valid to include this credit in the total life energy? To answer that we have to look at the energy accounting in a little more detail.

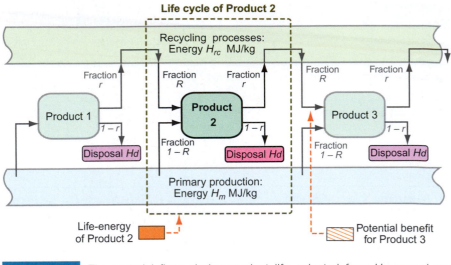

FIGURE 7.3 *The material flows during product life, adapted from Hammond and Jones (2011)*

Assigning recycle credits. Recycling passes material from one life cycle to the next. In general, it takes less energy and releases less carbon to recycle a unit of material than it takes to create the same quantity of virgin material from ores and feedstock—it is this that makes recycling attractive. But is the saved energy and CO_2 to be credited to the first life cycle or the second? It can't be credited to both, since that would be to count it twice.[1]

Figure 7.3 sets the scene. It shows the material flow through three successive product lives, labeled Product 1, 2, and 3. Part of the material used to make Product 1 is recycled at the end of life and becomes available to make Product 2, which is recycled, in turn providing part of the material for Product 3. Focus on Product 2. At the start of its life it draws on material created by the recycling of Product 1, supplemented by primary (virgin) material created from ores and feedstock. Product 2 enters service and performs its function. At the end of its life, part of its material is recycled and passes to Product 3, where the cycle repeats itself. If Product 2 is awarded the credit for receiving recycled material from Product 1 it can't also have the credit for delivering it to Product 3 because that belongs to Product 3. If instead Product 1 claims the credit for creating the recyclable material used by Product 2, then Product 2 loses its start-of-life credit but can now claim one at end-of-life.

---

[1]Hammond and Jones (2010) offer a helpful explanation of the alternatives and their relative merits, paraphrased here.

There are two methods for dealing with this allocation problem. The *recycled content method* gives Product 2 the full credit for recycled content at the start of life but none for recycling at life's end. We model it here for energy; the modeling for emissions follows the same path. Product 2 uses a fraction R of recycled materials with an embodied energy H_{rc} per kg supplemented by a fraction $(1 - R)$ of virgin materials with embodied energy H_m per kg. At the end of life a fraction r is (or could be) recycled, but Product 2 gets no credit for that. It must, however, accept a small penalty for the fraction $(1 - r)$ that must be disposed of at end of life, costing a disposal energy H_d per kg. The net embodied energy H_{net} per kg of material is then

$$(H_{net})_{recycle\ content} = R\,H_{rc} + (1 - R)H_m + (1 - r)H_d \qquad (7.1)$$

The alternative (the *substitution method*) is to award the full benefit of recycling at the end of life, $r(H_m - H_{rc})$, to Product 2 but give it no credit for recycled content at the start of life, giving the net embodied energy H_{net} per kg of material as

$$(H_{net})_{substitution} = H_m - r\,(H_m - H_{rc}) + (1 - r)H_d = rH_r + (1 - r)(H_m + H_d) \qquad (7.2)$$

Equations (7.1) and (7.2) become identical when $R = r$, that is, when the recycled fraction entering at the start of life is the same as that exiting at the end, r. But if the two fractions differ, so too do the two energies E_1 and E_2.

Recycling at end of life is a future benefit, one that may not be realized for many years or, indeed, at all. If the concern is for *present* resources, energy demands, and climate-changing emissions, then it does not make sense to use the substitution method. We therefore adopt the recycled content method in calculating the life-cycle energy and carbon emissions (equation 7.1). It deals with the present, not the future, it avoids double counting, and it conforms to the European guidelines on assessing the carbon footprint known as PAS 2050 and BSI, 2008 (Chapter 5).

But this choice leaves us with a difficulty. One purpose of an eco-audit is to guide *design* decisions. Designers who strive to design products using recycled materials will wish the eco-audit to reflect this, as the recycled content method does. On the other hand, designers who strive to make disassembly easy and to use materials that recycle well would also want the audit to reflect that, and the recycled content method fails to do so. To overcome this, we show bars for the energy and carbon contributions to the first life as bars of solid colors and add the potential energy and carbon saving (or penalty) arising from the EoL choice as a separate, cross-hatched bar, as in Figure 7.4. Thus if a fraction r of a product is recycled at end of life, the potential EoL credit is $H_{EoL} = r(\tilde{H} - H_{rc})$.

The outputs. The bar charts of Figure 7.4 show the final eco-audit for the PET bottles. What do we learn? The largest contribution to energy consumption and CO_2

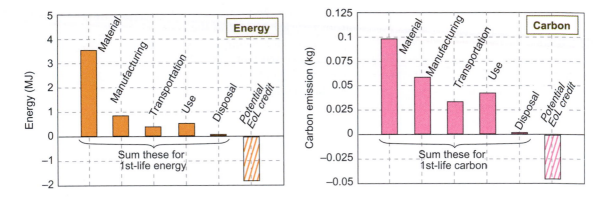

FIGURE 7.4 *The way of displaying end-of-life (EoL) data for energy and carbon for the PET bottle based on the "recycle fraction" method*

generation derives from the production of the polymers used to make the bottle. The second largest is that of the short, two-day refrigeration. The seemingly extravagant part of the life cycle—that of transporting water, 1 kg per bottle, 550 km from the French Alps to the diners' table in London—contributes 10% of the total energy and 17% of the total carbon. If genuine concern is felt about the eco-impact of drinking Alpure water, then (short of giving it up) it is the bottle that is the primary target. Could it be made thinner, using less PET? (Such bottles are 30% lighter today than they were 15 years ago.) Is there a polymer that is less energy intensive than PET? Could the bottles be reusable and of sufficiently attractive design that people would wish to reuse them for other purposes? Could recycling of the bottles be made easier? These are design questions, the focus of the lower part of Figure 3.13. Methods for approaching them are detailed in Chapters 9 and 10.

An overall reassessment of the eco-impact of the bottles should, of course, explore ways of reducing energy and carbon in all phases of life, not just one, seeking the most efficient molding methods, the least energy intensive transport mode (32-metric-ton truck, barge), and—an obvious step—minimizing the refrigeration time.

7.3 Computer-aided eco-auditing

An eco-audit guides decision making during design or redesign of a product and it points to the aspects of design that a fuller life-cycle assessment (LCA) should examine. Early in the design process the detailed information required for a rigorous LCA is not available, and even if it were, the formal LCA methods are sufficiently burdensome to inhibit their repeated use for rapid "what if?" explorations of alternatives. As we have seen, an eco-audit, though approximate, frequently reveals differences that are sufficiently large to be significant. LCA tools such as SimaPro,

GaBi, MEEUP, and the Boustead Model (see Appendix to Chapter 3) can be used in this way, but they were not designed for it. Others are much simpler to use. One—the CES eco-audit tool[2]—is described in the Appendix to this chapter. It implements the procedure shown in Figure 7.1, enabling the inputs and delivering the outputs shown there.

7.4 Summary and conclusions

Eco-aware product design has many aspects, one of which is the choice of materials and manufacturing route. Materials are energy intensive—they have high embodied energies and associated carbon footprints. Seeking to use low-energy materials might appear to be one way forward, but this can be misleading. Material choice influences the choice of the manufacturing process, it influences the weight of the product and its mechanical, thermal, and electrical characteristics and thus the energy it consumes during use, and it influences the potential for recycling or energy recovery at the end of life. It is the full-life energy that we should seek to minimize.

Doing so requires a two-part strategy developed in Chapter 3 and Figure 3.14. The first part is an *eco-audit*: a quick, approximate assessment of the distribution of energy demand and carbon emission over life. This provides inputs to guide the second part: that of *material selection to minimize the energy and carbon over the full life*, balancing the influences of the choice over each phase of life—the subject of Chapters 9 and 10. The eco-audit method described here is fast and easy to perform, and although approximate, it delivers information with sufficient precision to enable strategic decision making. This chapter has introduced the procedure and tools; the next illustrates their use with case studies. The Exercises at the end of both chapters provide opportunities for exploring the methods further.

7.5 Further reading

GaBi 4 (2008), PE International, www.gabi-software.com/. *(An LCA tool that complies with European legislation. It has facilities for analyzing cost, environment, social, and technical criteria and optimization of processes.)*

Hammond, G. and Jones, C. (2010), "Inventory of carbon and energy (ICE), Annex A: methodologies for recycling," The University of Bath, Bath, UK. *(A well-documented compilation of embodied energy and carbon data for building materials, with appendices explaining the alternative ways of assigning recycling credits between first and second lives)*

ISO 14040 (1998), Environmental management—Life-cycle assessment—Principles and framework.

---

[2]The CES Eco Audit tool is a standard part of the CES Edu package developed by Granta Design, Cambridge, UK (www.Grantadesign.com).

ISO 14041 (1998), Goal and scope definition and inventory analysis.

ISO 14042 (2000), Life-cycle impact assessment.

ISO 14043 (2000), Life-cycle interpretation, International Organization for Standardization, Geneva, Switzerland. *(The set of standards defining procedures for life cycle assessment and its interpretation)*

Matthews, B. (2011), "Java Climate Model," with UCL Louvain-la-neuve, KUP Bern, DEA Copenhagen, UNEP/GRID Arendal, http://chooseclimate.org. Accessed December 2011. *(An interactive tool to explore the effects of greenhouse gasses on global temperature and sea level)*

MEEUP method (2005), VHK, Delft, Netherlands, www.pre.nl/EUP/. *(The Dutch Methodology for Ecodesign of Energy-using Products (MEEUP) is a response to the EU directive on Energy Using Products (the EuP Directive) described in Chapter 5. It is a tool for the analysis of products—mostly appliances—that use energy, following the ISO 14040 series of guidelines.)*

PAS 2050 (2008), "Specification for the assignment of the life-cycle greenhouse gas emissions of goods and services," ICS code 13.020.40, British Standards Institution, London, UK. ISBN 978-0-580-50978-0. *(A proposed European Publicly Available Specification (PAS) for assessing the carbon footprint of products)*

SimaPro (2008), Pré Consultants, www.pre.nl. *(An LCA tool for analyzing products following the ISO 14040 series standards)*

The Nature Conservancy (2011), "The carbon footprint calculator," www.nature .org/initiatives/climatechange/calculator/. Accessed December 2011. *(A little online tool to estimate your personal carbon footprint)*

7.6 Appendix: eco-audit tools

Eco-auditing is made faster and simpler by eco-audit software tools. One, the CES Edu Eco-audit package, is described here. The double spread of Figure 7.5 is a mock-up of the user interface. It shows the user actions and the consequences. There are four steps, labelled 1 to 4. Actions and inputs are shown on the left in red.

> *Step 1, material, manufacture* allows entry of the bill of materials and processes: the *material*, the *primary shaping process*, the *mass*, the *process*, and an *end-of-life choice* for each component. The component name is entered in the first box. The material is chosen from the pull-down menu of box 2, opening the CES database of materials attributes.[3] Selecting a material from the tree-like hierarchy of materials causes the tool to retrieve and store its embodied energy and CO_2 footprint per kg. The component mass in kg is then entered in box 3. Box 4 allows a choice of the primary

---

[3]One of the CES Edu Materials databases depending on which was chosen when CES was opened.

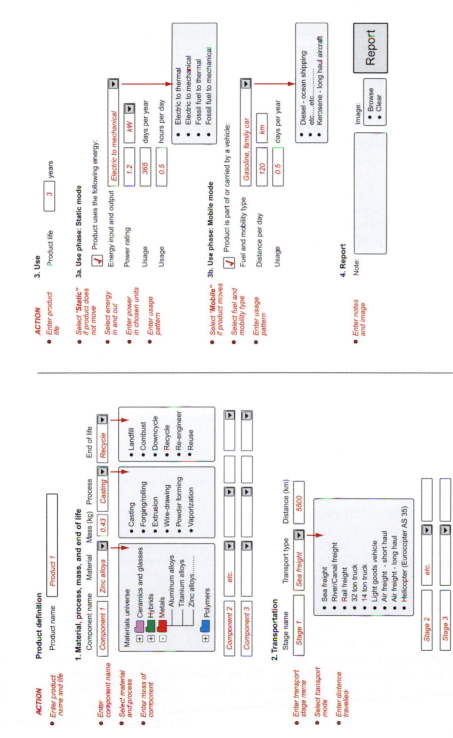

FIGURE 7.5 *Eco-audit tool*

shaping process from the pull-down menu of processes that are compatible with the material choice, again retrieving energy and carbon footprint per kg. The last box allows the choice of end of life. On completing a row-entry, a new row appears for the next component.

As explained earlier, for a first appraisal of the product it is frequently sufficient to enter data for the components with the greatest mass, accounting for perhaps 95% of the total. The residue is included by adding an entry for "residual components" giving it the mass required to bring the total to 100% and selecting a proxy material and process: "polycarbonate" and "molding" are good choices because their energies and CO_2 lie in the mid range of those for commodity materials. Dead weight (weight that is not part of the product but must be transported with it—like the water in the example of this chapter) is entered without choosing a material or process.

The tool multiplies the energy and CO_2 per kg of each component by its mass and totals them.

Step 2, Transportation allows for transportation of the product from the manufacturing site to the point of sale. The tool allows multistage transportation (e.g., shipping followed by delivery by truck). To use it, the stage is given a name, a transportation type is selected from the pull-down "Transportation type" menu, and a distance is entered in km or miles. The tool retrieves the energy/metric ton · km and the CO_2/metric ton · km for the chosen transportation type from a look-up table like that of Table 6.8 and multiplies them by the product mass and the distance travelled, finally summing the stages.

Step 3, the use phase allows entry of the information to evaluate the two contributions that were described in Section 7.2—the static mode contribution, Step 3(a), and the mobile contribution, Step 3(b).

Clicking the Static mode check box activates part 3(a). Selecting an energy conversion mode causes the tool to retrieve the efficiency (from a look-up table like that of Table 2.1) and to multiply it by the power and the duty cycle, entered here as *use-days per year* times *hours per day* times *product life* in years.

As explained earlier, products that are part of a transportation system carry an additional energy and CO_2 penalty by contributing to the system weight. Clicking the Mobile mode check box enables part 3(b). A pull-down menu allows selection of the type of transportation, listed by fuel and mobility type. Once a daily distance is entered the tool calculates the energy and CO_2 by multiplying product weight and distance carried by the energy or CO_2 per metric ton · km drawn from the same look-up table as that for the transportation phase.

Step 4, the final step, allows the user to enter notes to annotate the audit and to insert an image of the product as a JPEG or BMP file that will then appear at the head of the Report. Clicking on Report then completes the calculations, presenting the results in tabular form and as bar charts like those of Section 7.3 and in tabular form. A summary table presents the final result in the way indicated in Figure 7.4: as a total first-life energy and carbon emission, together with a potential end-of-life credit for both.

An advanced version of the Eco-audit tool allows secondary processing steps (machining, joining, finishing) to be included in the life-cycle audit.

7.7 Exercises

E7.1. If the embodied energies and CO_2 used in the Alpure water case study in the chapter are uncertain by a factor of $\pm 20\%$, do the conclusions change? Mark approximate $\pm 20\%$ error margins onto the top of each bar in a copy of the audit for the Alpure bottle, Figure 7.2 (you are free to copy it), and then state your case.

E7.2. Alpure water has proved to be popular. The importers now wish to move up-market. To do so they plan to market their water in 1-liter glass bottles of an appealing design instead of the rather ordinary PET bottles with which we are familiar. A single 1-liter glass bottle weighs 430 grams, much more than the 40 grams of those made of PET. Critics argue that this marketing strategy is irresponsible because of the increased weight. The importers respond that glass has a lower embodied energy than PET.

Use the methods of this chapter and the data available in Chapter 15 for glass and PET to compare the embodied energy of 100 1-liter glass bottles with that of 100 PET bottles of the same capacity, calculated in the text as 348 MJ. What do you conclude? Does the conclusion change if the glass bottle is made from recycled glass?

E7.3. Health and safety regulations prevent the use of recycled PET for bottling Alpure water, resulting in a surplus of waste PET bottles. The economics of recycling are unattractive; it is decided instead to burn the bottles, using the recovered heat to generate electricity.

(a) If the energy conversion efficiency from heat to useful electric power is 0.33, how much useful energy is recovered and how much CO_2 is released by the combustion of 100 PET bottles weighing, in total, 4.0 kg?

(b) Compare this with the end-of-life credit for 100 PET bottles if they are, instead, recycled. The table lists the data you will need.

Virgin PET		Recycled PET		Combustion of PET	
Embodied energy (MJ/kg)	CO_2 footprint (kg/kg)	Embodied energy (MJ/kg)	CO_2 footprint (kg/kg)	Combustion heat (MJ/kg)	Combustion CO_2 (kg/kg)
85	2.35	39	2.3	23.5	2.35

E7.4. A top London restaurant, we shall call it the *Extravaganza*, is impatient to receive its consignment of Alpure water and decides to have it air-freighted from France rather than wait for delivery by the 14-metric-ton truck. Compare the transportation energy for 100 bottles, with a total mass 104 kg including the water, over 550 km, by 14-metric-ton truck and by short-haul air freight. The relevant data are contained in Table 6.8. Is the embodied energy of the PET still the largest contribution to life energy?

E7.5. Rolled steel I-beams used in the construction of commercial buildings have a recycle content of 45%. The design life of commercial buildings is 60–100 years, though most are demolished after a shorter time. On demolition today, 80% of the steel is recycled. Take the embodied energy of primary steel to be $H_m = 32$ MJ/kg and that for recycled steel to be $H_{rc} = 9$ MJ/kg.

(a) What are the material life energies using the recycle fraction method and using the substitution methods (equations (7.1) and (7.2) of the text)?

(b) Report instead the material life energy using the recycle fraction method and the potential EoL credit of the steel if it were recycled at today's recycle energy and a recovery rate of 80%.

E7.6. Wine is sold in glass bottles. The average recycle content in a wine bottle is 38%. The overall recycling rate for wine bottles at end of life is 42%. Take the embodied energy of primary glass to be $H_m = 15$ MJ/kg and that for recycled glass to be $H_{rc} = 7$ MJ/kg.

(a) What are the material life energies using the recycle fraction method and using the substitution method?

(b) Report instead the material life energy using the recycle fraction method and the potential EoL credit of the glass of the bottle if it were recycled at today's recycle fraction and energy.

E7.7. Bottled water is sold in PET bottles. The overall recycling rate for PET at end of life is 22%. One maker of bottled water proposes to use 50% of recycled PET in their bottles. Retrieve the carbon footprint of primary PET, CO_2, and that for recycled PET, $CO_{2,rc}$, from the data sheet in Chapter 15. Use them to estimate the material life carbon footprint using the recycle fraction method and the Future EoL carbon credit of the PET of the bottles if 22% of them were recycled.

Exercises using the CES Edu software

E7.8. Create an eco-audit for the scenario described in the text: 100 blow-molded 1-liter PET bottles, each weighing 40 g when empty, filled with water in France, and transported 550 km to the United Kingdom by a 14-metric-ton truck. Select virgin PET, assign the manufacturing process to "Polymer molding," and select "Recycle" from the end-of-life options. Compare the output bar charts with those of Figure 7.2 to check that everything is entered correctly.

(a) What is the material contribution to the CO_2 footprint of the life cycle?

(b) Change the end-of-life choice to "Combust." How much CO_2 is released to the atmosphere by this end-of-life choice?

(c) If the filled PET bottles are transported by short-haul air transport rather than a 14-metric-ton truck, how much more carbon is released to the atmosphere?

(d) Change the material of the bottles from PET to soda glass, each bottle now weighing 430 grams instead of 40 grams. What is now the material contribution to the CO_2 footprint of the life cycle?

CHAPTER 8

Case studies: eco-audits

8.1 Introduction and synopsis

Here are 12 diverse case studies using the methods of Chapter 7. Each describes a product, lists the bill of materials and processes, and suggests a transportation route, a duty cycle, and an end-of-life choice. They illustrate well the features of eco-audits. They are fast and allow "what if?" exploration. But they are also approximate. Are they so approximate that they are meaningless? Comparison of eco-audits with more exhaustive analyses using life-cycle assessment (LCA) methods, described here, show that eco-audits capture the important features of the energy and carbon emissions of a life cycle, or at least part of a life cycle. They provide the tool we need to perform the first part of the strategy suggested in Chapter 3 and Figure 3.14.

8.2 Reusable and disposable cups

Disposable cups seem like an engine of waste—fill them once and then throw them away. Would not reusable cups (Figure 8.1) be more eco-benign? The answer is not as simple as it might seem. Reusable cups are more energy-intensive to make and

FIGURE 8.1 *A disposable polystyrene cup and a reusable polycarbonate cup*

Table 8.1	Material, manufacturing, and use of 1,000 cups		
Attribute		**Disposable**	**Reusable**
Material		Polypropylene	Polycarbonate
Mass of 1,000 cups (kg)		16	113
Shaping process		Molding	Molding
Mass of cardboard packaging for 1,000 cups (kg)		0.6	2.3
Electrical energy for single wash of 1,000 cups (kWh)		—	20

they have to be collected and washed in ways that meet health standards. They become a more energy-efficient choice only if reused a sufficient number of times.

Here we compare the energy profiles of 0.33 liter disposable cups with reusable plastic cups of the same volume, calculating the break-even number. We take as the functional unit 1,000 cups plus the associated cardboard packaging. The disposable cups are made of polypropylene, the reusable ones of polycarbonate. Tables 8.1 and 8.2 list the attributes of the cups and of the virgin materials from which they are made (health requirements preclude the use of recycled materials), taking the data from Chapter 15. Transport requirements are negligible. At end of life all cups are sent to the landfill. Washing 1,000 cups once requires 20 kWh of electrical energy. To convert this to $MJ_{oil\ equivalent}$ we first multiply by 3.6 to convert kWh to MJ and then divide by the conversion efficiency from oil to electricity, which we set at the European average of 0.38, giving a final value of $E_{wash} = 195\ MJ_{oe}$.

Using these data, the energy of material and manufacturing of 1,000 disposable cups plus cardboard packaging is

$$E_{disposable} = 16 \times (95 + 21) + 0.6 \times 28 = 1,873\ \text{MJ}$$

That for 1,000 reusable cups is

$$E_{reusable} = 113 \times (110 + 18.5) + 2.3 \times 28 = 14,580\ \text{MJ}$$

Table 8.2	Eco-attributes of the materials		
Attribute	**Polypropylene**	**Polycarbonate**	**Cardboard**
Embodied energy, virgin material (MJ/kg)*	95	110	28
Carbon footprint, virgin material $CO_{2,equiv}$ (kg/kg)*	2.7	5.6	1.4
Molding energy (MJ/kg)*	21	18.5	—
Molding carbon footprint, $CO_{2,equiv}$ (kg/kg)*	1.6	1.4	—
Oil-equiv. energy for single wash, 1,000 cups (MJ)	—	195	—

From the data sheets of Chapter 15

Suppose the reusable cups are used, on average, n times. Then they become more energy efficient when

$$n\, E_{disposable} > E_{reusable} + (n-1)E_{wash}$$

Inserting the data and solving for n reveal that reusable cups are more energy efficient only if reused at least 15 times.

This might be an achievable target in a restaurant, but for outdoor events it is totally unrealistic. Garrido and del Castillo, investigating a major outdoor event at the Barcelona Universal Forum of Culture in 2004, report the gloomy statistic that only 20% of reusable cups were actually returned. The rest ended up in trashcans or simply disappeared.

Further reading

Garrido, N. and del Castillo, M.D.A. (2007), "Environmental evaluation of single-use and reusable cups," *Int. J. LCA*, 12, pp. 252–256.

Imhoff, D. (2005), *Paper or plastic: searching for solutions to an overpackaged world*, University of California Press, Berkeley, CA. ISBN-13: 978-1578051175. *(What the title says: a study of packaging taking a critical stance)*

8.3 Grocery bags

Few products get a worse press than plastic grocery bags. They are distributed free, and in vast numbers. They are made from oil. They don't degrade. They litter the countryside, snare water birds, and choke turtles. Add your own gripe.

Paper bags are made from natural materials, and they biodegrade. Surely it's better to use paper? And come to think of it, why not use bags made out of jute—it's a renewable resource—and use them over and over? That must be the best of all?

A lot of questions. Let's see what answers the data sheets of Chapter 15 can give, using CO_2 footprint as the measure of goodness or badness. First we must look at some real bags (Figure 8.2 and Table 8.3). The function of an eco-bag is "low-carbon containment," but there is more to it than that. Bag 1 is a typical one-use supermarket container. It is made of polyethylene (PE) and it weighs just 7 grams. Bag 2 is also PE but it is 3 times heavier and the designer graphics tell you something else: the bag is a statement of the cultural and intellectual standing of the store from which it comes (it is a bookshop). It is attractive and strong, too good to throw away, at least not straight away.

FIGURE 8.2   *Carrier bags. The lightest weighs 7 grams, the heaviest weighs 257 grams.*

Table 8.3   The characteristics of the carrier bags shown in Figure 8.2

Bag	Material	Mass (g)	Material CO_2 footprint (MJ/kg)*	CO_2 footprint, 100 bags (kg)	How many reuses?
1	Polyethylene (PE)	7	2.1	1.5	1
2	Polyethylene (PE)	20	2.1	4.3	3
3	Paper	46	1.4	6.4	5
4	Paper	54	1.4	7.6	6
5	Polypropylene (PP)	75	2.7	20.3	14
6	Juco-75% jute 25% cotton	257	1.1	28.3	19

*From the data sheets of Chapter 15.

Bags 3 and 4 are made of paper. Paper bags suggest a concern for the environment, a deliberate avoidance of plastic, good for company image. But there is more mass of material here—about seven times more than that of Bag 1.

And finally, reusable bags—"Bags for life" as one supermarket calls them. Bag 5 is an example. It is robust and durable and looks and feels as if it is made from a woven fabric, but it's not—it's a textured polypropylene sheet. The color, the "Saving Australia" logo, and the sense that it really *is* green propelled this bag into near-universal popularity there. Here is one Aussi paper: "Forget the little black dress. The hot new item around town is the little green bag."

But isn't murmuring "Green" and "Save the planet" a little bit, well, yesterday? Today is Bag 6. Discrete, understated, almost—but not quite—unnoticeable. Those who have one have the quiet satisfaction of knowing that it is made of *Juco*, a mix of 75% jute and 25% cotton. But it uses a great deal of material—36 times more than Bag 1.

So you see the difficulty. We have wandered here into a world that is not just about containment but also about self-image and company branding. Our interest here is eco-analysis, not psychoanalysis. So consider the following question. If the 7 gram plastic bags are really used only once, how many times do you have to use the others to do better in eco-terms? The data sheets of Chapter 15 help here. The fourth column of Table 8.3 lists their values for the carbon footprints of PE, PP, paper, and jute (that for jute includes spinning and weaving). If you multiply these values by the masses, taking 100 bags as the unit of study, you get the numbers in the fifth column. Divide these by the value for the single-use bag, and you get the number of times the others must be used to provide containment at lower carbon per use than Bag 1—last column.

Now you must make your own judgment. Would you re-use a paper bag six or more times? Unlikely—they tear easily and get soggy when wet. If you don't, Bag 1 wins. Would you use the green Bag 5 more than fourteen times? I have one and it has already been used more than that, so it looks like a winner. Finally Bag 6, the thinking person's eco-bag, is less good than plastic until you've used it 19 times. Not impossible, provided nothing leaks or breaks inside it, causing terminal contamination.

So from a carbon and energy point of view, single-use bags are not necessarily bad—it depends how meticulous you are about reusing any of the others. The real problem with plastic is its negligible value (so people discard it without a thought) and its long life, causing it to accumulate on land, in rivers and lakes, and in the sea where it disfigures the countryside and harms wildlife.

Further reading

Edwards, C. and Fry, J.M. (2011), "Life-cycle assessment of supermarket carrier bags," Report: SC030148, The Environment Agency, Bristol, UK. www.environment-agency.gov.uk/static/documents/Research/Carrier_Bags_final_18-02-11.pdf. Accessed January 2011. (*An exemplar of LCA at its most LCA-like*)

González-García, S., Hospido, A., Feijoo, G., and Moreira, M.T. (2010), "Life cycle assessment of raw materials for non-wood pulp mills: Hemp and flax resources," *Conservation and Recycling*, 54, pp. 923–930.

Imhoff, D. (2005), *Paper or plastic: searching for solutions to an overpackaged world*, University of California Press, Berkeley, CA. ISBN-13: 978-1578051175. *(A study of packaging taking a critical stance)*

Shen, L. and Patel, M.K. (2008), "Life cycle assessment of polysaccharide materials: a review," *J. Polymer Environ.*, 16, pp. 154–167. *(A survey of the embodied energy and emissions of natural fibers)*

8.4 An electric kettle

Figure 8.3 shows a typical 2-kW electric kettle. The kettle is manufactured in Southeast Asia and transported to Europe by air freight, a distance of 12,000 km. Table 8.4 lists the materials. The kettle boils 1 liter of water in 3 minutes. It is used for this purpose, on average, twice per day for 300 days per year over a life of 3 years. At end of life the kettle is sent to the landfill. How are energy consumption and carbon emissions distributed across the phases of life?

Figure 8.4 shows the energy breakdown. The first two bars—materials (120 MJ) and manufacture (10 MJ)—are calculated from the data in the table by multiplying the embodied energy by the mass for each component, and summing them. Air freight consumes 8.3 MJ/metric ton·km (Table 6.8), giving 129 MJ/kettle for the 12,000 km transport. The duty cycle (6 minutes per day, 300 days for 3 years) at full power consumes 180 kWh of electrical energy. The corresponding consumption of fossil fuel and emission of CO_2 depends on the energy mix and conversion efficiency of the host country (Table 6.7). Taking the energy mix of Australia, for

FIGURE 8.3   *A 2-kW kettle*

example, gives a 1,800 MJ oil equivalent and 128 kg CO_2 for the use-phase. At the end of life the kettle is dumped, at an energy cost of 0.2 MJ.

The use-phase of life consumes far more energy and liberates far more carbon than all the others put together. Despite its low duty cycle of just 6 minutes per day, the electric power (or, rather, the oil equivalent of the electric power) accounts for 88% of the total. Improving eco-performance here has to focus on this use energy—even a large reduction in any of the others makes little difference. Heat is lost through the kettle wall. Selecting a polymer with lower thermal conductivity or using an insulated double wall could help here—it would increase the embodied energy of the material bar, but even doubling this leaves it small. A full vacuum insulation would be the ultimate answer—the water not used when the kettle is boiled would then remain hot for long enough to be useful the next time it is

Table 8.4	Electric kettle life 3 years						
Bill of materials and processes				**Material and process properties***			
Component	Material	Process	Mass (kg)	Material energy (MJ/kg)	Process energy (MJ/kg)	Material CO_2 kg/kg	Process CO_2 kg/kg
Kettle body	Polypropylene	Polymer molding	0.86	95	21	2.7	1.6
Heating element	Ni-Cr alloy	Wire drawing	0.026	133	22	8.3	1.7
Casing, heating element	Stainless steel	Def. processing	0.09	81	7.9	5.1	0.59
Thermostat	Nickel alloys	Wire drawing	0.02	133	22	8.3	1.7
Internal insulation	Alumina	Power forming	0.03	52	—	2.8	—
Cable sheath, 1 meter	Natural rubber	Polymer molding	0.06	66	7.6	1.6	1.3
Cable core, 1 meter	Copper	Wire drawing	0.015	71	15	5.2	1.1
Plug body	Phenolic	Polymer molding	0.037	90	13	3.0	2.2
Plug pins	Brass	Extrusion	0.03	72	3.1	6.3	0.23
Packaging, padding	Polymer foam	Polymer molding	0.015	107	11	3.7	1.6
Packaging, box	Cardboard	Construction	0.125	28	—	1.4	—
Other small components	*Proxy material:* Polycarbonate	*Proxy process:* Polymer molding	0.04	110	11	5.6	1.4
		Total mass	**1.3**				

*From the data sheets of Chapter 15

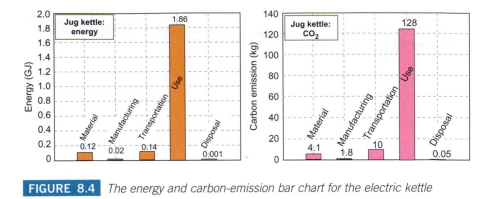

FIGURE 8.4 *The energy and carbon-emission bar chart for the electric kettle*

needed. The seeming extravagance of air freight accounts for only 6% of the total energy. Using sea freight instead increases the distance to 17,000 km but reduces the transportation energy per kettle to a mere 0.2% of the total.

This dominance of the use-phase of energy and of CO_2 emission is characteristic of small electrically powered appliances. Further examples can be found in the next case study and the exercises at the end of this chapter.

8.5 A coffee maker

The 640-watt coffee maker shown in Figure 8.5 makes 4 cups of coffee in 5 minutes (requiring full power) and then keeps the coffee hot for a subsequent 30 minutes, consuming one sixth of full power. It is manufactured in Southeast Asia and shipped 17,000 km to Europe, where it is sold and used. At the end of life it is sent to the landfill.

Table 8.5 summarizes the bill of materials. The housing is injection molded polypropylene, the carafe is glass, there are a number of small steel and aluminum parts, a heating element, and, of course, a cable and plug. The control system has some simple electronics and an LED indicator. Each brew requires a filter paper that is subsequently discarded. We take the life of the coffee maker to be 5 years, over which time it is used once per day.

Figure 8.6 shows the breakdown, calculated as in the previous case study. Here electronics (Table 6.5) are included, assigning them an energy of 3,000 MJ/kg. The first three bars—materials (150 MJ), manufacture (27 MJ), and transport (6 MJ)—are all small compared to the energy of use. A single use-cycle uses the equivalent of 10 minutes of full 640-watt power. Over 5 years, this consumes 194 kWh of electrical power, which, if generated in the United States (Table 6.6), corresponds to an equivalent oil consumption of 1,480 MJ and CO_2 emission of 105 kg. Each use also consumes one filter paper—1,825 of them over life—each weighing 2 grams, making 3.65 kg of paper. The average embodied energy of paper, from the

FIGURE 8.5 *A coffee maker*

Table 8.5 Coffee maker, life: 5 years

	Bill of materials and processes			Material and process properties*			
Component	Material	Process	Mass (kg)	Material energy (MJ/kg)	Process energy (MJ/kg)	Material CO₂ kg/kg	Process CO₂ kg/kg
Housing	Polypropylene	Molding	0.91	95	21	2.7	1.6
Small steel parts	Steel	Rolling	0.12	32	2.7	2.5	0.2
Small aluminum parts	Aluminum	Rolling	0.08	209	5.5	12	0.4
Glass carafe	Glass (Pyrex)	Molding	0.33	25	8.2	1.4	0.7
Heating element	Ni-Cr alloy	Wire drawing	0.026	133	22	8.3	1.7
Electronics and LED	Electronics	Assembled	0.007	3000	—	130	—
Cable sheath, 1 meter	PVC	Extrusion	0.12	66	7.6	1.6	1.3
Cable core, 1 meter	Copper	Wire drawing	0.035	71	15	5.2	1.1
Plug body	Phenolic	Molding	0.037	90	13	3.0	2.2
Plug pins	Brass	Extrusion	0.03	72	3.1	6.3	0.23
Packaging, padding	Polymer foam	Molding	0.015	107	11	3.7	1.6
Packaging, box	Cardboard	Construction	0.125	28	—	1.4	—
Other components	*Proxy material:* Polycarbonate	*Proxy process:* Molding	0.04	110	11	5.6	1.4
		Total mass	**1.9**				

*From the data sheets of Chapter 15 and Table 6.5

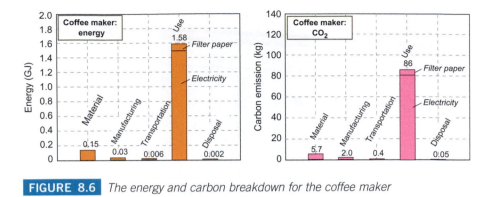

FIGURE 8.6 *The energy and carbon breakdown for the coffee maker*

data sheet in Chapter 15, is 28 MJ/kg, with an associated emission of 1.4 kg/kg, so the filter papers add 100 MJ to the use energy and 5.1 kg to the use-carbon. The figure shows these at the top of the Use bar of the bar charts.

There is nothing that can be done to recover the electrical power once it is used, but it *is* possible to reduce energy consumption by replacing the glass carafe with a stainless steel vacuum one, thereby eliminating the need for a heater to keep the coffee warm, reducing the electric power consumption by a factor of 2. The embodied energy of stainless steel is three times greater than that of glass, so it is necessary to check that this redesign really does save energy over life—a task left to the exercises at the end of this chapter.

8.6 An A-rated washing machine

Household appliances are major consumers of energy. In this case study we examine the energy demands and carbon emissions associated with the life of a generic washing machine (Figure 8.7). German studies suggest that, on average, a washing machine is used for 220 wash cycles per year and lasts for 10 years. The energy it consumes during use depends on the wash temperature, on whether it is hot-filled or cold-filled, and on the choice of cycle: an A-rated machine requires 1.22 kWh for a 90°C wash and 0.56 kWh for a 40°C wash, averaging out at 0.85 kWh for a typical mix between washes (National Energy Foundation, 2011). For illustration, we assume that the washing machine is manufactured in Germany and delivered to the United Kingdom, a distance of 1,000 km, by a 32-metric-ton truck. At end of life washing machines are shredded, and the metallic materials are recovered, roughly 49% of the total weight. The left side of Table 8.6 lists the bill of materials for a washing machine, taken from the study by Stahel (1992). The eco-audit is based on this bill of materials, using the data shown on the right of Table 8.6.

Figure 8.8 shows the results, using the average wash-spin cycle energy of 0.85 kWh. The use phase outweighs all others by a factor of at least 6. The

FIGURE 8.7 *A washing machine*

recycling of metals at end of life allows about a third of the material energy to be recovered. The transportation of the washing machine over 1,000 km makes a negligible difference to the totals.

Further reading

National Energy Foundation (2011), www.nef.org.uk/energysaving/labels.htm. *(Use energy for A to E rated washing machines)*

Stahel, W.R. (1992), *Langlebigkeit und Materialrecycliing*, 2nd edition, Vulkan Verlag, Essen, Germany. ISBN 3-8027-2815-7. www.productg-life.org/en/archive/case-studies/washing-machines.

8.7 Ricoh imagio MF6550 copier

Ricoh provides Environmental Product Declarations (EPDs) for its products. The one for their MF6550 digital copier (Figure 8.9) includes an approximate bill of materials, shown on the left of Table 8.7. The typical service life of the copier is 5 years, during which time it is used 8 hours per day for 20 days per month, consuming, on average, 0.35 kW of electrical power over that time. During this time, the copier consumes 2,880,000 sheets of paper weighing 12,215 kg. At end of life, 85%

Table 8.6	The washing machine, life: 10 years				
Bill of materials			**Material properties***		
Component	**Recycle content (%)**	**Part mass (kg)**	**Embodied energy (MJ/kg)**	**CO_2 footprint (kg/kg)**	
Mild steel parts	42%	23	22	1.7	
Cast iron parts	69%	3.8	8.9	0.5	
HSLA steel parts	Virgin (0%)	6.2	35	2.1	
Stainless steel parts	38%	5.4	59	3.7	
Aluminum parts	43%	1.9	134	7.6	
Copper, brass parts	43%	1.8	48	3.5	
Zinc parts	Virgin (0%)	0.1	72	3.8	
Polystyrene parts	Virgin (0%)	2.1	92	2.9	
Polyolefin parts (PP)	Virgin (0%)	1.3	94	2.7	
PVC parts	Virgin (0%)	0.7	80	2.4	
Nylon parts	Virgin (0%)	0.4	128	5.6	
Other polymers (ABS)	Virgin (0%)	1.9	96	3.4	
Rubber parts	Virgin (0%)	1.6	66	1.6	
Concrete blocks[+]	Virgin (0%)	22	1.1	0.1	
Cardboard packaging	Virgin (0%)	2.3	28	1.4	
Wood parts	Virgin (0%)	2.5	7.4	0.4	
Borosilicate glass	Virgin (0%)	0.1	25	1.4	
	Total mass	**77**			

*From the data sheets of Chapter 15

+ The 22 kg of concrete suppress vibration during the spin cycle.

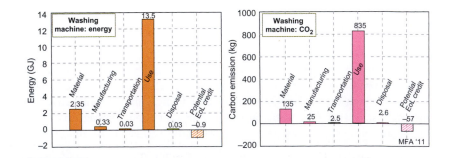

FIGURE 8.8 *The energy and carbon footprint of an A-rated washing machine, based on data from Stahel (1992) and the National Energy Foundation (2011)*

FIGURE 8.9   *A Ricoh imagio MF6550 copier*

Table 8.7   The copier, life: 5 years

	Bill of materials		Material properties*	
Component	Recycled content* (%)	Part mass (kg)	Embodied energy (MJ/kg)	CO_2 footprint (kg/kg)
Steel press parts	42%	1.1e +02	22	1.7
Zinc parts	22%	3.7	59	3.2
Copper/brass parts	43%	1.5	48	3.5
PS parts	0%	8.5	92	2.9
ABS parts	0%	4	96	3.4
PC parts	0%	2.4	110	5.6
POM parts	0%	1.4	105	4.0
PP parts	0%	1.1	95	2.7
PET parts	0%	0.8	84	2.3
Glass parts	0%	2.2	25	1.4
Neoprene parts	0%	5	101	3.7
Other materials (PC proxy)	0%	15	110	5.6
Total mass		**180**		

These data was paraphrased from Ricoh, 2010.

*From the data sheets of Chapter 15

of copiers are returned to Ricoh, where they are recycled, and 95% of the material content is recovered. The residue goes to the landfill. Transportation was set at 200 km by a 14-metric-ton truck for initial delivery, plus 800 km by a light goods vehicle for four annual services.

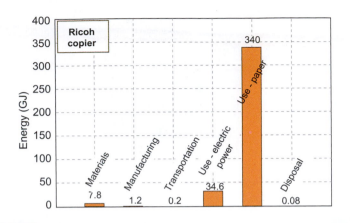

The energy-audit for the Ricoh copier. The contribution of the paper is shown as a green bar.

The bar chart of Figure 8.10 shows the output of the eco-audit with the use-phase split into two parts, one for the electrical energy, the other for the paper. The eco-audit matches the EPD well in energy of material and manufacturing, in use electric power, and in the strikingly large energy associated with the paper. Ricoh points out that double-sided copying, or, better, double-sided with two pages on each face, is the obvious way to save energy.

Further reading

Ricoh (2010), "Environmental Product Declaration for the Ricoh imagio MF6550 digital copier," www.ricoh.com.

8.8 A portable space heater

The space heater shown in Figure 8.11 is carried as part of the equipment of a light goods vehicle used for railway repair work. It burns 0.66 kg of propane gas per hour, delivering an output of 9.3 kW (32,000 BTU). The air flow is driven by a 38-W electric fan. The heater weighs 7 kg. The (approximate) bill of materials is listed in Table 8.8. The product is manufactured in India and shipped to the United States by sea freight (15,000 km); it is then carried by a 32-metric-ton truck for 600 km farther to the point of sale. It is anticipated that the vehicle carrying it will travel, on average, 700 km per week, over a 3-year life, releasing 0.1 kg CO_2/metric ton $\cdot$ km (Table 6.8). The heater itself will be used for 2 hours per day for 10 days per year. At end of life the carbon steel components are recycled.

FIGURE 8.11 *A space heater powered by liquid propane gas (LPG)*

Table 8.8	LPG space heater, life: 3 years					
Bill of materials and processes				**Material properties***		
Component	**Material**	**Process**	**Mass (kg)**	**Mat. CO_2 kg/kg**	**Proc. CO_2 kg/kg**	
Heater casing	Low C steel	Def. processing	5.4	2.5	0.19	
Fan	Low C steel	Def. processing	0.25	2.5	0.19	
Heat shield	Stainless steel	Def. processing	0.4	5.1	0.27	
Motor, rotor, and stator	Iron	Def. processing	0.13	1.0	0.21	
Motor, conductors	Copper	Def. processing	0.08	5.7	0.17	
Motor, insulation	Polyethylene	Polymer extrusion	0.08	2.1	0.51	
Connecting hose, 2 meter	Natural rubber	Polymer molding	0.35	1.5	0.61	
Hose connector	Brass	Def. processing	0.09	6.3	0.18	
Other components	*Proxy material:* polycarbonate	*Proxy process:* polymer molding	0.22	5.6	0.86	
		Total mass	**7.0**			

*From the data sheets of Chapter 15

This is a product that uses energy during its life in two distinct ways. First there is the electricity and LPG required to make it function. Second, there is the energy penalty that arises because it increases the weight of the vehicle that carries it by

7 kg. How much CO_2 does the product release over its life? And which phase of life releases the most?

Figure 8.12 shows the CO_2 emission profile. The first two bars—materials (19 kg) and manufacturing (1.7 kg)—are calculated from the data in the table. Transportation releases only 3.5 kg per unit. The power consumed by burning LPG for heat (9.3 kW) far outweighs that used to drive the small electric fan motor (38 W), so it is the CO_2 released by burning LPG that we evaluate here. It is less obvious how the static use for generating heat, drawn for only 20 hours per year, compares with the extra fuel consumed by the vehicle because of the product weight—remembering that, as part of the equipment, it is lugged over 36,000 km per year, releasing 76 kg of carbon. The figure shows the CO_2 of use outweighs all other contributions, here accounting for 93% of the total, as it does with most energy-using products. Of this, 65% derives from burning gas, and 35% comes from the additional fuel consumed by carrying the heater to the sites where it is used. The CO_2 burden to recycle steel, from the data sheet in Chapter 15, is about 0.7 kg/kg, saving the difference between this and that for primary production (2.5 kg/kg)—a net saving of 10 kg.

8.9 Ceramic pottery kilns

Kilns for firing pottery are simple structures, but they are large and heavy, and that means a lot of material goes into making them. Is this a case in which the embodied energy of the materials dominates the life energy? An energy audit will tell us.

At its most basic, a kiln is a steel frame lined with refractory brick that encloses the chamber in which the ceramics are fired. Nichrome heating elements are embedded in the inner face of the brick and connected to a power source with

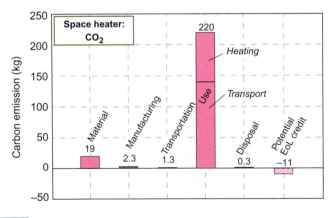

FIGURE 8.12 *The energy breakdown for the space heater. The use-phase dominates.*

insulated copper cable. The chamber is closed during firing by a brick-lined door (Figure 8.13). Table 8.9 gives a bill of the principal materials in a typical small pottery kiln capable of firing a chamber of 0.28 m^3 (10.5 cubic feet) at up to 1,200°C.

FIGURE 8.13 | *An electric kiln*

Table 8.9 | A pottery kiln, life: 10 years

| Component | Bill of materials | | Material properties* | |
	Recycle content (%)	Part mass (kg)	Embodied energy (MJ/kg)	CO$_2$ footprint (kg/kg)
Mild steel frame and casing	42%	65	22	1.7
Refractory brick	0%	500	2.8	0.2
Nichrome furnace windings	0%	5	133	8.3
Copper connectors	43%	2	48	3.5
Alumina high-temp insulation	0%	1	52	2.8
	Total mass	**572**		

*From the data sheets of Chapter 15, using "Brick" as a proxy for "Refractory brick"

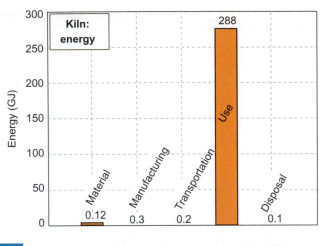

FIGURE 8.14 *The energy audit for the kiln, assuming a life of 10 years*

It is rated at 12 kW. We will suppose that the kiln is installed in an art department of a school, where it is fired once per week for 40 weeks per year. To get there, it had to be transported by a large truck over a distance of 750 km. The kiln's life is 10 years.

The data in Table 8.9 allow the embodied energies of the materials of the kilns to be summed: they amount to 3.6 GW. The transport energy data of Table 6.8 gives the energy consumption of a 32-metric-ton truck as 0.46 MJ/metric ton · km, so transporting this half-metric-ton kiln over 750 km consumes a trifling 0.2 GJ.

In a typical firing cycle the kiln operates at full power (12 kW) for 4.5 hours to heat the interior to 1,200°C, and then it is cut to 4.9 kW to maintain this temperature for another 2.5 hours, after which point power is cut, using a total of 66 kWh. Multiplying this by the duty cycle of 40 firings per year for 10 years gives the electrical energy used over life as 26.4 GWh. When you multiply this by 3.6 to convert the units to MJ, and then divide it by 0.33, the approximate conversion efficiency of fossil fuel to electric power gives the oil-equivalent use energy of 288 GJ.

Figure 8.14 displays the outcome of the audit. In 1 year the kiln consumes nearly 10 times more energy in electrical heating power than in material energy; in the 10-year life (short, for a kiln) it consumes nearly 100 times more. This is the most extreme example so far of the great energy commitment of the use-phase of devices that consume energy during use.

It has profound implications for material selection. The embodied energy of the material of the kiln hardly matters—it is conserving heat that is the central objective. That's where material selection comes in, but we'll keep that for Chapter 10.

8.10 Auto bumpers—exploring substitution

The bumpers of a car (Figure 8.15) are heavy; reducing their mass can save fuel. Here we explore the replacement of a low-carbon steel bumper with one of equal performance made from an age-hardened aluminum or of CFRP on a gasoline-powered family car. The steel bumper weighs 14 kg; the aluminum substitute weighs 11 kg, the CFRP substitute only 8 kg. But the embodied energies of aluminum and of CFRP are much higher than that of steel. Is there a net savings? We assume a vehicle life of 10 years, and that it is driven 25,000 km per year. Table 8.10 lists the energies involved. We take the energy penalty of weight for a gasoline-powered car (Table 6.8) to be 2.06 MJ/metric ton · km.

The bar charts of Figure 8.16 compare the material energy, the use energy, and the sum of the two, assuming virgin material is used for both the steel and the aluminum bumper. The substitution of aluminum or CFRP for steel results in a large increase in material energy but a drop in use energy. The right-hand columns in Table 8.10 list values and the totals: the aluminum substitute has a lower total energy than the steel, but not by much—the break-even comes at about 200,000 km. The CFRP substitute reduces the life-energy a little more. Both aluminum and CFRP cost more than steel. A final decision needs a cost-benefit trade-off. Methods for doing that come in the next chapter.

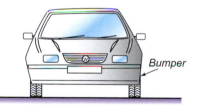

FIGURE 8.15 *Car and bumper*

Table 8.10	The analysis of the material substitution				
Material and mass		**Material properties**	**Eco-audit of bumper**		
Material of bumper	Mass (kg)	Embodied energy (MJ/kg)*	Material energy (MJ)	Use energy (MJ)	Total: material plus use (MJ)
Low-carbon steel	14	32	448	7,210	7,660
Age-hardened Alu	10	209	2,090	5,150	7,240
CFRP	8	272	2,180	4,120	6,300

*From the data sheets of Chapter 15

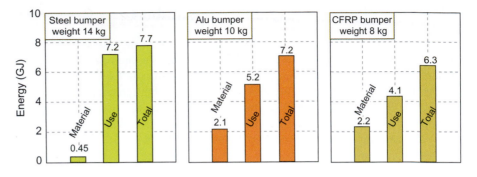

The comparison of the energy audits of a steel, an aluminum, and a CFRP bumper for a family car driven 250,000 km over its life

But this comparison has not been quite fair. A product like the bumper would incorporate recycled as well as virgin material. It is possible, using the data given in the data sheets of Chapter 15, to correct the material energies for the recycled content. This is left to the exercises at the end of this chapter.

8.11 Family car—comparing material energy with use energy

Argonne National Laboratory, working with the US Department of Energy, has developed a model (GREET) to evaluate energy and emissions associated with vehicle life. Table 8.11 lists the bill of materials for two of the vehicles they analyze: a conventional steel-bodied mid-sized family car with internal combustion engine (ICE), and a vehicle of similar size made of lightweight materials, aluminum and CFRP (Figure 8.17). The biggest differences in data values are bold and italicized. The total mass is shown at the bottom of the columns. Using the lightweight material reduces the total mass by 39%.

The data sheets of Chapter 15 provide the embodied energies of the materials. Fuel consumption scales with weight in ways that are analyzed in Chapter 6; for now, we use the results that claim a conventional car of this weight consumes 3.15 MJ/km, and the lighter one consumes 2.0 MJ/km.[1] There is enough information here to allow an approximate comparison of embodied energy and the use of the two vehicles, assuming both are driven 25,000 km per year for 10 years.

---

[1] 1 MJ/km = 2.86 liters/100 km = 95.5 miles per UK gallon = 79.5 miles per US gallon.

Table 8.11	Material content of a conventional family car and one made of lightweight materials		
Material	**Material energy H_m MJ/kg***	**Conventional ICE vehicle (kg)**	**Lightweight ICE vehicle (kg)**
Carbon steel	32	*839*	*254*
Stainless steel	81	0.0	5.8
Cast iron	17	*151*	*31*
Wrought aluminum (10% recycled content)	200	*30*	*53*
Cast aluminum (35% recycled content)	149	*64*	*118*
Copper/Brass	72	26	45
Magnesium	380	0.3	3.3
Glass	15	39	33
Thermoplastic polymers (PU, PVC)	80	94	65
Thermosetting polymers (Polyester)	88	55	41
Rubber	110	33	17
CFRP	273	*0.0*	*134*
GFRP	110	0.0	20
Platinum, catalyst (*Table 6.3*)	117,000	0.007	0.003
Electronics, emission control etc. (*Table 6.4*)	3,000	0.27	0.167
Other (*proxy material: Polycarbonate*)	110	26	18
	Total mass	**1,361**	**836**

*From the data sheets of Chapter 15 and Tables 6.4 and 6.5

 FIGURE 8.17 *A body in white. The audit examines the potential of alternative materials.*

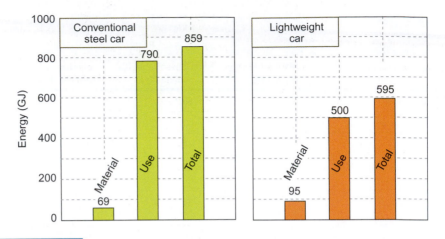

FIGURE 8.18 *The comparison of the energy audits of a conventional and a light-weight family car.*

The bar charts of Figure 8.18 show the comparison. The input data are of the most approximate nature, but it would take very large discrepancies to change the conclusion: the energy consumed in the use-phase of both vehicles greatly exceeds that embodied in their materials. The use of lightweight materials increases the embodied energy by 38% but reduces the much larger fuel-energy consumption by 37%. The result is a net gain: the sum of the material and use energies for the lightweight vehicle is 31% less than that of the conventional one.

Further reading

Burnham, A., Wang, M., and Wu, Y. (2006), "Development and applications of GREET 2.7," ANL/ESD/06-5, Argonne National Laboratory, Argonne, IL. USA, www.osti.gov/bridge. *(A report describing the model developed by ANL for the US Department of Energy to analyze life emissions from vehicles)*

8.12 Computer-assisted audits: a hair dryer

Audits of energy and emissions are greatly simplified by the use of computer-based tools, one of which was described at the end of Chapter 7. As an example of its use, consider one further case study—an eco-audit of a 2,000-watt electric hair dryer.

A hair dryer is a product found in almost every European and American household. Figure 8.19 shows a contemporary 2,000-W dryer, made in Southeast Asia and shipped by sea to Europe, roughly 20,000 km. The bill of materials and

A 2000-W "ionic" diffuser hairdryer

processes as entered into the Eco-audit tool is reproduced in Table 8.12. The dryer has an expected life of 3 years (it is guaranteed for only 2), and will be used, on average, for 3 minutes per day for 150 days per year. At end of life the polymeric housing and nozzle subsystem are recycled. The rest is dumped.

Figure 8.20 and Table 8.13 show part of the output. The tool provides more detail: tables listing the energy and CO_2 emission that can be attributed to each component, and the details of the transportation and use-phase calculations.

As with other energy-using products, it is the use-phase that dominates the audit. It may be possible to reduce the Material bar of Figure 8.20 a little by substituting with less energy-intensive materials or by using a greater recycled content, but the effect of this on the life energy will be small. So the target has to be the heater. When a hair dryer is being operated, most of the heat goes straight past the head. The fraction that is functionally useful is not known, but it is small. Anything that increases this fraction contributes to energy efficiency. The diffuser (a standard accessory) does that, but a lot of hot air is still lost. The makers of this hair dryer claim a dramatic development. Incorporated within it is a gas-discharge ionizer. It ionizes the air flowing past it, which, it is speculated, breaks down the water in the hair into smaller droplets, allowing them to evaporate faster. The manufacturers of ionic hairdryers claim that this dries the hair *twice as fast as non-ionic hair dryers* (it is printed on the box). If true, it gives an impressive total life-energy reduction of nearly 40%—it is marked on the figure. Although I have found no scientific basis for the claim, customer reviews express a market preference for ionic hairdryers, so maybe it is real.

Table 8.12	The hair dryer, life: 3 years			
Bill of materials and processes				
Subsystem	**Component**	**Material**	**Shaping process**	**Mass (kg)**
Housing and nozzle	Housing	ABS	Polymer molding	0.177
	Inner air duct	Nylons (PA)	Polymer molding	0.081
	Filter	Polypropylene	Polymer molding	0.011
	Diffuser	Polypropylene	Polymer molding	0.084
Fan and motor	Fan	Polypropylene	Polymer molding	0.007
	Casing	Polycarbonate	Polymer molding	0.042
	Motor—iron	Low carbon steel	Def. processing	0.045
	Motor—windings	Copper	Def. processing	0.006
	Motor—magnet	Nickel	Def. processing	0.022
Heater	Heating filament	Nickel-chrome alloys	Def. processing	0.008
	Insulation	Alumina	Ceramic power forming	0.020
	Support	Low carbon steel	Def. processing	0.006
Circuit board and wiring	Conductors	Copper	Def. processing	0.006
	Board	Phenolics	Polymer molding	0.007
	Insulators	Phenolics	Polymer molding	0.012
	Cable sheathing	Polyvinylchloride	Polymer molding	0.005
Cable and plug	Main cable, core	Copper	Def. processing	0.035
	Cable sleeve	Polyvinylchloride	Polymer molding	0.109
	Plug body	Phenolics	Polymer molding	0.021
	Plug pins	Brass	Def. processing	0.023
Packaging	Rigid foam padding	Rigid polymer foam	Polymer molding	0.011
	Box	Paper and cardboard	Construction	0.141
Residual components	Residual components	*Proxy material:* Polycarbonate	*Proxy process:* Polymer molding	0.010
			Total mass	***0.89***

8.13 Summary and conclusions

What do these case studies tell us? That, for energy-using products (household appliances, cars, copiers), it is the use-phase that dominates, frequently consuming more energy and creating more emissions than all the other phases combined, even when the duty cycle appears modest. For products that do not use energy or use

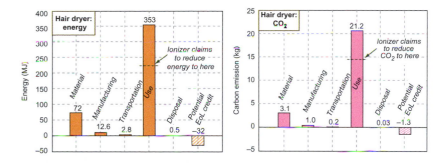

FIGURE 8.20 *The energy break-down for the hair dryer, as delivered by the CES Edu Eco-audit tool*

Table 8.13 The energy and CO_2 breakdown for the hair dryer

Phase of life	Energy (MJ)	CO_2 (kg)
Material	72.2	3.1
Manufacturing	12.6	1.0
Transportation	2.8	0.2
Use	353	21.2
Disposal	0.5	0.03
Total, end of first life	**441**	**25.2**
End of life potential	−32	−1.3

very little (the PET bottle, the disposable cups, the carrier bags) it is the embodied energy and associated CO_2 of the materials that dominate.

Eco-aware product design has many aspects, one of which is the choice of materials. Materials are energy-intensive, with high embodied energies and associated carbon footprints. Seeking to use low-energy materials might appear to be one way forward, but this can be misleading. Material choice has consequences for manufacturing; it influences weight, thermal performance, and electrical characteristics. All change the energy consumed by the product during use and the potential for recycling or energy recovery at the end of life. It is full-life energy that we seek to minimize.

Doing so requires a two-part strategy developed in Chapter 3. The first part is an *eco audit*: a quick, approximate assessment of the distribution of energy demand and carbon emission over life. This provides inputs to guide the second part: that of material selection to minimize the energy and carbon over the full life, thus balancing the influences of the choice over each phase of life—the subject of Chapters 9 and 10. The eco-audit method described here is fast and easy to perform, and although approximate, it delivers information with sufficient precision to enable strategic decision making. The exercises that follow provide opportunities for exploring them further.

8.14 Exercises

E8.1. If the glass container (weight 0.33 kg) of the coffee maker audited in Section 8.5 of the text is replace by a double-walled stainless steel one weighing twice as much, how much does the total embodied energy of the product change? If this reduces the electric power consumed over life (1,480 MJ) by 10%, does the energy balance favor the substitution?

E8.2. Figure 8.21 shows a 1,700-watt steam iron. It weighs 1.3 kg, 98% of which is accounted for by the seven components listed in the table. The iron heats up on full power in 4 minutes, and then it is used, typically, for 20 minutes on half power. At end of life the iron is dumped in a landfill. Create an eco-audit for the iron assuming that it is used once per week over a life of 5 years, using data from the following table. Do not take transportation and end of life into account.

FIGURE 8.21 *A steam iron*

What conclusions can you draw? How might the life-energy be reduced?

Steam iron: bill of materials

Component	Material	Shaping process	Mass (kg)	Material energy* (MJ/kg)	Process energy* (MJ/kg)
Body					
Polypropylene	Molded	0.15		97	21.4
Heating element	Nichrome	Wire drawn	0.03	134	22.1
Base	Stainless steel	Cast	0.80	82	11.3
Cable sheath, 3 meter	Polyurethane	Molded	0.18	82	18.8
Cable core, 3 meter	Copper	Wire drawn	0.05	71	14.8
Plug body	Phenolic	Molded	0.037	90	28
Plug pins	Brass	Rolled	0.03	71	1.7
		Total mass	**1.28**		

*From the data sheets of Chapter 15

E8.3. Figure 8.22 shows a 970-watt toaster. It weighs 1.2 kg including 0.75 m of cable and a plug. It takes 2 minutes and 15 seconds to toast two slices of bread. It is used to toast, on average, 8 slices per day, so it draws its full electrical power for 9 minutes (0.15 hours) per day for 300 days per year over its design life of 3 years. The toasters are made locally—transportation energy and CO_2 are negligible. At end of life it is dumped. Create an eco-audit for the toaster using data in the following table.

FIGURE 8.22 *An electric toaster*

Toaster: bill of materials

Component	Material	Shaping process	Mass (kg)	Material energy* (MJ/kg)	Process energy* (MJ/kg)
Body	Polypropylene	Molded	0.24	97	21.4
Heating element	Nichrome	Drawn	0.03	134	22.1
Inner frame	Low carbon steel	Rolled	0.93	32	2.7
Cable sheath, 0.75 meter	Polyurethane	Molded	0.045	82	18.8
Cable core, 0.75 meter	Copper	Drawn	0.011	71	14.8
Plug body	Phenolic	Molded	0.037	90	28
Plug pins	Brass	Rolled	0.03	71	1.7
		Total mass	**1.32**		

*From the data sheets of Chapter 15

E8.4. It is proposed to replace the low-carbon steel bumper set discussed in Case Study 8.10 in the text (the auto bumper) by one made of GFRP. It is anticipated that the GFRP set will weigh 9.5 kg. Following the procedure of the text, draw data from the data sheets of Chapter 15 to estimate whether, over the life pattern used in the text (250,000 km at 2.06 MJ/metric ton · km), there is a net energy saving compared with steel and aluminum, each with 50% recycled content. Is one choice significantly better if the embodied energy data could be in error by ±10%?

Material of fender	Mass (kg)	Embodied energy (MJ/kg)*
Low-carbon steel	14	32
Age-hardened Al	10	209
GFRP	9.5	112
50% recycled steel	14	21
50% recycled Al	10	114

E8.5. It is reported that the production of a small car such as that shown in Figure 8.23 (mass 1,000 kg) requires materials with a total embodied energy of 70 GJ, and a further 25 GJ for the manufacturing phase. The car is manufactured in Germany and delivered to the US show room first by sea freight (distance 10,000 km), followed by a heavy truck over another 1,500 km (Table 6.8 of the text gives the energy per metric ton · km for both). The car has a useful life of 10 years, and will be driven on average 20,000 km per year, consuming 2 MJ/km. Assume that recycling at end of life consumes 0.5 GJ but recovers 25 GJ per vehicle.

Make an energy-audit bar chart for the car with bars for material, manufacturing, distribution, use, and disposal. Which phase of life consumes the most energy?

FIGURE 8.23   *A small car*

The inherent uncertainty of current data for embodied and processing energies are considerable—if both of these were in error by up to 20% either way, can you still draw firm conclusions from the data? If so, what steps would do the most to reduced life-energy requirements?

E8.6. The following table lists one European automaker's summary of the material content of a mid-sized family car (such as that seen in Figure 8.23). Material proxies for the vague material descriptions are given in brackets and italicized. The vehicle is gasoline-powered and weighs 1,800 kg. The data sheets of Chapter 15 provide the embodied energies of the materials: mean values are listed in the table. Table 6.8 gives the energy of use: 2.1 MJ/metric ton · km, equaling 3.6 MJ/km for a car of this weight. Use this information to make an approximate comparison of embodied and use energies of the car, assuming it is driven 25,000 km per year for 10 years.

Material content of a family car, total weight 1,800 kg

Material	Mass (kg)
Steel (low-alloy steel)	950
Aluminum (cast aluminum alloy)	438
Thermoplastic polymers (PU, PVC)	148
Thermosetting polymers (polyester)	93
Elastomers (butyl rubber)	40
Glass (borosilicate glass)	40
Other metals (copper)	61
Textiles (polyester)	47
Total mass	**1,800**

E8.7. Conduct a CO_2 eco-audit for the patio heater shown in Figure 8.24. It is manufactured in Southeast Asia and shipped 12,000 km to the United States where it is sold and used. It weighs 24 kg, of which 17 kg is rolled stainless steel, 6 kg is rolled carbon steel, 0.6 kg is cast brass, and 0.4 kg is unidentified injection-molded plastic (so we need a proxy to deal with this). The masses and the material and process carbon footprints, from Chapter 15, are assembled in the following table.

When used the patio heater delivers 14 kW of heat ("enough to keep 8 people warm") consuming 0.9 kg of propane gas (LPG) per hour, and releasing 0.059 kg of CO_2/MJ. The heater is used for 3 hours per day for 30 days per year, over 5 years, at which time the owner tires of it and dumps it. We will ignore end-of-life.

FIGURE 8.24 *A patio heater*

Use these data to construct a bar chart for CO_2 emission over life.

Bill of materials		Carbon footprints	
Material	Mass (kg)	Material (kg/kg)*	Process (kg/kg)*
Stainless steel, rolled	17	5.1	0.6
Carbon steel, rolled	6	2.5	0.2
Brass, cast	0.6	6.3	0.6
Polymer molded (polypropylene)	0.4	2.7	1.6
Totals	24		

*From the data sheets of Chapter 15

E8.8. Carry out an eco-audit for a product of your choosing. Either pick something simple (a polypropylene dish washing bowl) or, more ambitiously, something more complex, and be prepared to dismember it, weigh the parts, and use your best judgment to decide what they are made of and how they were shaped.

Exercises using the CES eco-audit tool

The first four of these exercises repeat or extend those of the text or of the preceding exercises, using the eco-audit tool of the CES Edu software (found in the pull-down menu under "Tools"). The output and bar charts in some cases differ slightly from those of the earlier exercises because of the way the software creates means of the property ranges (it uses geometric rather than arithmetic means, more logical when ranges are large) and because it calculates the oil equivalent of electrical energy using a country-dependent energy mix (the mix of fossil fuel, hydro, wind, and nuclear in the grid supply).

E8.9. Carry out the eco-audit for the 2-kW electric kettle of Section 8.4, now using the fact that the PP kettle body and the cardboard packaging are recycled. Set transport at 12,000 km by air freight, and the duty cycle at 6 minutes (0.1 hours) per day, for 300 days per year, for 3 years, using the electricity mix for Australia.

E8.10. Carry out the eco-audit for the 640-watt coffee maker of Section 8.5, now using the fact that it is made entirely from recycled materials (select the grade "100% recycled") and that the PP housing and the steel parts are themselves recycled at end of life. Set transport at 17,000 km by sea freight, and the duty cycle at 10 minutes (0.17 hours), for 365 days per year, for 5 years.

E8.11. Carry out the eco-audit for CO_2 for the portable space heater using the bill of materials from Section 8.8's Table 8.8 of the text, assuming that at end of first life the fan and heat shield are re-engineered to incorporate them into a new product. Use the transportation and duty cycle described there.

E8.12. Carry out Exercise E8.3 (the toaster) using the CES eco-audit tool, using the bill of materials listed there. Make bar charts both for energy and for CO_2. Then select the grade "Standard" in the calculation, and repeat ("Standard" grade is material with a recycle content equal to the recycle fraction in current supply, data for which are included in the data sheets). Compare the result with using virgin material to make the toaster.

E8.13. Automobile door "skins" are the outermost surfaces of the doors (such as seen in Figure 8.25). They are usually made from sheet steel, but the pressure to reduce weight leads car makers to consider aluminum or sheet molding compound (SMC) as alternatives. A set of four steel door skins for a GM Holden mid-sized sedan weighs 17.0 kg. It is estimated that a replacement set made from aluminum would weigh 10.5 kg, and one made from GFRP would weigh 11.7 kg. If both the steel and the aluminum contain 50% recycle content, and the vehicle (with steel doors) weighs 1,400 kg and will be driven 150,000 km over its life, will the substitution of aluminum or SMC for steel save energy? Use the CES Eco-audit tool to find out.

FIGURE 8.25 *A door skin*

FIGURE 8.26 *A Gamesa wind turbine*

E8.14. Wind turbines (Figure 8.26) have been the subject of many LCAs, most of them motivated by a wish to demonstrate that the turbine quickly pays back the energy and saves the carbon emissions associated with its construction.

(a) An approximate bill of materials is available for a Gamesa 2-MW wind turbine installed in a Spanish wind farm. Use the CES Edu Eco-audit tool to assess the energy and carbon associated with the materials of the turbine, assuming virgin materials are used.

(b) If the turbine has a capacity factor of 25%, how long will it take to return the energy of the materials of which it is made?

2 MW wind turbine: bill of materials

Component	Material	Part mass (kg)
Blades	Epoxy/glass fiber	19,500
Blade hub	Cast iron	14,000
Nose cone	Epoxy/glass fiber	310
Foundation footing	Concrete	700,000
Foundation, steel	Low-carbon steel	25,000

Ferrule	Low-carbon steel	15,000
Tower	Low-carbon steel	143,000
Nacelle frame	Cast iron	11,000
Nacelle main shaft	Low-alloy steel	6,100
Nacelle transformer	Copper	1,500
Nacelle transformer	Low-carbon steel	3,300
Generator	Copper	2,000
Generator	Low-alloy steel	4,290
Gearbox, shafts	Low-alloy steel	8,000
Gearbox, gears	Cast iron	8,000
Nacelle cover	Epoxy/glass fiber	2,000
	Total mass	**9.6e +05**

CHAPTER 9

Material selection strategies

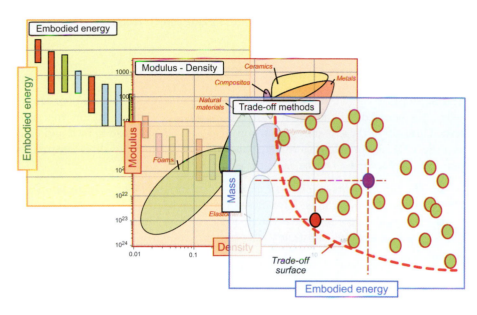

9.1 Introduction and synopsis

Life is full of decisions. Which shoes to buy? Which restaurant to eat at? Which camera? Which bike? Which car? Which university? Most of us evolve strategies for reaching decisions, some involving emotional response ("It's cool/I just couldn't resist the color, and besides, Joe/Joanna has one"), others based on cold logic.

It is cold logic that we want here. The strategy then takes the following form:

- Assemble data for the characteristics of the thing you wish to select—make a *database*, mental or physical.
- Formulate the characteristics that the thing must have to satisfy your requirements—list the *constraints*. Those that meet the constraints become candidates for selection.
- Decide on the criterion you will use to rank the candidates that meet your constraints—choose and apply the *objective*.
- Research the top-ranked candidates more fully to satisfy yourself that nothing has been overlooked—seek *documentation*.

This chapter is about selecting materials using this constraints-objectives-documentation strategy. It follows naturally from the eco-audits of the previous

227

chapters—first the audit to identify the phase of life that matters most, then the selection to find the material that most effectively minimizes the eco-impact of that phase of life. We start by illustrating the strategy with a product rather than a material—the ideas are the same, but the material has added complications. We base the discussion around the selection of a car to meet given constraints and with two objectives, one of them that of minimizing carbon footprint.

9.2 The selection strategy: choosing a car

You need a new car. To meet your needs it must be a mid-sized four-door family sedan, gas-powered, and deliver at least 150 horsepower—enough to tow your sailboat. Given all of these requirements, you wish it to cost as little to own and emit as little CO_2 as possible (Figure 9.1, left-hand side). There are three *constraints* here, but they are not all of the same type.

- The requirements of *four-door family sedan* and *gasoline power* are *simple constraints*—a car **must** have these to be a candidate.
- The requirement of *at least* 150 hp places a lower limit but no upper one on power; it is a *limit constraint*—any car with 150 hp or more is acceptable.

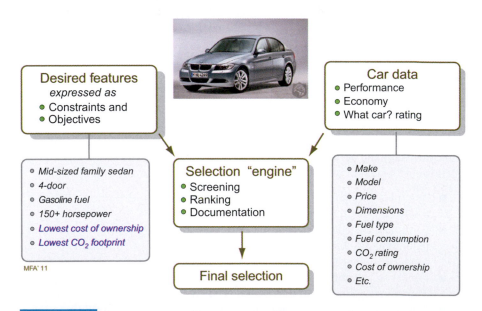

FIGURE 9.1 *Selecting a car. The requirements are expressed as constraints and objectives (objectives in blue). Records containing data for cars are* screened *using the* constraints and ranked *by the objectives to find the most attractive candidates. These are then explored further by examining* documentation.

The wish for minimum cost of ownership is an *objective*, a criterion of excellence. The most desirable cars, from among those that meet the constraints, are those that minimize this objective. The wish to minimize the CO_2 emission is a second objective, one that may not be compatible with the first.

To proceed, you need information about available cars (Figure 9.1, right-hand side). Car makers' web sites, dealers, car magazines, and advertisements in the national press list such information. It includes car type and size, number of doors, fuel type, engine power, and price; car magazines go further and estimate the cost of ownership, meaning the sum of running costs, taxes, insurance, servicing, and depreciation, listing it as \$/mile or €/km.

Now: decision time (Figure 9.1, central box). The selection engine (you, in this example) uses the constraints to *screen out*, from all the available cars, those that are not four-door gasoline-powered family sedans with 150 hp. Many cars meet these constraints; the list is still long. You need a way to order it so that the best choices are at the top. That is what the objective is for: it allows you to *rank* the surviving candidates by cost of ownership—those with the lowest values are ranked most highly. Rather than just choosing the one at the top, it is better to keep the top three or four and seek further *documentation*, exploring their other features in depth (delivery time, size of trunk, service frequency, security . . .) weighing the small differences in cost against the desirability of these features.

But we have overlooked a second objective, listed in blue on the left of Figure 9.1. You are an environmentally responsible person; you wish to minimize the CO_2 rating as well as the cost of ownership. Choosing to meet two objectives is more complicated than to meet just one. The problem is that the car that best satisfies one objective—minimizing cost, for example—may not be the one that minimizes the other—CO_2—and vice versa, so it is not possible to minimize both at the same time. A compromise has to be reached, and that needs *trade-off methods*.

Figure 9.2 shows the method. Its axes are the two objectives: cost of ownership and CO_2 rating. Suppose a friend has recommended a particular car, shown as an orange dot at the center of the diagram. Your research has revealed cars with combinations of cost and carbon shown by the other dots. Several—the purple ones in the upper right—have higher values of both; they lie in the "unacceptable" quadrant. Several—the blue ones—either have lower cost or lower carbon, but not both. One—the green one—is both cheaper *and* produces less carbon; it ranks more highly than the orange dot by both objectives. It is the obvious choice.

Or is it? That depends on the value you attach to a low carbon footprint. If you think it is a good idea so long as you don't have to pay a premium for it, then the car marked "Choice if cost matters most" is the best. If instead you are ready to pay whatever it takes to minimize CO_2 emission, then the car marked "Choice if carbon matters most" is the one to go for.

All three choices lie on the boundary of the occupied region of the figure together with several others that are compromises between them. The envelope of these—the broken line—is called the *trade-off line*. Cars that lie on or near this line have the best compromise combination of cost and carbon. So even if we can't reach a single

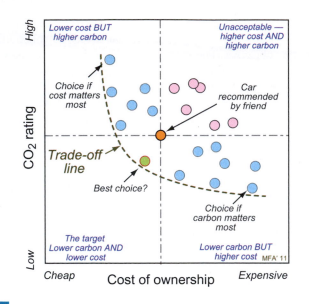

FIGURE 9.2 *The trade-off between carbon footprint and cost of ownership*

definitive choice (at least not yet) we have made progress. The viable candidates are those on or close to the trade-off line. All others are definitely less good.

Methods like this are used as a tool for decision making in many fields: in deciding between design options for new products, in optimizing the operating methods for a new plant, in guiding where the site of a new town will be—and in selecting materials. We turn to that next.

9.3 Principles of materials selection

Selecting materials involves seeking the best match between design requirements and the properties of the materials that might be used to make it. Figure 9.3 shows the strategy of the last section applied to selecting materials for a portable bike shed. On the left is the list of requirements that the material must meet, expressed as constraints and objectives. The constraints include the ability to be molded, weather resistance, adequate stiffness, and strength. The objectives include that it needs to be as light and as cheap as possible. On the right is the database of material attributes, drawn from suppliers' data sheets, handbooks, web-based sources, or software specifically designed for materials selection or from Chapter 15. The comparison "engine" applies the constraints on the left to the materials on the right and ranks the survivors using an objective, delivering a short list of viable candidates, just as with the cars. If two or more objectives are active, trade-off methods like the one in Figure 9.2 resolve the conflict.

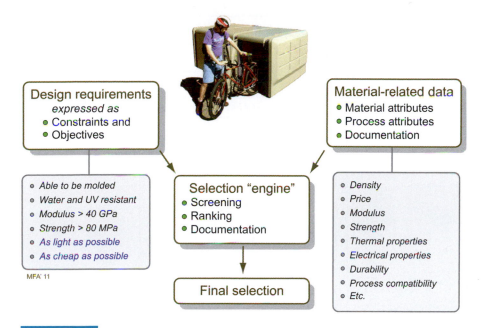

Design requirements
expressed as
● Constraints and
● Objectives

Material-related data
● Material attributes
● Process attributes
● Documentation

● *Able to be molded*
● *Water and UV resistant*
● *Modulus > 40 GPa*
● *Strength > 80 MPa*
● *As light as possible*
● *As cheap as possible*

MFA' 11

Selection "engine"
● Screening
● Ranking
● Documentation

● *Density*
● *Price*
● *Modulus*
● *Strength*
● *Thermal properties*
● *Electrical properties*
● *Durability*
● *Process compatibility*
● *Etc.*

Final selection

FIGURE 9.3 *Selecting a material for a portable bike shed. The requirements are expressed as constraints and objectives (objectives in blue). Records containing data for materials are* screened *using the constraints and* ranked *by the objectives to find the most attractive candidates. These are then explored further by examining documentation.*

There is, however, a complication. The requirements for the car were straightforward—doors, fuel type, power—all of these are explicitly listed by the manufacturer. The design requirements for a component of a product specify what it should *do* but not what *properties* its materials should have. So the first step is one of *translation*: converting the design requirements into constraints and objectives that can be applied to the materials database (Figure 9.4). The next task is that of *screening*, as with cars, eliminating the materials that cannot meet the constraints. This is followed by the *ranking* step, ordering the survivors by their ability to meet a criterion of excellence, such as that of minimizing cost, embodied energy, or carbon footprint. The final task is to explore the most promising candidates in depth, examining how they are used at present, case histories of failures, and how best to design with them, the step we called *documentation*. Now let's take a closer look at each step.

9.3.1 Translation. There is a story about a man—his name was Claude Shannon—who invented a product with only one function. It was a box with a switch on the front. When you pressed the switch, the box opened, a hand came

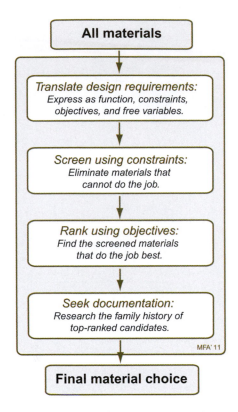

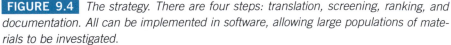

MFA' 11

FIGURE 9.4 *The strategy. There are four steps: translation, screening, ranking, and documentation. All can be implemented in software, allowing large populations of materials to be investigated.*

out, switched off the switch, and went back inside. If you pressed the button after that, nothing happened. Just one function. Once.

Some products are like that: air bags, for example, or disposable diapers. But most of the things engineers design function more than once, and many have more than one function. Typical functions are to support a load, to contain a pressure, to transmit heat, to provide electrical insulation, and so forth. This must be achieved subject to *constraints*: that certain dimensions are fixed, that the component must carry the design loads without failure, must insulate against or conduct heat or electricity, must function safely in a certain range of temperature and in a given environment, and many more. In designing the component, the designer has one or more *objectives*: to make it as cheap as possible, perhaps, or as light, or as environmentally benign, or some combination of these. Certain parameters can be adjusted in order to optimally meet the objective—the designer is free to vary dimensions that are not constrained by design requirements and, most importantly, free to

Table 9.1	Function, constraints, objectives, and free variables
Function	What does the component do?
Constraints	What non-negotiable conditions must be met?
Objective	What is to be maximized or minimized?
Free variables	What parameters of the problem is the designer free to change?

Table 9.2	Examples of common constraints and objectives*	
Common constraints		**Common objectives**
Must be		**Minimize**
■ Electrically conducting		■ Cost
■ Optically transparent		■ Mass
■ Corrosion resistant		■ Volume
■ *Non-toxic*		■ Thermal losses
■ *Non-restricted substance*		■ Electrical losses
■ *Able to be recycled*		■ *Resource depletion*
■ *Biodegradable*		■ *Energy consumption*
		■ *Carbon emissions*
Must meet a target value of		■ *Waste*
■ Stiffness		■ *Environmental impact*
■ Strength		■ *Water use*
■ Fracture toughness		
■ Thermal conductivity		
■ Service temperature		

Environment-related constraints and objectives are italicized.

choose the material for the component. We call these *free variables*. Constraints, objectives, and free variables (Table 9.1) define the boundary conditions for selecting a material and—in the case of load-bearing components—the choice of shape for its cross-section.

It is important to be clear about the distinction between constraints and objectives. A constraint is an essential condition that must be met, usually expressed as an upper or lower limit on a material property. An objective is a quantity for which an extreme value (a maximum or minimum) is sought, frequently the minimization of cost, mass, volume, or—of particular relevance here—environmental impact (Table 9.2).

The outcome of the translation step is a list of the design-limiting properties and the constraints they must meet. The first step in relating design requirements to material properties is therefore a clear statement of function, constraints, objectives, and free variables.

Translation of the design requirements for the helmet visor

Example: A material is required for the visor of a safety helmet to provide maximum facial protection.

Translation: To allow clear vision the visor must be *optically transparent*. To protect the face from the front, from the sides, and from below, it must be doubly curved, requiring that the material can *be molded*. We thus have two constraints: transparency and ability to be molded.

Fracture of the visor would expose the face to damage: "maximizing facial protection" therefore translates into maximizing resistance to fracture. The material property that measures resistance to fracture is the *fracture toughness*, K_{1c}. The objective is therefore to maximize K_{1c}.

9.3.2 Screening. Constraints are gates: meet the constraint and you pass through the gate, fail to meet it and you are shut out. Screening does just that: it eliminates candidates that cannot do the job at all because one or more of their attributes lies outside the limits set by the constraints. As examples, the requirement that "the component must function in boiling water" or that "the component must be non-toxic" imposes obvious limits on the attributes of maximum service temperature and toxicity that successful candidates must meet. The left-hand column of Table 9.2 lists common constraints.

9.3.3 Ranking: material indices. To rank the materials that survive the screening step we need criteria of excellence—what we have called objectives. The right-hand column of Table 9.2 lists common objectives. Each is a measure of performance. Performance is sometimes limited by a single property, sometimes by a combination of them. Thus the best materials to minimize thermal losses (an objective) are the ones with the smallest values of the thermal conductivity, λ. The best materials to minimize DC electrical losses (another objective) are those with the lowest electrical resistivity ρ_e—provided, of course, that they also meet all other constraints imposed by the design. Here the objective is met by selecting the material with an extreme (here, the lowest) value of a single property. Often, though, it is not one property but a group of properties that are relevant. Thus the best materials for a light stiff tie-rod are those with the smallest value of the group, ρ/E, where ρ is the density and E is Young's modulus. Those for a strong beam of lowest embodied energy are those with the lowest value of $H_m\rho/\sigma_y^{2/3}$ where H_m is the embodied energy of the material and σ_y is its yield strength. The property or property group that maximizes performance for a given design is called its *material index*.

Table 9.3 lists indices for stiffness and strength-limited design for three generic components: a tie, a beam, and a panel, for each of five objectives. The first three relate to design for the environment. Selecting materials with the objective of minimizing volume uses as few materials as possible, conserving resources. Selecting with the objective of minimizing mass is central to the eco-design of transportation systems (or indeed of anything that moves) because fuel consumption for transportation scales with weight. Selecting with the objective of minimizing embodied energy is important when large quantities of material are used, as they are in construction of buildings, bridges, roads, and other infrastructure. The fifth column, selection with the objective of minimizing cost, is always with us. Table 9.4 lists indices for thermal design. The first is a single property, the thermal conductivity, λ; materials with the lowest values of λ minimize heat loss at steady state, that is, when the temperature gradient is constant. The other two guide

Table 9.3	Indices for stiffness and strength-limited design					
	Configuration		**Objective: to minimize**			
		Volume	**Mass**	**Embodied energy**	**Carbon footprint**	**Material cost**
Stiffness-limited design	Tie	$1/E$	ρ/E	$H_m\rho/E$	$CO_2.\rho/E$	$C_m\rho/E$
	Beam	$1/E^{1/2}$	$\rho/E^{1/2}$	$H_m\rho/E^{1/2}$	$CO_2.\rho/E^{1/2}$	$C_m\rho/E^{1/2}$
	Panel	$1/E^{1/3}$	$\rho/E^{1/3}$	$H_m\rho/E^{1/3}$	$CO_2.\rho/E^{1/3}$	$C_m\rho/E^{1/3}$
Strength-limited design	Tie	$1/\sigma_y$	ρ/σ_y	$H_m\rho/\sigma_y$	$CO_2.\rho/\sigma_y$	$C_m\rho/\sigma_y$
	Beam	$1/\sigma_y^{2/3}$	$\rho/\sigma_y^{2/3}$	$H_m\rho/\sigma_y^{2/3}$	$CO_2.\rho/\sigma_y^{2/3}$	$C_m\rho/\sigma_y^{2/3}$
	Panel	$1/\sigma_y^{1/2}$	$\rho/\sigma_y^{1/2}$	$H_m\rho/\sigma_y^{1/2}$	$C_m\rho/\sigma_y^{1/2}$	$CO_2.\rho/\sigma_y^{1/2}$

Density, ρ (kg/m³); Elastic (Young's) modulus, E (GPa); Yield strength, σ_y (MPa); Carbon footprint CO_2 (kg/kg); Price, C_m ($/kg); Embodied energy/kg of material, H_m (MJ/kg)

Table 9.4	Indices for thermal design		
	Objective: to minimize		
Objective	Steady-state heat loss	Thermal inertia	Heat loss in a thermal cycle
Index	λ	$C_p\rho$	$(\lambda C_p\rho)^{1/2}$

Thermal conductivity λ (W/m·K); Specific heat, C_p (J/kg·K); Thermal diffusivity, $a = \lambda/C_p\rho$ (m²/s)

material choice when the temperature fluctuates. The symbols are defined under each table.

There are many such indices, each associated with maximizing some aspect of performance. They provide criteria of excellence that allow ranking of materials by their ability to perform well in the given application. Their derivation is described more fully in the appendix to this chapter and their use is illustrated by case studies in Chapter 10. All can be plotted on material property charts that identify the best candidates. The charts for the indices of Tables 9.3 and 9.4 appear later in this chapter (Section 9.6).

To summarize, then: *screening* uses constraints to isolate candidates that are capable of doing the job; *ranking* uses an objective to identify the candidates that can do the job best.

Screening and ranking for the helmet visor

Example: A search for transparent materials that can be molded delivers the following list. The first four are thermoplastics, the last two, glasses. Fracture toughness values come from the data sheets of Chapter 15.

Material	Average fracture toughness K_{1c} MPa $\cdot$ m$^{1/2}$
Polycarbonate (PC)	3.4
Cellulose acetate (CA)	1.7
Polymethyl methacrylate (Acrylic, PMMA)	1.2
Polystyrene (PS)	0.9
Soda-lime glass	0.6
Borosilicate glass	0.6

The constraints have reduced the number of viable materials to six candidates. When ranked by fracture toughness, the top-ranked candidates are PC, CA, and PMMA.

9.3.4 Documentation. The outcome of the steps so far is a ranked short list of candidates that meet the constraints and are ranked most highly by the objective. You could just choose the top-ranked candidate, but what hidden weaknesses might it have? What is its reputation? Has it a good track record? To proceed further, we seek a detailed profile of each: its documentation (Figure 9.4, bottom).

What form does documentation take? Typically, it is descriptive, graphical, or pictorial: case studies of previous uses of the material, failure analyses, details of its corrosion behavior in particular environments, of its availability and pricing, warnings of its environmental impact or toxicity, or descriptions of how it is recycled.

Such information is found in handbooks, suppliers' data sheets, web sites of environmental agencies, and other high-quality web sites. Documentation helps narrow the short list to a final choice, allowing a definitive match to be made between design requirements and material choice.

Why are all these steps necessary? Without screening and ranking, the candidate pool is enormous and the volume of documentation is overwhelming. Dipping into it, hoping to stumble on a good material, gets you nowhere. But once a small number of potential candidates have been identified by the screening and ranking steps, detailed documentation can be sought for these few alone, and the task becomes viable.

Documentation for materials for the helmet visor

Example: At this point it helps to know how the three top-ranked candidates listed in the previous Examples box are used. A quick web search reveals the following.

Polycarbonate: Safety shields and goggles; lenses; light fittings; safety helmets; laminated sheet for bulletproof glazing.

Cellulose acetate: Eyeglasses frames; lenses; goggles; tool handles; covers for television screens; decorative trim and steering wheels for cars.

PMMA, Plexiglas: Lenses of all types; cockpit canopies and aircraft windows; containers; tool handles; safety spectacles; lighting; automotive tail lights.

This is encouraging: all three materials have a history of use for goggles and protective screening. The one that ranked highest in our list—polycarbonate—has a history of use for protective helmets. We select this material, confident that, with its high fracture toughness, it is the best choice.

9.4 Selection criteria and property charts

Material property charts were introduced in Chapter 6. They are of two types: *bar charts* and *bubble charts*. A bar chart is simply a plot of one or a group of properties—Chapter 6 has several of them. Bubble charts plot two properties or groups of properties. Constraints and objectives can be plotted on them.

Screening: constraints on charts. As we have seen, design requirements impose non-negotiable demands ("constraints") on the material of which it is made. These limits can be plotted as horizontal or vertical lines on material property charts. Figures 9.5 and 9.6 show two schematic examples. The first is a bar chart of embodied energy. A selection line has been placed to impose the limit *embodied energy* <10 MJ/kg; all the materials below the line meet the constraint. The second shows a schematic of the *modulus-density* chart. We suppose that the design imposes limits on these of *modulus* >10 GPa and *density* $<2,000$ kg/m^3, shown

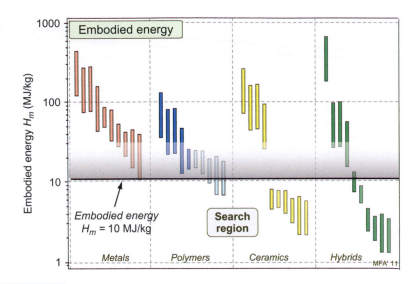

FIGURE 9.5 *Screening using a bar chart. Here we seek materials with embodied energies less than 10 MJ/kg. The materials in the "Search region" below the selection line meet the constraint.*

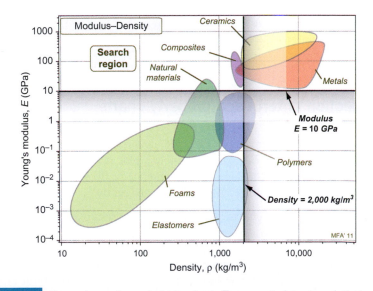

FIGURE 9.6 *Screening using a bubble chart. Two constraints are plotted: modulus >10 GPa and density <2,000 kg/m³. The materials in the "Search region" at the upper-left meet both constraints.*

on the figure. All materials in the window defined by the limits, labeled "Search region," meet both constraints.

Ranking: indices on charts. Material indices measure performance; they allow ranking of the materials that meet the constraints of the design. We use the design of light, stiff components as examples; the other material indices are used in a similar way.

Figure 9.7 shows a schematic of the $E-\rho$ chart shown earlier. The logarithmic scales allow all three of the indices—$M = \rho/E$, $\rho/E^{\frac{1}{3}}$ and $\rho/E^{\frac{1}{2}}$—listed in Table 9.3 of the previous section to be plotted onto it. Consider the first of these:

$$M = \frac{\rho}{E} \tag{9.1}$$

taking logs

$$\log(E) = \log(\rho) - \log(M) \tag{9.2}$$

For a given value of M, this is the equation of a straight line of slope 1 on a plot of $\log(E)$ against $\log(\rho)$, as shown on the figure. Similarly, the condition

$$M = \frac{\rho}{E^{\frac{1}{3}}} = \text{constant}, \ C \tag{9.3}$$

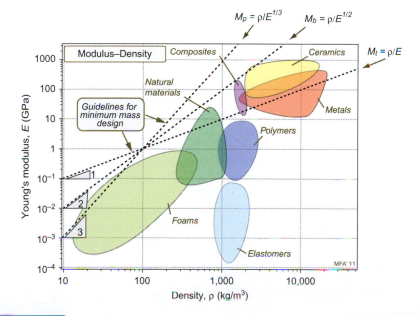

FIGURE 9.7 A schematic E−ρ chart showing guidelines for three material indices for stiff, lightweight structures

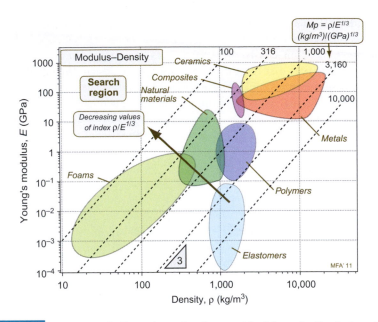

FIGURE 9.8 *A schematic* E–ρ *chart showing a grid of lines for the index* $\rho/E^{1/3}$. *The units are* $(kg/m^3)/(GPa)^{1/3}/(kg/m^3)$.

becomes, on taking logs,

$$\log(E) = 3\log(\rho) - 3\log(C) \tag{9.4}$$

This is another straight line, this time with a slope of 3, also shown. By inspection, the third index, $\rho/E^{1/2}$, will plot as a line of slope 2. We refer to these lines as *selection guidelines*. They give the slope of the family of parallel lines belonging to that index, each line corresponding to a different value of the index, M. Selection guidelines are marked on the charts that appear later in this chapter.

It is now easy to read off the subset of materials that maximize performance for each loading geometry. For example, all the materials that lie on a line of constant $M = \rho/E^{1/3}$ perform equally well as a light, stiff panel; those above the line perform better, those below, less well. Figure 9.8 shows a grid of lines corresponding to values of $M = \rho/E^{1/3}$ from $M = 100$ to $M = 10,000$ in units of $(kg \cdot m^{-3})/GPa^{1/3}$. A material with $M = 100$ in these units gives a panel that has one tenth the weight of one with $M = 1,000$. The first two texts listed under Further reading develop numerous case studies illustrating the use of the method.

9.5 Using indices for scaling

Most of the products we use today were designed when the dominant objectives were cost, performance, and safety. It is only now that the objective of minimizing environmental impact has been added to this list. An eco-audit or full LCA of these products identifies the phase of life that is causing the most damage, and it often suggests that replacing the materials with which the product is made by a set that are lighter, or stronger, or have lower embodied energies, or are easier to recycle would reduce the eco-burden.

But substitution is not that simple. The density of aluminum is one third of that of steel, so you might think that the weight of an aluminum car body-in-white (BiW) would be one third of that of a steel BiW. But aluminum is less than half as stiff and (depending on the alloy and heat treatment) about half as strong as steel. If the BiW is to function as well after the substitution as it did before, the section thickness of the aluminum components must be increased to compensate for the lesser properties. Thick sections are heavier than thin ones, so the reduction in weight is not nearly as large as it at first appeared. And there is something else to remember: aluminum costs, per kg, three times more than steel. Substitution is seldom cost-neutral. What, then, are the scaling laws when one material is replaced by another? And what is the gain in eco-performance when property-compensation is properly included? Indices can tell us.

The factor by which the mass of a tie, beam, or panel is changed by substitution is given by the ratio of the index for the new material to that of the old one. The factors by which embodied energy, carbon footprint, or material cost are changed by substitution are similarly given by the ratios of the relevant indices, giving the scaling factors in Table 9.5. The index describing the performance of a material as

Table 9.5	Scaling laws for stiffness- and strength-limited design				
	Configuration	**Eco-performance gain by substitution**			
		Volume*	**Mass***	**Embodied energy***	**Material cost***
Stiffness-limited design	Tie	$\left(\dfrac{E_o}{E_1}\right)$	$\dfrac{\rho_1}{\rho_o}\cdot\left(\dfrac{E_o}{E_1}\right)$	$\dfrac{H_{m,1}\rho_1}{H_{m,o}\rho_o}\cdot\left(\dfrac{E_o}{E_1}\right)$	$\dfrac{C_{m,1}\rho_1}{C_{m,o}\rho_o}\cdot\left(\dfrac{E_o}{E_1}\right)$
	Beam	$\left(\dfrac{E_o}{E_1}\right)^{1/2}$	$\dfrac{\rho_1}{\rho_o}\cdot\left(\dfrac{E_o}{E_1}\right)^{1/2}$	$\dfrac{H_{m,1}\rho_1}{H_{m,o}\rho_o}\cdot\left(\dfrac{E_o}{E_1}\right)^{1/2}$	$\dfrac{C_{m,1}\rho_1}{C_{m,o}\rho_o}\cdot\left(\dfrac{E_o}{E_1}\right)^{1/2}$
	Panel	$\left(\dfrac{E_o}{E_1}\right)^{1/3}$	$\dfrac{\rho_1}{\rho_o}\cdot\left(\dfrac{E_o}{E_1}\right)^{1/3}$	$\dfrac{H_{m,1}\rho_1}{H_{m,o}\rho_o}\cdot\left(\dfrac{E_o}{E_1}\right)^{1/3}$	$\dfrac{C_{m,1}\rho_1}{C_{m,o}\rho_o}\cdot\left(\dfrac{E_o}{E_1}\right)^{1/3}$
Strength-limited design	Tie	$\left(\dfrac{\sigma_{y,o}}{\sigma_{y,1}}\right)$	$\dfrac{\rho_1}{\rho_o}\cdot\left(\dfrac{\sigma_{y,o}}{\sigma_{y,1}}\right)$	$\dfrac{H_{m,1}\rho_1}{H_{m,o}\rho_o}\cdot\left(\dfrac{\sigma_{y,o}}{\sigma_{y,1}}\right)$	$\dfrac{C_{m,1}\rho_1}{C_{m,o}\rho_o}\cdot\left(\dfrac{\sigma_{y,o}}{\sigma_{y,1}}\right)$
	Beam	$\left(\dfrac{\sigma_{y,o}}{\sigma_{y,1}}\right)^{2/3}$	$\dfrac{\rho_1}{\rho_o}\cdot\left(\dfrac{\sigma_{y,o}}{\sigma_{y,1}}\right)^{2/3}$	$\dfrac{H_{m,1}\rho_1}{H_{m,o}\rho_o}\cdot\left(\dfrac{\sigma_{y,o}}{\sigma_{y,1}}\right)^{2/3}$	$\dfrac{C_{m,1}\rho_1}{C_{m,o}\rho_o}\cdot\left(\dfrac{\sigma_{y,o}}{\sigma_{y,1}}\right)^{2/3}$
	Panel	$\left(\dfrac{\sigma_{y,o}}{\sigma_{y,1}}\right)^{1/2}$	$\dfrac{\rho_1}{\rho_o}\cdot\left(\dfrac{\sigma_{y,o}}{\sigma_{y,1}}\right)^{1/2}$	$\dfrac{H_{m,1}\rho_1}{H_{m,o}\rho_o}\cdot\left(\dfrac{\sigma_{y,o}}{\sigma_{y,1}}\right)^{1/2}$	$\dfrac{C_{m,1}\rho_1}{C_{m,o}\rho_o}\cdot\left(\dfrac{\sigma_{y,o}}{\sigma_{y,1}}\right)^{1/2}$

*The subscript "o" refers to the original material; the subscript "1" refers to the substitute. The scaling laws for carbon footprint are the same as those for embodied energy with H_m replaced by CO_2.

Table 9.6	Scaling laws for thermal design		
Objective	**Eco-performance gain by substitution**		
	Fixed steady-state heat loss*	**Volume for given heat capacity***	**Heat loss in a thermal cycle***
Scaling law	$\dfrac{t_1}{t_o} = \dfrac{\lambda_1}{\lambda_o}$	$\dfrac{V_1}{V_o} = \dfrac{C_{p,1}\,\rho_1}{C_{p,o}\,\rho_o}$	$\dfrac{Q_1}{Q_o} = \left(\dfrac{\lambda_1\,C_{p,1}\rho_1}{\lambda_o\,C_{p,o}\rho_o}\right)^{1/2}$

*Wall thickness t, (m); Volume V, (m³); Heat Q (kJ)

thermal insulation is simply the thermal conductivity, λ (Table 9.4). The factor by which the insulation thickness must change to maintain the same level of heat loss when a new material is chosen is given by the ratio of the index for the new and the old. Change in thermal mass and thermal loss in a heat cycle, similarly, scale with the ratios of the indices, giving the scaling laws in Table 9.6.

Weight savings by materials substitution

Example: A steel beam, loaded in bending, is to be replaced by an aluminum one to save weight. The beam stiffness must remain unchanged. What is the maximum potential weight savings that this substitution allows? Here are the material properties.

Material	Density ρ (kg/m³)	Modulus E (GPa)
Steel	7,850	210
Aluminum	2,710	70

Answer: From Table 9.5, the ratio of the mass after substitution to that before is

$$\frac{m_1}{m_o} = \frac{\rho_1}{\rho_o}\cdot\left(\frac{E_o}{E_1}\right)^{1/2} = 0.6$$

Here the subscript "o" refers to steel, the subscript "1" to aluminum. Inserting the data gives the ratio 0.6, meaning that the maximum possible weight saving is 40% rather than the factor of 3 that the ratio of the densities suggests.

Volume savings by material substitution

Example: Standard polystyrene foam is used as thermal insulation for a small refrigerator. The foam has a thermal conductivity $\lambda_o = 0.035$ W/m · °C. It is suggested that the same thermal performance can be had with thinner walls (increasing the useful volume) by using a polymethacrylimide foam instead, which has a thermal conductivity $\lambda_1 = 0.028$ W/m·°C. By what factor can the wall thickness be reduced by this substitution while still maintaining the same thermal performance as before?

Answer: From Table 9.6 the ratio of wall thicknesses that will give the same heat loss per unit area is

$$\frac{t_1}{t_o} = \frac{\lambda_1}{\lambda_o} = 0.8$$

Thus the substitution allows the walls to be made 20% thinner than before.

9.6 Resolving conflicting objectives: trade-off methods

Just as with cars, real-life materials selection almost always requires that a compromise be reached between conflicting objectives. Table 9.2 lists nine of them, and there are more. The choice of materials that best meets one objective will not usually be that which best meets the others; the lightest material, for instance, will generally not be the cheapest or the one with the lowest carbon footprint. To make any progress, the designer needs a way of trading mass against cost and both against carbon footprint. This section describes ways of resolving this and other conflicts of objective.

Such conflicts are not new; engineers have sought methods to overcome them for at least a century. The traditional approach is that of using experience and judgment to assign *weight factors* to each constraint and objective, using them to guide choice in the way summarized below.

Weight factors. Weight factors seek to quantify judgment. The method works like this. The key properties or indices are identified and their values, M_i, are tabulated for promising candidates. Since their absolute values can differ widely and depend on the units in which they are measured, each is first scaled by dividing it by the largest index of its group, $(M_i)_{max}$, so that the largest, after scaling, has the value 1. Each is then multiplied by a weight factor, w_i, with a value between 0 and 1, expressing its relative importance for the performance of the component. This gives a weighted index W_i:

$$W_i = w_i \frac{M_i}{(M_i)_{max}} \tag{9.5}$$

For properties that are to be minimized, like corrosion rate, the scaling uses the minimum value $(M_i)_{min}$, expressed in the form

$$W_i = w_i \frac{(M_i)_{min}}{M_i} \tag{9.6}$$

The weight factors w_i are chosen so that they add up to 1, that is: $w_i < 1$ and $\Sigma w_i = 1$. The most important property is given the largest w, the second most important, the second largest, and so on. The W_i's are calculated from equation (9.5) and (9.6) and summed. The best choice is the material with the largest value of the sum

$$W = \Sigma_i W_i \tag{9.7}$$

Sounds simple, but there are problems, some obvious, like that of subjectivity in assigning the weights, some more subtle. Experienced engineers can be good at assessing relative weights, but the method nonetheless relies on judgment, and judgments can differ. For this reason, the rest of this section will focus on systematic methods.

Systematic trade-off strategies. Consider the choice of material to minimize both mass (performance metric P_1) and cost (performance metric P_2) while also meeting a set of constraints such as a required strength or durability in a certain environment. Following the standard terminology of optimizations theory, we define a *solution* as a viable choice of material, meeting all the constraints but not necessarily optimal by either of the objectives. Figure 9.9 is a plot of P_1 against P_2 for alternative solutions, each bubble describing a solution. The solutions that minimize P_1 do not minimize P_2, and vice versa. Some solutions, such as that at **A**, are far from optimal—all the solutions in the box attached to it have lower values of both P_1 and P_2. Solutions like **A** are said to be *dominated* by others. Solutions like those at **B** have the characteristic that no other solutions exist with lower values of *both* P_1 and P_2. These are said to be *non-dominated* solutions. The line or surface on which they lie is called the non-dominated or optimal *trade-off line*. The values of P_1 and P_2 corresponding to the non-dominated set of solutions are called the *Pareto set*.

Just as with cars (see Figure 9.2), the solutions on or near the trade-off line offer the best compromise; the rest can be rejected. Often, this is enough to identify a short list, using intuition to rank them. When it is not, the strategy is to define a *penalty function*.

Penalty functions. Consider first the case in which one of the objectives to be minimized is cost, C (units: $), and the other is mass, m (units: kg). We define a locally linear penalty function[1] Z:

$$Z = C + \alpha_m\, m \tag{9.8}$$

---

[1]Also called a *value function* or *utility function*. The method allows a local minimum to be found. When the search space is large, it is necessary to recognize that the values of the exchange constants, α_i, may themselves depend on the values of the performance metrics, P_i.

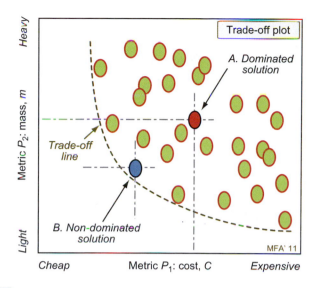

FIGURE 9.9 *Multiple objectives. Mass and cost for a component made from alternative material choices. The trade-off surface links non-dominated solutions.*

Here α_m is the change in $Z(\$)$ associated with unit increase in m (kg) and has the units of $/kg. It is called an *exchange constant*. Rearranging gives:

$$m = -\frac{1}{\alpha_m} C + \frac{1}{\alpha_m} Z \qquad (9.9)$$

This defines a linear relationship between m and C that plots as a family of parallel penalty lines, each for a given value of Z, as shown in Figure 9.10. The slope of the lines is the negative reciprocal of the exchange constant, $-1/\alpha_m$. The value of Z decreases toward the bottom left: the best choices lie there. The optimum solution is the one nearest the point at which a penalty line is tangential to the trade-off line, since it is the one with the smallest value of Z.

If instead the two objectives were cost ($) and carbon footprint CO_2 (kg), the penalty function becomes

$$Z = C + \alpha_c\, CO_2 \qquad (9.10)$$

Here the exchange constant, α_c, is the change in Z for unit increase in CO_2, and thus again has the units of $/kg. Contours of Z can be plotted on a chart of mass versus carbon footprint, just as in the previous example.

When all three objectives are active, the penalty function becomes

$$Z = C + \alpha_m \cdot m + \alpha_c\, CO_2 \qquad (9.11)$$

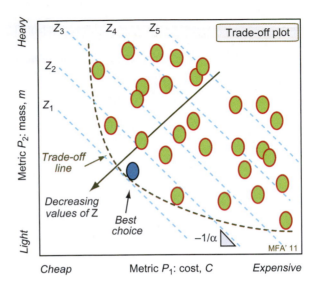

FIGURE 9.10 *The penalty function Z superimposed on the trade-off plot. The contours of Z have a slope of −1/α. The contour that is tangent to the trade-off surface identifies the optimum solution.*

It can no longer be plotted as contours on a two-dimensional chart, but it can be evaluated for candidate materials, choosing the one that minimizes Z. To do this we need values for α_m and α_c.

Values for the exchange constants α_m and α_c. An exchange constant is the value or "utility" of a unit change in a performance metric. In the example we have just seen, α_m is the utility ($) of saving 1 kg of weight. Its magnitude depends on the application. Thus the utility of weight savings in a family car is small, though significant; in aerospace it is much larger. The utility of heat transfer in house insulation is directly related to the cost of the energy used to heat the house; that in a heat-exchanger for electronics can be much higher because high heat transfer allows faster data processing, something worth far more. The utility can be real, meaning that it measures a true saving of cost. But it can also, sometimes, be perceived, meaning that the consumer, influenced by scarcity, advertising, or fashion, will pay more or less than the true value of the performance metric.

In many engineering applications the exchange constants can be derived approximately from technical models for the life-cost of a system. Thus the utility of weight savings in transportation systems is derived from the value of the fuel saved or that of the increased payload, evaluated over the life of the system. Table 9.7 gives approximate values for α for various modes of transportation. The most striking thing about them is the enormous range: the exchange constant depends in a dramatic way on the application in which the material will be used. It

Table 9.7	Exchange constants α_m for the mass–cost trade-off for transport systems	
Sector: transport systems	**Basis of estimate**	**Exchange constant, α_m (US\$/kg)**
Family car	Fuel saving	1–2
Truck	Payload	5–20
Civil aircraft	Payload	100–500
Military aircraft	Payload, performance	500–1,000
Space vehicle	Payload	3,000–10,000

is this that lies behind the difficulty in adopting aluminum alloys for cars despite their universal use in aircraft, it explains the much greater use of titanium alloys in military than in civil aircraft, and it underlies the restriction of beryllium (a very expensive metal) to use in space vehicles.

Exchange constants can be estimated approximately in various ways. The cost of launching a payload into space lies in the range \$3,000 to \$10,000/kg; a reduction of 1 kg in the weight of the launch structure would allow a corresponding increase in payload, giving the ranges of α_m shown in the table. Similar arguments based on increased payload or decreased fuel consumption give the values shown for civil aircraft, commercial trucks, and automobiles. The values change with time, reflecting changes in fuel costs, legislation to increase fuel economy, and the like.

These values for the exchange constant are based on engineering criteria. More difficult to assess are those based on perceived value. That for the performance/cost trade-off for cars is an example. To the enthusiast, a car that is able to accelerate rapidly is alluring. He (or she) is prepared to pay more to go from 0 to 60 mph in 5 seconds than to wait around for 10, as we will see in Chapter 10.

There are other circumstances in which establishing the exchange constant can be more difficult. An example is that for *environmental impact*—the damage to the environment caused by manufacture, or use, or disposal of a given product. Minimizing environmental impact has now become an important objective, almost as important as minimizing cost. Ingenious design can reduce the first without driving the second up too much. But how much is a unit decrease in impact worth?

Exchange constants for eco-design. One outcome of the Kyoto Protocol (see Chapter 5) was the creation of a market in carbon permits. A carbon permit to emit 1 metric ton of CO_2 per year sells today (March 2011) for €17 (\$24, or £15). Not all emitters of carbon have to hold carbon permits—ordinary households, for example, are exempt—but those that do not are now threatened with a carbon tax, at present aiming for €20 per metric ton). The permit price or tax set a value for the carbon-to-currency exchange constant, α_c, important because the penalty function can now be evaluated, allowing a properly balanced trade-off.

Using penalty functions

Example (1): The exchange constant for weight savings in a light goods vehicle is $\alpha_m = \$12/kg$, meaning that the value of weight reduction over the life of the vehicle is \$12 for each kilogram saved. A maker of such vehicles offers three models. The first uses steel panels for the body work. The second uses aluminum, costs \$2,500 more, but weighs 300 kg less. The third offers carbon-fiber paneling, costs \$8,000 more and weighs 500 kg less. Which is the best buy?

Answer: The penalty functions for the steel (1) and aluminum (2) vehicles are

$$Z_1 = C_1 + \alpha_m\, m_1$$

and

$$Z_2 = C_2 + \alpha_m\, m_2$$

The aluminum vehicle is attractive only if its value of Z is lower than that of the steel one. Writing

$$\Delta Z = Z_2 - Z_1 = C_2 - C_1 + \alpha_m(m_2 - m_1)$$
$$= 2,500 - 12 \times 300 = -\$\,1,100$$

The aluminum-paneled vehicle offers a life saving of \$1,100—it is a good buy. Repeating the comparison for the composite-paneled vehicle gives a value of $\Delta Z = + \$2,000$. It is not a good buy.

Example (2): A European-wide carbon tax is planned. It will be set initially at around \$20/metric ton (\$0.02/kg) of carbon emitted. If this is applied to material production, will it result in a significant rise in material prices? Express the result as a % increase above the current untaxed price.

Answer: Form a penalty function using $\alpha_c = 0.02$ \$/kg as the carbon-to-price exchange constant:

$$Z = C_m + 0.02 \times CO_2$$

where C_m is the material price in \$/kg and CO_2 is its carbon footprint in kg/kg. Rearranging to express Z as a % increase in C_m gives

$$\left(\frac{Z - C_m}{C_m}\right) \times 100 = \frac{2 \times CO_2}{C_m}$$

This quantity is plotted in the chart in Figure 9.11. The tax results in a 20% increase in the price of cement, a 10% increase in the price of aluminum and magnesium and smaller, but still significant increases in the price of the other materials.

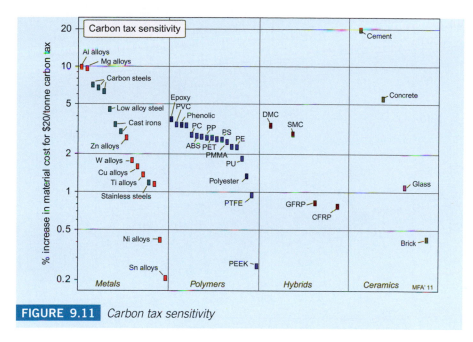

FIGURE 9.11 *Carbon tax sensitivity*

9.7 Seven useful charts

Seven material property charts guide materials selection to minimize mass, material embodied energy, material carbon footprint, and thermal losses using the indices of Tables 9.3 and 9.4. They are five of a much larger collection that can be found in the texts listed under Further reading at the end of this chapter.[2]

The Modulus–Density chart (Figure 9.12). The modulus E of engineering materials spans seven decades,[3] from 0.0001 GPa to nearly 1000 GPa; the density ρ spans a factor of 2,000, from less than 0.01 to 20 Mg/m^3. Members of each family cluster together and can be enclosed in envelopes, each of which occupies a characteristic part of the chart. The members of the ceramics and metals families have high moduli and densities; none has a modulus less than 10 GPa or a density less than 1.7 Mg/m^3. Polymers, by contrast, all have moduli below 10 GPa and densities that are lower than those of any metal or ceramic—most are close to 1 Mg/

---

[2]Some of charts can be downloaded, free, from grantadesign.com/education.

[3]Very low-density foams and gels (which can be thought of as molecular-scale, fluid-filled foams) can have lower moduli than this. As an example, gelatin (as in Jello) has a modulus of about 10^{-5} GPa.

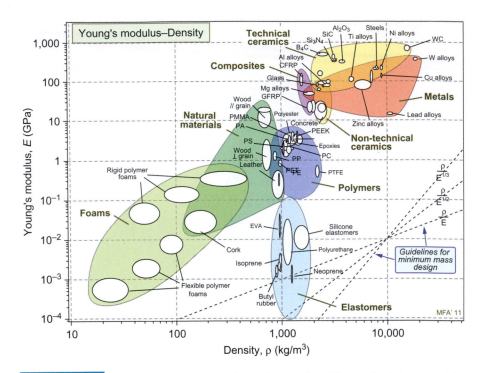

FIGURE 9.12 *The Modulus–Density chart: the one for stiffness at minimum weight*

m^3. Elastomers have roughly the same density as other polymers but their moduli are lower by a further factor of 100 or more. Materials with a lower density than polymers are porous: man-made foams and natural cellular structures like wood and cork.

This chart lets you select materials to minimize the mass of stiffness-limited structures. To do that, you need the three indices for the lightweight, stiffness-limited design in Table 9.3. Guidelines showing the slope of each of these are plotted on the chart. You might think that most structures are strength-, not stiffness-, limited, but that is wrong. Stiffness determines not only elastic deflection under load, but also vibration frequencies and resistance to buckling. When a hard-top vehicle line is augmented with an open-top model, its structure is beefed up to maintain stiffness, not strength.

The Strength–Density chart (Figure 9.13). The range of the yield strength σ_y or elastic limit σ_{el} of engineering materials, like that of the modulus, spans about six decades: from less than 0.01 MPa for foams, used in packaging and energy-absorbing systems, to 10^4 MPa for diamond, exploited in diamond tooling for machining and polishing. Members of each family again cluster together and can be enclosed in envelopes.

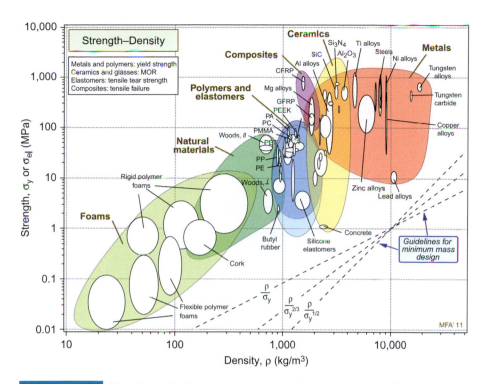

FIGURE 9.13 *The Strength–Density chart: the one for strength at minimum weight.*

Comparison with the Modulus-Density chart (Figure 9.12) reveals some marked differences. The modulus of a solid is a well-defined quantity with a narrow range of values. The yield strength is not. The strength range for a given class of metals, such as stainless steels, can span a factor of 10 or more, depending on its state of work hardening and heat treatment—it is this that leads to the elongated strength bubbles for metals. Polymers cluster together with strengths between 10 and 100 MPa. The composites CFRP and GFRP have strengths that lie between those of polymers and ceramics, as one might expect since they are mixtures of the two.

This chart is the one to select materials to minimize the mass of strength-limited structures. To do that you need the three indices for lightweight, strength-limited design in Table 9.3. Guidelines showing the slope of each of these are plotted on the chart.

The Modulus–Embodied energy and Strength–Embodied energy charts (Figures 9.14 and 9.15). The two charts just described guide design to minimize mass. If the objective becomes to minimize the energy embodied in the material of the product, we need equivalent charts for these.

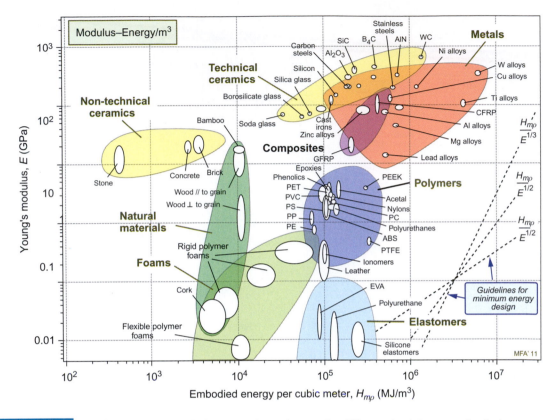

FIGURE 9.14 *The Modulus–Embodied energy chart: the one for stiffness at minimum embodied energy*

Figure 9.14 shows modulus E plotted against $H_m\rho$; the guidelines give the slopes for three of the commonest performance indices for stiffness-limited design at minimum embodied energy. Figure 9.15 shows strength σ_y plotted against $H_m\rho$. Guidelines give the slopes for strength-limited design at minimum embodied energy. They are used in exactly the same way as the $E-\rho$ and $\sigma_y-\rho$ charts for minimum mass design.

The Modulus–Carbon footprint and Strength–Carbon footprint charts (Figures 9.16 and 9.17). Two further charts allow optimized choice of when the objective is to minimize the carbon footprint of the material. The first, Figure 9.16, shows modulus plotted against carbon footprint per unit volume, $CO_2 \cdot \rho$, where CO_2 is the carbon footprint per kg of the material. The second, Figure 9.17, does the same for strength. Guidelines show the slopes associated with the indices of Table 9.3.

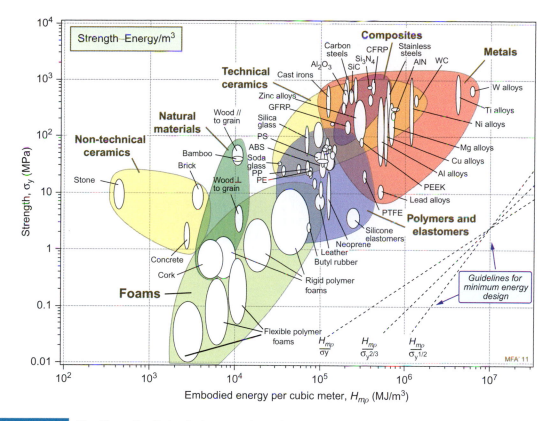

FIGURE 9.15 *The Strength–Embodied energy chart: the one for strength at minimum embodied energy*

The Thermal conductivity–Thermal diffusivity chart (Figure 9.18). The thermal conductivity, λ, is the material property that governs the flow of heat, q (W/m^2), in a steady temperature gradient dT/dx:

$$q = -\lambda \frac{dT}{dx} \qquad (9.12)$$

The thermal diffusivity, a (m^2/s), is the property that determines how quickly a thermal front diffuses into a material. It is related to the conductivity

$$a = \frac{\lambda}{\rho C_p} \qquad (9.13)$$

where ρC_p is the specific heat per unit mass ($J/kg \cdot K$). The contours show the volumetric specific heat, ρC_p, equal to the ratio of the two, λ/a. The data span almost five decades in λ and a. Solid materials are strung out along the line

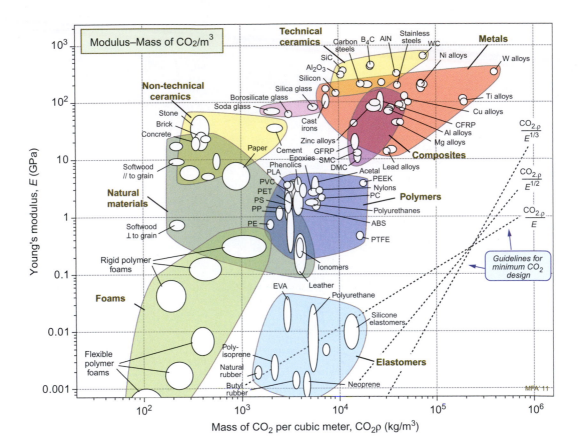

FIGURE 9.16 *The Modulus–Carbon footprint chart: the one for stiffness at minimum carbon release*

$$\rho C_p \approx 3 \times 10^6 \; J/m^3 \cdot K \qquad (9.14)$$

meaning that the heat capacity per unit volume, ρC_p, is almost constant for all solids, something to remember for later. As a general rule, then,

$$\lambda = 3 \times 10^6 \, a$$

(λ in W/m·K and a in m²/s). Some materials deviate from this rule: they have lower-than-average volumetric heat capacity. The largest deviations are shown by porous solids: foams, low-density firebrick, woods, and the like. Because of their low density they contain fewer atoms per unit volume and, averaged over the volume of the structure, ρC_p is low. The result is that, although foams have low *conductivities* (and are widely used for insulation because of this), their *thermal diffusivities* are not necessarily low. This means that they don't transmit much heat, but they do heat up or cool down quickly.

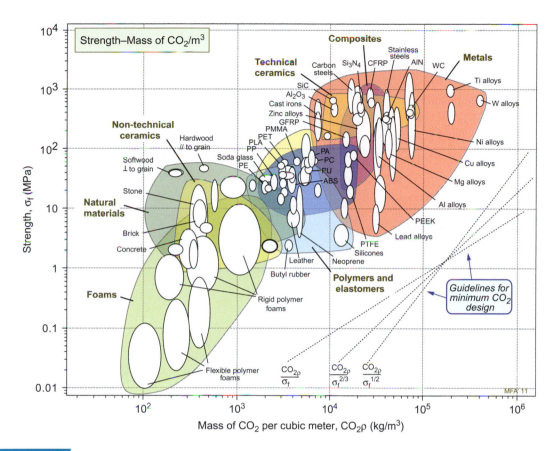

FIGURE 9.17   *The Strength–Carbon footprint chart: the one for strength at minimum carbon release*

Use of the eco-charts

Example: Which polymer, in sheet form, offers the lowest carbon footprint per unit of bending stiffness?

Answer: Table 9.3 gives the index for stiffness at a minimum carbon footprint for a panel in bending

$$M = \frac{CO_2 \cdot \rho}{E^{1/3}}$$

Selecting the corresponding guideline and displacing it to the left until only one polymer remains above it identifies PLA (polylactide—a biopolymer) as the best choice.

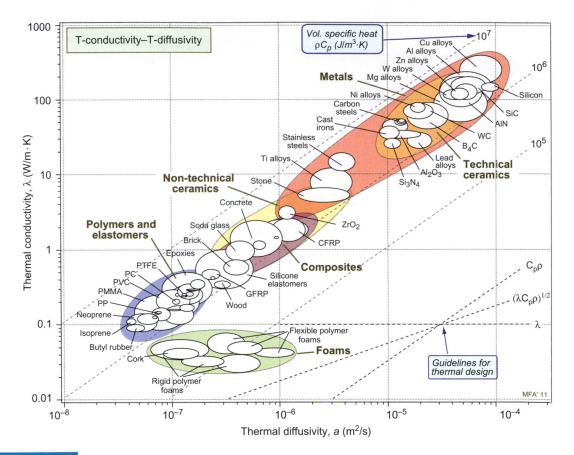

FIGURE 9.18 *The Thermal conductivity–Thermal diffusivity chart with contours of volumetric specific heat: the one for minimum thermal loss*

9.8 Computer-aided selection

The charts give an overview, but the number of materials that can be shown on any one of them is obviously limited. Selection using them is practical when there are very few constraints, but when there are many—as there usually are—checking that a given material meets them all is cumbersome. Both problems are overcome by a computer implementation of the method.

The *CES* material selection software is one such implementation (Figure 9.19). Its databases contain records for materials. It has a search and selection engine. It has plotting facilities to make material property charts and the ability to plot material indices on the charts and use them to optimize material selection, implementing the strategy of this chapter. More information can be found at grantadesign .com/education.

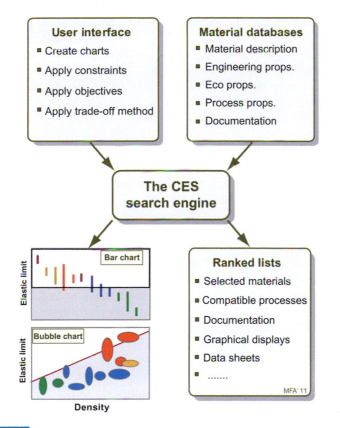

FIGURE 9.19   *The operation and outputs of typical selector software*

9.9 Summary and conclusions

A broad strategy exists that works for selecting almost anything anything—products, services, ... and materials. First decide on the attributes the material must (or must not) have, defining a set of constraints. Then apply the constraints, leaving a list of candidate-materials that meet them. Decide on an objective—a measure of excellence. It could be price (the cheaper the better), or weight (the lighter the better), or eco-impact (the lower the better), or some other measure of performance. Use this to rank the list of surviving candidates. Then get to work researching the top three or four on the list, gathering as much information as possible to make a well-informed final choice.

There are almost always many constraints, but this does not create a difficulty: simply apply them sequentially, retaining only those entities that meet them all. Often, too, there are two or more objectives and that does create a difficulty: the choice that best satisfies one is rarely the best choice for the other. Then trade-off methods become useful, either graphical ones, which allow you to plot the alternatives, identify the trade-off line, and then use your judgment to select an entity on or near the line; or analytical ones, where you formulate a penalty function and seek the entities that carry the lowest penalty.

Now we have a set of tools. In Chapter 10 they are used to analyze and select materials for design for the environment.

9.10 Further reading

Ashby, M. F. (2011), *Materials selection in mechanical design*, 4th edition, Butterworth Heinemann, Oxford, UK. ISBN 978-1-85617-663-7. *(A text that develops the ideas presented here in more depth, including the derivation of material indices, a discussion of shape factors, and a catalog of simple solutions to standard problems)*

Ashby, M. F., Shercliff, H. R., and Cebon, D. (2009), *Materials: engineering, science, processing and design*, 2nd edition, Butterworth Heinemann, Oxford, UK. ISBN 978-1-85617-895-2. *(An elementary text introducing materials through material property charts, and developing the selection methods through case studies)*

Bader, M. G. (1977), "Composites applications and design," Proc. of ICCM-11, Gold Coast, Australia, Vol. 1: ICCM, London. *(An example of trade-off methods applied to the choice of composite systems)*

Bourell, D. L. (1997), "Decision matrices in materials selection," *ASM Handbook Vol. 20, Materials Selection and Design*, editor G.E. Dieter, ASM International, Materials Park, Ohio, USA, pp 291−296. ISBN 0-87170-386-6. *(An introduction to the use of weight factors and decision matrices)*

Dieter, G. E. (2000), *Engineering design, a materials and processing approach*, 3rd edition, McGraw-Hill, New York, USA, pp 150−153, 255−257. ISBN 0-07-366136-8. *(A well-balanced and respected text, now in its 3rd edition, focusing on the role of materials and processing in technical design)*

Field, F. R. and de Neufville, R. (1988), "Material selection—maximizing overall utility," *Metals and Materials*, June, pp 378−382. *(A summary of utility analysis applied to material selection in the automobile industry)*

Goicoechea, A., Hansen, D. R., and Druckstein, L. (1982), *Multi-objective decision analysis with engineering and business applications*, Wiley, New York, NY, USA. *(A good starting point for the theory of multi-objective decision making)*

Keeney, R. L. and Raiffa, H. (1993), *Decisions with multiple objectives: preferences and value tradeoffs*, 2nd edition, Cambridge University Press, Cambridge, UK. ISBN 0-521-43883-7. *(A notably readable introduction to methods of decision making with multiple, competing objectives)*

9.11 Appendix: deriving material indices

This Appendix describes how material indices are derived. You can find out more about them and their use in the first two texts listed under Further reading. The Appendix has five sections:

(a) Material indices for stiffness and strength at minimum mass—simple section shapes
(b) Using shape to increase stiffness and strength at minimum mass
(c) Arches and shells
(d) Indices for stiffness and strength at minimum material embodied energy or carbon footprint
(e) Indices for stiffness and strength at minimum material cost

The performance of a component is characterized by a performance equation called the *objective function*. The performance equation contains a group of material properties. This group is the material indices of the problem. Sometimes the "group" is a single property. Thus if the performance of a uniform beam is measured by its stiffness, the performance equation contains only one property, the elastic modulus E. More commonly, the performance equation contains a group of two or more properties. Familiar examples are the *specific stiffness*, E/ρ (where E is Young's modulus, and ρ is the density), and the *specific strength*, σ_y/ρ (where σ_y is the yield strength or elastic limit), but there are many others. For reasons that will become apparent, we express the indices in a form for which a *minimum*, not a *maximum* is sought.

Recall that the life-energy and emissions for transport systems are dominated by the fuel consumed during use. The lighter the system is made, the less fuel it consumes and the less carbon it emits. So a good starting point is *minimum weight design*, subject, of course, to the other necessary constraints of which the most important, here, have to do with stiffness and strength. We consider the generic components shown in Figure 9.20: ties, panels, beams and shells, loaded as shown. The derivation when the objective is that of minimizing embodied energy, carbon footprint, or material cost follows in a similar way.

(a) Material indices for stiffness and strength at minimum mass— simple section shapes. **A light, stiff tie-rod.** A material is sought for a cylindrical tie rod that must be as light as possible (Figure 9.20(a)). Its length L_o is specified and it must carry a tensile force F without extending elastically by more than δ. Its stiffness must be at least $S^* = F/\delta$. We are free to choose the cross-section area A and, of course, the material. The design requirements, translated, are listed in Table 9.8.

We first seek an equation that describes the quantity to be minimized, here the mass m of the tie. This equation, the *objective function*, is

$$m = A\, L_o\, \rho \qquad (9.15)$$

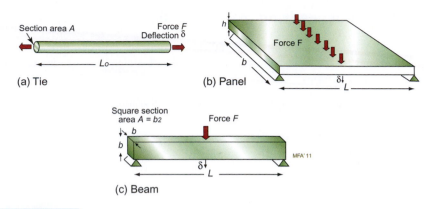

FIGURE 9.20 *Generic components with simple section shapes. (a) A tie, a tensile component. (b) A panel, loaded in bending. (c) A beam, loaded in bending.*

Table 9.8	Design requirements for the light stiff tie
Function	Tie rod
Constraints	Stiffness S^* specified (A functional constraint)
	Length L_0 specified (A geometric constraint)
Objective	Minimize mass
Free variables	Choice of material
	Choice of cross-section area A

where ρ is the density of the material of which the tie is made. We can reduce the mass by reducing the cross-section, but there is a constraint: the section area A must be sufficient to provide a stiffness $S^* = F/\delta$. The stiffness of a tie is

$$S = \frac{A E}{L_o} \qquad (9.16)$$

where E is Young's modulus. If the material has a low modulus, E, a large cross-section A is needed to give the necessary stiffness; if E is high, a smaller A is needed. The area required to provide at least the stiffness S^* is

$$A \geq \frac{L_o S^*}{E}$$

Inserting this into equation (9.15) gives

$$m \geq S^* L_o^2 \left(\frac{\rho}{E}\right) \qquad (9.17)$$

Both S^* and L_o are specified. The lightest tie that meets the design requirements is that made of the materials with the smallest values of the index

$$M_{t_1} = \frac{\rho}{E} \qquad (9.18,a)$$

provided that they also meet all other constraints of the design. If the constraint is not stiffness but strength, the index becomes

$$M_{t_2} = \frac{\rho}{\sigma_y} \qquad (9.18,b)$$

where σ_y is the yield strength. The best choice of material for the lightest tie that can support a load F without yielding is that with the smallest value of this index.

The mode of loading that most commonly dominates in engineering is not tension, but bending—think of the floor joists of buildings, of wing spars of aircraft, of shafts of golf clubs and racquets. The index for bending differs from that for tension, and this (significantly) changes the optimal choice of material. We start by modeling a panel.

A light, stiff panel. A panel is a flat slab, like a table top. Its length, L, and width, b, are specified but its thickness, h, is free. It is loaded in bending by a central load, F (Figure 9.20(b)). The stiffness constraint requires that it must not deflect more than δ. The objective is to achieve this with minimum mass, m. Table 9.9 summarizes the design requirements.

The objective function for the mass of the panel is the same as that for the tie:

$$m = A L \rho = b h L \rho$$

Its bending stiffness S must be at least S^*:

$$S = \frac{C_1 E I}{L^3} \geq S^* \qquad (9.19)$$

Table 9.9	Design requirements for the light stiff panel
Function	Panel
Constraints	Stiffness S^* specified (A functional constraint)
	Length L and width b specified (A geometric constraint)
Objective	Minimize mass
Free variables	Choice of material
	Choice of panel thickness h

Here C_1 is a constant that depends only on the distribution of the loads and I is the second moment of area, which, for a rectangular section, bh, is

$$I = \frac{b\,h^3}{12}$$

(9.20)

We can reduce the mass by reducing h, but only so far that the stiffness constraint is still met. Using the last two equations to eliminate h in the objective function gives

$$m \geq \left(\frac{12\,S^*}{C_1\,b}\right)^{1/3} (b\,L^2)\left(\frac{\rho}{E^{1/3}}\right)$$

(9.21)

The quantities S^*, L, b, and C_1 are all specified; the only freedom of choice left is that of the material. The best materials for a light, stiff panel are those with the smallest values of

$$M_{p1} = \frac{\rho}{E^{1/3}}$$

(9.22,a)

Repeating the calculation with a constraint of strength rather than stiffness leads to the index

$$M_{p2} = \frac{\rho}{\sigma_y^{1/2}}$$

(9.22,b)

These don't look much different from the previous indices, ρ/E and ρ/σ_y, but they are: they lead to different choices of material, as we saw in Section 9.4. For now, note the procedure. The in-plane dimensions of the panel were specified but we were free to vary the thickness h. The objective is to minimize its mass, m. Use the stiffness constraint to eliminate the free variable, here h. Then read off the combination of material properties that appears in the objective function—the equation for the mass. It sounds easy, and it is—as long as you know from the start what the constraints are, what you are trying to maximize or minimize, and which parameters are specified and which are free.

Now for another bending problem, in which the freedom to choose shape is greater than for the panel.

A light, stiff beam. Consider first a beam with a very simple cross-section, one that is square with an edge-length b and an area $A = b^2$. It is loaded in bending over a span of fixed length L with a central load F (Figure 9.20(c)). The stiffness constraint is again that it must not deflect more than δ under the load F, with the objective that the beam should again be as light as possible. Table 9.10 summarizes the design requirements.

Table 9.10	Design requirements for the light stiff beam
Function	Beam
Constraints	Stiffness S^* specified (A functional constraint)
	Length L ⎱
	Section shape square ⎰ (Geometric constraints)
Objective	Minimize mass
Free variables	Choice of material
	Area A of cross-section

Proceeding as before, the objective function for the mass is

$$m = A L \rho = b^2 L \rho$$

The bending stiffness S of the beam must be at least S^*

$$S = \frac{C_1 EI}{L^3} \geq S^* \tag{9.23}$$

where C_1 is a constant—we don't need its value. The second moment of area, I, for a square section beam is

$$I = \frac{b^4}{12} = \frac{A^2}{12} \tag{9.24}$$

For a given length, L, the stiffness, S^*, is achieved by adjusting the size of the square section. Now eliminating b (or A) in the objective function for the mass gives

$$m = \left(\frac{12 \, S^* \, L^3}{C_1}\right)^{1/2} (L) \left(\frac{\rho}{E^{1/2}}\right) \tag{9.25}$$

The quantities S^*, L, and C_1 are all specified or constant. The best materials for a light, stiff beam are those with the smallest values of the index M_b where

$$M_{b_1} = \frac{\rho}{E^{1/2}} \tag{9.26,a}$$

Repeating the calculation with a constraint of strength rather than stiffness leads to the index

$$M_{b_2} = \frac{\rho}{\sigma_y^{2/3}} \tag{9.26,b}$$

If the objective is to minimize mass, a square section is not a good choice. So we must examine the gains that can be made by using more efficient section shapes.

(b) Using shape to increase stiffness and strength at minimum mass.

Lightweight design is carried a step further by combining material choice with the choice of section shape and configuration. Figure 9.21 shows three examples. All three increase bending stiffness and strength by introducing stretching where before bending predominated.

Beams with efficient shapes. The mechanical efficiency of beams is increased by increasing the second moment of area, I, without increasing the cross-section A or mass m. This is achieved by locating the material of the beam as far from the neutral axis as possible, as in thin-walled tubes or I-beams (Figures 9.21(a) and 9.22). Some materials are more amenable than others to being made into efficient shapes. Comparing materials on the basis of the index M_b (equation 9.26) therefore requires some caution—materials with lower values of the index may "catch up" by being made into more efficient shapes. So we need to get an idea of the effect of shape on bending performance.

Figure 9.22 shows a solid square beam, of cross-section area A. If we turned the same area into an I beam or a tube in the way shown below each square of the figure, the mass of the beam is unchanged but the second moment of area, I, and thus the stiffness S (equation 9.23) are greater. We define the ratio of I for the shaped section to that for a solid square section with the same area (and thus

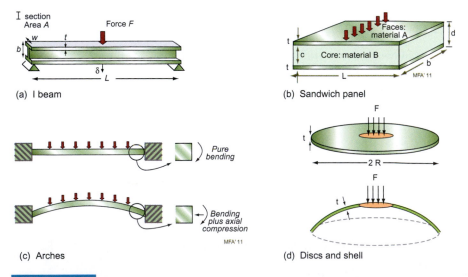

(a) I beam

(b) Sandwich panel

(c) Arches

(d) Discs and shell

FIGURE 9.21 *Generic components with more complex section shapes. (a) An I beam. (b) A sandwich panel. (c) An arch. (d) A shell. All are stiffer and stronger than the corresponding simple shape.*

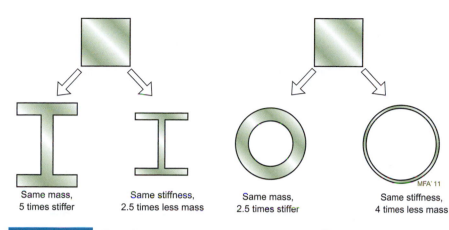

| Same mass, 5 times stiffer | Same stiffness, 2.5 times less mass | Same mass, 2.5 times stiffer | Same stiffness, 4 times less mass |

FIGURE 9.22   *The effect of section shape on bending stiffness* EI: *a square-section beam compared, left, with I and O shapes of the same area and thus mass per unit length as the square section, and right, with shapes that have the same stiffness as the square section, but less area and mass per unit length.*

mass) as the *shape factor* Φ. The more slender the shape, the larger is Φ, but there is a limit—make it too thin and the flanges will buckle or the tube will kink—so there is a maximum shape factor for each material that depends on its properties. Table 9.11 lists some typical values.

Shaping is used to make structures lighter: it is a way to get the same stiffness with less material. The mass ratio is given by the reciprocal of the square root of the maximum shape factor, $\Phi^{-\frac{1}{2}}$ (that is because C_1, which is proportional to the shape factor, appears as $(C_1)^{-\frac{1}{2}}$ in equation (9.23)). Table 9.11 lists the factor by which a beam can be made lighter, for the same stiffness, by shaping. Metals and composites can all be improved significantly (though the metals do a little better), but wood has a more limited potential because it is more difficult to shape it into efficient, thin-walled shapes. So, when comparing materials for light, stiff beams using the indices of equation (9.26), remember that the performance of wood is not as good as it looks because other materials can be made into more efficient shapes. Composites (particularly CFRP) have attractive (i.e., low) values of all the indices, M_t, M_p, and M_b, but this advantage relative to metals is reduced a little by the greater shape efficiency of which metal are capable.

Sometimes a beam of a given shape is to be replaced with one of a different material but with the same shape. If we constrain the shape to be *self-similar* so that all dimensions change in proportion as we vary the overall size, the indices derived in the previous section still hold. This is a consequence of equation (9.24)—for a given shape, the second moment of area I can always be expressed as a constant times A^2, so changing the shape just changes the constant C_1 in equation (9.23), not the resulting index.

Sandwich structures. A sandwich panel combines two materials configured so that one forms the faces, the other the core, to give a structure of high bending stiffness

Table 9.11	The effect of shaping on stiffness and mass of beams in different structural materials	
Material	Typical maximum shape factor (stiffness relative to that of a solid square beam)	Typical mass ratio by shaping (mass relative to that of a solid square beam)
Steels	64	1/8
Al alloys	49	1/7
Composites (GFRP, CFRP)	36	1/6
Wood	9	1/3

and strength at low weight (Figure 9.20 (b)). The separation of the faces by the core increases the moment of inertia of the section, I, producing a structure that resists bending and buckling loads well. Sandwiches are used when weight saving is critical: in aircraft, trains, trucks, and cars, in portable structures, and in sports equipment.

The faces, each of thickness t, carry most of the load, so they must be stiff and strong, and they form the exterior surfaces of the panel so they must also tolerate the environment in which it operates. The core, of thickness c, occupies most of the volume; it must be light, and stiff and strong enough to carry the shear stresses necessary to make the whole panel behave as a load-bearing unit. If the core is much thicker than the faces, these stresses are small.

Very approximately, if the monolithic panel of thickness h of Figure 9.20(b) is split in half to make the two faces of the sandwich panel of thickness d in Figure 9.21(b), the bending stiffness rises by the factor

$$\frac{S_{sandwich}}{S_{mono}} = 1 + 3\frac{d}{h}\left(\frac{d}{h} - 1\right) \qquad (9.27,a)$$

provided the faces remain thick enough that they do not buckle. The bending strength rises by

$$\frac{P_{f,\ sandwich}}{P_{f,\ mono}} = 3\frac{d}{h} + \frac{h}{d} - 3 \qquad (9.27,b)$$

subject to the same provision. Thus if the sandwich is twice as thick as the monolithic panel, it is 4 times stiffer and 3.5 times stronger. If the sandwich is three times thicker, it is 19 times stiffer and 6 times stronger. There is more to the

Table 9.12	Design requirements for the light, stiff, strong shell
Function	Support distributed load F without excessive deflection or failure
Constraints	Stiffness S^* specified $\quad\}$ (Functional constraints) Failure load F^* specified Radius R specified $\quad\}$ (Geometric constraints) Load distribution specified
Objective	Minimize mass
Free variables	Thickness of shell wall, t Choice of material

design of sandwich structures than this,[4] but it does illustrate the large gains that sandwich construction makes possible.

*(c) **Arches and shells.*** Singly curved plates and doubly curved panels ("shells") are stiffer and stronger than flat ones because the external load is carried, in large part, by membrane stresses. A *membrane stress* is a stress that acts in the plane of the plate or shell, tending to stretch or compress it (Figure 9.21(c)). Thin shapes are much stiffer and stronger in simple compression or tension than they are in pure bending, so the curvature allows a real increase in performance at the same mass, or a mass saving with the same performance.

Figure 9.21(d) shows a circular disc and a shell of radius R, both of thickness t carrying a distributed load F. The load induces a deflection, δ, and a maximum membrane stress, σ. Define the stiffness S as F/δ. The stiffness constraint then becomes $S \geq S^*$, where S^* is the desired stiffness. The strength constraint is simply $\sigma \leq \sigma_y$ where σ_y is the yield strength of the material. Table 9.12 summarizes the requirements.

The mass of the disc and the shell segment (the objective function) is

$$m \approx \pi R^2 t \, \rho \qquad (9.28)$$

where ρ is the density of its material. The deflection δ and the maximum bending stress σ created by a distributed load like that in the figure are standard results.[5] The stiffness of the disc is

$$S = \frac{F}{\delta} = \frac{E\,t^3}{AR^2} \geq S^* \qquad (9.29)$$

The maximum membrane stress σ is

$$\sigma = B\frac{F^*}{t^2} \leq \sigma_y \qquad (9.30)$$

---

[4]See Chapter 11 of the text "Materials Selection in Mechanical Design," referenced under Further reading, for details.

[5]See the compilation by Young listed under Further reading.

where E is Young's modulus, ν is Poisson's ratio, S^* is the desired stiffness, F^* is the desired load-bearing capacity, and $A \approx 0.27$ and $B \approx 0.95$ are constants that depend weakly on how the load is distributed on the surface and the value of Poisson's ratio.

The corresponding values for the shell segment are

$$S = \frac{F}{\delta} = \frac{E\,t^2}{CR} \geq S^* \qquad (9.31)$$

The maximum membrane stress σ is

$$\sigma = D\frac{F^*}{t^2} \leq \sigma_y \qquad (9.32)$$

where E is Young's modulus, ν is Poisson's ratio, S^* is the desired stiffness, F^* is the desired load-bearing capacity, and $C \approx 0.4$ and $D \approx 0.65$ are constants that, like A and B, depend weakly on how the load is distributed and on Poisson's ratio. Solving each of these for t and substituting the result into the objective function gives for the disc

$$m_1 = \pi\,R^2 (A\,R\,S^*)^{1/3}\left[\frac{\rho}{E^{1/3}}\right] \quad \textit{(stiffness constraint)} \qquad (9.33)$$

and

$$m_2 = \pi\,R^2 (BF^*)^{1/2}\left[\frac{\rho}{\sigma_y^{1/2}}\right] \quad \textit{(strength constraint)} \qquad (9.34)$$

The corresponding masses for the shell are

$$m_1 = \pi\,R^2 (C\,R\,S^*)^{1/2}\left[\frac{\rho}{E^{1/2}}\right] \quad \textit{(stiffness constraint)} \qquad (9.35)$$

and

$$m_2 = \pi\,R^2 (DF^*)^{1/2}\left[\frac{\rho}{\sigma_y^{1/2}}\right] \quad \textit{(strength constraint)} \qquad (9.36)$$

Everything in these equations is specified except for the material properties in square brackets, defining the indices. The shell is stiffer and stronger than the plate by the factors

$$\frac{S_{shell}}{S_{disc}} = \frac{A\,R}{C\,t} \approx 0.8\frac{R}{t} \quad \text{and} \quad \frac{\sigma_{shell}}{\sigma_{disc}} = \frac{D}{B} \approx 1.5 \qquad (9.37)$$

Extremely light, stiff, strong structures are made possible by combining sandwich construction with a shell-like shape (*monocoque construction*). We will encounter this further in the Case Studies of Chapter 10.

(d) Indices for stiffness and strength at minimum material embodied energy or carbon footprint. We've done all the hard work. Extending the ideas to minimizing the contribution of material embodied energy H_m or carbon footprint CO_2 to product life is just a case of replacing density ρ in the indices derived so far by $H_m\rho$ or $CO_2 \cdot \rho$. Here is the argument.

Minimizing embodied energy or carbon footprint. When the objective is to minimize embodied energy rather than mass the indices change. If the embodied energy of the material is H_m MJ/kg, the energy embodied in a component of mass m is just mH_m. The objective function for the energy H embodied in the tie, panel, or beam then becomes

$$H = m\, H_m = A\, L\, H_m\, \rho \tag{9.38}$$

Proceeding along the same steps as for minimum mass then leads to indices that have the form of equations (9.18), (9.22), and (9.26) with ρ replaced by $H_m\rho$, as in Table 9.3. Minimizing the carbon footprint follows the same procedure, replacing H_m MJ/kg by the material footprint CO_2 kg/kg in the indices.

(e) Indices for stiffness and strength at minimum material cost. **Minimizing material cost**. When, instead, the objective is to minimize cost rather than mass, the indices change again. If the material price is C_m \$/kg, the cost of the material to make a component of mass m is just mC_m. The objective function for the material cost C of the tie, panel, or beam then becomes

$$C = m\, C_m = A\, L\, C_m\, \rho \tag{9.39}$$

Proceeding as before then leads to indices that have the form of equations (9.16), (9.20), and (9.24) with ρ replaced by $C_m\rho$, as in Table 9.3. It must be remembered that the material cost is only part of the cost of a shaped component; there is also the manufacturing cost—the cost to shape, join, and finish it.

9.12 Exercises

E9.1. What is meant by an *objective* and what is meant by a *constraint* in the requirements for a design? How do they differ?

E9.2. Describe and illustrate the "Translation" step of the material selection strategy. Materials are required to make safe, eco-friendly swings and climbing frames for a children's playground. How would you translate these design requirements into a specification for selecting materials?

E9.3. Bicycles come in many forms, each aimed at a particular sector of the market:

- Sprint bikes
- Touring bikes
- Mountain bikes
- Shopping bikes
- Children's bikes
- Folding bikes

Use your judgment to identify the primary objective and the constraints that must be met for each of these.

E9.4. You are asked to design a fuel-saving cooking pan with the goal of wasting as little heat as possible while cooking. What objective would you choose, and what constraints would you recommend should be met?

E9.5. Formulate the constraints and objective you would associate with the choice of material to make the forks of a racing bicycle.

E9.6. What is meant by a *material index*?

E9.7. The objective in selecting a material for a panel of given in-plane dimensions for the lid casing of an ultra-thin portable computer is that of minimizing the panel thickness, h, while meeting a constraint on bending stiffness, S^*, to prevent damage to the screen. What is the appropriate material index?

E9.8. Plot the index for a light, stiff panel on a copy of the *Modulus–Density* chart of Figure 9.11, positioning the line so that six materials are left above it, excluding ceramics because of their brittleness. Which six do you find? What material classes do they belong to?

E9.9. Panels are needed to board up the windows of an unused building. The panels should have the lowest possible embodied energy but be strong enough to deter an intruder who, in attempting to break in, will load the panels in bending. (a) Which index would you choose from Table 9.3 to guide your choice?

(b) Plot the index on the *Strength–Embodied energy* chart of Figure 9.14, positioning the line to find the best choice, excluding ceramics because of their brittleness. Which six do you find? What material classes do they belong to?

E9.10. A material is required for disposable forks for a fast-food chain. List the objectives and the constraints that you would see as important in this application.

E9.11. Use the $E–H_m\rho$ chart of Figure 9.14 to find the metal with a modulus E greater than 100 GPa and the lowest embodied energy per unit volume.

E9.12. *Weight saving by materials substitution.* A steel beam, loaded in bending, is to be replaced by an aluminum one to save weight. The bending strength of the beam must remain unchanged. What is the maximum potential weight savings that this substitution allows? Here are the material properties.

Material	Density ρ (kg/m^3)	Strength σ_y (MPa)
Steel YS260	7,850	288
Aluminum 6061-T4	2,710	113

E9.13. *Carbon penalty of material substitution.* The Guggenheim Museum in Bilbao, Spain, is clad with titanium sheet. It is suggested that a stainless steel sheet of the same thickness would be mechanically just as good and have a much lower carbon footprint. Is this statement accurate? To figure this and the next question out, you will need the data in this table.

Material	Density ρ (kg/m^3)	Carbon footprint CO_2 (kg/kg)	Young's modulus E (GPa)
Stainless steel	7,800	5.0	200
Titanium	4,600	46.4	115

E9.14. *Carbon penalty of material substitution again.* The person who suggested that stainless steel might be a more environmentally thoughtful choice now points out further that stainless steel has a higher modulus than titanium, so its substitution by a stainless steel sheet of the same bending stiffness, rather than the same thickness, would make more sense. What is the ratio of carbon footprints per unit area when this comparison is made? The previous question has the data you need.

E9.15. A maker of polypropylene (PP) garden furniture is concerned that the competition is stealing part of his market by claiming that the "traditional" material for garden furniture, cast iron, is much less energy- and CO_2-intensive than the PP. A typical PP chair weighs 1.6 kg; one made of cast iron weighs 11 kg. Use the data sheets for these two materials in Chapter 15 to find out who is right—are the differences significant? (Remember the warning about precision at the start of Chapter 6.)

If the PP chair lasts 5 years and the cast iron chair lasts 25 years, does the conclusion change?

E9.16. Show that the index for selecting materials for a strong panel with the dimensions shown here, loaded in bending, with minimum embodied energy content is

$$M = \frac{\rho\, H_m}{\sigma_y^{1/2}}$$

where H_m is the embodied energy of the material, ρ is its density, and σ_y is its yield strength. To do so, rework the panel derivation in Section 9.11 by replacing the stiffness constraint with a constraint on failure load F requiring that it exceed a chosen value F^* where

$$F = C_2 \frac{I\, \sigma_y}{h\, L} > F^*$$

where C_2 is a constant and I is the second moment of area of the panel: $I = \frac{b\, h^3}{12}$.

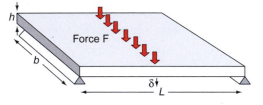

Exercises using CES Edu Level 2 Eco

E9.17. Use a "Limit" stage to find materials with modulus $E > 180$ GPa and embodied energy $H_m < 30$ MJ/kg.

E9.18. Use a "Limit" stage to find materials with yield strength $\sigma_y > 100$ MPa and a carbon footprint $CO_2 < 1$ kg/kg.

E9.19. Make a bar chart of embodied energy H_m. Limit the selection to polymers alone by creating a subset that only contains the "Polymer" folder. Which polymers have the lowest embodied energy?

E9.20. Make a chart showing modulus E and density ρ. Apply a selection line of slope 1, corresponding to the index ρ/E, positioning the line so that six materials are left above it. What families do they belong to?

E9.21. A material is required for a tensile tie to link the front and back walls of a barn to stabilize both. It must meet a constraint on strength and have as low an embodied energy as possible. To be safe, the material of the tie must have fracture toughness $K_{1c} > 18$ MPa $\cdot$ m$^{1/2}$. The relevant index is

$$M = \frac{H_m \rho}{\sigma_y}$$

Construct a chart of σ_y plotted against $H_m\rho$. Add the constraint of adequate fracture toughness, meaning $K_{1c} > 18 \ \mathrm{MPa \cdot m}^{V_2}$, using a "Limit" stage. Then plot an appropriate selection line on the chart and report the three materials that are the best choices for the tie.

E9.22. A company wishes to enhance its image by replacing oil-based plastics in its products by polymers based on natural materials. Use the "Search" facility in CES to find *biopolymers*. List the material you find. Are their embodied energies and CO_2 footprints less than those of conventional plastics? Make bar charts of embodied energy and CO_2 footprint to find out.

CHAPTER 10

Eco-informed materials selection

CONTENTS

10.1 Introduction and synopsis

Audits like those of Chapters 7 and 8 point the finger, directing attention to the life phase that is of most eco-concern. If you point fingers you invite the response: what do you propose to do about it? That means moving from *auditing* to *selection*—from the top part of the strategy of Figure 3.11 to the bottom.

Chapter 9 introduced selection methods. Here we illustrate their use with case studies. The first two are simple, showing how selection methods work. Remember, in reading them, that there is always more than one answer to environmental questions. Material substitution guided by eco-objectives is one way forward but it is not the only one. It may sometimes be better to abandon one way of doing things altogether (the IC engine, for example) and replace it with another (fuel-cell/electric

---

Super light-weight vehicles: Shell eco-marathon contester, Cal Poly Supermilage team vehicle, "Microjoule" by the students of the Lycee La Joliverie, France, and "Pivo2" electric car by Nissan. All have shell bodies made from materials chosen for stiffness and strength at minimum mass.

275

power, perhaps). So while change of material is one option, change of concept is another. And of course there is a third: change of lifestyle (no vehicle at all).

The case studies are in four groups.

- Materials for drink containers (10.2, 10.3)
- Materials for buildings (10.4, 10.5)
- Heating and cooling (10.6–10.8)
- Materials for transport (10.9–10.12)

The groups are self-contained. You can jump to the group that interests you without needing to read those that come before it. They are, however, arranged so that the simplest are at the beginning and the more complicated at the end.

10.2 Which bottle is best? Selection per unit of function

Drink containers co-exist that are made from many different materials: glass, polyethylene, PET, aluminum, steel—Figure 10.1 shows examples of them. Surely one material must be a better environmental choice than the others? The audit of a PET bottle in Chapter 7 delivered a clear message: the phase of life that dominates energy consumption and CO_2 emission is that of producing the material of which the bottle is made. Embodied energies per kg for the five materials are plotted in the upper part of Figure 10.2; a plot of their carbon footprints shows a similar distribution. Glass has the lowest values of both. It would seem that glass is the best choice.

But hold on. These are energies *per kg of material*. The containers differ greatly in weight and volume. What we need are *energies per unit of function*. So let's start again and do the job properly, listing the design requirements. The material must not corrode in fluids that are mildly acidic (fruit juice) or those that are mildly alkaline (milk). It must be easy to shape, and—given the short life of a container—it must be recyclable. Table 10.1 lists the requirements, including the objective of minimizing embodied energy *per unit volume of fluid contained*.

The volumes and masses of the five competing container types, the material of which they are made, and the embodied energy of each are listed Table 10.2. All

Glass   PE   PET   Aluminum   Steel

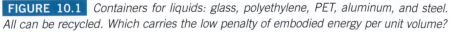

FIGURE 10.1 *Containers for liquids: glass, polyethylene, PET, aluminum, and steel. All can be recycled. Which carries the low penalty of embodied energy per unit volume?*

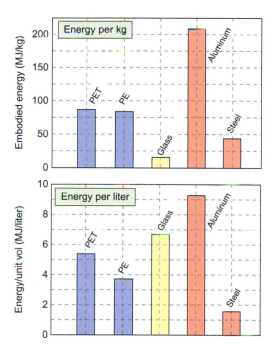

FIGURE 10.2 *Above: the embodied energy of the bottle materials. Below: the material energy per liter of fluid contained.*

Table 10.1	Design requirements for drink containers
Function	■ Drink container
Constraints	■ Must be immune to corrosion in the drink ■ Must be easy and fast to shape ■ Must be recyclable
Objective	■ Minimize embodied energy per unit capacity
Free variable	■ Choice of material

five materials can be recycled. For all five, cost-effective processes exist for making containers. All but one—steel—resist corrosion in the mildly acidic or alkaline conditions characteristic of bottled drinks. But steel is easily protected with lacquers.

That leaves us with the objective. The last column of the table lists the embodied energies per liter of fluid contained, calculated from the numbers in the other columns. The results are plotted in the lower part of Figure 10.2. The ranking is now very different: steel emerges as the best choice, polyethylene the next best. Glass (because so much is used to make one bottle) and aluminum (because of its high embodied energy) are the least good.

Table 10.2	Data for the containers with embodied energies for virgin material			
Container type	Material	Mass (grams)	Embodied energy (MJ/kg)	Energy/liter (MJ/liter)
PET 400 ml bottle	PET	25	84	5.3
PE 1 liter milk bottle	High density PE	38	81	3.8
Glass 750 ml bottle	Soda glass	325	15.5	6.7
Al 440 ml can	5,000 series Al alloy	20	208	9.5
Steel 440 ml can	Plain carbon steel	45	32	3.3

Postscript. In all discussion of this sort there are issues of primary and of secondary importance. There is cost—we have ignored this because eco-design was the prime objective. There is the economics of recycling: the value of recycled materials depends to differing degrees on impurity pick-up. There is the fact that real cans and bottles are made with some recycle content, reducing the embodied energies of all five to varying degrees, but not enough to change the ranking. There is the extent to which current legislation subsidizes or penalizes one material or another. And there is appearance: transparency is attractive for some containers but irrelevant for others. But we should not let these cloud the primary finding: that the containers differ in their life-energy, dominated by material, and that steel is by far the least energy intensive.

News-clip: Small savings

Lighter bottle tops.
It's all change at the top for ASB Miller, brewer of Grolsch and Peroni. The company has developed a bottle cap that uses less steel, reducing its raw material costs and cutting carbon dioxide emissions, thanks to the lighter loads on delivery trucks.... The group uses 42 billion tops a year.
***The Sunday Times**, July 3, 2011*

How impressed should we be by this headline? The crown caps referred to in the article weigh 2.5 grams. Let us suppose a 20% weight saving: 0.5 grams per cap. Multiplied by 42 billion this gives 21,000 metric tons of steel. The first claim—that of reducing raw material costs for the cap-makers—appears justified. What about the transport? A 500 ml (half liter) bottle of beer weighs 310 grams when empty, 810 grams when full. Saving 0.5 grams of steel per unit reduces its mass, and thus the transport energy and CO_2 release, by 0.06%. Much greater savings are possible by asking the driver of the delivery truck to drive slightly more slowly.

10.3 Systematic eco-selection: carbonated-water bottles

Bottles for bottled water are used only once and then recycled or combusted. The audit of Section 7.2 revealed that the largest contribution to the life-energy and

CO_2 release derived from the material of the bottle itself; subsequent manufacture, transport, use, and disposal all contributed much less. Here we consider material selection to minimize life-energy of a bottle for carbonated water, requiring it to support the pressure of the gas without failing (Figure 10.3). Further requirements are that the bottle be moldable, transparent, and able to be recycled (Table 10.3).

The internal pressure p in the bottle creates tensile stresses in its walls. The circumferential stress is $\sigma_c = pr/t$ and the axial stress is $\sigma_a = pr/2t$, where r is the radius of the bottle and t its wall thickness. The wall must be thick enough to support the larger of these stresses without failing, requiring

$$t = S\frac{pr}{\sigma_y} \tag{10.1}$$

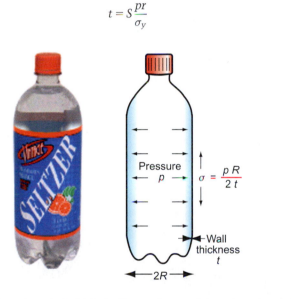

FIGURE 10.3 *A carbonated drink bottle carries an internal pressure that creates stresses in its walls.*

Table 10.3	Design requirements for drink containers
Function	■ Drink container
Constraints	■ Must contain the pressure of dissolved CO_2 safely ■ Must be moldable ■ Must be transparent or translucent ■ Must be recyclable
Objective	■ Minimize embodied energy per unit capacity ■ Minimize the cost per unit capacity
Free variables	■ Choice of material

where σ_y is the yield strength of the wall material and S is a safety factor. The embodied energy of the material of the wall per unit area, H_A (the quantity we wish to minimize) is

$$H_A = tH_m \rho = Spr\frac{H_m\rho}{\sigma_y} \qquad (10.2)$$

where ρ is the density of the bottle material and H_m is its embodied energy per kg. The best choice of material is one with the smallest value of the index

$$M_1 = H_m\rho/\sigma_y \qquad (10.3)$$

Cost, of course, is an issue in a product like this one. The cost of the bottle material per unit area is calculated as above, simply replacing the embodied energy/kg, H_m, with the material cost/kg C_m. Thus the cheapest bottle is that made from the material with the smallest value of the index

$$M_2 = C_m\rho/\sigma_y \qquad (10.4)$$

Table 10.4 shows the properties of transparent thermoplastics.

Figure 10.4 is a trade-off plot with M_1 on one axis and M_2 on the other. The choice with the lowest embodied energy (and carbon footprint) is polylactide, PLA. The least expensive is polyethylene terephthalate, PET. Both lie on the trade-off surface, making them better choices than any of the others.

Postscript. Today most carbonated drink containers are PET. They are likely to remain so. Economic benefits outweigh environmental benefits until the eco-gain is a lot larger than that suggested by this case study. The costs and uncertainties in changing from a material that is well-tried, attractive to consumers, and has an

Table 10.4   Transparent thermoplastics and their properties

Material (transparent thermoplastics)	Density (kg/m^3)	Embodied energy (MJ/kg)	Price ($/kg)	Energy index ($M_1 \times 10^4$)	Cost I index ($M_2 \times 10^2$)
Polylactide (PLA)	1,230	53	2.5	**8.2**	1.7
Polyhydroxyalkanoates (PHA, PHB)	1,240	54	2.4	5.6	1.2
Polyurethane (tpPUR)	1,180	119	5.4	3.3	0.7
Polystyrene (PS)	1,040	92	2.2	4.2	1.8
Polymethyl methacrylate (PMMA)	1,190	102	2.2	5.2	1.8
Polyethylene terephthalate (PET)	1,340	84	1.7	5.3	2.6
Polycarbonate (PC)	1,170	110	4.0	5.0	1.4

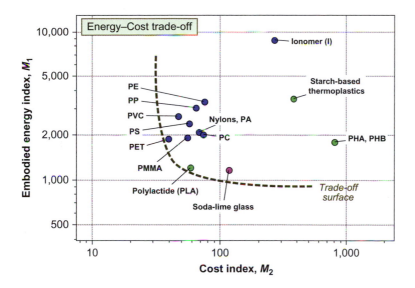

FIGURE 10.4   *A carbonated drink bottle must support an internal pressure. The trade-off plot for moldable, transparent, or translucent polymers shows that the PET is the cheapest, but PLA is the lowest in embodied energy.*

established recycling infrastructure to another that is less well-tired, more expensive, and more difficult to recycle are too great to accept.

Saving material in drink containers

Example: The working pressure in a Pepsi bottle is about 5 atmospheres (0.5 MPa, 75 psi). The bottle has a diameter $2r = 64$ mm and is made of PET. How thick must the wall of the bottle be to carry this pressure safely? Use a safety factor S of 2.5. The tensile strength of PET is approximately 70 MPa. The wall thickness of Pepsi bottles is about 0.5 mm. Is there scope for making them thinner while retaining a safety factor of 2.5?

Answer: The required wall thickness t is

$$t = S\frac{pr}{\sigma_y} = 2.5\frac{0.5 \times 0.032}{70} = 0.00057 \ \text{m} = 0.57 \ \text{mm}$$

No. It seems that Pepsi bottles already use as little PET as is practical.

10.4 Structural materials for buildings

The built environment is the largest of all consumers of materials. The aggregated embodied energy of the materials in buildings, too, is large. We are talking GJ now, not MJ, and the functional unit is "per unit area (m^2) of floor space." The embodied energy of a building is the energy used to acquire raw materials, manufacture building products, and transport and install them when it is first built. Frequently the embodied energy is a large fraction of a building's life-energy, so it is architects and civil engineers who look most closely at the embodied energies of the materials they use. What are they?

A local realtor (real estate agent) advertises "an exceptional property boasting wood construction with delightful concrete car-parking space with exquisite steel-framed roof that has to be seen to be appreciated." Filter out the noise and you are left with three words: wood, concrete, steel. These are, indeed, the principal materials of the structure of buildings. The *structure* is just one of the material-intensive parts of a building. It provides the frame, meaning the structure that carries the self-weight and working loads, resists the wind forces, and, where needed, supports the dynamic loads of earthquakes. The structure is clad and insulated by the *envelope*. It provides weather protection, thermal insulation, radiation screening, acoustic separation, and the color, texture, and short-term durability of the building. The building has to work and therefore it needs *services*: internal dividers, water, gas, electricity, heating and cooling, ventilation, control of light and sound, and disposal of waste. And there is the *interior*: the materials that the occupants see, use, and feel—the floor and wall coverings, furnishings, and fittings. The four different groups—structure, envelope, services, interior—have different primary functions. All four are material-hungry (Table 10.5).

The initial embodied energy per unit area of floor space of a building depends on what it is made of and where. An approximate figure is $4.5\,GJ/m^2$; we'll get more specific in a moment. Figure 10.5 shows where it goes: about a quarter each into the materials of the structure, those of the envelope, those of the services, and those of site preparation, building work, and interior lumped together. They differ most in the choice of materials for the structure.

Table 10.5 Embodied energy per m^2, concrete frame building

	Embodied energy (GJ/m^2)	% of total
Site work	0.29	6
Structure	0.93	21
Envelope	1.26	28
Services	1.11	23
Construction	1.37	7
Interior finishes	0.30	14
Total	**4.52**	

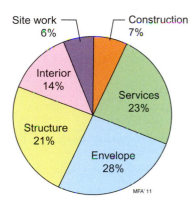

FIGURE 10.5 *The relative energies associated with the construction of a typical three-story office building.*

Table 10.6	Embodied energy/m^2 of alternative building structures	
Structural type	**Mass of materials (kg/m^2)**	**Embodied energy* (GJ/m^2)**
Steel frame	86 steel, 625 concrete	1.2
Reinforced concrete frame	68 steel, 900 concrete	0.9
Wood frame	80 timber	0.67

*Cole and Kernan (1996). Underground parking, add 0.26 GJ/m^2.

Table 10.6 compares the structural embodied energy per m^2 of a steel-framed, a reinforced concrete, and a wood-framed building. The wood frame has the lowest value, the steel frame the highest. The steel frame is 72% more energy intensive than wood and 33% more so than concrete.

Postscript. If wood is the most energy-efficient material for building structures, why are not all buildings made of wood? As always, there are other considerations. There is the obvious constraint of scale: wood is economic for small structures but for buildings above four stories in height, steel and concrete are more practical. There is availability: where wood is plentiful (Massachusetts, USA), wood is widely used, but elsewhere (London, UK) it is not. There are issues of recyclability: steel is easily recycled, but re-using wood or concrete at end of life is more difficult. The trade-off between all of these determines the final choice.

10.5 Initial and recurring embodied energy of buildings

The life cycle of a building is more complex than that of a short-lived product. There are recurring contributions to the embodied energy from repairs and upgrading over life. Thus it is useful to distinguish between five aspects of life and their associated energy:

- The embodied energies of the materials used to construct the building in the first place
- The energy to erect the building
- The recurring embodied energy incurred each time the building is upgraded during its life
- The use-energy to operate, condition (heat, cool, light, and ventilate), and power the building
- The end-of-life energy to demolish the building and dispose of its materials

Consider, as an example, the life cycle of a small office block with a 60-year design-life. Such a building, if built of reinforced concrete, uses about 968 kg of material per m^2 of floor area (Table 10.6) and has a total embodied energy of materials of about 4.5 GJ/m^2 (Table 10.5). The structure of the building is usually designed to last for its full life. The envelope, interior, and services, for functional or aesthetic reasons, might be upgraded every 15 years. This means replacing interior walls, floors, doors, finishes, and mechanical and electrical services. Estimates for the energy of upgrades vary between 2 and 4 GJ/m^2—we shall take 3 GJ/m^2—incurred once every 15 years, as typical. There will be three such upgrades over the 60-year life of the building, absorbing 9 GJ/m^2, twice as much as the initial embodied energy.

Environmental design in the building industry over the past 30 years has focused on reducing the operating energy, which is now much lower than it used to be: values of 0.5 to 0.9 GJ/m^2 per year are now possible. We will use the value 0.7 GJ/m^2/year, which gives us a 60-year use energy of 42 GJ/m^2. If the building materials are transported 500 km to the site by HGV, consuming 0.8 MJ/metric ton/km (Table 6.9), the transport energy is $0.8 \times 500 \times 0.95 = 0.38$ GJ/m^2. Finally, at end-of-life the building is demolished. Demolition and subsequent transport of rubble to the landfill are estimated to require 0.13 GJ/m^2.

These data are plotted in Figure 10.6(a). The significant terms are the initial embodied energy (8% of the total), the recurring embodied energy (16%), and the energy of use (74%). Figure 10.6(b) shows how the energy accumulates over life. The use-energy exceeds the initial embodied energy after only about 7 years, despite the extremely efficient heat conservation implied by the value we chose for the use-energy per year. The recurring embodied energy first appears after 15 years, and steps up after each subsequent 15-year interval.

Postscript. Some constructors now offer "carbon-neutral" or "zero-energy" housing, meaning zero use-energy. Heat is captured by passive solar heating: high heat-capacity walls oriented so they are warmed by sunlight during the winter day and release heat into the house at night. A large roof overhang shades the walls when the sun is high in the summer. Cooling is provided by airflow and the use of the underlying earth as a heat sink. Electricity is derived from solar panels.

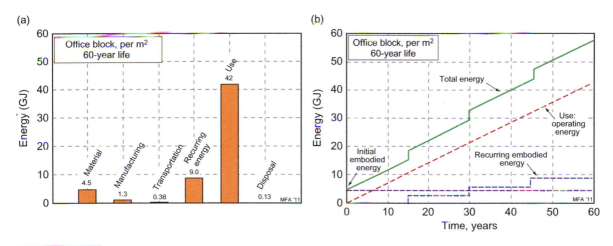

FIGURE 10.6 *(a) The contributions to the life energy per m^2 of floor area of three-story office building with a life of 60 years. (b) The way the energy evolves over the life.*

The use-energy is the largest column in Figure 10.6(a). If it is reduced to zero, the initial and recurring embodied energies of the materials dominate the picture. The priority is then to seek materials with low embodied energy per unit of function—"function" meaning bending strength for the structure, thermal resistance for the walls, durability for the floor covering, and so forth.

Further reading. Cole, R.J. and Kernan, P.C. (1996), "Life-cycle energy use in office buildings," *Building and Environment*, 31: 4, pp 307−317. *(An in-depth analysis of energy use per unit area of steel, concrete, and wood-framed construction)*

10.6 Heating and cooling (1): refrigeration

Heating and cooling are among the most energy-gobbling, CO_2-belching things we do. Refrigerators, freezers, and air-conditioners keep things cold. Central heating, ovens, and kilns keep things hot. For all of these appliances it is the use-phase of life that contributes most to energy consumption and emissions. Some, like refrigerators and incubators, aim to hold temperatures constant over long periods of time—the fridge or incubator is cooled or heated once and then held like that. Others, like ovens and kilns, heat up and cool down every time they are used, zig-zagging up and down in temperature over the span of a few hours. Commercial office space lies somewhere in between, heated or cooled during the day, but not at night or on weekends. The best choice of material to minimize heat loss depends on the form of the use-cycle.

Refrigerators. To get into the topic, take a look at refrigerators. The function of a fridge is to provide cold space. A fridge is an energy-using product (EuP), and like most EuPs it is the use-phase of life that dominates energy consumption and emission release. Thus the eco-objective for a fridge is to minimize the *use-energy per year per cubic meter of cold space*, H_f^*, (kWh/m³ · year). The "*" signifies "per cubic meter" and the "*f*" means "frigid."

But suppose the fridges that are good by this criterion are expensive and the ones that are not so good are cheap. Then economically minded consumers will perceive a second objective in choosing a fridge: that of minimizing *the initial cost per cubic meter of cold space*, C_f^*. The fridge with the lowest H_f^* is probably not the one with the lowest C_f^*, and vice versa. Resolving the conflict needs the trade-off methods of Chapter 9. The global objective is to minimize the *life-cost per cubic meter of cooled space* (Table 10.7).

Figure 10.7 plots the two measures of excellence for 95 contemporary (2008) fridges. The trade-off line is sketched (remember that it is just the convex-down envelope of the occupied space). The fridges that lie on or near the line are the best

Table 10.7	Design requirements for refrigerator
Function	■ Provide long-term cooled space
Objective	■ Minimize life-cost per cubic meter of cooled space
Free variable	■ Choice of refrigerator from those currently available

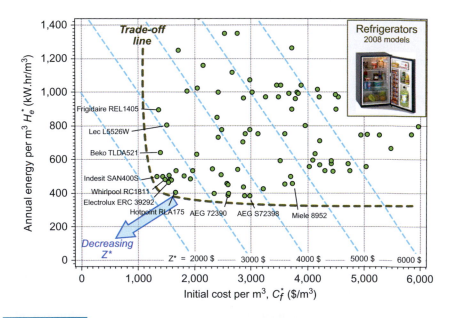

FIGURE 10.7 *A trade-off plot for fridges with contours of the penalty function Z*. Data from 2008 advertising.*

choices: several are identified. They are "non-dominated solutions," offering lower energy for the same price or lower price for the same energy than any of the others.

That still leaves many, and they differ a lot. To get further we need a *penalty function*. The easiest unit of penalty is that of cost, in whatever currency you choose (here US$). We wish to minimize life-cost, which we take to be the sum of the initial cost of the fridge and the cost of the energy used over its life. So define the penalty function

$$Z^* = C_f^* + \alpha_e H_f^* t \qquad (10.5)$$

where α_e, the exchange constant, is the cost of energy per kWh and t is the service life of the fridge in years, making Z^* the life-cost of the fridge per cubic meter of cold space. The global objective is to minimize Z^*.

Take the service life to be 10 years and the cost of electrical power to be 0.2 US$ per kWh. Then the penalty function becomes

$$Z^* = C_f^* + 2H_f^* \qquad (10.6)$$

or, solving for H_y^*:

$$H_f^* = \frac{1}{2} Z^* - \frac{1}{2} C_f^* \qquad (10.7)$$

The axes of Figure 10.7 are H_f^* and C_f^*, so this equation describes a family of straight lines with a slope of $-1/2$, one for any given value of penalty Z^*. Five are shown for Z^* values between $2,000 and $6,000. The best choices are the fridges with the lowest value of Z^*. They are the ones where the Z^* contour just touches the trade-off line.

If by some miracle the cost of energy dropped by a factor of 10, the Z^* contours get 10 times steeper—almost vertical—and the best choice becomes the cheapest fridge, regardless of power consumption. If, more probably, it rose by a factor of 10, the contours become almost flat and the best choices shift to those that use the least energy, regardless of initial cost.

You could argue that this purely economic view of selection is misguided. The environment is more important than that; reducing use-energy and emissions has a greater value than 0.2 $ per kWh. Fine. Then you must define what you believe to be a fair value for α_e, basing your selection on the Z^* contours that result. But if you want the rest of the world to follow your example, you must persuade them to use—or, if you are in government, make them use—the same value. Systematic choice is not possible until a value for the exchange constant is agreed upon.

The trade-off plot of Figure 10.7 has a sharp curve at the lower left. When the plots are like this, the optimal choice is not very sensitive to the value of the exchange constant α_e. Unless α_e changes a lot, the Z^* contour continues to touch the trade-off line at more or less the same place. But when the curve is more rounded, its value influences choice more strongly. We will see an example in a later case study.

Postscript. Why would anyone buy one of the fridges from the upper right? The life-cost of some of them is three times greater than it is for those that are most economical. But consumers don't always think in purely economic terms. There are other considerations, notably the perceived value of quality, aesthetics, brand image, and prestige, all of which can be manipulated by clever marketing.

10.7 Heating and cooling (2): materials for passive solar heating

There are a number of schemes for capturing solar energy for home heating: solar cells, liquid-filled heat exchangers, and solid heat reservoirs. The simplest of these is the heat-storing wall: a thick wall, the outer surface of which is heated by exposure to direct sunshine during the day and from which heat is extracted at night by blowing air over its inner surface (Figure 10.8). An essential of such a scheme is that the time-constant for heat flow through the wall be about 12 hours; then the wall first warms on the inner surface roughly 12 hours after the sun first warms the outer one, giving out at night what it took in during the day. We will suppose that, for architectural reasons, the wall must not be more than 400 mm thick. What materials maximize the thermal energy captured by the wall while retaining a heat-diffusion time of up to 12 hours? Table 10.8 summarizes the requirements.

The heat content, Q, per unit area of wall, when heated through a temperature interval ΔT, gives the objective function

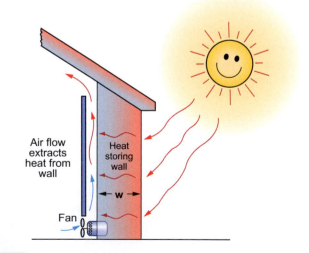

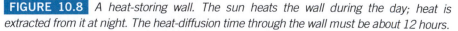

FIGURE 10.8 *A heat-storing wall. The sun heats the wall during the day; heat is extracted from it at night. The heat-diffusion time through the wall must be about 12 hours.*

Table 10.8	Design requirements for materials for passive solar heating
Function	■ Heat storing medium
Constraints	■ Heat diffusion time through wall $t \approx 12$ hours ■ Wall thickness ≤ 0.4 m ■ Adequate working temperature $T_{max} > 100°C$
Objective	■ Maximize thermal energy stored per unit material cost
Free variables	■ Wall thickness, w ■ Choice of material

$$Q = w \, \rho \, C_p \, \Delta T \qquad (10.8)$$

where w is the wall thickness, and ρC_p is the volumetric specific heat (the density ρ times the specific heat C_p). The 12-hour time constant is a constraint. It is adequately estimated by the approximation for the heat-diffusion distance in time t

$$w = \sqrt{2at} \qquad (10.9)$$

where a is the thermal diffusivity. Eliminating the free variable w gives

$$Q = \sqrt{2t} \, \Delta T \, a^{1/2} \, \rho \, C_p \qquad (10.10)$$

or, using the fact that $a = \lambda/\rho C_p$ where λ is the thermal conductivity,

$$Q = \sqrt{2t} \, \Delta T \left(\frac{\lambda}{a^{1/2}} \right)$$

The heat capacity of the wall is maximized by choosing material with a high value of

$$M = \frac{\lambda}{a^{1/2}} \qquad (10.11)$$

The restriction on thickness w requires (from equation (10.19)) that

$$a \leq \frac{w^2}{2t}$$

with $w \leq 0.4$ m and $t = 12$ hours $(4.3 \times 10^4 \text{ s})$, we obtain an attribute limit

$$a \leq 1.9 \times 10^{-6} \text{ m}^2/\text{s} \qquad (10.12)$$

Figure 10.9 shows thermal conductivity λ plotted against thermal diffusivity a with M and the limit on a plotted on it. It identifies the group of materials, listed in Table 10.9: they maximize M_1 while meeting the constraint on wall thickness. Solids are good; porous materials and foams (often used in walls) are not.

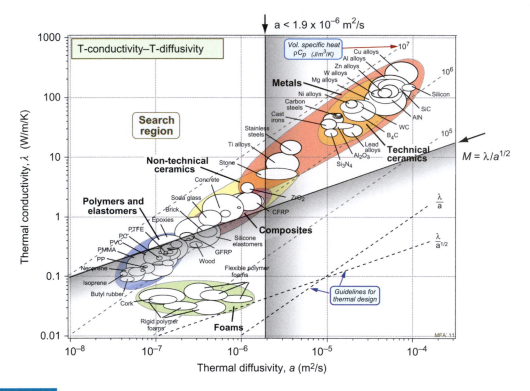

FIGURE 10.9 *Materials for heat-storing walls. Concrete, brick, stone, and glass are practical choices.*

Table 10.9	Materials for passive solar heat-storage		
Material	$M_1 = \lambda / a^{1/2}$ $(W \cdot s^{1/2}/m^2 \cdot K)$	**Approx. cost $/m³**	**Comment**
Concrete	2.2×10^3	200	The best choice—good performance at minimum cost
Stone	3.5×10^3	1,400	Better performance than concrete because specific heat is greater, but more expensive
Brick	10^3	1,400	Less good than concrete
Glass	1.6×10^3	10,000	Useful—part of the wall could be glass

Postscript. All this is fine, but what about the cost? If this scheme is to be used for housing, cost is an important consideration. The approximate cost per unit volume, read from the data sheets of Chapter 15, is listed in the table—it points to the selection of concrete, with stone and brick as alternatives.

10.8 Heating and cooling (3): kilns and cyclic heating

When space is held at a constant temperature, energy loss is minimized by choosing materials with as low a thermal conductivity λ as possible. But when space is heated and cooled in a cyclic way, the choice of material for insulation is more subtle. Take, as a generic example, the oven or kiln sketched in Figure 10.10—we will refer to it as "the kiln." The design requirements are listed in Table 10.10.

When the kiln is fired, the internal temperature rises from ambient, T_o, to the operating temperature, T_i, where it is held for the firing time t. The energy consumed in one firing has two contributions. The first is the heat absorbed by the kiln wall in raising it to T_i. Per unit area, it is

$$Q_1 = C_p \rho w \left(\frac{T_i - T_o}{2} \right) \tag{10.13}$$

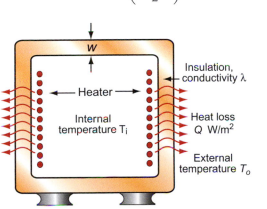

FIGURE 10.10 *A heated chamber with heat loss by conduction through the insulation.*

Table 10.10	Design requirements for kiln wall
Function	■ Thermal insulation for kiln (cyclic heating and cooling)
Constraints	■ Maximum operating temperature 1,000°C) ■ Upper limit on kiln-wall thickness for space reasons
Objective	■ Minimize energy consumed in a heat-cool cycle
Free variables	■ Kiln wall thickness, w ■ Choice of material

where C_p is the specific heat of the wall per unit mass (so $C_p\rho$ is the specific heat per unit volume) and w is the insulation wall thickness; $(T_i - T_o)/2$ is just the average temperature of the kiln wall. Q_1 is minimized by choosing a wall material with a low heat capacity $C_p\rho$ and by making it as thin as possible.

The second contribution is the heat lost by conduction through the wall, Q_2, per unit area. It is given by the first law of heat flow. If held for time t it is

$$Q_2 = -\lambda\frac{dT}{dx}t = \lambda\frac{(T_i - T_o)}{w}t \tag{10.14}$$

It is minimized by choosing a wall material with a low thermal conductivity λ and by making the wall as thick as possible.

The total energy consumed per unit area is the sum of these two:

$$Q = Q_1 + Q_2 = \frac{C_p\,\rho\,w\,\Delta T}{2} + \frac{\lambda\,\Delta T}{w}t \tag{10.15}$$

where $\Delta T = (T_i - T_o)$. Consider first the limits when the wall thickness w is fixed. When the heating cycle is short the first term dominates and the best choice of material is that with the lowest volumetric heat capacity $C_p\rho$. When instead the heating cycle is long, the second term dominates and the best choice of material is that with the smallest thermal conductivity, λ.

A wall that is too thin loses much energy by conduction, but little is used to heat the wall itself. One that is too thick does the opposite. There is an optimum thickness, which we find by differentiating equation (10.15) with respect to wall thickness w and equating the result to zero, giving

$$w = \left(\frac{2\lambda t}{C_p\rho}\right)^{1/2} = (2at)^{1/2} \tag{10.16}$$

where $a = \lambda/\rho C_p$ is the thermal diffusivity of the wall material. The quantity $(2at)^{1/2}$ has dimensions of length and is a measure of the distance heat can diffuse in time t. Substituting equation (10.16) back into equation (10.15) to eliminate w gives

$$Q = (\lambda\,C_p\,\rho)^{1/2}\Delta T(2t)^{1/2} \tag{10.17}$$

This is minimized by choosing a material with the lowest value of the quantity

$$M = (\lambda\,C_p\,\rho)^{1/2} = \frac{\lambda}{a^{1/2}} \tag{10.18}$$

Figure 10.11 shows the $\lambda - a$ chart of Chapter 9, expanded to include more materials that are good thermal insulators. All three of the criteria we have derived—minimizing $C_p\rho$, λ, and $(\lambda C_p\rho)^{1/2}$—can be plotted on it; the "guidelines" show the slopes. For long heating times it is λ we wish to minimize and the best

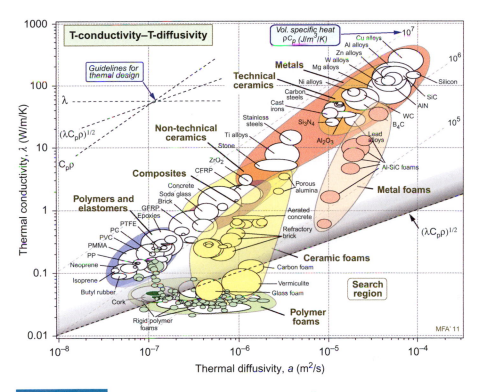

FIGURE 10.11 *The thermal conductivity—thermal diffusivity chart with contours of volumetric specific heat. It guides selection for efficient thermal design.*

choices are the materials at the bottom of the chart: polymeric foams or, if the temperature T_i is too high for them, foamed glass, vermiculite, or carbon. But if we are free to adjust the wall thickness to the optimum value of equation (10.16), the quantity we wish to minimize is $(\lambda C_p \rho)^{1/2}$. A selection line with this slope is plotted on Figure 10.11. The best choices are the same as before, but now the performance of vermiculite, foamed glass, and foamed carbon is almost as good as that of the best polymer foams. Here the limitation of the hard-copy charts becomes apparent: there is not enough room to show a large number of specialized materials such as refractory bricks and concretes. The limitation is overcome by the computer-based methods mentioned in Chapter 9, allowing a search over a much greater number of materials.

Postscript. It is not generally appreciated that, in an efficiently designed kiln, as much energy goes in heating up the kiln itself as is lost by thermal conduction to the outside environment. It is a mistake to make kiln walls too thick; a little is saved in reduced conduction loss, but more is lost in the greater heat capacity of the kiln itself.

That, too, is the reason that foams are good: they have a low thermal conductivity λ *and* a low volumetric heat capacity $C_p \rho$. Centrally heated houses in which the heat is turned off at night suffer a cycle like that of the kiln. Here (because T_i is lower) the best choice is a polymeric foam, cork, or fiberglass (which has thermal properties like those of foams). But as this case study shows—turning the heat off at night doesn't save you as much as you think, because you have to supply the heat capacity of the walls in the morning.

10.9 Transportation (1): introduction

Transportation by sea, road, rail, and air together account for 32% of all the energy we use and 34% of all the emissions we generate (Figure 2.4). Cars contribute a large part to both. The primary eco-objective in car design is to *provide mobility at minimum environmental impact*, which we will measure here by the CO_2 rating in grams per kilometer (g/km). The audits of Chapter 8 confirmed what we already knew: that the energy consumed during the life-phase of a car exceeds that of all the other phases put together. If we are going to reduce it, we first need to know how it depends on the vehicle mass and propulsion system.

But first let us look at another aspect of car data by performing the selection that was set up earlier (Figure 9.1): the selection of a car to meet constraints on power, fuel type, and number of doors, and subject to two objectives: one environmental (CO_2 rating), the other economic (cost of ownership). The schematic of Figure 9.2 showed how to set up the trade-off between the objectives. Real data are plotted in this way in Figure 10.12(a). It shows carbon rating and cost of ownership for 2,600 cars plotted on the same axes as the schematic.[1] We are in luck: the trade-off line has a sharp curve; the cars with the lowest CO_2 rating also have the lowest cost of ownership. The best combinations are the cars that lie nearest to the sharpest part of the curve of the trade-off line at the lower left of the plot. Not surprisingly they are all very small.

That is the picture before the constraints are applied. If we now screen out all models that are not gasoline powered, have less than 150 hp, and fewer than 4 doors, the trade-off plot looks like Figure 10.12(b). Again, the trade-off line has a sharp curve. The cars that lie there have the best combination of cost and carbon and meet all the constraints—we don't need a penalty function to find them. The Honda Civic 2.0 is a winner, but the Toyota Corolla 1.8 and the Renault Laguna 1.8 lie close. It is worth exploring these in more depth. What are the service intervals? How close is the nearest service center? What do consumer magazines say about them? It is this documentation that allows a final choice to be reached.

Not all trade-offs are so straightforward. Performance, to many car owners, is a matter of importance. For those who wish to be responsible about carbon yet drive a high performance vehicle, the compromise is more difficult: high performance generally means high carbon. Figure 10.13 shows a second trade-off, using

---

[1]Data from *What Car?* (2005). The database used to construct Figures 6.6, 6.7, 10.10, and 10.11 is available for use with the CES Edu 2012 software.

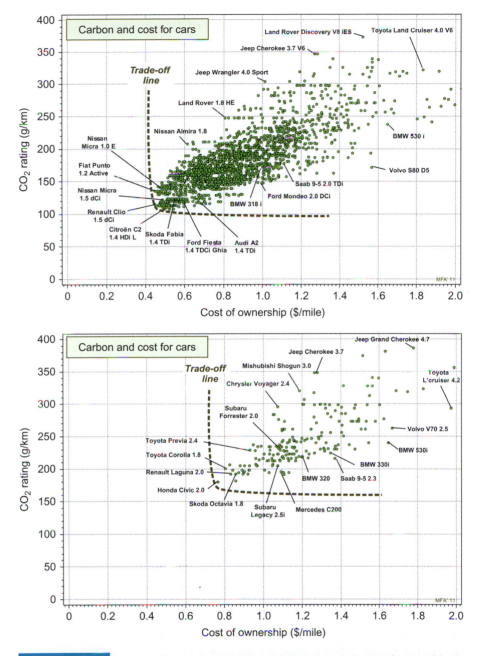

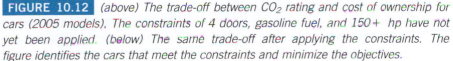 **FIGURE 10.12** *(above) The trade-off between CO$_2$ rating and cost of ownership for cars (2005 models). The constraints of 4 doors, gasoline fuel, and 150+ hp have not yet been applied. (below) The same trade-off after applying the constraints. The figure identifies the cars that meet the constraints and minimize the objectives.*

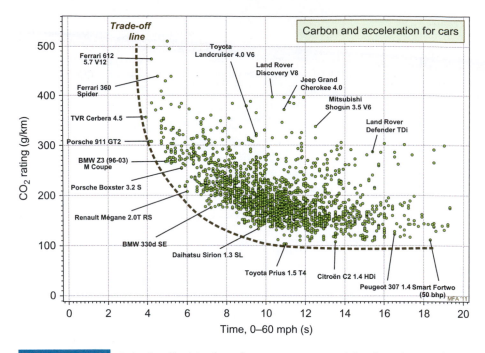

FIGURE 10.13 *A trade-off plot of performance, measured by time to accelerate from 0 to 60 mph (100 km/hour), against CO_2 rating for 1,700 cars.*

acceleration, measured by the time it takes the car to go from 0 to 60 mph, as one objective (the shorter the time, the greater the performance) and carbon rating as the second. Now the trade-off line is a broad curve. As before, the best choices are those that lie on or near this line; all others (and there are many) can be rejected. To get further it is necessary to assign relative values to performance and carbon rating. If the first is the most highly valued, it is the cars at the upper left that become the prime candidates. If it is carbon, it is those at the lower right. The compromise is a harsh one—any choice with low carbon has poor performance.

That was a digression. Now back to the energy consumption and carbon emission of road transportation. The fuel consumption and emission of cars increase with their mass. Data presented in Figures 6.6 and 6.7 identified the energy and CO_2 emission per km as a function of the mass of the car for gasoline, diesel, liquid propane gas (LPG), and hybrid-engine vehicles. The resulting energy and CO_2 emission per kg per km for a typical 1,000 kg vehicle are listed in Table 10.11. These are the penalty in energy and carbon for a 1 kg increase in the mass of the car, or the savings in both if the car mass is reduced by 1 kg.

Modeling: where does the energy go? Energy is dissipated during transportation in three ways: as the energy needed to accelerate the vehicle up to its cruising speed, giving it kinetic energy that it lost by braking; as drag exerted by the air or water

Table 10.11	The energy and CO_2 penalty of mass for cars	
Fuel type	**Energy (MJ/km · kg)** **(m = 1,000 kg)**	**CO_2 (kg/km · kg)** **(m = 1,000 kg)**
Gasoline power	2.1×10^{-3}	1.4×10^{-4}
Diesel power	1.6×10^{-3}	1.2×10^{-4}
LPG power	2.2×10^{-3}	0.98×10^{-4}
Hybrid power	1.3×10^{-3}	0.92×10^{-4}

through which it is passing; and as rolling friction in bearings and the contact between wheels and road. Imagine (following MacKay (2008)) that a vehicle, with mass m accelerates to a cruising velocity v acquiring kinetic energy

$$E_{ke} = \frac{1}{2}mv^2 \tag{10.19}$$

It continues over a distance d for a time d/v before braking, losing this kinetic energy as heat and dissipating energy per unit time (*power*) of

$$\frac{dE_{ke}}{dt} = \frac{E_{ke}}{d/v} = \frac{1}{2}\frac{mv^3}{d} \tag{10.20}$$

While cruising, the vehicle drags behind it a column of air with a cross-section proportional to its frontal area A. The column created in time t has a volume c_dAvt where c_d is the drag coefficient, typically about 0.3 for a car, 0.4 for a bus or truck. This column has a mass $\rho_{air}c_dAvt$ and it moves with velocity v, so the kinetic energy imparted to the air is

$$E_{drag} = \frac{1}{2}m_{air}v^2 = \frac{1}{2}\rho_{air}c_dA\,v^3t \tag{10.21}$$

where ρ_{air} is the density of air. The drag is the rate of change of this energy with time (= power)

$$\frac{dE_{drag}}{dt} = \frac{1}{2}\rho_{air}c_dA\,v^3 \tag{10.22}$$

If this cycle is repeated over and over again, the power dissipated is the sum of these two:

$$Power = \frac{1}{2}\left(\frac{m}{d} + \rho_{air}\,c_d\,A\right)v^3 \tag{10.23}$$

Rolling resistance adds another small term that is proportional to the mass, which we will ignore.

The first term in the brackets is proportional to the mass m of the vehicle and inversely proportional to the distance d it moves between stops. The second depends only on the frontal area A and the drag coefficient c_d. Thus for a short haul, stop-and-go, or city driving, the way to save fuel is to make the vehicle as light as possible. For long haul, steady cruising, or highway driving, mass is less important than minimizing drag, making frontal area A and drag coefficient c_d of central importance.

Efforts to reduce fuel consumption and emissions from cars seek to reduce all three of the power-dissipating mechanisms listed at the start: recuperative braking to reuse kinetic energy, a low drag coefficient shape to reduce air-drag, low rolling resistance tires to reduce road-surface losses, and low overall mass because of its direct link to fuel consumption. The data for average combined[2] energy consumption of Table 6.10 show a near-linear dependence of energy per km on mass, meaning that, in normal use, it is the kinetic energy term, equation (10.20), that dominates. Thus design to minimize the energy and CO_2 of vehicle use must focus on material selection to minimize mass.

10.10 Transportation (2): crash barriers—matching material to purpose

Barriers to protect the driver and passengers of road vehicles are of two types: those that are static—the central divider of a freeway, for instance—and those that move—the bumper of the vehicle itself (Figure 10.14). The static type lines tens of

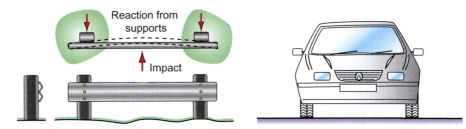

FIGURE 10.14 *Two crash barriers, one static, the other—the bumper of the car—attached to something that moves. In action both are loaded in bending. The criteria for material selection for each differ.*

---

[2]"Combined energy consumption" means that for a typical mix of urban and highway driving.

Table 10.12	Design requirements for crash barriers
Function	■ Crash barrier: transmit impact load to absorbing elements
Constraints	■ High strength ■ Adequate fracture toughness ■ Recyclable
Objectives	■ Minimize embodied energy for given bending strength (static barrier) ■ Minimize mass for a given bending strength (mobile barrier)
Free variables	■ Choice of material ■ Shape of cross-section

thousands of miles of road. Once in place this type consumes no energy, creates no CO_2, and lasts a long time. The dominant phases of its life cycle are those of material production and manufacturing. The bumper, by contrast, is part of the vehicle; it adds to its mass and thus to its fuel consumption. If eco-design is the objective, the criteria for selecting materials for the two sorts of barriers differ: minimizing embodied energy for the first, minimizing mass for the second (Table 10.12).

In an impact the barrier is loaded in bending (Figure 10.14). The function of the barrier is to transfer the load from the point of impact to the support structure where reaction from the foundation or from the crush elements in the vehicle support or absorb it. To do this the material of the barrier must have high strength, σ_y, be adequately tough, and able to be recycled. That for the static barrier must meet these constraints with *minimum embodied energy* as the objective, since this will reduce the overall life energy most effectively. We know from Chapter 9 that this means materials with low values of the index

$$M_1 = \frac{H_m \rho}{\sigma_y^{2/3}} \tag{10.24}$$

where σ_y is the yield strength, ρ is the density, and H_m is the embodied energy per kg of material. For the car bumper it is mass, not embodied energy, that is the problem. If we change the objective to that of *minimum mass*, we require materials with low values of the index

$$M_2 = \frac{\rho}{\sigma_y^{2/3}} \tag{10.25}$$

These indices can be plotted onto the charts of Figures 9.12 and 9.14. We leave that as an exercise at the end of this chapter and show here an alternative: simply plotting the index itself as a bar chart. Figures 10.15 and 10.16 show the results for metals, polymers, and polymer-matrix composites. The first guides the selection for static barriers. It shows that embodied energy for a given bending-load bearing capacity (equation (10.24)) is minimized by making the barrier from carbon steel or

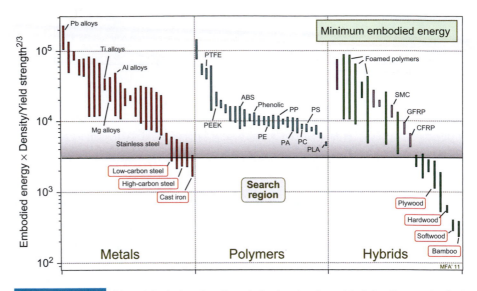

FIGURE 10.15 *Material choice for the static barrier is guided by the embodied energy per unit of bending strength, $H_m\rho/\sigma_y^{2/3}$, here in units of (MJ/m^3)/ MPa$^{2/3}$. Cast irons, carbon steels, or low-alloy steels are the best choices. (Here the number of materials has been limited for clarity).*

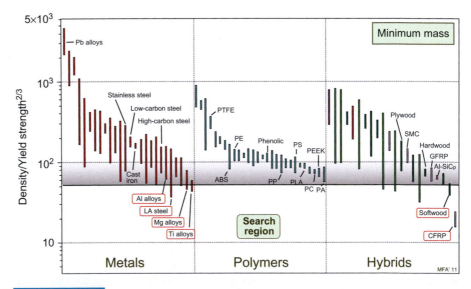

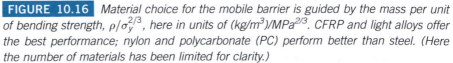

FIGURE 10.16 *Material choice for the mobile barrier is guided by the mass per unit of bending strength, $\rho/\sigma_y^{2/3}$, here in units of (kg/m^3)/MPa$^{2/3}$. CFRP and light alloys offer the best performance; nylon and polycarbonate (PC) perform better than steel. (Here the number of materials has been limited for clarity.)*

cast iron or wood; nothing else comes close. The second figure guides selection for the mobile barrier: it is a bar chart of equation (10.25). Here CFRP (continuous fiber carbon-epoxy, for instance) excels in its bending strength per unit mass, but it is not recyclable. Heavier, but recyclable, are alloys of magnesium, titanium, and aluminum, and low-alloy steel.

Postscript. Roadside crash barriers have a profile like that shown on the left of Figure 10.14. The "3"-shaped profile increases the second moment of area of the cross-section, and through this, the bending stiffness and strength. This is an example of combining material choice and section shape (Section 9.11 and Table 9.11) to optimize a design. A full explanation of the co-selection of material and shape can be found in the first text listed in Further reading.

10.11 Transportation (3): materials for light weight structures

The energy and carbon plots for cars, Figures 6.6 and 6.7, illustrate the striking dependence of both on vehicle mass. Every 10% reduction in passenger vehicle mass reduces fuel consumption by about 8%. The challenge is to minimize mass while still meeting the structural integrity and safety standards provided by the material we use today.

Material selection to support tensile or compressive loads at minimum mass is straightforward: select materials with the smallest values of the indices ρ/E and ρ/σ_y. The challenge more usually is that of dealing with bending loads. Then minimizing mass is not just a question of material choice, but one of choice of material, shape, and configuration. An example will bring this to life.

A material is required to support a force F that is cantilevered a distance L perpendicular to F. The support must be as light as possible. The simplest solution is a solid square section beam, built in at its left end, as in Figure 10.17(a). The method described in the appendix to Chapter 9 gives its mass as

$$m_1 = 6\, F^{2/3}\, L^{5/3}\, \frac{\rho}{(\sigma_y)^{2/3}} \tag{10.26}$$

The best materials are those with the smallest values of the index $\rho/\sigma_y^{2/3}$. Table 10.13 lists data for four competing materials with this index in column 5.

A solid square section is a bad choice for a light beam. When a beam is loaded in bending, most of the load is carried by the outer elements; the elements along the central neutral axis of bending carry no load at all. You get more bending strength and stiffness by redistributing the material to where it works best, far from the neutral axis (Figure 10.17(b)). Doing so increases the section modulus Z of the beam by the shape-factor Φ^f introduced in Section 9.11—it measures the efficiency of use of material. I-section and hollow tube-sections can have large values of Φ^f.

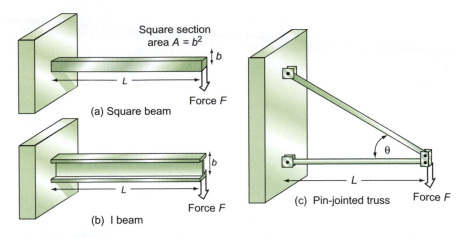

FIGURE 10.17 *A square-section beam, an I-beam, and a truss, all supporting the same load* F *at the same overhang* L

Table 10.13	Properties of materials for light weight structures						
Material	ρ **(kg/m³)**	σ_y **(MPa)**	$\dfrac{\rho}{\sigma_y}$	$\dfrac{\rho}{\sigma_y^{2/3}}$	Φ^f	$\dfrac{\rho}{(\Phi^f \sigma_y)^{2/3}}$	$\dfrac{m_2}{m_1}$
1020 Steel, normalized	7,850	330	23.8	164	13	30	0.18
6061-T4 Al	2,710	113	24	116	10	25	0.22
GFRP SMC 30% glass	1,770	83	21	93	6	28	0.30
Wood (oak), along grain	760	40	19	65	3	31	0.48

The maximum shape factor depends on the material—Table 10.13 lists values for the four materials in column 6.

The mass of the shaped section required to carry the force F is now

$$m_2 = 6\,F^{2/3}\,L^{5/3}\,\frac{\rho}{(\Phi^f \sigma_y)^{2/3}} \tag{10.27}$$

(it is exactly the same as equation (10.26) except for the factor Φ^f). The best materials are now those with the smallest $\rho/(\Phi^f \sigma_y)^{2/3}$. Shape has reduced the mass by the factor

$$\frac{m_2}{m_1} = \frac{1}{(\Phi^f)^{2/3}} \tag{10.28}$$

The reduction in mass that is made possible through shaping is listed in the last column of Table 10.13.

Can we do better still? To do so we have to think not only of shape but also of configuration. A beam is a configuration. There are others. One is a pin-jointed truss (Figure 10.17(c)). The truss can support bending loads, but neither of its members is loaded in bending: the upper one is in simple tension, the lower one in simple compression. If both have the same cross-section and are made of the same material, the mass of the truss needed to support a load F without failing is

$$m_2 = FL\frac{\rho}{\sigma_y}\frac{1}{\sin\theta}\left(1 + \frac{1}{\cos\theta}\right) \qquad (10.29)$$

There is a new variable: it is the angle θ between the two members of the truss. If the lower one is normal to the loading direction the angle that minimizes the mass is $52°$, and for this value, the terms involving θ reduce to the factor 3.3, giving

$$m_2 = 3.3\,FL\,\frac{\rho}{\sigma_y} \qquad (10.30)$$

The best material for the lightest truss is that with the smallest value of ρ/σ_y, the same as that for simple tension and compression. Not surprising—it is because the truss turns bending loads into axial loads. The truss is lighter than the original square beam by the factor

$$\frac{m_3}{m_1} = 0.55\left(\frac{F}{L^2\sigma_y}\right)^{1/3} \qquad (10.31)$$

The gain now depends on the magnitudes of the force F and length L, and it can be large.

Can we do yet better? Yes, by combining material, configuration, *and* shape. The upper member of the truss is loaded in simple tension—only the area of its cross-section is important, not its shape. But the lower one is loaded in compression; if it is slender, it will fail not by yielding but by elastic buckling. Giving it a tubular section can reduce the mass further. The optimization now has to be done numerically, but it is not difficult.

Lightweight design

Example: A material is required to support a load 100 kg (1,000 N = 0.001 MN) that is cantilevered a distance $L = 1$ m perpendicular to F. Four materials are available with the properties and shapes listed in the following table. Evaluate the mass of the square section beam, I-beam, and truss to support this load. Which material/shape/configuration offers the lowest mass solution?

Answer: The table shows that, for the combination of load and span given, the truss is much lighter than beams of any shape. The titanium alloy truss is the lightest of all.

Material	ρ (kg/m³)	σ_y (MPa)	Φ^f	m_1 (kg)	m_2 (kg)	m_3 (kg)
1020 Steel, normalized	7,850	330	13	9.8	1.8	0.078
6061-T4 Al	2,710	113	10	6.9	1.5	0.079
GFRP SMC 30% glass	1,770	83	6	5.6	1.7	0.069
Ti-6%Al-4%V, aged	4,430	1050	11	2.6	0.53	0.025

Shells and sandwiches. The four eco-cars pictured on the opening page of this chapter all have casings that are thin, doubly curved sheets, or *shells*. The designers wanted a casing that was adequately stiff and strong, and as light as possible. The double curvature of the shell helps with this: a shell, when loaded in bending, is stiffer and stronger than a flat or singly curved sheet of the same thickness because any attempt to bend it creates *membrane stresses*: tensile or compressive stress in the plane of the sheet. Sheets support tension or compression much better than they support bending.

So what is the best material for an adequately stiff and strong shell that is as light as possible? Table 10.14 lists the requirements. To meet them we need material indices for shells. They are derived in Section 9.11c. The results we need are that the minimum mass of a shell element of diameter $2R$ (Figure 9.20d) with prescribed stiffness S^* and failure load F^* is

$$m_1 = \pi R^2 (C\,R\,S^*)^{1/2} \left(\frac{\rho}{E^{1/2}} \right) \qquad \text{(stiffness constraint)} \qquad (10.32)$$

or

$$m_2 = \pi R^2 (DF^*)^{1/2} \left(\frac{\rho}{\sigma_y^{1/2}} \right) \qquad \text{(strength constraint)} \qquad (10.33)$$

Table 10.14	Monocoque construction: design requirements for the light, stiff, strong shell
Function	■ Doubly curved shell
Constraints	■ Stiffness S^* specified ■ Failure load F^* specified (functional constraints) ■ Radius R specified ■ Load distribution specified (geometric constraints)
Objective	■ Minimize mass
Free variables	■ Thickness of shell wall, t ■ Choice of material

with $C \approx 0.4$ and $D \approx 0.65$, depending on the dominant constraint. If both stiffness and strength are important, then it is the greater of m_1 and m_2 that determines the mass. Everything in these two equations is specified except for the material properties in square brackets, so the two indices are

$$M_1 = \frac{\rho}{E^{1/2}} \qquad (\textit{light, stiff shell}) \qquad (10.34)$$

and

$$M_2 = \frac{\rho}{\sigma_y^{1/2}} \qquad (\textit{light, strong shell}) \qquad (10.35)$$

The shell is more efficient than a flat panel, offering lower mass for the same bending stiffness and strength.

The best materials for shells are found by evaluating the indices for competing materials or by plotting them onto appropriate charts. Take the index for adequate strength and minimum mass, M_2, as an example. Taking logs of equation (10.35) and rearranging gives

$$Log\ \sigma_y = 2log\rho - 2logM_2 \qquad (10.36)$$

This is the equation of a family of lines of slope 2 on the $\sigma_y - \rho$ chart. Figure 10.18 shows it with such a line positioned to leave three materials with the low values of M_2 in the search area: CFRP (carbon fiber reinforced polymer) and two grades of rigid polymer foam. Certain ceramics come close. A similar selection line for M_1, plotted on the Modulus–Density chart of Figure 9.11, gives the same result. Ceramics are ruled out by their brittleness. Foams are eliminated for a different reason: they are light, but to achieve the necessary stiffness and strength, a foam shell has to be very thick, increasing the frontal area and drag of the car. That leaves CFRP as the unambiguous best choice.

Postscript. CFRP, of course, is what the mileage-marathon cars use. But there is more to it than that. The best performance is achieved by a combination of material and shape. CFRP offers exceptional stiffness and strength per unit mass; making it into a doubly curved shell adds shape-stiffness, further enhancing performance. Yet higher efficiency is possible by making the shell from a sandwich with CFRP faces separated by a high-performance foam core, combining both shape and configuration.

10.12 Transportation (4): material substitution for eco-efficient design

Making cars lighter means replacing heavy steel and cast iron components by those made of lighter materials: light alloys based on aluminum, magnesium, or

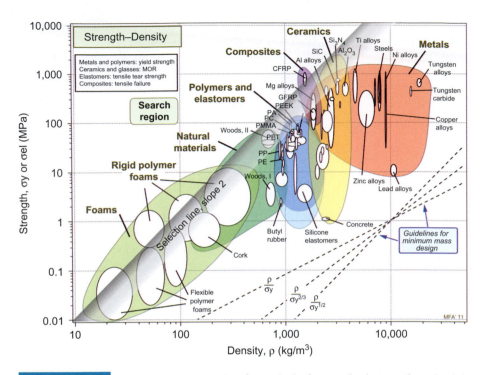

FIGURE 10.18 *A chart-based selection for a shell of prescribed strength and minimum mass. Carbon-fiber reinforced polymer (CFRP) is the best choice.*

titanium, or polymers and composites reinforced by glass or carbon fibers. All these materials have greater embodied energy per kg than steel, introducing another sort of trade-off: that between the competing energy demands or emissions of different life phases. There is a net saving of energy and CO_2 only if that saved in the use-phase by reducing mass exceeds that invested in the material-phase as extra embodied energy. Figure 10.19 illustrates the problem. It shows what happens when an existing material (which we take to be steel for this example) for a vehicle component is replaced by a substitute. The horizontal axis plots the change in material embodied energy, ΔH_{emb}. The vertical one plots the change in use energy, ΔH_{use}. The black circle at (0,0) is the steel; the other circles, enclosed by a trade-off line, represent substitutes. The diagonal contours show the penalty function which, in this instance, is particularly simple: it is just the sum of the two energies or the two emissions.

$$Z = \Delta H_{emb} + \Delta H_{use} \qquad (10.37)$$

The best materials are those with the lowest (most negative) value of Z. Any substitute that lies on the contour of $Z = 0$ passing through steel offers neither

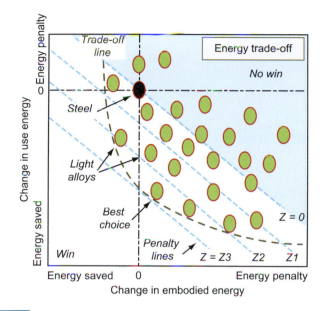

FIGURE 10.19   *The use energy change and the embodied energy change when a steel component is replaced by one made of a light alloy. There is a net energy savings only if the sum of the two is negative. The diagonal lines are contours of constant (negative) sum.*

reduction nor increase in life-energy. Those lying in the blue "No win" zone increase the life-energy. Those in the white "Win" zone offer a savings. The best choice of all is that nearest the point that a penalty contour is tangent to the trade-off line. It is indicated on the figure.

All very straightforward. Well, not quite. Material substitution in a component that performs a mechanical function requires that the component be rescaled to have the same stiffness or strength (whichever is design-limiting) as the original. In making the substitution, the mass of the component changes both because the density of the new material differs from that of the old and because the scaling changes the volume of material that is used. Steel has a larger density than—say—aluminum, but it is also stiffer and stronger. We must scale in such a way that the *function* is maintained. The scaling rules are known: they are given by material indices developed in Chapter 9. They predict the change in *mass* Δm and in *embodied energy* ΔH_{emb} when one material is replaced by a substitute that must perform the same function. Multiplying Δm by the energy per kg·km from Table 10.11 and the distance traveled over life, which we will take as 200,000 km, gives the change in *use-phase energy* ΔH_{use} resulting from substitution.

Consider, then, the replacement of the pressed steel bumper set of a car by one made of a lighter material. The function and mass scaling of the bumper were described earlier: it is a beam of given bending strength with a mass, for a given bending strength, that scales as $\rho/\sigma_y^{2/3}$ where ρ is the density and σ_y is the yield strength of the material of which it is made. The mass change of the bumper set, on replacing one made of steel (subscript "o") with one made of a lighter material (subscript "1"), is thus:

$$\Delta m = B\left(\frac{\rho_1}{\sigma_{y,1}^{2/3}} - \frac{\rho_o}{\sigma_{y,o}^{2/3}}\right) \tag{10.38}$$

where B is a constant. If the mass m_o of the steel bumper set is 20 kg, then

$$m_o = B\frac{\rho_o}{\sigma_{y,o}^{2/3}} = 20 \text{ kg} \tag{10.39}$$

defining B. Substituting this value for B into the previous equation gives

$$\Delta m = 20\left(\frac{\rho_1}{\sigma_{y,1}^{2/3}}\left[\frac{\sigma_{y,o}^{2/3}}{\rho_o}\right] - 1\right) \tag{10.40}$$

The property group in the square brackets is that for the steel of the original bumper; we will use data for an AISI 1022 rolled steel with 0.18% C 0.7% Mn, a yield strength of 295 MPa, and a density of 7,900 kg/m³, giving the value of the group in square brackets as 5.6×10^{-3} in these units. From Table 10.11, a gasoline-engine vehicle weighing 1,000 kg requires 2.1×10^{-3} MJ/km·kg. Thus the change in use-energy, ΔH_{use}, found by multiplying this by Δm and the distance traveled over a life of 200,000 km, is

$$\Delta H_{use} = 8.4 \times 10^3 \left(5.6 \times 10^{-3}\left(\frac{\rho_1}{\sigma_{y,1}^{2/3}}\right) - 1\right) \text{ MJ} \tag{10.41}$$

The change in embodied energy ΔH_{emb} is found in a similar way. The embodied energy of steel is $H_{mo} = 33$ MJ/kg. The embodied energy of the initial 20 kg steel bumper set is

$$(H_{emb})_o = m_o H_{mo} = B\frac{H_{mo}\rho_o}{\sigma_{y,o}^{2/3}} = 20 \times 33 = 660 \text{ MJ} \tag{10.42}$$

defining the value of the constant C. The change on substituting a new material is then:

$$\Delta H_{emb} = C\left(\frac{H_{m1}\rho_1}{\sigma_{y,1}^{2/3}} - \frac{H_{mo}\rho_o}{\sigma_{y,o}^{2/3}}\right) = 660\left(\frac{H_{m1}\rho_1}{\sigma_{y,1}^{2/3}}\left[\frac{\sigma_{y,o}^{2/3}}{H_{mo}\rho_o}\right] - 1\right) \tag{10.43}$$

As before, the property group in the square brackets is that for the steel of the original bumper. Its value, using steel data given above, is 1.7×10^{-4}, giving the final expression for change in embodied energy

$$\Delta H_{emb} = 660\left(1.7 \times 10^{-4}\frac{H_{m1}\rho_1}{\sigma_{y,1}^{2/3}} - 1\right) \text{ MJ} \qquad (10.44)$$

The trade-off between ΔH_{use} and ΔH_{emb} is plotted in Figure 10.20. It explores the life energy change when the carbon steel bumper set is replaced by one made of low-alloy steels (colored orange), 6,000 series aluminum alloys (green), wrought magnesium alloys (blue), composites (blue), or titanium alloys (yellow). The original 1022-grade steel lies at the origin (0,0). The other bubbles, labeled, show the changes in ΔH_{use} and ΔH_{emb} calculated from equations (10.41) and (10.44). The "No win" area is shaded blue. For titanium alloys the energy saving is slight. For magnesium and aluminum alloys it is considerable. The greatest saving is made possible by using composites. The trade-off line is sketched in. The penalty contours show the total energy saved over 200,000 km. The contour that is tangent to

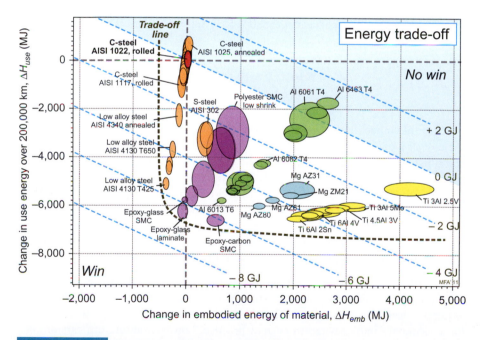

FIGURE 10.20 *The trade-off between embodied energy and use energy. Energy savings is shown as a negative value, energy penalty as positive.*

the trade-off line has (by interpolation) a value of about 7 GJ. It identifies the best choice: here, the epoxy-glass laminate.

Before a final decision is reached, it is helpful to have a feeling for the likely change in cost. The change in use-cost ΔC_{use} is the use energy change ΔH_{use} (equation (10.41)) multiplied by the price of energy; gasoline at $0.8/liter ($3 per US gallon) gives an energy price of approximately 0.025 $/MJ. Thus

$$\Delta C_{use} = 210\left(5.6\times 10^{-3}\left(\frac{\rho_1}{\sigma_{y,1}^{2/3}}\right)-1\right)\ \$ \tag{10.45}$$

The change in material cost ΔC_{mat} might seem (in parallel with that of embodied energy) to be the change in the product of component mass m and the price per unit mass C_m, and we shall use this here:

$$\Delta C_{mat} = C\left(\frac{C_{m1}\rho_1}{\sigma_{y,1}^{2/3}} - \frac{C_{mo}\rho_o}{\sigma_{y,o}^{2/3}}\right) = 20C_{mo}\left(\frac{C_{m1}\rho_1}{\sigma_{y,1}^{2/3}}\left[\frac{\sigma_{y,o}^{2/3}}{C_{mo}\rho_o}\right]-1\right) \tag{10.46}$$

Taking the cost C_{mo} of steel to be $0.8/kg makes the value of the steel property group in the square brackets equal to 7.0×10^{-3} and the equation becomes

$$\Delta C_{mat} = 16\left(7.0\times 10^{-3}\left(\frac{C_{m1}\rho_1}{\sigma_{y,1}^{2/3}}\right)-1\right)\ \$ \tag{10.47}$$

Cost, of course, has other contributions than just that of the materials. Part of the cost of a component is that of manufacturing. If it takes longer to shape, join, and finish with the new material than with the old, there is an additional cost penalty, so the material cost of equation (10.47) must be regarded as approximate only.

With this simplifying approximation, the cost trade-off (following the pattern used for energy) appears as in Figure 10.21. Titanium alloys lie well inside the "No win" zone. Low-alloy steels, aluminum alloys, and magnesium alloys are the most attractive from an economic point of view, though epoxy-glass is almost as good. The most striking result is that the total sum saved over life is so small.

Postscript. We have focused here on *primary* mass savings that material substitution brings. In reality the saving is—or can be—larger because the lighter vehicle requires a less heavy suspension, lighter tires, less powerful brakes, which allows for a secondary mass saving. In current practice, however, aluminum cars are not much lighter than the steel ones they replace because manufacturers tend to load them with more extras (such as air conditioning and double glazing), adding mass back on.

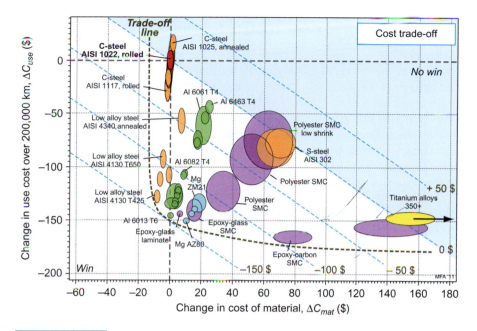

FIGURE 10.21 *The trade-off between material cost and use cost. Cost savings is shown as a negative value, cost penalty as positive.*

News-clip: Ultra efficient transport

Volkswagen reveals its 313 mpg hybrid.

The Sunday Times, January 30, 2011

A fuel efficiency of 313 mpg is 1 liter/100 km. Auto makers' claims for fuel efficiency seldom quite match the experience of the purchaser of the vehicle. But even allowing for that, this car will still go five times farther on a liter of fuel than any ordinary one. How has Volkswagen achieved this? Aerodynamics, low-loss tires, a tiny 800 cc diesel power source, hybrid technology, and above all, low mass—it is only 795 kg all added up, about half the weight of a mid-sized car—achieved through the use of CFRP monocoque construction.

As we have seen, the energy consumption and carbon emission of an ordinary car are dominated by those of the use-phase. But if you divide these by 5 (as VW claimed to do) and use materials with 4 times greater embodied energy materials for the structure (CFRP), the balance changes. The dominant contribution to life energy and emissions now becomes those of the materials, which changes the criteria for material selection.

> ### Weight saving by substituting
>
> *Example:* All-CFRP construction allows for great savings of mass. Both stiffness and strength are central to the light weight mechanical design of the body shell of cars, but assume for a moment that in extreme light weight design it is strength that is the first concern—if it were stiffness instead there would be little point in replacing mild steel with high-strength steel to save mass, as some car makers are doing since both materials have the same modulus and density. Bending, not tension, is the critical mode of loading. Given the material properties listed in the following table, by what factor can mass be reduced if an all-steel shell is replaced by an all-CFRP shell of the same strength?
>
> | Material | Density (kg/m^3) | Yield strength (MPa) | $\rho/\sigma_y^{2/3}$ |
> |---|---|---|---|
> | Mild steel | 7,850 | 310 | 171 |
> | CFRP | 1,550 | 540 | 23 |
>
> *Answer:* The mass savings in material substitution is measured by the index $\rho/\sigma_y^{2/3}$. The last column lists the value of this index for steel and CFRP. The CFRP structure has the potential to be lighter by a factor of about 7. All that assumes that the composite shell is as efficiently designed as present-day steel shells and that the problems of joining it to the power train do not add mass.

10.13 Summary and conclusions

Rational selection of materials to meet environmental objectives starts by identifying the phase of product life that causes greatest concern: production, manufacturing, transportation, use, or disposal. Dealing with all of these requires data not only for the obvious eco-attributes (energy, emissions, toxicity, ability to be recycled, and the like) but also data for mechanical, thermal, electrical, and chemical properties. Thus if material production is the phase of concern, selection is based on minimizing embodied energy and associated emissions and using as little material as possible. But if it is the use-phase that is of concern, selection is based instead on light weight, excellence as a thermal insulator, or as an electrical conductor. These define the objective; the idea is to minimize this while meeting all the other constraints of the design: adequate stiffness, strength, durability, and the like.

Almost always there is more than one objective and, almost always, they conflict. They arise in more than one way. One is the obvious conflict between eco-objectives and cost, illustrated both in Chapter 8 and here. Then trade-off methods offer a way forward provided a value can be assigned to the eco-objective, something that is not always easy. Another conflict is the one between the energy demands and emissions of different phases of life: the conflict between increased embodied energy of material and reduced energy of use, for example. Trade-off methods work particularly well for this type of problem because both energies can be quantified.

The case studies of this chapter illustrate how such problems are tackled. The exercises of Section 10.15 present more.

10.14 Further reading

Ashby, M.F. (2011), *Materials selection in mechanical design*, 4th edition, Butterworth Heinemann: Oxford, UK, Chapter 4. ISBN 978-1-85617-663-7. *(A text that develops the ideas presented here in more depth, including the derivation of material indices, a discussion of shape factors, and a catalog of simple solutions to standard problems)*

Ashby, M.F., Shercliff, H.R., and Cebon, D. (2009), *Materials: engineering, science, processing and design*, 2nd edition, Butterworth Heinemann: Oxford, UK. ISBN 978-1-85617-895-2. *(An elementary text introducing materials through material property charts and developing the selection methods through case studies)*

Caceres, C.H. (2007), "Economical and environmental factors in light alloys automotive applications," Metallurgical and Materials Transactions, A. Vol 18, No. 7, pp. 1649–1662. *(An analysis of the cost-mass trade-off for cars)*

Calladine, C.R. (1983), *Theory of Shell Structures*, Cambridge University Press: Cambridge, UK. ISBN 0-521-36945-2. *(A comprehensive text developing the mechanics of shell structures)*

Carslaw, H.S. and Jaeger, J.C. (1959), *Conduction of Heat in Solids*, 2nd Edition, Oxford University Press: Oxford, UK. ISBN 0-19-853303-9. *(A classic text dealing with heat flow in solid materials)*

Hollman, J.P. (1981), *Heat Transfer*, 5th Edition, McGraw-Hill: New York, U.S.A. ISBN 0-07-029618-9. *(An introduction to problems of heat flow)*

MacKay, D.J.C. (2008), "Sustainable energy—without the hot air," Department of Physics, Cambridge University, Cambridge, UK. www.withouthotair.com/ *(MacKay brings common sense into the discussion of energy use)*

Young, W.C. (1989), *Roark's Formulas for Stress and Strain*, 6th Edition, McGraw-Hill, New York, USA. ISBN 0-07-072541-1. *(A "Yellow Pages" for results for calculations of stress and strain in loaded components)*

10.15 Exercises

E10.1. *Drink containers with recycled content.* The drink containers of Figure 10.1 use recycled materials to different degrees. How does the ranking in Table 10.2 of the text change if the contribution of recycling is included? To figure this out, multiply the energy per liter in the last column of the table by the factor

$$1 - f_{rc}\left(1 - \frac{H_{rc}}{H_m}\right)$$

where f_{rc} is the recycle fraction in current supply, H_m is the embodied energy for primary material production, and H_{rc} is that for recycling of the material. The following table lists values for the drink-container materials.

Container type	Material	Embodied energy* (MJ/kg)	Recycle energy* (MJ/kg)	Recycle fraction in current supply*
PET 400 ml bottle	PET	84	38.5	0.21
PE 1 liter milk bottle	High density PE	81	34	0.085
Glass 750 ml bottle	Soda glass	15.5	6.8	0.24
Al 440 ml can	5000 series Al alloy	208	19.5	0.44
Steel 440 ml can	Plain carbon steel	32	9.0	0.42

* From the data sheets of Chapter 15

E10.2. *Allowing for recycled content*. Derive the correction factor to allow for recycled content cited in Exercise E10.1.

E10.3. *Estimating embodied energies for building structures*. The following table lists the approximate mass of structural materials per m² of floor space required for a steel and concrete-framed building and a wood-framed building. Assuming that the steel has 100% recycled content and that the concrete and wood have none, what are the embodied energies per m² for the two alternative frames? To evaluate these take data for embodied energies of 100% recycled carbon steel and for virgin concrete and softwood from the data sheets of Chapter 15. Compare the resulting energies per m² with the values from the LCA quoted in Table 10.6 of the text.

Structural type	Mass of materials (kg/m²)
Steel frame	86 steel
	625 concrete
Wood frame	80 timber

E10.4. *Estimating embodied energies for building envelopes*. The cladding of a framed building provides an envelope to protect it against the elements. It must do so while imposing as little additional dead load on the structural frame as possible, and it should enhance the architectural concept and appearance of the building. Alternative cladding materials include 1.5 mm aluminum sheet, 10 mm plywood, or 3 mm PVC. Use data from Chapter 15 to estimate the embodied energy per m^2 of clad surface using each of these materials. Assume that the aluminum has a recycled content of 50% and that the other two materials have none.

E10.5. *Thermal mass.* Define specific heat. What are its usual units? How would you calculate the specific heat per unit volume from the specific heat per unit mass? If you wanted to select a material for a compact heat storing device, which of the two would you use as a criterion of choice?

E10.6. *Importance of the value of exchange constants.* In a far-away land, fridges cost the same as they do here and last just as long (10 years), but electrical energy costs 10 times more—that is, it costs $2/kWh. Make a copy of the trade-off plot for fridges (Figure 10.7) and plot a new set of penalty lines onto it, using this value for the exchange constant, α_e. If you had to choose just one fridge to use in this unfortunate land, which would it be? In another 20 years, it may be us.

E10.7. You are asked to design a large heated workspace in a cold climate, making it as eco-friendly as possible by attaching polystyrene foam insulation to the inside wall. Polystyrene foam has a density of 50 kg/m^3, a specific heat capacity of 1220 J/kg·K, and a thermal conductivity of 0.034 W/m·K. The space will be heated during the day (12 hours) but not at night. What is the optimum thickness, w, of foam to minimize the energy loss?

E10.8. Use the indices for the crash barriers (equations 10.19 and 10.20) with the charts for strength and density (Figure 9.12) and strength and embodied energy (Figure 9.14) to select materials for each of the barriers. Position your selection line to include one metal for each. Reject ceramics and glass on the grounds of brittleness. List what you find for each barrier.

E10.9. Complete the selection of materials for light, stiff shells of Section 10.4 by plotting the stiffness index

$$M_1 = \frac{\rho}{E^{1/2}}$$

onto a copy of the modulus-density chart of Figure 9.11. Reject ceramics and glass on the grounds of brittleness, and foams on the grounds that the shell would have to be very thick. Which materials do you find? Which of these would be practical for a real shell?

E10.10. The makers of a small electric car wish to make bumpers out of a molded thermoplastic. Which index is the one to guide this selection if the aim is to maximize the range for a given battery storage capacity? Plot it on the appropriate chart from the set shown as Figures 9.11–9.14, and make a selection.

E10.11. Car bumpers used to be made of steel. Most cars now have extruded aluminum or glass-reinforced polymer bumpers. Both materials have a much higher embodied energy than steel. Take the weight of a set of steel bumpers to be 20 kg, and that of an aluminum one to be 14 kg. Table 10.11 gives the equation for the energy consumption per kg per km for a 1,000 kg car as 2.1×10^{-3} MJ/km·kg.

(a) Work out how much energy is saved by changing the bumper set of a 1,000 kg car from steel to aluminum, over an assumed life of 200,000 km.

(b) Calculate whether the switch from virgin steel to virgin aluminum has saved energy over life. You will find the embodied energies of steel and aluminum in the datasheets of Chapter 15. Ignore the differences in energy in manufacturing the two bumpers—it is small.

(c) Repeat the calculation for the bumpers made from 100% recycled steel and aluminum. The datasheets of Chapter 15 give the necessary energies

(d) The switch from steel to aluminum increases the price of the car by $60. Fuel cost depends on country; it varies at present between $0.5/liter and $2/liter. Using a pump price of $1 for gasoline, work out whether, over the 200,000 km life, it is cheaper to have the aluminum bumper or the steel one.

Exercises using the CES Edu software

E10.12. Refine the selection for shells (Section 10.11) using Level 2 of the CES software. Make a chart with the two shell indices

$$M_1 = \frac{\rho}{E^{1/2}} \quad \text{and} \quad M_2 = \frac{\rho}{\sigma_y^{1/2}}$$

as axes, using the "Advanced" facility to make the combination of properties. Then add a "Tree" stage, selecting only metals, polymers, and composites and natural materials. Which ones emerge as the best choice? Why?

E10.13. Tackle the crash barrier case study using CES Level 2 following the requirements set out in Table 10.12. Use a Limit stage to apply the constraints on fracture toughness $K_{1c} \geq 18$ MPa·m$^{1/2}$ and the requirement of recyclability. Then make a chart with yield strength α_y on the y-axis and density ρ on the x-axis and apply a selection line with the appropriate slope to represent the index for the mobile barrier:

$$M_2 = \frac{\rho}{\sigma_y^{2/3}}$$

List the best candidates. Then replace Level 2 with Level 3 data, limiting the search to metals, and explore what you find.

E10.14. Repeat the procedure of Exercise E10.13, but this time make a chart using CES Level 2 on which the index for the static barrier

$$M_1 = \frac{H_m \rho}{\sigma_y^{2/3}}$$

can be plotted—you will need the Advanced facility to make the product $H_m \rho$. List what you find to be the best candidates. Then, as before, dump in Level 3 data and explore what you find.

CHAPTER 11

Sustainability: living within our means

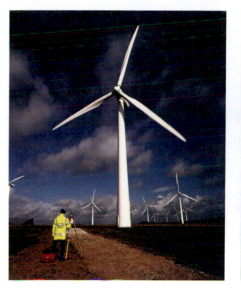

11.1 Introduction and synopsis

Sustainable development is development that meets the needs of the present without compromising the ability of future generations to meet their own needs.

This much-quoted definition from the Brundtland Report of the World Council on Economic Development (WCED, 1987) captures what most people would agree is the principle of sustainability. But agreeing how to apply the principle is not so easy.

Sustainability is a word that has come to mean whatever the user wants it to mean. To a political party it means votes. To large corporations it means staying in business and continuing to grow. To an oil company it is the future after oil.

Renewable and non-renewable energy. (Image of wind turbine courtesy of Leica Geosystems, Switzerland. Image of oil rig courtesy of the US National Park Service.)

To those who think on a broader scale, sustainability means using technology to decouple gross domestic product (GDP)[1] from environmental damage (carbon emissions, for instance), allowing the first to grow while reducing the second. There are those, however, who have no faith in this approach, seeing the free market as the cause of environmental problems and not their solution; it is the relentless drive for growth that threatens a sustainable future. Sustainability, to them, means equilibrium, not growth. And there are many shades in between.

In the following pages, we examine the meaning of sustainability. The view you take of it depends on scale—local or global—and on the time frame—weeks, years, centuries. We start by discussing these and then explore sustainability from a materials perspective.

11.2 The concept of sustainable development

As engineers and scientists it is natural that we should wish to tackle problems at a level at which we can bring our skills to bear. The methods described in the previous chapters are examples of this. They address immediate environmental issues, ones that are already evident and identified. But they do little to tackle the deeper problem—that of long-term sustainability. Figure 11.1 introduces the concept. The horizontal axis describes the time scale, ranging from that of the life of a product to that of the span of a civilization. The vertical axis describes the spatial scale, again ranging from that of the product to that of society as a whole. It has four nested boxes, expanding outward in conceptual scale, each representing an approach to thinking about the environment.

The least ambitious of these—the smallest box—is that of *pollution control and prevention* (PC and P). This is intervention on the scale and lifetime of a single product and is frequently a clean-up measure. Take transportation as an example: we added catalytic converters to cars; this was a step to mitigate an identified problem associated with an existing product or system.

The next box is *design for the environment* (DFE)—the techniques we discussed in Chapters 7 to 10. Here the time and spatial scales include the entire design process; the strategy is to foresee and minimize the environmental impact of product families at the design stage, balancing them against the conflicting objectives of performance, reliability, quality, and cost. Retaining the example of the car: it is to redesign the vehicle, giving emphasis to the objectives of minimizing emissions by reducing weight and adopting an alternative propulsion system—hybrid, perhaps, or electric.

---

[1]The annual gross domestic product (GDP) is an indicator used to gauge the health of a country's economy. It is the total value of all goods and services produced over a given year—think of it as the size of the economy. GDP per capita is used as a measure of average individual income. However, as Dasgupta (2010) points out, this measure fails to include the loss of natural capital as resources are consumed or damaged.

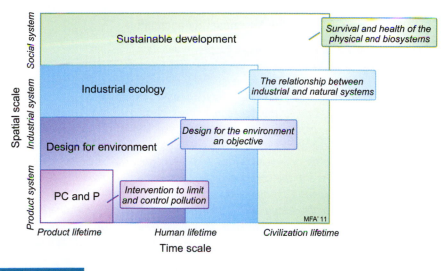

Approaches, differing in spatial and temporal scale of thinking, about industrialization and the natural ecosystem (adapted from Coulter et al. 1995).

The third box, that of *industrial ecology*, derives from the perception of human activities as part of the global ecosystem. Here the idea is that the processes and balances that have evolved in nature might suggest ways to reconcile the imbalance between the industrial and the natural systems, an idea known as the "ecological metaphor." We return to this idea in a moment as a way of structuring thinking about the last box, that of *sustainable development*.

A set of operating principles for true sustainable development

Example: Daly (1990) suggests the following as a set of operating principles for sustainable development:

- The rate of use of renewable resources (air, water, biomass) must be no greater than the rate of regeneration.
- The rate of use of nonrenewable resources (high-quality minerals, fossil fuel) must be no greater than the rate at which renewable resources, used sustainably, can be substituted for them.
- The rate of emission of pollutants (gas, liquid, and solid emissions) must be no greater than the rate at which they can be recycled, assimilated, or degraded by the environment.

These principles require changes in the way we use materials that appear, today, to be unachievable. But they remain an ideal, something against which sustainability measures can be judged.

News-clip: Sustainability accounting

Triple bottom line. It consists of three Ps: profit, people and planet.
The phrase "the triple bottom line" was first coined in 1994 by John Elkington, the founder of a British consultancy called SustainAbility. His argument was that companies should be preparing three different (and quite separate) bottom lines. One is the traditional measure of corporate profit—the "bottom line" of the profit and loss account. The second is the bottom line of a company's "people account"—a measure in some shape or form of how socially responsible an organisation has been throughout its operations. The third is the bottom line of the company's "planet" account—a measure of how environmentally responsible it has been. The triple bottom line (TBL) thus consists of three Ps: profit, people and planet. It aims to measure the financial, social and environmental performance of the corporation over a period of time. Only a company that produces a TBL is taking account of the full cost involved in doing business.

The Economist, **November 17, 2009**

This is a concept that feels right: a recognition of corporate social and environmental responsibility as well as of their responsibility to shareholders. It is often shown as three overlapping circles with a "sustainability" sweet-spot in the middle where social duty, environmental stewardship, and profitability coincide. Has it happened? Many enterprises have introduced environmental and social accounting into their business process. But many others, driven by pressures to cut costs, have transferred production and services to low-cost countries where excessive use of hydrocarbons and the exploitation of cheap labor are commonplace. Getting the three circles to overlap is not easy.

11.3 The ecological metaphor

Natural and industrial systems have certain features in common and others that are strikingly different. Consider three.

- Both the natural and the industrial systems transform resources (meaning materials and energy)—nature through biological growth, industry through manufacturing. The plant kingdom captures energy from the sun, carbon dioxide from the atmosphere, and minerals from the earth to create carbohydrates; the animal kingdom derives its energy and essential minerals from those of plants or from each other. The industrial system, by contrast, acquires most of its energy from fossil fuels and its raw materials from those that occur naturally in the earth's crust, in the oceans, and in the natural world.

- Both systems generate waste, the natural system through the metabolism and death of organisms, the industrial system through the emissions of manufacturing and through the obsolescence and finite life of the products it produces. The difference is that the waste of nature is recycled with 100% efficiency, drawing on renewable energy (sunlight) and natural decay to return it to the ecosystem, allowing a steady state to be established. The waste of industry is recycled much less effectively, requires nonrenewable energy (fossil fuels), and leaves a legacy of depleted resources and contaminated ecosystems.

- Both the natural and the industrial systems exist within the ecosphere, which provides the raw materials and other primary resources; acts as a reservoir for waste, absorbing and, in nature, recycling it; and provides the essential environment for life, meaning fresh water, a breathable atmosphere, tolerable temperatures, and protection from UV radiation. The natural system manages, for long periods, to live in balance with the ecosphere. Our present industrial system, it appears, does not. Are there lessons to be learned about managing industrial systems from the balances that have evolved in nature? Can nature give guidance, or at least provide an ideal?

Of the 92 usable elements of the Periodic Table, four are of central importance for life: carbon, nitrogen, hydrogen, and oxygen. They make up the carbohydrates, fats, and proteins on which life depends. To allow a steady state, these elements must be constantly recycled around the ecosystem. The most important of these circular paths are the *carbon cycle*, the *nitrogen cycle*, and the *hydrological (water) cycle*. Subsystems have evolved that provide the links in the cycles and do so at rates that manage to avoid bottle necks at which waste accumulates. Figure 11.2 is a sketch of one of these, the carbon cycle. Carbon dioxide in the atmosphere is captured by green plants and algae on land, and by phytoplankton and other members of the aquatic biomass in water. Fungal and bacterial action enables decomposition of plants and animals when they die, returning much of the carbon to the atmosphere but also sequestering some as carbon-rich deposits of peat, gas, oil, coal, and, in the oceans, limestone, $CaCO_3$.

The elements important for man-made products, by contrast, are far more numerous. Today they include most of the periodic table (Table 11.1). Carbon is one. As in nature, the products that use it (in the form of coal, oil, or gas) eject most of it into the atmosphere, but the natural subsystems that recycle carbon

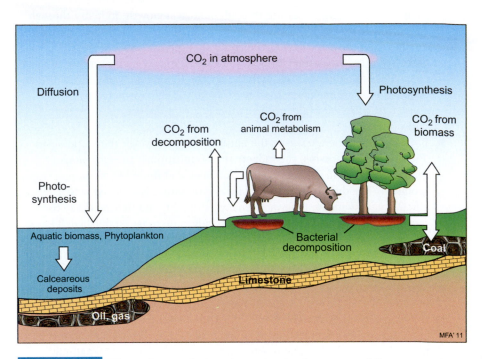

FIGURE 11.2 *The carbon cycle in nature, showing how some of the many subsystems that have evolved transform resources. They do so in such a way as to give balance and long-term stability.*

Table 11.1	The increasing diversity of elements used in materials and devices over the past 75 years	
Sector	**Elements**	
	75 years ago	**Today**
Iron-based alloys*	Fe, C	Al, Co, Cr, Fe, Mn, Mo, Nb, Ni, Si, Ta, Ti, V, W
Aluminum alloys*	Al, Cu, Si	Al, Be, Ce, Cr, Cu, Fe, Li, Mg, Mn, Si, I, V, Zn, Zr
Nickel alloys*	Ni, Cr	Al, B, Be, C, Co, Cr, Cu, Fe, Mo, Ni, Si, Ta, Ti, W, Zr
Copper alloys*	Cu, Sn, Zn	Al, Be, Cd, Co, Cu, Fe, Mn, Nb, P, Pb, Si, Sn, Zn
Magnetic materials*	Fe, Ni, Si	Al, B, Co, Cr, Cu, Dy, Fe, Nd, Ni, Pt, Si, Sm, V, W
Displays	W	Eu, Ge, Ne, Si, Tb, Xe, Y
(Micro) electronics	Cu, Fe, W	As, Ga, In, Sb, Si
Low-C energy (Solar, Wind)	Cu, Fe	Ag, Dy, Ga, Ge, In, Li, Nd, Pd, Pt, Re, Se, Sm, Te, Y

*Data from the composition fields of records in the CES Edu '12 Level 3 database, Granta Design (2012)

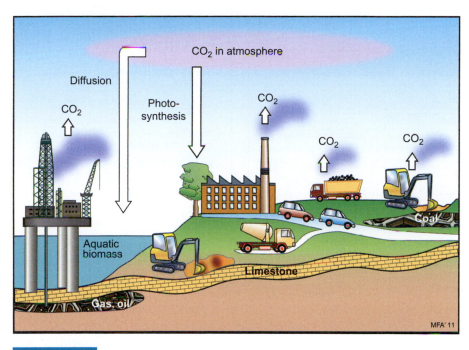

FIGURE 11.3   *The additional burden placed on the carbon cycle by large-scale industrialization. The subsystems that evolved to balance the cycle still exist but work at rates that do not begin to replace the resources that are consumed.*

have not evolved to remove it at a matching rate (Figure 11.3). The problem is more acute with other elements, the heavy metals, for example, for which no natural subsystems exist to provide recycling. When rates don't match, stuff piles up somewhere. Focusing on carbon again, this imbalance is evident in the steep rise of atmospheric carbon since 1850 visible in Figure 3.8. Burning fossil fuels and calcining limestone for cement generates large masses of CO_2. Reduced forestation, rising water temperatures, and soil contamination reduce the rate at which it is reabsorbed by the bio-subsystems. When the production rate exceeds the absorption rate, the atmospheric concentration of carbon rises.

What do we learn? Figure 11.4 summarizes the differences between the natural and industrial systems: one is sustainable over long periods of time; the other, in its present form, is not. Of the many aspects of sustainability, two relate directly to materials. The first is the lack of appropriate subsystems to close many of the recycling paths, and the second is that, where subsystems exist, there is an imbalance of rates. However, there is a more fundamental difference: it has to do with metrics of well-being. In nature this metric is achieving balance so that the system is in *equilibrium*. In the industrial system, well-being is achieved by *growth*. An economy that is growing is healthy, one that is static is sick. Economic growth, our metric for the well-being of businesses, nations, and society as a whole, carries with it

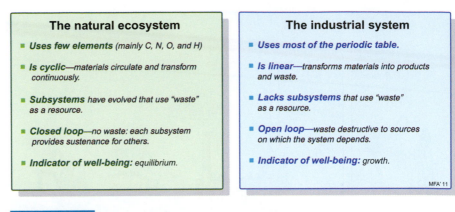

The natural ecosystem

- **Uses few elements** *(mainly C, N, O, and H)*

- **Is cyclic**—*materials circulate and transform continuously.*

- **Subsystems** *have evolved that use "waste" as a resource.*

- **Closed loop**—*no waste: each subsystem provides sustenance for others.*

- **Indicator of well-being:** *equilibrium.*

The industrial system

- **Uses most of the periodic table.**

- **Is linear**—*transforms materials into products and waste.*

- **Lacks subsystems** *that use "waste" as a resource.*

- **Open loop**—*waste destructive to sources on which the system depends.*

- **Indicator of well-being:** *growth.*

MFA' 11

FIGURE 11.4   *The comparison of the natural and the industrial ecosystems*

the need for an ever-increasing consumption of materials and energy and of creating waste. The present system of industrial production has been likened to an organism that ingests resources, produces goods, and expels waste. The characteristic that makes this organism devastating for the environment is its insatiable appetite: the faster it ingests resources, the greater the output of products, and the better is its health, even though this does not coincide with the health of the biosphere.[2] The comparison, then, highlights the ideal: an industrial system in which the consumption of materials and energy and the production of waste are minimized, and the discarded material from one process becomes the raw material for another, ultimately closing the loop.

But are natural systems really at equilibrium? Over long periods of time, yes. The forces for change are minimal, allowing optimization at an ever more refined and detailed level. This leads to an interdependence on a scale that even now we do not fully grasp, but which man-made activities too frequently disturb. But on a geological time scale there have been massive disruptions of the natural system. Most derived from sudden climate change. Are there lessons in the way the natural system then adapted? When dinosaurs—reptiles—succumbed to one of these disruptions, some small, furry mammal—a mouse, perhaps—survived, evolved, and multiplied because it was better adapted to the changed climate and habitat. Where, in the technical world of today, are the post-industrial mice, and what do they look like?

We don't know. But there is a message here. The life raft, the Noah's Ark, so to speak, of the natural world, allowing continuity in times of change, lies in its diversity: in other words, new mice, ready and waiting. The nuclei of the new system existed within the old and could emerge and grow when circumstances changed. Without knowing what we will need to do, can we create a society—a scientific

[2]Frosch and Gallopoulos, 1989; Regge and Pallante, 1996; cited by Guidice et al., 2006.

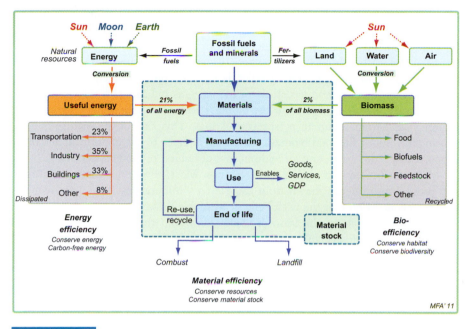

MFA' 11

FIGURE 11.5 *The energy, biomass, and material flows*

society—with sufficient diversity that, painful though change will be, the nuclei of the next phase pre exists in it?

11.4 Material sustainability

As we have seen, the word "sustainability" has no meaning without context. So let's give "material sustainability" a context. Figure 11.5 is an attempt to do so. It has a lot on it. The boxes across the top itemizes the five families of natural resources available to us: *energy, minerals, land, water,* and *air*. All except minerals have a renewable component and so can be drawn upon indefinitely provided they are managed. Minerals are a finite resource, but a large one (Chapter 2), sufficient to meet present needs but a likely concern for the future.

Focus first on the top-left of the figure. Renewable energy reaches us from the sun via radiation, from wind and wave power, from the moon via the tides, and from the earth via geothermal heat. At present we get about 8% of our energy from these, relying instead on the mineral reserves of oil, coal, and gas for all the rest. Before we can use any of these, we have to harvest them and convert them into a form that can be distributed—*useful energy* (the orange box), which fuels the activities listed on the center left. Once used, almost all this energy decays to low-grade heat and is lost. This side of the picture is truly sustainable only if the energy consumed by transportation, industry, and daily life equates to the useful energy derived directly from the sun, moon, and earth. Measures to promote *energy efficiency* (bottom-left) seek to reduce the demand for energy per unit of GDP and to draw as much energy as possible from low carbon, renewable sources.

Now focus on the right-hand side of Figure 11.5. The biosphere draws energy from the sun, carbon from the atmosphere, water from rain, and minerals from the land to create *biomass* (the green box). The biomass provides fuel, fodder for animals, food for humans, and feedstock for industry, as indicated on the center right. Once burned, eaten, or processed, the biomass reverts to atmospheric carbon and minerals, and from a human perspective, is lost. This side of the picture is truly sustainable only if the total biomass consumed equates to the quantity that can be supported by land, water, and air without loss of habitat or biodiversity.

Materials occupy the center of the picture. Most, today, are made by mining and processing minerals drawn from the planet's mineral reserves. This requires energy—about 21% of all the useful energy available to us is consumed in material production. Materials in the form of wood, fiber, and the raw materials for bio-derived animal and vegetable products of engineering are drawn from biomass— about 2% of all biomass is used in this way. The materials enter the supply chain and are processed into products that provide services that contribute to the GDP and, with it, prosperity. At the end of product life the material may be rejected and sent to the landfill or burned for energy, or it may recirculate through the supply chain via recycling, remanufacturing, or reuse. The box with the broken blue line encloses the *material stock*: the material currently in use or held for reuse.

Measures to promote *material efficiency* seek to maintain sufficient material stock to meet present needs while minimizing new inputs of minerals, energy, and biomass. "Conservation" here has a double meaning: conserving mineral, energy, and biomass resources, and conserving the material stock currently in circulation.

Materials, in this sense, differ from energy and biomass. The services that energy and biomass provide degrade both to such an extent that they become value-less. Materials, by contrast, need *not* be degraded by use. There are difficulties, described in Chapter 5, but in principle at least, a material, once created, can provide service and survive to be used again and again. For this reason the central *material stock* box is not grayed out.

We return to material efficiency in Chapter 13. First we must explore renewable materials.

11.5 Renewable materials

No matter how you look at it, using materials costs energy. Let us put that aside and examine the degree to which the materials themselves are sustainable. To be so the material must

- be drawn from a resource so vast that its use by man has no effect on supply (like sea water);
- be re-created in perfect form after use (like ice in the arctic); or
- regrow as fast as it is used (the biosphere, if properly managed).

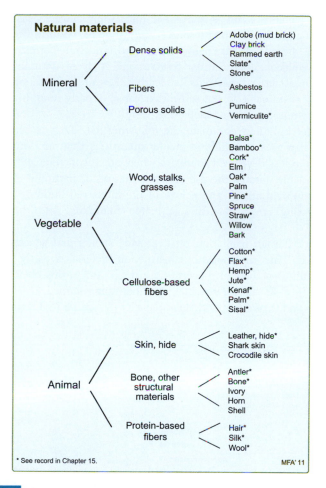

FIGURE 11.6 *A hierarchy of natural structural materials*

Examples of the first two are rare. It is the third, the management of the bio-sphere, on which we must depend for truly renewable materials. But our present appetite for materials is large. Which natural materials can satisfy it?

Not many. Figure 11.6 is a list of possibilities. You will find data sheets for most of them in Chapter 15. Some come from vegetable sources, some from animal, and some from mineral. We examine them in more depth in a moment, but it is first important to know the constraints. Wood, as an example, is a renewable resource. Provided the total tree stock is constant, so that wood is harvested at the same rate as it is grown, wood could be seen as a sustainable material. Today some softwoods are managed sustainably in developed nations (an essential for the paper industry), but globally wood is harvested (or simply burned) faster than it is replaced, making it a diminishing resource. And wood in the form we use it for construction has to be cut, dried, chemically treated, and transported, all with some nonrenewable consequences. Similar reservations apply to natural fibers, natural rubber, and animal-derived materials like leather. Few of the materials we use today qualify as truly sustainable in the sense that those of early man, himself part of a natural ecosystem and its closed-loop cycle, were.

Materials for buildings dominate the following list. That is because the construction industry uses materials in greater quantities than any other, and much construction is in less developed countries where steel, concrete, and fired brick are not readily available, forcing the use of those that occur naturally. Architects with concern for the environment and a love of pre-industrial building materials now use an interesting range of near-sustainable materials, and these are worth exploring.

Buildings for human habitation must provide shelter and generally have to be heated or cooled. Thus the choice of construction material is partly conditioned by the local climate. Two thermal properties are important in this choice: the thermal conductivity λ (J/m·K) and the thermal mass per unit volume, ρC_p (J/m^3·K), where ρ is the density and C_p is the specific heat capacity (Figure 11.7). The first, λ, controls

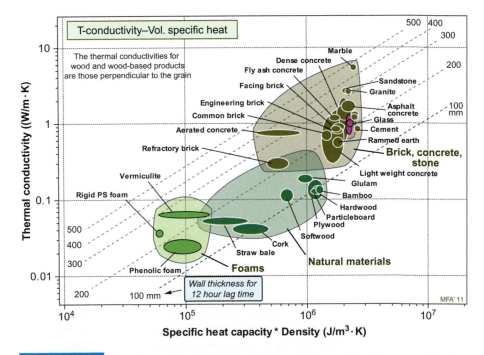

FIGURE 11.7   *The thermal conductivity and thermal mass of building materials. The contours show the wall thickness required to give a thermal cycle time of 12 hours.*

the loss or gain per unit area through the wall. The second, ρC_p, measures the heat energy absorbed to raise the temperature of unit volume of the wall by 1°C, or the heat released when it cools by 1°C. A wall with a high thermal mass is slow to heat up and slow to cool. The time lag for heating or cooling depends on how rapidly heat diffuses into or out of a structure, measured by the thermal diffusivity, a (m²/s)

$$a = \frac{\lambda}{\rho C_p} \qquad (11.1)$$

The thermal lag time t_l (seconds) for heat to diffuse into or out of a wall of thickness w is well approximated by

$$t_l = \frac{w^2}{2a} \qquad (11.2)$$

Thermal lag

Example: The following table lists the thermal properties of ice. If an ice house has walls 500 mm thick, how long will it take the interior temperature to start changing after the exterior is exposed to a sudden drop in temperature?

Ice property	Value
Density	925 kg/m³
Specific heat	2,100 J/kg · K
Thermal conductivity	2.1 W/m · K

Answer: The thermal diffusivity of ice, calculated from the data in the table, is $a = 1.08 \times 10^{-6}$ m²/s. The thermal lag time for a 500 mm wall is then

$$t_l = \frac{(0.5)^2}{2.16 \times 10^{-6}} = 1.2 \times 10^5 \text{ s} = 32 \text{ hours}$$

The thick ice wall is very effective in screening out changes of external temperature.

Passive thermal management is best achieved by a thermal lag time of 12 hours. Then the wall can absorb heat for most of the day before its inner surface becomes warm, and it can release heat to the interior for most of the night when the outside is cool. The wall thickness that gives a 12-hour (4,300 second) thermal lag time is thus

$$w_{12} = \sqrt{2a t_l} = \sqrt{8,600\,a} = 294 \sqrt{\frac{\lambda}{\rho C_p}} \qquad (11.3)$$

This wall thickness appears as a family of parallel lines in Figure 11.7. A wood or cork wall 100 mm is enough to give a 12-hour lag time, but it does not store much heat (the values of ρC_p are low). Rammed earth, concrete, or brick walls that are 200 to 250 mm thick give the same 12-hour lag time, but they store almost 10 times more thermal energy.

With that background we can now explore specific materials.

11.5.1 Mineral-based materials. Ice. When the temperature drops below 0°C, water freezes into ice. Ice is easy to cut into blocks and stack into walls, arches, and domes—it is a practice thousands of years old. Today ice hotels lure winter tourists to Norway, Finland, and Sweden. While the temperature remains below zero, the structure is permanent. When it rises above zero, the entire structure reverts to pure water, ready to be frozen again. It's hard to think of a material more renewable than that.

Stone and lime. Stone may not be renewable, but one might think the resource from which it is drawn is near infinite. True in general, but not in particular: Carrera marble, Sydney sandstone, Portland stone, Welsh slate—all are now scarce. But more generally, stone is an eco-friendly material, durable and reliable. "Dressed" stone—the material of city banks, venerable universities, and corporate head-offices—is expensive. Much labor and quite some energy go into the dressing. Fieldstone, by contrast, is there to be picked up. Dry-stone walls are made by skilled stacking of stones as found. They work well as field boundaries but they are not load-bearing. Stone bonded with a lime mortar, however, is robust and durable. Roman arenas, amphitheatres, and aqueducts built with this combination still stand today.

Rammed earth and adobe. Soil is available almost everywhere. Mix it with straw or hair and a little lime-cement, stomp it down between wooden shuttering, let it set, and you have a wall that will support a light roof. If the earth wall is thick enough its thermal mass keeps the room cool as the outside temperature rises but warms it at night when the outside cools.[3] Adobe is the traditional building material of Mexico and parts of Africa and is still widely used today.

Brick, fired and unfired. Ordinary bricks are fired at between 800 and 1,200°C. Clay bricks made from compressed clay-rich earth are not fired so their embodied energy is low. Their compressive strength, too, is low (on the order of 0.1 MPa), so they are used for non–load-bearing walls or infill for timber frame constructions. Unfired bricks have a high density ($1,500/m^3$), giving them good sound-damping properties, a high thermal mass, and a low thermal conductivity. They help regulate temperature and humidity by *breathing*, absorbing and releasing moisture. This and their high thermal mass make buildings with thick brick walls cool in

---

[3]Typical rammed earth walls have a density of 1,950 kg/m^3, a compressive strength of 0.6 MPa, a thermal mass of 1.8 MJ/m$^3 \cdot$ K, and a thermal conductivity of 0.6 W/m $\cdot$ K (Maniatidis and Walker 2003).

summer and warm in winter. Clay brick walls can be assembled with a light clay mortar (a mixture of clay and sand), plastered with a clay plaster, or left bare on internal surfaces but external walls have to be rendered to make them water resistant.

Mineral fibers—asbestos. Asbestos is a hydrated form of magnesium silicate with a sheet silicate structure that rolls up to form hollow tubes about 4 microns in diameter and a few mm long. It is used to reinforce cement to provide flat sheets, tiles, pipes, and guttering, and as reinforcement in phenolic resin for brake and clutch linings. Its use is now restricted because some grades are carcinogenic.

Mineral foams—pumice and vermiculite. Vermiculite is a lamellar form of hydrated magnesium-aluminum-iron-silicate that resembles mica. It is used in its exfoliated (expanded) form, made by heating it at 1,000°C, which reduces its density from 2,300 kg/m^3 to about 100 kg/m^3. It is used for non-flammable thermal insulation, for fire protection, and for sound absorption.

11.5.2 Vegetable-derived materials—wood and wood-like materials.

Wood and wood-like materials offer a remarkable combination of properties. They are light, and, parallel to the grain, they are stiff, strong, and tough—as good, per unit weight, as almost any man-made material except carbon fiber reinforced polymer (CFRP). They are cheap, easily machined, carved, and joined, and—when laminated—they can be molded into complex shapes. And they are aesthetically pleasing, warm both in color and feel, and with associations of craftsmanship and quality.

All wood and wood-like materials rely on cellulose for their mechanical stiffness and strength. The data sheets in Chapter 15 list properties of clear wood samples, meaning samples with no knots or other defects. These are not, however, the data needed for mechanical design. All engineering materials have some variability in quality and properties. To allow for this design handbook's list, "allowables" will have property values that will be met or exceeded by, say, 99% of all samples (meaning the mean value of the property minus 2.3 standard deviations). Natural materials like wood show greater variability than man-made materials like steel, with the result that the allowable values for mechanical properties may be only 50% of the mean. There is a second problem: large structures made of wood contain knots, shakes, and sloping grain, all of which degrade properties. To deal with this, the wood is "stress-graded" by visual inspection or by automated methods, assigning each piece a stress grading G between 0 and 100. A grading of G means that properties listed for clear wood are knocked down by the factor $G/100$. Finally, in building construction, there is the usual requirement of sound practice, requiring an overall safety factor, typically 2.25. The result is that the permitted stress for design may be as low as 20% of the values listed in the data sheets of Chapter 15.

Mechanical design with wood

Example: Pine with a stress grading G of 45 is to be used to construct a bridge. Using data for Softwood from Chapter 15 and a typical safety factor of 2.25, what is the acceptable maximum design stress for the material of the bridge?

Answer: The yield strength of pine parallel to the grain is 40 MPa (the mean of the value in the Softwood in Chapter 15). Allowance for the stress grading and the safety factor gives an acceptable maximum design stress σ_D of

$$\sigma_D = 40 \times \frac{45}{100} \times \frac{1}{2.25} = 8 \text{ MPa}$$

Tree wood. All tree wood has broadly the same chemical composition: 40–50 wt% crystalline cellulose fibers in a matrix consisting of 20–25 wt% partly crystalline hemi-cellulose and 25–30 wt% amorphous lignin; in addition there is 0–10% oily extractives (terpenes and polyphenols) that give the wood its color and smell. The great range of properties exhibited by different woods is largely due to the difference in structure and relative density (the fraction of solid in the wood), which ranges from 0.1 to 0.85, giving the wood a density between 140 and 1,050 kg/m^3.

Wood can be grown sustainably, but many hardwoods are now classified as endangered and their harvesting and export are restricted or banned.

News-clip: Banned species

Gibson frets over claims that guitars made from illegal wood.
Gibson, the legendary guitar maker, is the target of a federal investigation. The authorities indicated that they may charge the company for illegally importing ebony from India.

The Times, August 29, 2011

Ebony and mahogany are the traditional materials for the bodies of electric guitars. Both species are endangered. It has proved very difficult to persuade musicians to switch to instruments made from sustainable materials.

Palms are monocots, more closely related to grasses and ferns than to trees. Some palms grow to a considerable height, supporting it by increasing the thickness and amount of lignin in the older cell walls and giving the stems a gradient structure, thickest at the outside. These differences give palm wood a structure and properties that differ from those of tree wood.

Bamboos are grasses. The hollow, cylindrical stem derives its strength and stiffness from its tube-like shape, reinforced by parallel fiber-bundles that occupy about 50%

of the cross-section. Like palms, bamboos do not thicken by secondary growth and show no growth rings.

Cork. All trees have a thin layer of cork just below the outer bark. The cork-oak, *Quercus suber*, is unique in that, at maturity, the cork forms a layer several centimeters thick around the trunk. Structurally, it is a low-density, polymeric closed-cell foam. Cork contains suberin, a fatty acid, and waxes that make it impervious to air, water, and alcohol (think of the cork in a wine bottle). Its function in nature appears to be to protect the tree from fire and from loss of moisture, and perhaps to discourage damage by animals (cork is pretty indigestible stuff).

Straw and reed. Straw, a by-product of agriculture, has long been used for roofs and for wall insulation. Straw bales, a product of more recent technology, are like building blocks: stack them up and you create a wall. Surface them with earth plaster or wood and they become durable. The walls have low thermal conductivity and low thermal mass (quite different from rammed earth) so they insulate well. Thatch is a reed, one with a long history of use as a roofing material. The reed grows with its base in water so it has evolved to resist it. Like straw, it insulates and is surprisingly durable: a well-thatched roof lasts for 80 years.

Thermal mass of renewable building materials

Example: Compare the thermal masses per unit area of walls made of brick, adobe, softwood, and straw bales, drawing data from the datasheets for these materials in Chapter 15. Typical wall thicknesses, w, are brick: 89 mm, adobe: 200 mm, wood: 51 mm, straw bale: 450 mm.

Answer: The following table shows the data. The wood and straw bale walls have the lowest thermal mass, the adobe wall has the highest.

Material	Density, ρ (kg/m^3)	Specific heat, C_p (J/kg·K)	Thermal mass /unit area, $\rho C_p w$ (MJ/m^2·K)
Brick	1,800	800	0.12
Adobe	1,850	710	0.26
Softwood	510	1,700	0.04
Straw bale	123	1,680	0.09

Thermal management with renewable building materials

Example: Compare the thermal optimum wall thicknesses for a 12-hour thermal lag time for the same four materials of the previous example. If the walls are built to these thicknesses, how much energy is required to heat the unit area of the wall by 10°C or is recovered from it by a drop in temperature of 10°C?

Answer: The following table lists the results. The brick wall, because of its relatively high thermal conductivity, has to be thicker than the others to have a 12-hour thermal lag time. It stores and releases the most heat. The straw bale wall insulates well but stores and releases very little heat because its thermal mass is very low.

Material	T-conductivity λ(W/m·K)	T-diffusivity $a = \lambda/\rho C_p$ (m²/s)	Wall thickness x for t = 12-hour lag time $w = \sqrt{2at}$ (mm)	Energy recovered per unit area (MJ/m²) $\rho C_p w$ (MJ/m²·K)
Brick	0.9	0.64×10^{-6}	235	3.3
Adobe	0.57	0.44×10^{-6}	195	2.5
Softwood	0.26	0.3×10^{-6}	160	1.4
Straw bale	0.05	0.25×10^{-6}	147	0.3

11.5.3 Vegetable-derived materials—fibers.

Natural fibers have been used to make ropes, fabrics, and shelters for thousands of years. They still are. Protein fibers (wool, hair) are derived from animals. Vegetable fibers are based on cellulose and are found in stem (bast), leaf, seed, and fruit. Many derive from fast growing plants, making them potentially renewable. Their properties are compared with those of man-made fibers in Figures 11.8 and 11.9. No natural fiber matches carbon, Kevlar, or drawn steel for absolute strength and modulus, but because the densities of natural fibers are low, their specific properties are much more attractive. This has led to efforts to use them as reinforcement for panels for cars and trucks, an application we return to in a later section. First, some background on natural fibers. They are listed below in alphabetical order.

Coir (from Malayalam *kayar*, cord) is a coarse fiber extracted from the hairy outer shell of a coconut. There are two sorts. White coir, harvested from unripe coconuts, is spun to yarn that is used in mats or rope. Brown coir, from fully ripened coconuts, is thicker and stronger than white coir. It is used in mats, brushes, and sacking. Pads of brown coir sprayed with latex (rubberized coir) are used as upholstery padding in cars. The coir fiber is relatively waterproof and is one of the few natural fibers resistant to damage by salt water.

Cotton is fiber from the blossom of the plant *Gossypium*. There are many species, all chemically identical (95% cellulose with a little wax) but differing in fiber length and slenderness. Cotton is used to make a great range of different fabrics with wonderful names, many deriving from the country or town of their origin: voile, calico, gingham, muslin, crepe, damask, cambric, organdie, crinoline, twill, sailcloth, canvas. They differ in the quality of the fiber and the nature of the weave.

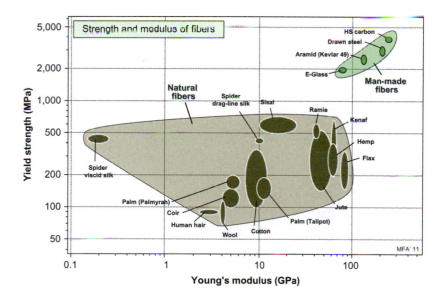

FIGURE 11.8 *The tensile strength and modulus of natural and man-made fibers. Records for most of these can be found in Chapter 15.*

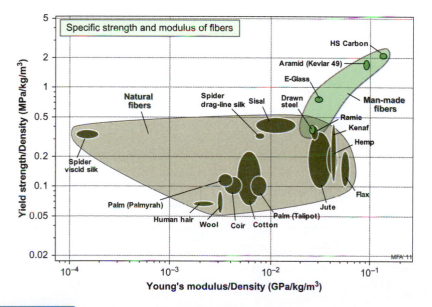

FIGURE 11.9 *The specific tensile strength and modulus for the same natural and man-made fibers that appear in Figure 11.8*

Flax is a fiber from the stalk of the linseed plant, *Linum usitatissimum*. As with hemp, the fibers are freed from the stalk by retting (for "retting" read "rotting"), an exceptionally smelly process. Flax is used to make linen and rope, valued for their strength and durability. Flax has been used as a reinforcement for polymer matrix composites.

Hemp comes from the weed *Cannabis sativa*. The hemp fibers, like those of flax, are freed from the stalk by a retting process. The properties of hemp resemble those of cotton. It is used for rope and coarse fabric, and is a potential reinforcement for natural fiber–reinforced composites.

Jute is a long, soft, shiny vegetable fiber made from the stems of plants in the genus *Corchorus*, family *Malvaceae*. It is one of the cheapest of the natural fibers and is second only to cotton in the amount produced and variety of uses. It is spun into coarse, strong threads and woven to make sack-cloth, hessian, or burlap. There is interest in using jute as a reinforcement in composites, replacing glass.

Kenaf (*Hibiscus cannabinus*) is a fast growing stem-fiber used to make industrial textiles, ropes, and twines. Emerging uses of kenaf fiber include engineered wood, insulation, and clothing-grade cloth, and as a reinforcement in polymer-matrix composites.

Palmyra palm (*Borassus*) is a genus of fan palms, native to tropical regions of Africa, Asia, and New Guinea. Palm fiber (sometimes called vegetable horsehair) is derived from its leaves. The fibers are springy and strong, making it good for stuffing furniture and mattresses.

Ramie (*Boehmeria nivea*) comes from the stem of a flowering plant in the nettle family. It is one of the strongest natural fibers. It is used for cordage and thread and is woven into fabrics for household furnishings (upholstery, canvas) and clothing, frequently in blends with other textile fibers.

Sisal fiber is derived from an agave, *Agave sisalana*. It is valued for cordage because of its strength, durability, ability to stretch, affinity for certain dyestuffs, and, like coir, it is resistant to deterioration in saltwater. The higher-grade fiber is converted into yarns for the carpet industry. Sisal is now also used as a reinforcement in polymer-matrix composites.

Fiber strengths

Example: Which vegetable fibers have the highest specific strengths? Use Figure 11.9 to find out.

Answer: The three vegetable fibers with the greatest strengths are sisal, ramie, and kenaf. All three are candidates for natural fiber–reinforced composites.

11.5.4 Animal-derived materials—wool, silk, skin, and bone.
Animal life, from an energy point of view, is high maintenance. That means that materials derived from animals have a larger embodied energy than those of the vegetable world.

Wools are keratin, the hair or fleece of the sheep, alpaca, rabbit (angora), camel (camel hair), certain goats (mohair, cashmere) and llamas (vicuna). It is hard wearing (as in wool carpets) and an excellent thermal insulator (as in wool sweaters and blankets). Wool is woven in many different ways to produce cloth, tweed, plaid, flannel, or worsted. Wool fabrics are dirt-resistant, flame-resistant, and, in many weaves, resistant to wear and tear. But keratin is a protein, easily digested by the larvae of moths if steps are not taken to stop them.

Silks, too, are proteins, the fibers spun by the silkworm to wrap and insulate its cocoon and by the spider to trap its prey. Silk is remarkable for its strength and extendibility. The combination imparts energy-absorbing qualities that make spiders webs so effective. It is the material of high-altitude balloons, parachutes, and the clothing of film stars and empresses.

Bone, tooth, and shell. Bone is a composite of hydroxyapatite (hydrated calcium phosphate) and collagen. It provided early man with tools and ornaments. Today it is calcined for bone china. Tooth, too, is hydroxyapatite in a dense form, familiar as ivory. Shell is calcium carbonate in the form of calcite or aragonite. Where it is plentiful it is calcined to make lime for cement or crushed as a filler in concrete.

Skin and hide. Real leather has become a symbol of luxury; most of its uses in earlier times are now capably met by polymers such as polyvinylchloride (PVC) and polyurethane (PU). To find real leather you need to explore the interiors of high-fashion boutiques, expensive furniture shops, and high-end automobiles.

News-clip: Engineered wood

Rising high: the timber tower block.
Is concrete so last century? Some architects are starting to think so. Wood is usually seen as a decorative material but a new type of reinforced timber is changing the picture. The so-called cross-laminated timber takes the form of spruce cut into sheets, which are stacked and glued under high pressure. This material is robust enough to use as walls, floors and even lift shafts.
The Sunday Times, January 30, 2011

Cross-laminated timber is a sort of glulam. It has low density (480 kg/m^3), good thermal and sound insulation, and build-time is half that of a concrete and steel structure of the same size. The tallest building so far made of it is a nine-story residential tower in east London called the Stadthaus.

11.6 Bio-derived materials

Environmental prerogatives stimulate interest in using natural materials in more innovative ways. Some have been with us for 2,000 years. Some are new. Figure 11.10 lists some of them.

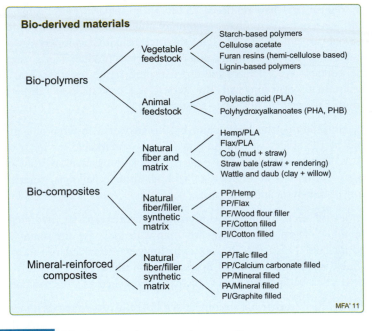

Bio-derived materials

Bio-polymers
- Vegetable feedstock
  - Starch-based polymers
  - Cellulose acetate
  - Furan resins (hemi-cellulose based)
  - Lignin-based polymers
- Animal feedstock
  - Polylactic acid (PLA)
  - Polyhydroxyalkanoates (PHA, PHB)

Bio-composites
- Natural fiber and matrix
  - Hemp/PLA
  - Flax/PLA
  - Cob (mud + straw)
  - Straw bale (straw + rendering)
  - Wattle and daub (clay + willow)
- Natural fiber/filler, synthetic matrix
  - PP/Hemp
  - PP/Flax
  - PF/Wood flour filler
  - PF/Cotton filled
  - PI/Cotton filled

Mineral-reinforced composites
- Natural fiber/filler synthetic matrix
  - PP/Talc filled
  - PP/Calcium carbonate filled
  - PP/Mineral filled
  - PA/Mineral filled
  - PI/Graphite filled

MFA' 11

FIGURE 11.10 *Bio-derived polymers and composites*

Cob and wattle and daub. Cob (mud and straw) and wattle and daub (mud mixed with clay over a slender wood lattice) are among the oldest of building materials. Both have high thermal mass and thermal lag, smoothing the sometimes extreme day-to-night temperature variations. And repair and replacement are simple. Even today, about half the world's population live in earth-walled houses.

Straw-bale construction. Straw has been used as a building material for centuries. Today, straw is used as both the structure and primary building enclosure material. Typically the straw is molded into bales and cut to dimension before being stacked to form thick insulating walls of small to medium-sized buildings. In load-bearing straw-bale construction, bales are stacked and reinforced to provide structural walls that carry the roof load. With in-fill straw-bale construction, a wood, metal, or masonry structural frame supports the roof, and bales are stacked to provide non-structural insulating walls. With either alternative, the bale walls are plastered or stuccoed on both the interior and exterior. In a load-bearing design of a building up to two stories high, bales are stacked on a poured concrete stem wall that extends about 6 inches (150 mm) above the floor slab. Wooden frames are installed for windows and doors as the layers of bales are installed.

Bonded board and linoleum. Plywood, chip board, particle board, and cork board are reconstituted wood, bonded with synthetic adhesive. They can be made from low-quality wood and wood trimmings that would otherwise be burned, making them genuinely green though not totally carbon-free. The appealingly named meadow board (made from ryegrass stems that remain after seed harvest), sunflower seed board (compacted sunflower hulls), wheat board (what's left when the grain has been removed from the wheat grass by threshing), and peanut husk board (what it says) are substitutes for wood. These are new developments when compared with linoleum, invented in 1855, and are 100% natural in their ingredients. To make linoleum, linseed oil is filled with sawdust or cork-dust and allowed to oxidize. It is then catalyzed by tiny additions of lead acetate and zinc sulfate, to produce a material like hard rubber.

Hempcrete. The use of a mix of industrial hemp and lime ("hempcrete") for infill in a wood frame building is growing in Europe, but it is held back elsewhere because of its mistaken association with marijuana, derived from a different variety of hemp. The hemp content of hempcrete—75% by volume—is truly sustainable; it is grown as fast as it is used and it requires no fertilizer. Hempcrete sequesters 0.3 kg of carbon per kg—up to 20 metric tons of carbon in a typical house—giving an eco-credit provided the building lasts a long time.

Biopolymers. Most commercial polymers are derived from petro chemicals, but not all. Cellulose, one of the main structural polymers of wood, can be converted to a resin by acid treatment ("esterification"). The resin can be polymerized to give cellulose-based polymers: cellulose acetate (CA), cellulose acetate butyrate (CAB) or cellulose acetate propionate (CAP). CA and its variants are tough and transparent, familiar as cellophane and rayon. Starch-based polymers are derived from natural polysaccharides made up of glucose molecules. Starch itself softens and dissolves in water but treatment with complexing agents links the molecules more strongly to produce a bio-degradable polymer ("Mater-Bi") that resembles polystyrene. Lactic acid–based polymers are derived from annually renewable crops such as corn, maize, or milk. Polymerization gives polylactic acid or polylactide (PLA). Natural oil-based polymers arise from bacterial action on soybean oil, corn oil, or palm oil to yield polyhydroxyalkanoates (PHAs), the most-used of which is poly-3-hydroxybutyrate (PHB, "Biopol") with properties resembling polypropylene. Furan-based polymers are derived from furfuryl alcohol, made from hemi-cellulose and sugarcane. All of these are approved for food packaging, and several are biodegradable. The difficulty is the price: all biopolymers at present cost two to six times more than the commodity polymers with which they compete.

Bio-composites. There was a time when cars, like carriages, were largely made of natural materials—wood, fabric, natural rubber; only the engine and drive train were metal. By the 1970s man-made materials had replaced all of them. Today, however, natural materials are regaining ground, driven by the environmental agenda. Thus far their application has been limited to trim (flax-fiber reinforced epoxy or furan for interior door panels, for example), for sound insulation (recycled cotton fibers) and seat cushions (coconut fibers in latex), but interest in using plant

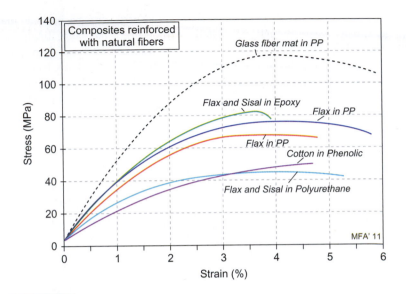

FIGURE 11.11 *Stress-strain curves for natural fiber–reinforced composites compared with that for a glass-polypropylene composite*

fibers as reinforcement is growing.[4] The fibers of most interest—flax, sisal, jute, hemp, kenaf, and coir—are annual crops that grow quickly. We saw in Figure 11.8 that single fibers have tensile strengths in the range 400–700 MPa. Most current composites use these as reinforcement in thermosetting phenolic or epoxy resins because their fluidity allows them to permeate the fiber array. The resulting properties are good (Figure 11.11), but recycling is not an option because the binder is a thermoset. This has shifted attention to plant fiber–reinforced thermoplastics, particularly polypropylene, produced in sheet form, that can be compression-molded to shape. The obstacles remain the cost, the variability of fiber quality, and the limit on processing temperature ($< 220°C$) required to avoid fiber damage.

News-clip: Traditional building materials

Council houses made of straw.

Local authorities (in Britain) are rushing to embrace traditional building materials.

The Sunday Times, July 31, 2011

---

[4]Using natural fibers as reinforcement for composites is not a new idea: Henry Ford developed Fordite, a composite of wheat straw, rubber, sulfur, and silica in the 1920s. It ended up in steering wheels.

Council houses are built by local authorities to provide low-rental accommodation for low-income families. Several authorities are experimenting with straw-bale construction. The bales are stacked into walls and surfaced with lime plaster that allows the straw to breathe. The construction materials are cheap and the straw provides excellent thermal insulation. Running costs for the tenants are reported as £320 ($500) per year for heating and electricity compared with about £1,100 ($1,800) for a traditional house of the same size.

News-clip: Bio-composites

An electric car made from hemp.
Canadian company Motive Industries, Inc. is testing a hybrid electric car, the Kestrel, made from hemp and other natural and synthetic fibers. If all goes according to plan, Motive will finish its prototype mid-2011 and make the car available to the public in late-2012, according to Nathan Armstrong, Motive's president.

The Audubon Magazine, **February 23, 2011**

Hemp has a long history with auto makers. In 1941 Henry Ford unveiled a car body made primarily out of organic fibers, hemp included. The Kestrel, a three-door hatchback, is made of a hemp composite with a strength comparable to that of fiberglass. It weighs just 2,500 pounds with its battery compared to 3,720 pounds of a comparably sized Ford Fusion. To make the bio-composite, hemp stalks are combed and rolled into a mat that is infused with a polymer resin.

11.6.1 Obstacles to the widespread use of biomaterials. The concept of replacing oil-, gas-, and coal-based materials by those of similar performance derived directly or indirectly from nature on a significant (and thus large) scale is an appealing one. There are the immediate problems that have been mentioned already:

- The price: bio-polymers and bio-composites with competitive property profiles are not cheap.
- The variability: the properties of biomaterials depend on geography and fluctuations of weather.
- The uncertainty of supply, which fluctuates with annual weather patterns.

There are probably ways of solving these problems, but they are not the ultimate difficulty. It is on a different scale. The world has now consumed itself into a corner in which there is not enough productive land to grow both the food we need and at the same time grow structural biomaterials or biofuels on a really large scale. Many studies conclude that the population carrying capacity of the planet is close to saturation. Space, water, and fertile land are the essentials for human habitation and activity. Large-scale replacement of man-made materials by those of nature no longer appears possible.

11.7 Summary and conclusions

Sustainable development, at the time of writing, is a vision, a grand ideal. Nature achieves it through balanced cycles—the carbon, nitrogen, and water cycles are examples—in which materials are used and, at the end of life, recycled to replenish the source from which they were drawn. That requires a closed loop in which the flows at each point match in rate, for if they do not, waste accumulates between them. We (the industrial "we") fail to achieve this because the ways in which we use materials either do not form a closed loop, or if they do, the units of the loop do not function at equal rates.

It is clear that there are emerging constraints to the way we draw on fossil fuels for energy and the earth's resources to create materials, both because these become depleted and because the way we use them damages the environment. The resources of many of the materials we use in the largest quantities are sufficiently abundant that we can continue to rely on them for a long time to come provided we have enough cheap, pollution-free energy to do so.

How is that to be achieved? Low carbon power, energy storage, and material efficiency are the subjects of the next two chapters.

11.8 Further reading

Australian government technical publications, building materials (2010), www.yourhome.gov.au/technical/pubs/fs49.pdf. *(A site providing practical advice on the choice and use of building materials. Enter section number to get the pages.)*

Azapagic, A., Perdan, S., and Clift, R. (Editors) (2004), *Sustainable development in practice*, John Wiley, Chichester, UK. ISBN 0-470-85609-2.

Baillie, C. (Editor) (2004), *Green composites*, Woodhead Publishing Ltd., London, UK. ISBN-13: 978-1855737396. *(A collection of papers dealing with natural fibers and composites based on them)*

Coulter, S., Bras, B., and Foley, C. (1995), A lexicon of green engineering terms, Proc. ICED 95, Prague, CZ, pp 1–11. *(Coulter presents the nested-box analogy of green design, here used as Figure 11.1.)*

Daly, H. (1990), "Toward some operational principles of sustainable development," *Ecological Economics*, 2, pp 1–6. *(Herman Daly, an economist at the University of Maryland, offers a simple set of operational principles to guide sustainable development.)*

Dasgupta, P. (2010), "Nature's role in sustaining economic development," *Phil. Trans. Roy. Soc. B*, 365, pp 5–11. *(Dasgupta argues that measuring well-being by GDP, education, and health ignores the loss of natural capital through damage to the ecosystem.)*

Furan (2010), www.biocomp.eu.com/uploads/04_THERMOSET_BIOPOLYMER_MATERIALS-TFC.pdf. *(Background on the furic acid–based bio-polymer, furan)*

Geiser, K. (2001), *Materials matter: towards a sustainable materials policy*, The MIT Press, Cambridge, MA, USA. ISBN 0-262-57148-X. *(A monograph examining the historical and present-day actions and attitudes relating to material conservation, with informative discussion of renewable materials, material efficiency, and dematerialization)*

Guidice, F., La Rosa, G., and Risitano, A. (2006), *Product design for the environment*, CRC/Taylor and Francis, London, UK. ISBN 0-8493-2722-9. *(A well-balanced review of current thinking on eco-design)*

Imhoff, D. (2005), *Paper or plastic: searching for solutions to an overpackaged world*, University of California Press, Berkeley, CA, USA. ISBN-13: 978-1578051175. *(What it says: a study of packaging taking a critical stance)*

Lovelock, J. (2000), *Gaia, a new look at life on earth*, Oxford University Press, Oxford, UK. ISBN 0-19-286218-9. *(A visionary statement of man's place in the environment)*

MacKay, D. J. C. (2008), "Sustainable energy—without the hot air," Department of Physics, Cambridge University, Cambridge, UK. www.withouthotair.com/ Accessed December 2011. *(MacKay brings a welcome dose of common sense into the discussion of energy sources and use. Fresh air replacing hot air.)*

Maniatidis, V. and Walker, P. (2003), "A review of rammed earth construction," Natural Building Technology Group, Department of Architecture & Civil Engineering, University of Bath, Bath, UK. http://people.bath.ac.uk/abspw/ rammedearth/review.pdf. Accessed December 2011. *(A comprehensive review of rammed earth properties and construction)*

Nielsen, R. (2005), *The little green handbook*, Scribe Publications Pty. Ltd., Carlton North, Victoria, Australia. ISBN 1-9207-6930-7. *(A cold-blooded presentation and analysis of hard facts about population, land and water resources, energy, and social trends)*

PROSPECT (Version 2.1) (1995), Oxford Forestry Institute, Department of Plant Sciences, Oxford University, Oxford, UK. The full database can be downloaded from www.plants.ox.ac.uk/ofi/prospect/indcx.htm. Accessed December 2011. *(A database of wood with illustrations of structure, information about uses, origins, habitat, and more)*

Schlösser, T., Gayer, U., and Karrer, G. (1999), Technischer Bericht 0003-98 Daimler-Chrysler Gmbh, Stuttgart, Germany. *(The Daimler-Chrysler Corporation has explored a number of natural fiber reinforced composites for their vehicles; this report presents the data shown in the figure in this chapter.)*

Schmidt-Bleek, F. (1997), *How much environment does the human being need—factor 10—the measure for an ecological economy*, Deutscher Taschenbuchverlag, Munich, Germany. ISBN 3-936279-00-4. *(Both Schmidt-Bleek and von Weizsäcker, referenced below, argue that sustainable development will require a drastic reduction in material consumption.)*

von Weizsäcker, E., Lovins, A. B., and Lovins, L. H. (1997), *Factor four: doubling wealth, halving resource use*, Earthscan Publications, London, UK. ISBN 1-85383-406-8; ISBN-13: 978-1-85383406-6. *(Both von Weizsäcker and Schmidt-Bleek, referenced above, argue that sustainable development will require a drastic reduction in material consumption.)*

WCED (1987), Report of the World Commission on the Environment and Development, Oxford University Press, Oxford, UK. *(The so-called Bruntland report launched the current debate and stimulated current actions on moving toward a sustainable existence)*

Wool, R. P. and Sun, X. S. (2005), *Bio-based polymers and composites*, Elsevier, Amsterdam, Netherlands. ISBN 0-12-763952-7. *(A text introducing the chemistry and production of bio-polymers)*

Woolley, T. (2006), *Natural building—A guide to materials and techniques*, The Crowood Press Limited, Ramsbury, UK. ISBN 1-861-26841-6. *(A well-illustrated introduction to traditional building materials and their present-day modifications)*

11.9 Exercises

E11.1. Distinguish *pollution control and prevention* (PC and P) from *design for the environment* (DFE). When would you use the first? When the second?

E11.2. What is meant by the *ecological metaphor*? What does it suggest about ways to use materials in a sustainable way?

E11.3. Use Figure 11.7 to estimate the thickness of a sandstone wall required to give a thermal delay time of 12 hours.

E11.4. What are the thermal characteristics of a straw-bale wall construction? Use Figure 11.7 to find out. The insulating ability or *thermal resistance* of a wall is measured by its R value, equal to its thickness (in meters) divided by its thermal conductivity (in W/m · K). What is the R value of a 200 mm straw-bale wall?

E11.5. The following table lists the thermal properties of rammed earth (adobe). If an adobe house has walls 300 mm thick, how long will it take the interior temperature to start changing after the exterior is exposed to a sudden drop in temperature?

Ice property	Value
Density	1,950 kg/m³
Specific heat	923 J/kg · K
Thermal conductivity	0.6 W/m · K

E11.6. Oak with a stress grading G of 60 is to be used to construct a bridge. Using data for Hardwood from Chapter 15 and a typical safety factor of 2.25, what is the acceptable maximum design stress for the material of the bridge?

E11.7. It is proposed to replace the steel door skins (out panels) of a particular model of car with composite skins made from 40% ramie fibers in a phenolic matrix. If all the fibers lie in the plane of the sheet and, on average, 50% of them contribute to stiffness and strength in any in-plane direction, what approximate in-plane modulus and strength could be expected? Use a simple rule of mixtures to calculate upper bounds for both. Data for ramie and phenolic can be found in Chapter 15.

CHAPTER 12

Materials for low-carbon power

12.1 Introduction and synopsis

If you want to make and use materials the first prerequisite is *energy*. The global consumption of primary energy today is approaching 500 exajoules (EJ),[1] derived principally from the burning of gas, oil, and coal. This reliance on fossil fuels will have to diminish in coming years to meet three emerging pressures:

- To adjust to diminishing reserves of oil and gas
- To reduce the flow of carbon dioxide and other greenhouse gases into the atmosphere
- To reduce dependence on imports of fossil fuels (where this is large) and the tensions this dependence creates

---

[1] 1 MJ = 0.28 kWh = 948 BTU. 1 quadrillion (10^{15}) BTU is called a "Quad," symbol Q. Thus 1 EJ ≈ 1 Q.

Clockwise from top left: land-based wind turbines in Brittany, a solar array (image courtesy of Voodo Solar, CA.), Pelamis wave machine in Portugal, Icelandic Geothermal plant (photo by Asegeir Eggertsson).

349

Table 12.1	Alternatives for power generations with current (2008) installed capacity and cost			
Power system	Current installed capacity (GW)	Growth rate (% per year)	Delivered cost ($/kWh)	Lifetime (years)
Conventional (gas)	960	1.5	0.01–0.03	30–40
Conventional (coal)	2,800	1.5	0.015–0.04	30–40
Fuel Cell	0.1	50	0.08–0.1	10–15
Nuclear (fission)	400	2.2	0.02–0.04	30–40
Wind	204	20–35	0.02–0.05	25–30
Solar Thermal	1.3	50	0.013–0.016	25–35
Solar PV	154	40	0.04–0.07	20–30
Hydro	675	4.5	0.003–0.014	75–100
Wave	0.004	50	0.03–0.07	20–30
Tide (current)	0.03	10	0.015–0.04	20–30
Tide (barrage)	0.26	10	0.009–0.015	75–100
Geothermal	8.9	20	0.01–0.02	30–40
Biomass	35	16	0.007–0.02	30–40

The world-wide demand for energy is expected to triple by 2050. Most of this energy will be electrical. How will it be generated in ways that relieve these pressures? And how much time will the transition take? The options are listed in Table 12.1.

We have history as a guide for the time it takes to replace one source of power with another. Figure 12.1 shows the way in which power sources have changed in the past 150 years. Past transitions have taken about 40 years for a 50% replacement. Speed, of course, depends on urgency and the ability to manage change; and both, in the coming years, may be greater than in the past. But the message of the figure is clear: a major shift in a vital underpinning technology such as power generation takes decades.

Renewable power systems draw their energy from natural sources: the sun, through solar; wind and wave power; the moon, through tidal power; and the earth's interior through geothermal heat. But it is a mistake to think that they are in any sense "free." Their construction incurs a capital cost, which can be large. They occupy land area. Materials and energy are consumed to construct and maintain them, and both construction and operation have an associated carbon footprint.

The best way to compare these alternative power systems is to examine their *resource intensities*.[2] By this we mean the quantity of capital, land area, energy, and

---

[2]Similar metrics are used by IAEA (1994) and San Martin (1989), reviewed by Rashad and Hammad (2000).

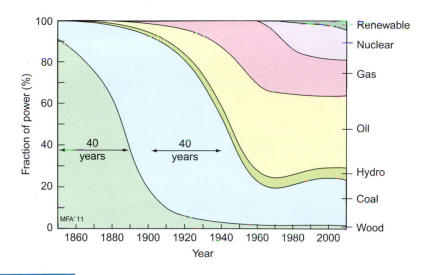

FIGURE 12.1 *The replacement of one source of power by another over the past 150 years. Transition times are on the order of 40 years. (Data from IEA, 2010)*

carbon-release per kW (kilowatt) of generating capacity. Equally important, each system has a *material intensity*, meaning the quantities of materials, in kg per rated kW of generating capacity, required to construct it. If a chosen power system were adopted on a scale that would make a major contribution to global power needs, its demand for materials could distort the materials supply chain. We explore this by comparing the demand of each power system for critical materials (material deemed by governments to be vital for their economy) with the current global production of these materials, highlighting where supply shortages might arise.

The main findings are introduced in Section 12.2. Subsequent sections develop the background and examine the implications. A warning before we start. The resource intensities of a given power system depend on many things—on the type and scale of the system, on its location, and on the way it is managed. The intensities are tabulated and plotted as representative ranges, but there is no guarantee that they enclose all members of a given system. There are also distinctions between energy and power, between the rated power of a system and the power it actually produces, between its efficiency and its capacity factor, and between energy and carbon of construction and those of operation. Appendix 1 defines all these terms fully.

12.2 The resource intensity of power sources—the big picture

The current world electric power–generating capacity is 2,200 GW (gigawatts). At present, about 66% of this derives from fossil fuels, about 16% from hydropower, 15% from nuclear power, and 3% from other renewable sources (IEA 2008). Electric

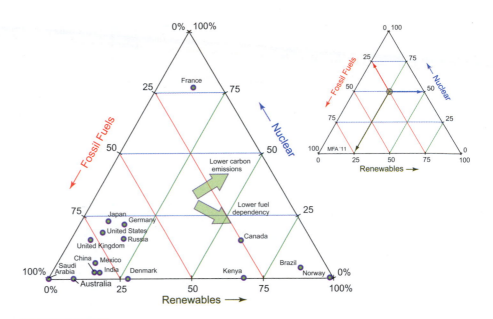

FIGURE 12.2 *A map of the distribution of electric power generation from fossil fuel, nuclear, and renewable sources. The map is read in the way shown in the smaller triangle: the energy mix is found from a vector at 3 o'clock for nuclear, one at 7.30 for renewables, and one at 11.00 for fossil fuels.*

power appears to be the future for almost everything except sea, air, and space transportation. Before examining individual systems for generating it, we should look at the electric energy mix to which individual nations have committed themselves. The fossil/nuclear/renewable power triangle of Figure 12.2 shows that this is diverse. The green arrows indicate the way changes of the mix reduce carbon emission or reduce dependence on nonrenewable fossil or nuclear fuels. The smaller triangle and the caption explain how to read the diagram.

In one way this is a reassuring picture. We are not all stuck in one corner. Nations that are endowed with natural energy sources (Norway, Brazil, Canada, Iceland) have developed cost-efficient ways of using them. The high commitment to nuclear power in France demonstrates that it is a viable option.

The flows of energy through the industrial system are complex. Energy is drawn from the sources listed in Table 12.1. Some energy is converted from one form to another—gas to electricity, for instance—before its final use to provide domestic, commercial, industrial, and transportation services. The flows can be visualized in what is called a Sankey diagram,[3] of which Figure 12.3, based on an original assembled by the Laurence Livermore National Laboratory, is an example. It shows the

---

[3]Riall Sankey, an Irish engineer, devised the diagram in 1898 to display losses in steam engines.

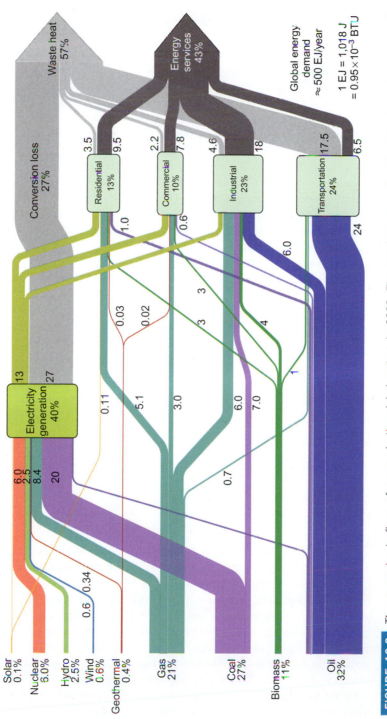

FIGURE 12.3 *The approximate flows of energy in the global system in 2009. The numbers show the % of the incoming energy of about 500 exajoules per year in each path. (Compiled from multiple sources)*

flows of energy in the US industrial system, which is typical of an industrialized nation.

The raw sources of energy enter on the left. Threads run from the left, through intermediate energy conversion steps, delivering energy to the sector listed in the boxes on the right. The width of a thread is proportional to the quantity of energy it carries per year. Colored threads represent the flow of useful energy. Light gray threads represent energy lost as low-grade waste heat. Dark gray threads represent energy used to provide useful services, which generally degrade it to the level of waste heat. Values for the flows are listed in exajoules (10^{18} joules).

The major mid-path conversion is that of coal, gas, oil, and nuclear energy to electricity at an overall efficiency of 33%, resulting in a "loss" of two-thirds of the incoming energy as low-grade heat. The electricity passes to the sectors on the right within which it is again converted to provide services: heat, light, manufacturing, transportation, and so on. Primary energy also enters these sectors in the form of gas, oil, coal, and biomass. They too provide services with varying efficiencies, all of which contribute to the light gray waste-heat threads exiting the sector boxes. The energy that provides final services is shown as the dark gray threads leaving the boxes.

The most striking feature of the diagram is that barely 40% of the incoming primary energy survives to provide useful service. The other 60% is lost on the way.

Energy conversion efficiencies

Example: Which two energy-conversion processes are the least efficient in the global energy flows of the US industrial system? Use the Sankey diagram for US energy flow to find out.

Answer: The two energy conversion processes with the lowest efficiencies are

- Generating electricity from coal, gas, oil, and nuclear sources with an overall efficiency of 33%
- Providing transportation (conversion of energy as oil to kinetic energy as motion) with an overall efficiency about 25%

Table 12.2 summarizes the resource intensities and typical capacity factors of alternative power systems. The data from which they are derived appear in subsequent sections of this chapter. We define "resource intensity" as the quantity of each resource per kW of nominal power generating capacity—*nominal* meaning the rated power of the system (for instance, a 5 kW solar array or a 600 MW power station). The actual averaged power output of the system over 1 year is less than the nominal rating because the *capacity factor*—the fraction of time that the system operates at full power—is less than 1. Thus for nuclear power, the capacity factor is typically above 75%. That for hydropower is about 55%; for off-shore wind, about 35%; for land-based wind, about 22%; and for photovoltaic solar power in Europe, about 10%. The table lists typical ranges of capacity factor.

Table 12.2	Average approximate global resource intensities for power generating systems					
Power system	Capital intensity (k$/kW$_{nom}$)	Area intensity (m^2/kW$_{nom}$)	Material intensity (kg/kW$_{nom}$)	Construction energy intensity (MJ/kW$_{nom}$)	Construction carbon intensity (kg/kW$_{nom}$)	Capacity factor (%)
Conventional—gas	0.6–1.5	1–4	605–1,080	1,730–2,710	100–200	75–85
Conventional—coal	2.5–4.5	1.5–3.5	700–1,600	3,580–9,570	100–700	75–85
Phosphoric acid fuel cell	3–4.5	0.1–0.5	80–120	5,000–10,000	600–1,000	>95
Solid oxide fuel cell	7–8	0.3–1	50–100	2,000–6,000	200–400	>95
Nuclear—fission	3.5–6.4	1–3	170–625	2,000–4,300	105–330	75–95
Wind—land-based	1.0–2.4	150–400	500–2,000	3,500–6,000	240–600	17–25
Wind—off-shore	1.6–3	100–300	300–900	5,000–10,000	480–1,000	30–40
Solar PV—single crystal	4–12	30–70	800–1,700	30,000–60,000	2,000–4,000	8–12*
Solar PV—poly-silicon	3–6*	50–80*	1,000–2,000	20,000–40,000	1,500–3,000	8–12*
Solar PV—thin-film	2–5	50–100	1,500–3,000	10,000–20,000	550–1,000	8–12*
Solar—thermal	3.9–8	20–100	650–3,500	19,000–40,000	1,500–3,500	20–35†
Hydro—earth dam	1–5	200–600	15,000-100,000	7,260–15,000	630–1,200	45–65
Hydro—steel reinforced	1–5	120–500	8,000–40,000	30,000–66,000	1,000–4,000	50–70
Wave	1.2–4.4	42–100	1,000–2,000	22,950–31,540	1,670–2,070	25–40
Tidal—current	10–15	150–200	350–650	12,000–18,000	800–1,130	35–50
Tidal—barrage	1.6–2.5	200–300	5,000–50,000	30,000–45,000	2,400–3,520	20–30
Geothermal—shallow	1.15–2	1–3	61–500	7,000–13,500	160–250	75–95
Geothermal— deep	2–3.9	1–3	400–1,200	20,000–40,700	1,700–3,900	75–95
Biomass—dedicated	2.3–3.6	10,000–33,000	500–922	5,000–19,800	600–1,800	75–95

*Estimated capacity factor for PV in the UK and equivalent latitudes in Europe. The capacity factor in central Australian, Sahara, or Mojave deserts could be four times greater.

†Typical capacity factor for solar thermal built in a suitable location, such as Spain, North Africa, Australia, or Southern USA.

Some of the data in Table 12.2 are easy to find but others are not. Some have been estimated from diagrams or schematics of the system, some deduced by analogy with other systems with similar structural requirements, some inferred from the physics on which the system depends. The material intensities vary greatly with the design—alternative choices exist for magnetic materials for generators and for the semiconductor panels for solar cells, for example—allowing wide variation. That means that the precision of these data is low. But the differences between the resource intensities of competing systems is sufficiently great that it is still possible to draw meaningful conclusions.

The remainder of this section examines what can be learned from the data in Table 12.2.

System efficiency and capacity factor

Example: What is meant by the *system efficiency* and by the *capacity factor* of a power system? How do they differ?

Answer: The system efficiency (%) is the efficiency of conversion of the primary energy source (coal, solar radiation, wind, wave, or tidal energy) into electrical power under ideal working conditions. Taking photovoltaic power as an example, up to 20% of the energy of the incident radiation is converted to electricity provided the incident intensity of the radiation is within the working range of the solar panel.

The capacity factor (%) is the fraction of time that a power system operates at its rated or nominal power. It is reduced by downtime for maintenance or fuel replacement and by the unavailability of the primary energy source. Photovoltaic power, for example, has a capacity factor as low as 10% because the sun does not shine at night, because of cloud cover, and because of the inclination of the panel to incoming radiation.

News-clip: Capacity factors

A gloomy winter.

The South East of England suffered the gloomiest winter on record, the Met Office revealed yesterday. Between December and the end of February London was particularly grey with only 98 hours of sun in three months.

The Times, **March 3, 2011**

The capacity factor of a solar panel in London was dismally low over this period. Three months is 90 days, or 2,160 hours. If the panel were oriented normal to those few rays of sunshine, its capacity factor C would have been $98/2,160 = 4.5\%$. If it were less happily oriented, C would be even less.

Resource intensities

Example: What is meant by the resource intensities of a power system?

Answer: Construction and commissioning of a power system requires resources: capital, materials, and energy and space, meaning land or sea area. A *resource intensity* is the amount of the resource required to create one unit of power generating capacity. Power systems have a rated nominal power, kW_{nom}, but none operates at full capacity all the time, so it is necessary to define also the average delivered power, kW_{actual}. Because of this we need two intensities for each resource. The first is the intensity of the resource per unit of rated nominal capacity (per kW_{nom})—a well-defined quantity since the rated power is a fixed characteristic of the system (e.g., a 1 kW solar panel). The second is the intensity of the resource per unit of delivered power when averaged over a representative period such as a year. It is equal to the nominal value divided by the capacity factor C expressed as a fraction (e.g., $C = 20\% = 0.2$). The resource intensity per kW_{actual} is always larger, sometimes much larger, than the intensity per kW_{nom}, and it is less well defined because it depends on how the system is operated, and, in the case of renewable systems, on the influence of the weather on sunshine, wind, wave, and tide.

Charts of resource intensities. The data of Table 12.2 are displayed in the next five figures. The first (Figure 12.4) shows the material and area intensities. For a meaningful comparison the nominal power will not do; instead we need the intensities associated with the actual power output averaged over a year, kW_{actual}. To calculate these we divide each nominal intensity in Table 12.2 by the capacity factor expressed as a fraction. The most striking thing about the figure is the enormous differences between the area intensities of different systems. Gas and coal-fired power stations, nuclear, and geothermal have small footprints of around $3 \ m^2/kW_{actual}$. All others require an area 50 to 500 times greater. This space-hungry characteristic

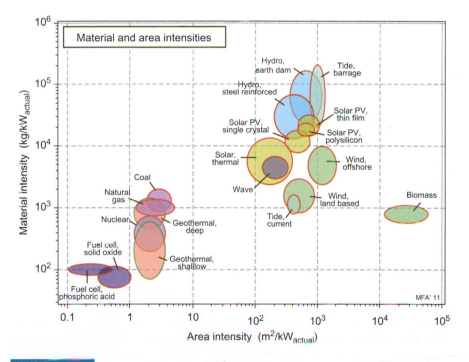

FIGURE 12.4 *The area and material intensities of power systems, based on actual power output during life*

may not be a problem for offshore wind and wave power, but for land-based systems the occupancy of land that could be used for other purposes presents difficulties. Conventional, nuclear, and geothermal systems also have lower material intensities than many of the others, but to understand the material implications of alternative power systems, we must examine their bills of materials in more depth: some systems, like hydropower, use materials that are cheap and readily available; others, like fuel cells, use materials that are scarce and expensive. This we do in subsequent sections of this chapter.

The demand for space

Example: Use mean values of the area intensities of Table 12.2 to compare the land area required to build a *nominal* 0.5 GW of new generating power using (a) nuclear, (b) single crystal solar PV power, and (c) land-based wind sources. How do these areas change if 0.5 GW of *actual*, not nominal, power is to be built?

Answer: For 0.5 GW of nominal power, the areas are (a) nuclear, 1 km^2, (b) single-crystal PV power, 50 km^2, and (c) land-based wind, 138 km^2.

For 0.5 GW for actual power, these values must be divided by the capacity factor, C. Using mean values from Table 12.2, the areas become (a) nuclear, $1/0.8 = 1.2$ km^2, (b) single-crystal PV power, $50/0.1 = 500$ km^2, and (c) land-based wind, $138/0.21 = 657$ km^2.

Material intensities

Example: The material intensity for the construction of offshore wind turbines averages about 825 kg per rated (nominal) kW of generating capacity. If the capacity factor for offshore wind is 0.35, what is the material intensity per kW$_{actual}$ of delivered power?

Answer: An offshore turbine rated at 1 kW actually delivers an average of 0.35 kW, a capacity factor of 35%. To deliver an average of 1 kW$_{actual}$ requires $1/0.35 = 2.5$ kW of nominal generating capacity, making the material intensity 2,063 kg/kW$_{actual}$.

The capital and energy intensities for the construction of power systems, plotted in Figure 12.5, are calculated in the same way as those for material and area—by dividing the nominal intensities by the capacity factor expressed as a fraction. The two actual intensities are approximately proportional. This arises partly because systems that are energy-intensive to construct are generally more expensive than those that are not, and partly because a low capacity factor (like that of solar photovoltaics) inflates both intensities.

Figure 12.6 shows the balance between the energy to construct the power system and the energy it generates per year in MJ of electrical energy, per nominal kW

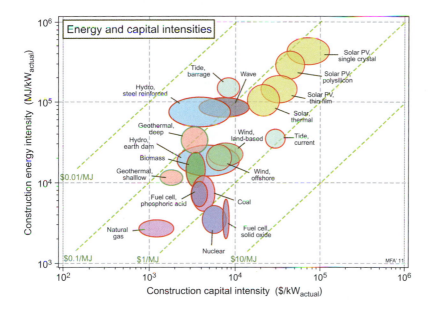

FIGURE 12.5 *The capital and energy intensities of construction of power systems, based on actual power output during life. The low capacity factor of solar PV systems in temperate climates makes them expensive in both capital and energy.*

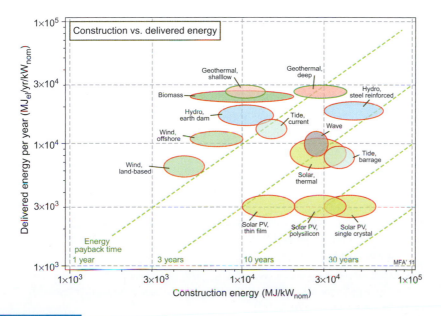

FIGURE 12.6 *Energy payback—the balance between construction and delivered energy*

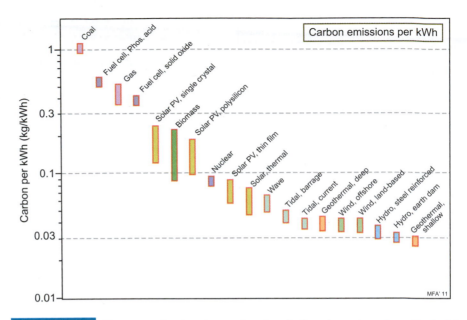

FIGURE 12.7 *The approximate release of carbon to the atmosphere from the building and operation of alternative power systems, assuming a 20-year life. None is carbon free, but all emit less than coal.*

of generating capacity.[4] It should be remembered that the construction energy is expressed in oil equivalent, whereas the delivered energy is electrical. Contours show the energy payback time, equal to the time in years before the delivered energy exceeds that invested in construction of the plant. The data suggest an energy payback time of 1 to 2 years for wind and hydro, rising to 3–10 years for solar and tidal barrier.

Figure 12.7 brings out the large differences in carbon emissions of power systems, measured in kg of CO_2 per kWh of delivered energy. Each bar describes the sum of three terms:

■ The construction carbon intensity (kg/kW_{nom}) prorated by the energy delivered over the system life in kWh/kW_{nom}, using the system lives listed in Table 12.1[5]

■ The carbon release associated with plant operation, estimated at 0.03 kg/kWh for coal and 0.02 kg/kWh for the others (estimates by White and Kulcinski, 2000)

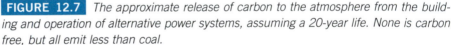

[4]$31,530 \times$ Capacity factor MJ_{elec}/kW_{nom} (31,530 is the number of hours in the year multiplied by 3.6 to convert kWh to MJ).

[5]Equal to (construction carbon intensity/8544 L C) where C is the capacity factor, L the system lifetime in years, and 8544 is the number of hours in a year.

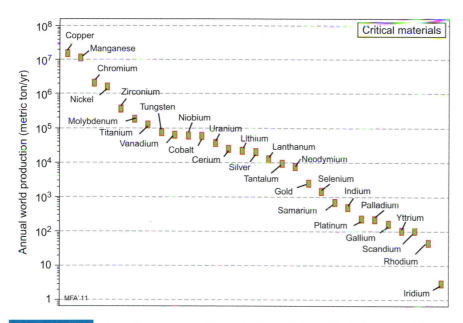

FIGURE 12.8 *Current annual world production of 29 critical elements. Many of them are used in power generation. A major shift from one power system to another can put pressure on their supply.*

- The release of CO_2 from hydrocarbon fuels, where they are used, per kWh of delivered electrical power[6] (fuel cells are assumed to burn methanol)

The figure demonstrates that no power system is completely carbon free because of the contributions from construction and maintenance, but renewable systems produce up to 30 times less carbon than those burning fossil fuels. The capacity factor for solar PV systems used in the calculation is that for Northern Europe. Solar panels in sunnier climates will have larger C and lower CO_2 emissions per kWh.

In subsequent sections we examine power systems in turn, focusing on the underlying physics of their operation and the implications for material supply if they are deployed on a large scale. One concern that emerges is the demands made on materials deemed to be critical, either because the global supply is limited, or because the main ore bodies are localized in such a way that a free market does not operate, or because they play a vital economic role for which no ready substitute exists (like copper for electrical conduction and manganese as an alloying element in steels). Figure 12.8 provides background for discussing this. It shows the 2008

---

[6]For coal this is equal to 0.088 kg/MJ $\times$ 3.6 MJ/kWh$_e$/0.33 (conversion efficiency from coal to electric power) = 0.96 kg/kWh. For gas it is equal to 0.055 kg/MJ $\times$ 3.6 MJ/kWh$_e$/0.38 (conversion efficiency of gas to electric power). For nuclear fuel it is approximately 0.022 kg/kWh$_e$.

world production of 29 elements that are considered to be critical.[7] Later sections compare the demand of each power system for these elements with the world production, highlighting where material constraints are likely. To do this we examine a hypothetical scenario: if 2,000 GW, which is roughly equal to the current world generating capacity or one third of what is projected for 2050, were to be replaced by a given alternative power system over a period of 10 years. From this we calculate the fraction of current world production of each critical material that would be required to make the replacement, revealing where material supply might be a problem.

Example: Differing reasons for criticality. Why are graphite, gallium and lithium listed as critical?

Answer:

Graphite. Over 95% of the world's supply of graphite comes from a single country (China). This makes graphite vulnerable to export tariffs or restrictions.

Gallium. Gallium is recovered from bauxite during aluminum refining; it cannot be economically mined on its own. Increase in supply is only possible if demand for aluminum increases or the efficiency of gallium recovery from bauxite is improved.

Lithium. Production of lithium is dominated by South American countries, not all of which are politically stable. Stable supply is essential to meet the anticipated surge in demand for lithium batteries, for which no viable substitute is currently known. This unique functionality is the reason lithium is classed as critical, despite its relative abundance.

12.3 Conventional fossil-fuel power: gas and coal

Fossil fuel power generation is the benchmark against which alternatives must be judged. The use of fossil fuels as a source of energy started in the 1700s and has grown in scale ever since. The high energy density of fossil fuels makes them easy to transport and allows a large amount of power to be generated in compact plants taking up little land area. Today, fossil fuels are our principal source of power, but they are also a source of political and social tensions as the limits to their supply become more apparent. And there is the concern about the emissions that result from their use.

Natural gas. It is said that natural gas, leaking from the ground and burning, so awed the people of Delphi that the place became both a shrine and an appropriately mysterious seat for an oracle. Be that as it may, natural gas is the star among fossil fuels. The low capital cost and the short lead times for building natural gas plants makes them the first choice for new capacity. Natural gas is primarily

---

[7]Many publications, such as that of the British Geological Survey (2011), list critical materials, though not all agree.

methane. It is the cleanest burning and least polluting hydrocarbon fuel. Gas was at one time made from coal, but today it is drawn from gas, oil, and shale reservoirs where it is found along with heavier hydrocarbons. Many gas reserves are now depleted, and while others remain to be exploited, most are deep and lie beneath water or ice.

For electricity generation, gas is either burned to produce steam or combusted in a gas turbine. In combined cycle units, a gas turbine produces electricity and the waste heat generates steam to drive a secondary steam turbine, giving conversion efficiencies above 50%. Natural gas fuel cells, which we meet in Section 12.6, allow small-scale electricity generation. In addition to its use as a fuel, natural gas is crucial for the manufacture of plastics, fabrics, and other chemicals and materials. It is, however, a finite resource, the production of which is expected to peak before 2050. The number of high-tech industries that rely on natural gas as a feedstock highlights the importance of conserving it for the future. This conservation imperative and the emissions associated with burning gas for energy motivate the efforts to find alternative sources of power.

Coal. Coal is the sun's energy in solid fossil form. There are four basic types, classed by their carbon content and calorific (heat) value. *Anthracite*, with 86−98% carbon, is the cleanest burning of the four. *Bituminous coal*, the most common type, contains 46−86% carbon The final two classes, *sub-bituminous* and *lignite*, both with a carbon content of 46−60% are soft and burn with a smoky flame. All forms of coal are used to generate electricity in large plants (at a typical conversion efficiency of 38%), but they also have important secondary uses: many plastics and organic chemicals rely on the distillation of coal for feedstock.

The global reserves of coal far exceed those of oil or gas. If we are to generate power from fossil fuels, then we will be driven to build coal-fired stations. Coal contains hydrogen, nitrogen, sulfur, and many other elements besides carbon. Combustion releases not only greenhouse gases such as carbon dioxide but also the oxides of sulfur and nitrogen that cause acid rain. These environmental concerns have prompted the development of clean-up technologies. *Washing* reduces the nitrogen content of coal. *Scrubbing*, spraying a lime-water mix into the smoke, removes acidic oxides of sulfur by neutralizing them. *Carbon capture and storage* (*CCS*), an emerging technology, captures the carbon dioxide, compresses it, and stores it in spent oil and gas reservoirs. This last possibility would allow coal-derived power to be included in the list of low-carbon power sources, although the material implications of CCS are not yet known.

Figure 12.9 shows the layout of a coal-fired power station and identifies the materials used in the largest quantities. An approximate bill of materials is given in Appendix 2 of this chapter, expressed as mass of material in kg per nominal kW of generating capacity, kg/kW_{nom}. As already explained, the actual output of any power source, averaged over life, is less than the nominal or rated value because the capacity factor C is less than 1, making the actual material intensities, kg/kW_{actual}, larger than the nominal ones by the factor $1/C$.

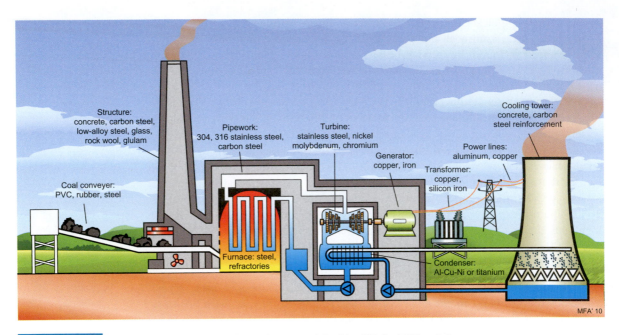

Structure:
concrete, carbon steel,
low-alloy steel, glass,
rock wool, glulam

Pipework:
304, 316 stainless steel,
carbon steel

Turbine:
stainless steel, nickel
molybdenum, chromium

Generator:
copper, iron

Cooling tower:
concrete, carbon
steel reinforcement

Power lines:
aluminum, copper

Transformer:
copper,
silicon iron

Coal conveyer:
PVC, rubber, steel

Furnace: steel,
refractories

Condenser:
Al-Cu-Ni or titanium

MFA' 10

FIGURE 12.9   *Coal-fired power station (based on an original by Bill C., Wikipedia)*

If a sufficient number of new coal stations were built over the next 10 years to provide 2,000 GW of additional capacity, would the drain on material supply be significant? As an indicator, we divide the annual material demand that this implies by the current annual global production of that material, expressing the resulting demand ratio in %. Thus a demand ratio of 1% means the construction would require a mere 1% of annual global production; a demand ratio of 100% means a quantity equal to the current global production would be needed. The results, for critical materials, are plotted in Figure 12.10. Only chromium might give cause for concern.

12.4 Nuclear power

Nuclear power is seen by some as a viable means of generating the electrical power needed to meet future needs. Others perceive it to be only an interim solution and one with the inherent risks of accidents and nuclear proliferation. Today, many governments take the view that, despite the risks, nuclear power offers the fastest, cheapest way to reduce dependence on imported hydrocarbons, cut carbon emissions, and assure energy supply to 2050.

Nuclear power derives from the energy released during the fission of a nuclear fuel—typically uranium-235—when it captures neutrons. The briefly formed uranium-236 is unstable and breaks into lighter nuclei, releasing more neutrons in a chain reaction. The energetic neutrons are slowed by a moderator, usually water, converting their kinetic energy to heat. Fuel consumption is roughly 1 mg of

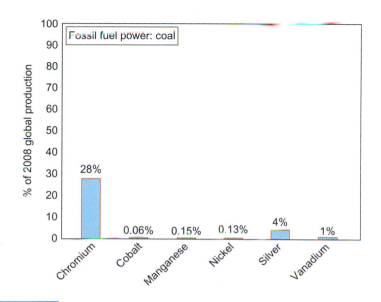

FIGURE 12.10 *Resource demands for strategic materials used in conventional power systems*

uranium per kWh of electrical energy. Fuel is replaced every 1 or 2 years, during which time the plant is shut down.

Currently there are 436 nuclear power stations worldwide, of which 60% are pressurized water reactors (PWRs) and 21% are boiling water reactors (BWRs). The remaining 19% include older CANDU and gas-cooled reactors and new, more advanced reactor designs. The most controversial issue surrounding the large-scale use of nuclear power is that of dealing with radioactive waste, which requires secure storage for up to 1,000 years.

The core of a pressurized water reactor (Figure 12.11) has some 200 tube assemblies containing ceramic pellets of enriched uranium dioxide (UO_2), or of a mixture of both uranium and plutonium oxides known as MOX (mixed oxide fuel). These are encased in a Cladding of a zirconium alloy, Zircaloy 4. Either B_4C-Al_2O_3 pellets or borosilicate glass rods are used as burnable poisons to limit the neutron flux when the fuel is new. Water, pumped through the core at a pressure sufficient to prevent boiling, acts as both a coolant and a moderator, slowing down high-energy neutrons. The power is controlled by control rods inserted from the top of the core and by dissolving boric acid into the reactor water. The boron carbide (B_4C) or Ag-In-Cd alloy control rods are clad in Inconel 627 or Type 304 stainless steel tubes. The primary pressurized water loop carries heat from the reactor core to a steam generator under a pressure of about 15 MPa, which is sufficient to allow the water in it to be heated to near 600°K without boiling. The heat is transferred to a secondary loop generating steam at 560°K and about

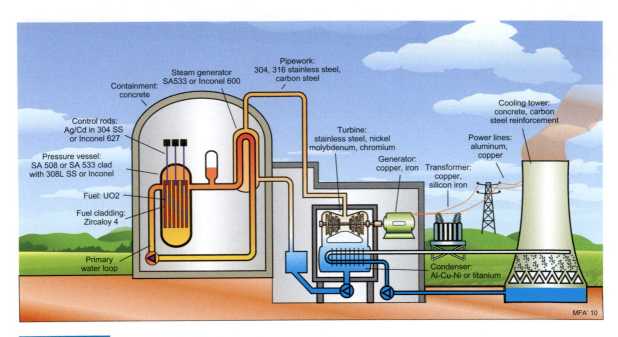

FIGURE 12.11 *The pressurized water reactor*

7 MPa that drives the turbine. An approximate bill of materials appears in Appendix 2 of this chapter.

Resource demands of nuclear power

Example **(1):** The energy density of uranium is 470,000 MJ/kg. If this energy is converted to electrical power at a conversion efficiency of 38%, how much uranium is required per year to provide a steady 1 GW of electrical power?

Answer: 1 kg of uranium delivers $470,000 \times 0.38/3.6 = 49,600$ kWh$_{electrical}$. (The factor 3.6 converts MJ to kWh.) There are $24 \times 365 = 8,760$ hours in a year, so a steady power of 1 GW over 1 year equates to an energy of $8,760 \times 10^6$ kWh. This requires $8,760 \times 10^6/49,600 = 1.77 \times 10^5$ kg = 177 metric tons of uranium per year.

Example **(2):** The annual global production of uranium (in 2008) was 40,000 metric tons per year. How many GW of power will that support? How does this compare with the anticipated demand in 2050?

Answer: The previous example showed that 177 metric tons of uranium is needed to provide a steady 1 GW of power for 1 year. The current annual global production of 40,000 metric tons of uranium could provide $40,000/177 = 226$ GW continuously for 1 year, sufficient to provide 15% of today's consumption, or 5% of the expected demand in 2050.

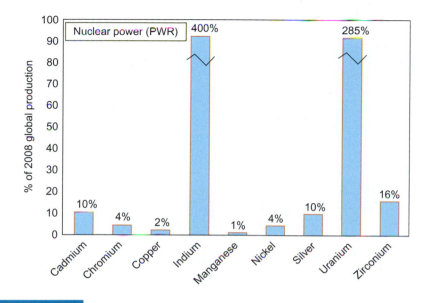

FIGURE 12.12 *Demand ratios for strategic materials used in nuclear power systems*

Installing 2,000 GW of additional nuclear capacity over the next 10 years carries the implications for critical materials plotted in Figure 12.12. The annual demand for indium for control rods and of uranium for fuel greatly exceeds the current annual production of these two materials.

12.5 Solar energy: thermal, thermoelectric, and photovoltaics

If you think of the earth as a flat disc facing the sun, then the energy that the sun beams onto the disc is a prodigious 1 kW/m². Multiplying this by the disc area (roughly 10^{14} m²) gives 100,000 TW (10^{17} W), more than a million times more power than we currently use. Not all of it is accessible: some is reflected, some absorbed in the atmosphere, and much falls where it can't be reached. Nor is it evenly distributed: the length of day and the angle that the surface presents to the sun differ between the poles and the tropics. When cloud cover and length of day are allowed for, the sun's energy per unit area in countries with a temperate climate averages 100 W/m²; in the tropics it can be three times larger.

Simple thermal systems. If a black panel is placed so that photons fall onto it, their energy is absorbed as phonons—lattice vibrations—raising its temperature. The energy can be harvested by passing water or air through the panel, providing low-grade heat for water or space heating.

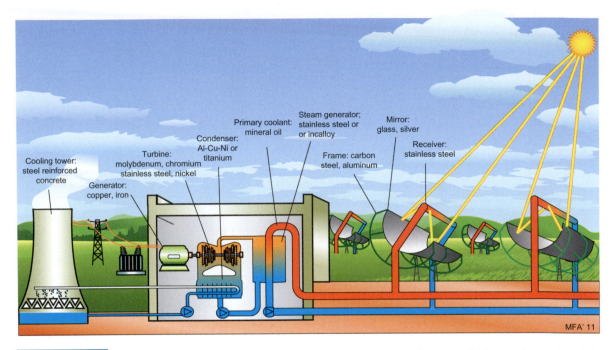

FIGURE 12.13 *A solar concentrating thermal system using parabolic reflectors, which can be replaced by Fresnel mirrors*

The main materials issues here are those of durability and cost. The materials of the panel must survive for the design life (30 years or more) without maintenance, and they must be sufficiently cheap that the cost of the panel is quickly offset by the value of the energy that is captured.

Concentrating thermal systems. Archimedes, it is said, incinerated enemy ships at the Siege of Syracuse by using polished shields to focus the sun's rays into lethal beams. Concentrating Solar Thermal (CST) plants use the same idea to generate high temperature steam to power turbines, using one of three schemes.

- Heliostatic mirrors track to follow the sun (Figure 12.13) and focus radiation on a tower receiver where it heats molten salt or pressurized water to above 400°C. The heated fluid is used to generate steam for a conventional turbine.
- Parabolic mirrors or liner Fresnel reflectors track the sun in one dimension, focusing radiation on a tube running down their length which contains the heat-transfer fluid. The fluid, typically mineral oil, passes to a heat exchanger where it is used to produce steam to drive the turbine.
- Parabolic dish reflectors resembling a satellite dish have a central receiver mounted in front of the mirror. The receiver is a Stirling engine coupled to a small generator which, combined with three-axis tracking of the sun, gives this design the highest efficiency. Expense limits their deployment.

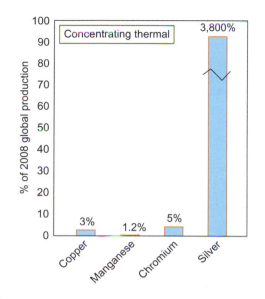

FIGURE 12.14 *Demand ratios for strategic materials used in solar concentrating systems*

Some of the incoming energy is lost by reflection, and some is lost by parasitic conduction or convection giving overall conversion efficiencies of 30–50%. If this heat is then used to generate electricity there is a further conversion loss, reducing the efficiency to 8–15%.

The parabolic trough is the cheapest and the most robust scheme for harvesting solar energy. Both it and the solar tower, which can reach higher temperatures, are compatible with thermal energy storage using molten nitrate salt as the storage medium. This energy is recovered when there is less sunlight but still high demand.

A bill of materials for a typical parabolic trough or solar tower plant appears in Appendix 2. Figure 12.14 shows the demand ratios. It is clear that supply constraints on the use of silver for the reflectors would be a concern for widespread use of CST. Low-cost polymer-based mirrors with aluminum reflective coatings are under trial. Molten salts used for storage are not critical materials and are already produced in large quantities for agriculture.

News-clip: Solar thermal power

Total (a French oil company) makes big commitment to desert sun.
Over the next few months the largest solar-power system in the world will spread over an area of sand the size of 290 football pitches, some 50 kilometers from Abu Dhabi.

Le Figaro, **April 25, 2011**

Total's project, called Shams 1 (*shams* is Arabic for sun) uses parabolic mirrors to heat oil-filled pipes to 400°C. The oil passes to a heat exchanger powering a steam turbine. The oil retains enough heat at night to keep the turbine running at half power. The plant covers 2.5 km^2 and costs 440 million euros. When fully operational, it will supply 1% of the needs of the Emirate.

Thermoelectric systems. Two dissimilar metal wires, joined at one end, develop an emf (a voltage difference) $\Delta V = S\Delta T$ between the two unjoined ends when the joined end is heated, where S is the Seebeck constant with units of volts/K, a characteristic of the materials of the couple. Thermoelectric capture can be used in combination with photovoltaics to generate useful power from otherwise wasted heat.

The efficiency of a thermoelectric system depends on the materials that form the junction and the temperature difference between the junction and the free ends of the wires. It is measured by a figure of merit, Z

$$Z = \frac{S^2 \kappa_e}{\lambda} \tag{12.1}$$

where κ_e is the electrical conductivity, and λ is the thermal conductivity. This is more commonly expressed as the *dimensionless figure of merit ZT* by multiplying Z by the average temperature $T = (T_1 + T_2)/2$ of the extremes of the wires. Larger values of ZT indicate greater thermodynamic efficiency. Values of $ZT = 1$ are considered high, but values in the 3 to 4 range are needed for thermoelectrics to compete with conventional power generation. Compounds of bismuth (Bi), selenium (Se), tellurium (Te), ytterbium (Yb), and antimony (Sb) have the highest values of ZT but none as high as 4. These are materials with small reserves and localized sources, making large-scale deployment of thermoelectric generation problematic.

Photovoltaic (PV) systems. Although photovoltaic power is expensive, world capacity is growing rapidly, spurred on by government subsidies to expand renewable electricity generation. In remote locations where transmission costs are high, solar power can compete with power from fossil fuels. But as Figure 12.5 showed, constructing PV systems is both capital and energy intensive.

The semiconductor that forms the active element of a PV collector has an energy gap (the gap between the conduction band and the valence band) comparable with the energy of the sun's photons. Solar radiation arrives as photons with wavelengths λ between 0.3 and 3 microns, the intensity peaking at 0.5 microns. The corresponding photon energy is hc/λ (here h is Planck's constant, 6.6×10^{-34} J/s and c is the velocity of light, 3×10^8 m/s). If an electron with a charge e absorbs a photon, it acquires a higher electric potential

$$\Delta V = h c/\lambda e \tag{12.2}$$

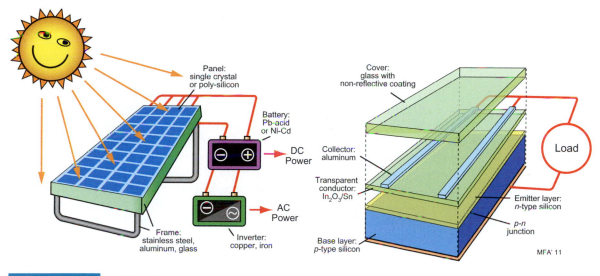

FIGURE 12.15 *A silicon-based photovoltaic panel*

For solar photons, ΔV is between 0.5 and 2.5 volts. If the energized electron now flows through an electric circuit across this potential difference, it can deliver electrical energy to an external load.

Most photovoltaic cells today use silicon as the semiconductor (Figure 12.15). A *base* layer of *p*-type silicon is joined to a thin *emitter layer* of *n*-type silicon that is exposed to the sun's rays, which are absorbed in a layer near the *p-n* junction. The *n*-layer is doped with electron donors (5-valent elements such as phosphorous or arsenic) that readily give up an electron. The *p*-layer is doped with electron receptors (3-valent elements such as boron or gallium) that readily accept an electron, creating an electron "hole." The mobile electrons in the *n*-layer and mobile holes in the *p*-layer provide charge-carriers that allow an electric current to flow through the cell. At the *p-n* junction a potential difference exists because of the excess electrons on one side and holes on the other. When solar photons with an energy greater than the band gap penetrate this junction, they create electron-hole pairs. The electrons move to the negative electrode and the holes to the positive one to provide the current in the external circuit.

Only a fraction of the incoming solar energy is captured because long-wavelength photons have too little energy to create electron-hole pairs and are simply absorbed as heat; short-wavelength photons have more than is needed and the difference, again, is absorbed as heat. The result is that the efficiency of conversion is low and it is reduced further if the panel does not face the sun but lies at an angle to it. Static thin-film devices made of amorphous silicon are the cheapest and give an efficiency of 8 to 9%. Poly-silicon crystals have efficiencies of 12–14%. Single crystal silicon cells (the most expensive) provide an efficiency of 15–17%, though they are more energy-intensive to make. The average output is increased further by tracking the panels so that they face the sun continuously, allowing up to 20% conversion efficiency. Concentrators in the form of lenses or mirrors can improve the efficiency further. Newer systems use cadmium telluride (CdTe) or copper selenide $Cu(In, Ga)Se_2$ semiconductors.

Once installed, solar PV power is effectively carbon-free, and although conversion efficiency declines over time, it requires little maintenance and has a long life. The difficulty with solar power in temperate climates is the low capacity factor, about 10%. This is because the rating of a panel (1 kW, for example) is the power it produces when the incoming solar power density is $1,000 \, W/m^2$, a value only reached at midday on a completely clear day. The average incoming power density, allowing for hours of darkness, cloud cover, and other factors, is about $100 \, W/m^2$. This low capacity factor drives up the capital and material intensities and stretches the energy payback time to between 3 and 10 years (Figures 12.5 and 12.6). When installed in a dry, tropical location, the capacity factor increases by 3 or more times, with a proportional drop in these intensities and times.

A bill of materials for a typical PV system is given in Appendix 2; details vary with panel type and manufacturer. Many of these materials, such as indium, gallium, and tellurium, are critical (Figure 12.16). A major expansion in photovoltaic power generation would put their supply under pressure. It is for this reason that current research focuses on cheaper, more plentiful alternatives such as copper or iron sulfides.

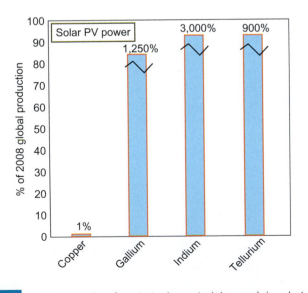

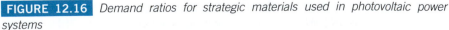

FIGURE 12.16 Demand ratios for strategic materials used in photovoltaic power systems

12.6 Fuel cells

Electrical conduction in solids can be electronic (a flow of electrons) or ionic (a flow
of ions). Many ionic conductors are electronic insulators, a characteristic exploited
in fuel cells. A fuel cell consists of an anode and a cathode separated by an electron-
ically insulating electrolyte. Oxidation takes place at the anode, releasing electrons,
while reduction at the cathode absorbs them:

$$2H_2 \rightarrow 4H^+ + 4e^- \qquad \text{(typical anode reaction)} \qquad (12.3,a)$$

$$O_2 + 4H^+ + 4e^- \rightarrow 2H_2O \qquad \text{(typical cathode reaction)} \qquad (12.3,b)$$

To allow the reaction, protons (H^+) diffuse through the electrolyte from the
anode to the cathode. Anode and cathode are connected by an external electron-
conducting circuit that includes the load. The reactions have to be catalyzed by
platinum, a critical material.

Fuel can be fed to the cell as hydrogen gas, but hydrogen supply is at present
limited. Most fuel cells create hydrogen by reforming methane with steam:

$$CH_4 + H_2O \rightarrow 3H_2 + CO \qquad (12.4,a)$$

$$CO + H_2O \rightarrow H_2 + CO_2 \qquad (12.4,b)$$

These reactions must be catalyzed with nickel and require high temperatures. Low-
temperature fuel cells use external reformers. High-temperature fuel cells can reform
the methane at the electrolyte. Both emit CO_2, generated by the reforming process.

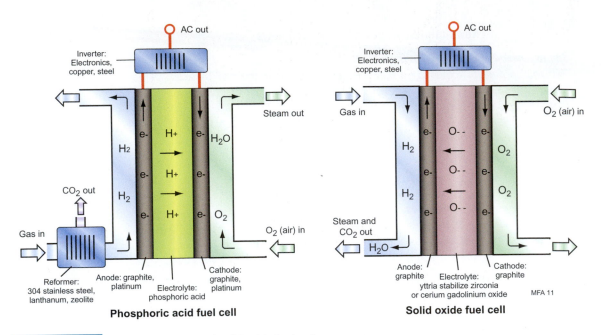

Phosphoric acid fuel cell

Solid oxide fuel cell

FIGURE 12.17 *Phosphoric acid and solid oxide fuel cells*

There are many types of fuel cells, which are generally classified by their electrolyte.

Phosphoric acid fuel cell (PAFC). The PAFC is the cheapest fuel cell and the one with the largest installed base (over 75 MW) and the longest useful life (10 years). It uses a liquid phosphoric acid electrolyte at relatively low temperatures of 150−200°C. Protons from the oxidation of hydrogen at the anode are transported through the phosphoric acid to the cathode where they react with oxygen from the air to form water, as in Figure 12.17. The reaction requires a platinum catalyst. The low temperatures limit efficiency to around 40%, which is not competitive with the most efficient gas power plants. Appendix 2 contains a bill of materials for a PAFC. Platinum is the only critical material that might constrain widespread deployment (Figure 12.18).

Solid oxide fuel cell (SOFC). The electrolyte of an SOFC, typically, is yttria-stabilized zirconia (YSZ). The high temperatures required for sufficient ionic mobility, typically 600−1,000°C, remove the need for expensive catalysts, like platinum, at the electrodes. Oxygen ions diffuse though the electrolyte, reacting with hydrogen at the anode to give water:

$$O_2 + 4e^- \rightarrow 2O^{2-} \qquad \text{(cathode reaction)} \qquad (12.5,b)$$

$$2O^{2-} + 2H_2 \rightarrow H_2O + 4e^- \qquad \text{(anode reaction)} \qquad (12.5,a)$$

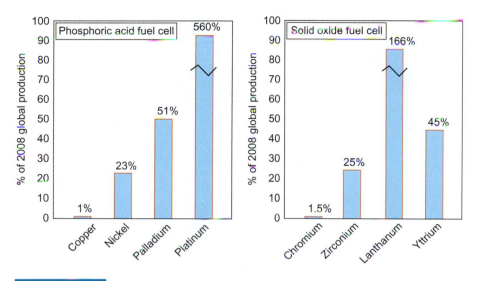

FIGURE 12.18   *Demand ratios for strategic materials used in fuel cells*

SOFC systems can run on natural gas, liquid propane gas (LPG), or biogas. The hydrocarbon fuel is steam reformed to hydrogen and carbon monoxide within the cell. The high temperatures give an efficiency of 50−60% but, because many components of a SOFC are ceramics, the start-up has to be slow to prevent thermal shock. Replacing the YSZ electrolyte by cerium gadolinium oxide reduces operating temperatures to 500−600°C, allowing the replacement of many structural ceramic components with stainless steel, improving thermal shock resistance and reducing start-up time.

The bill of materials for a YSZ electrolyte SOFC in Appendix 2 suggests that the only critical materials used in significant quantities are yttrium, zirconium, and lanthanum, a component of the oxide electrolyte.

12.7 Wind power

Wind has been used as a source of power for centuries. Today wind turbines are the fastest growing sector of the renewable power business driven, like solar PV power, by government subsidies (Figure 12.19). The problem with wind power, like that of most other renewable energy sources, is the low *power density*, that is, power per unit area. On land, it averages 2 W/m²; off-shore it is larger, about 3 W/m². The average land area per person in a country with a population density like that of the UK is about 3,500 m². That means that if the *entire country* were packed with the maximum possible number of wind turbines, it would generate just 7 kW per person (MacKay, 2009), approximately what we use today. Placing them off-shore helps solve the overcrowding problem, but maintenance costs are higher.

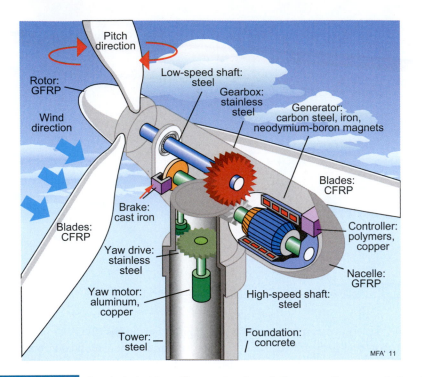

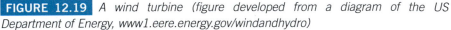

FIGURE 12.19 *A wind turbine (figure developed from a diagram of the US Department of Energy, www1.eere.energy.gov/windandhydro)*

Living on wind power alone

Example: The land area of the Netherlands (Holland) is 41,526 km.^{2} Its population is 16.5 million, and the average power consumption per capita there is 6.7 kW. Could Holland's power needs be met by land-based wind turbines operating at a (high) capacity factor of 0.3? Use mean values of the data in Table 12.2 to find out.

Answer: The land area occupied by wind turbines required to meet the Netherlands's needs is

$$A = \text{Population} \times \text{Power per person} \times \text{Area per unit power} / \text{Capacity factor}$$

From Table 12.2, the mean area intensity for land-based wind turbines is 275 m^2/kW$_{nom}$. The capacity factor is 0.21, giving

$$A = 1.01 \times 10^{11} \text{ m}^2 = 101,000 \text{ km}^2$$

This is 2.4 times the area of the Netherlands. There is no way land-based wind power alone can supply all the country's needs.

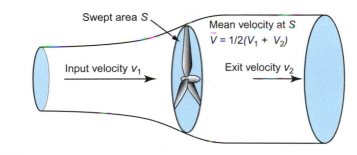

Control volume across a wind turbine

When wind comes into contact with the rotor of a wind turbine, some of its kinetic energy is imparted to the blades, driving their rotation. This rotation is transmitted through a gearbox to a generator, creating electric power. Wind speed v increases with height h above ground level

$$v(h) \approx v_{10}\left(\frac{h}{10}\right)^{0.14} \tag{12.6}$$

where v_{10} is the wind speed at a height $h = 10\,m$. We show in a moment that power depends on wind speed cubed, so increasing the height of a wind turbine by a factor of 2 increases power by 30%, as we show in a moment. Wind turbines have a cut-in and cut-out wind speed (typically 3 m/s^2 cut-in and 20 m/s^2 cut-out). They stop altogether when the wind speed is outside this range.

The maximum power generated by a turbine is limited to 58% of the kinetic energy of the wind stream passing through it (the Betz limit, symbol C_B). It is calculated by considering mass and energy conservation across the turbine, shown in Figure 12.20. Here the approaching wind velocity is v_1, the exit velocity is v_2, the swept area is S, and the mean velocity in the plane of the turbine is $\bar{v} = 1/2\,(v_1 + v_2)$. The energy flux per second (= power) of an uninterrupted air flow of velocity v_1 across an area S is

$$P_{in} = \frac{1}{2}\dot{m}\,v_1^2 = \frac{1}{2}\,(S\rho\,v_1)\,v_1^2 = \frac{1}{2}S\rho\,v_1^3 \tag{12.7}$$

where ρ is the density of air, assumed constant, and $\dot{m}$ is the mass per second passing through S.

Flow in an ideal (non-viscous) incompressible fluid is governed by Bernoulli's equation:

$$p + \frac{1}{2}\rho\,v^2 + \rho\,g\,h = \text{constant} \tag{12.8}$$

where p is the pressure, v the flow velocity, g the gravitational constant, and h the height. Thus the pressure difference Δp across the turbine caused by the drop in flow velocity from v_1 to v_2 at constant h is

$$\Delta p = \frac{1}{2}\rho\,(v_1^2 - v_2^2) \tag{12.9}$$

The air velocity in the plane of the turbine is $\bar{v}$ so the work is done by Δp per second, and thus the power P delivered to the turbine, is

$$P = \Delta p\, S\, \bar{v} = \frac{1}{4} S\,\rho\,(v_1 + v_2)\,(v_1^2 - v_2^2)$$

The power coefficient is found by dividing this by the kinetic energy per second of the uninterrupted air flow passing across an area S giving

$$\frac{P}{P_{in}} = \frac{1}{2}\left(1 + \frac{v_2}{v_1}\right)\left(1 - \left(\frac{v_2}{v_1}\right)^2\right)$$

Differentiating and equating to zero to find the maximum gives the value of the Betz limit $(P/P_{in})_{max} = C_B = 0.58$.

The power P is then

$$P = C_B\, P_{in} = \frac{1}{2}C_B\,\rho\,S\,v_1^3 \approx 0.3\,\rho\,S\,v_1^3 \tag{12.10}$$

Thus the peak (rated) power of a turbine varies as the swept area S times the cube of the incoming wind speed v_1.

The power rating of a wind turbine (e.g., 2 MW) is its peak power. The actual power output depends on its capacity factor and this depends on turbine design and location. Off-shore turbines have a capacity factor of about 35% but they are expensive to build and maintain. The average capacity factor for land-based turbines is closer to 21% but they are cheaper to build and easier to service.[8]

Wind energy is free, but harvesting it is not. There is an energy investment associated with the construction of a wind turbine. Figure 12.6 showed that the energy payback time is typically between 1 and 2 years. The problem with wind power is not energy payback time, however, but the small power output per unit. Even with a greatly over-optimistic capacity factor of 50%, about 1,000 2 MW wind turbines are needed to replace the power output of just one conventional coal-fueled

[8]The averaged capacity factor for all British wind turbines, land-based and off-shore, for the 12 months to April 2011 was a mere 21.7% because of unusually calm weather conditions (*The Times*, April 1, 2011).

power station, so the capital intensity and the intensity of land and materials, is high (Figures 12.4 and 12.5).

The blades are the most vulnerable parts of a wind turbine. The drive for ever longer blades (to increase S) places constraints on the choice of materials. The self-weight of the blade creates alternating bending loads at the blade root as the turbine rotates. Superimposed on these is an axial load caused by centrifugal force and a bending load caused by wind pressure. As we saw in Chapters 9 and 10, the most efficient configuration for a light, strong structure is a shell made from a material with the smallest value of the index ρ/σ_y, where σ_y is the yield strength and ρ is the density: laminated wood, GFRP, and, above all, CFRP, are the materials of choice for the blades of large wind turbines. The generator, too, is a critical component. Wind turbines typically rotate at 10–20 rpm. Some use induction generators (which do not use permanent magnets) that are efficient only at around 750 rpm, requiring a gearbox that needs regular maintenance. Permanent magnet generators operate efficiently at the low speeds at which the blades turn, allowing direct drive. Neodymium-iron-boron magnets (Nd-Fe-B) give the highest performance generators, with only the significantly more expensive samarium-cobalt (Sm-Co) coming close.

Appendix 2 has an approximate bill of materials for a wind turbine. The resource-demand ratios for critical materials, assuming CFRP blades and Nd-Fe-B magnets, appear in Figure 12.21. Both that for CFRP, the blade material, and for neodymium, a component of the permanent magnets of the generator, far exceed current production capability. It may be necessary to greatly increase production or find substitutes for both.

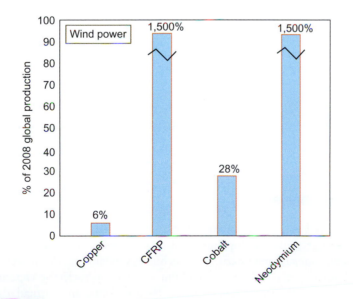

FIGURE 12.21 *Demand ratios for strategic materials used in wind power systems*

News-clips: The prizes and perils of wind power

It is time to believe in wind power.
Spanish energy group plans to make Britain a world leader. Britain is to become the industrial center for what is claimed will be the world's most advanced wind turbines after Gamesa committed itself to open three big installations in the UK.

The Times, February 8, 2011

Taller than the Gherkin (a London skyscraper), wider than the London Eye (a giant Ferris wheel): giant wind turbines face an epic challenge.
The Danish group Vestas yesterday unveiled plans to produce giant wind turbines with a generating capacity of 7 MW.

The Times, March 31, 2011

Lack of wind raises fears for future of green energy.
Britain's leading renewable energy company has reported a 20% fall in the amount of electricity produced by its wind turbines last year, which was unusually unwindy.

The Times, February 2, 2011

Wind farm "threatens heritage coast view."
An offshore wind farm with 240 turbines taller than the London Eye will be visible from the eastern part of the Jurassic Coast World Heritage Site. Eneco, the Dutch energy company, was granted development rights for a zone off the South Coast of Britain. "It's England's only geological World Heritage Site and it would spoil a view that is half as old as time itself," say protesters.

The Times, February 19, 2011

Exploiting wind power requires good meteorology and diplomacy as well as good engineering.

12.8 Hydropower

More energy is generated from hydropower (Figure 12.22) than from any other renewable energy source. It supplies most of the electricity for thirteen countries, among them, Norway and Brazil. The technology is simple, the power is always available (provided it rains), and dams built without hydroelectric capacity can be retrofitted with turbines and generators. Hydro plants are particularly flexible

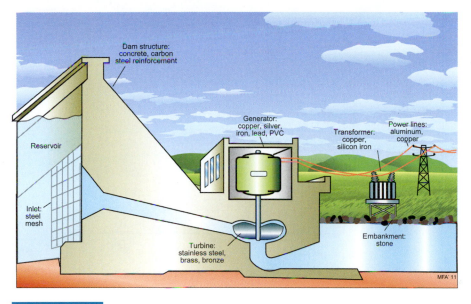

Dam structure:
concrete, carbon
steel reinforcement

Generator:
copper, silver,
iron, lead, PVC

Transformer:
copper,
silicon iron

Power lines:
aluminum,
copper

Reservoir

Inlet:
steel
mesh

Turbine:
stainless steel,
brass, bronze

Embankment:
stone

MFA' 11

FIGURE 12.22 *Schematic of a hydroelectric plant*

because they both store energy and generate power, either continuously or intermittently, to deal with spikes in demand.

The energy to fill reservoirs comes from the sun, but it is gravity that drives the water through the turbine. The flow of water through a water turbine, like that of air through a wind turbine, is governed by Bernoulli's equation. When the water flow is from an essentially stationary reservoir to an essentially stationary outflow pool, the pressure difference between the inlet and the outlet is $\Delta p = \rho g\,\Delta h$, where Δh is the difference in surface height between the reservoir and the outflow pool. The power captured by the turbine is then

$$P = \eta\, g\, \dot{Q}\,\Delta h \tag{12.11}$$

where $\dot{Q}$ is the volumetric flow rate in m^3/s and η is the efficiency of the system (over 90% for large hydro plants and falling to 50% for small ones). The capital cost of a hydro plant can be high, and its construction may damage natural and human habitat. But its lifetime is long, it requires little maintenance, it creates no emissions, and its fuel is free. Hydropower is a long-term investment.

The most material-intensive part of a hydro plant is the dam. Small dams can be made of earth. Larger ones are made of concrete, some requiring 5 metric tons of concrete for $1\ kW_{nom}$ of generating capacity. The demands on other materials are modest (Figure 12.23).

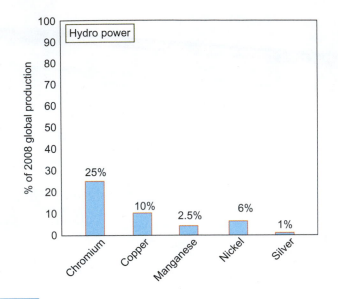

FIGURE 12.23 *Demand ratios for strategic materials used in hydropower systems*

News-clip: Power from glaciers

Plan to dam Arctic for hydro-power.
Water from the Arctic's melting glaciers could be harnessed to generate "green" electricity under a scheme proposed by one of the world's largest producers. Alcoa wants to build a network of industrial smelters around the Arctic powered by hydroelectric dams that capture meltwater. Greenland and Iceland are among its favored locations.

The Times, **November 20, 2010**

Aluminum consumption is expected to rise from 38 metric tons in 2009 to 74 metric tons in 2020. The demand from China alone is expected to reach 44 metric tons by that date. Aluminum has an exceptionally high embodied energy, all of it provided by electrical power. Alcoa's plan is to use hydro-power from melting glaciers to make aluminum.

12.9 Wave power

Energy can be captured from waves by placing something in their path—a fixed barrier with a turbine driven by the water whooshing in and out, for instance. Waves carry an energy per unit length rather than an energy per unit area, and it is large—as much as 40 kW per meter—as you will know if you've been hit by one. Capturing it, however, is not easy; it is unlikely that any wave machine would trap more than a quarter of this. Not many countries have long coastlines (some have

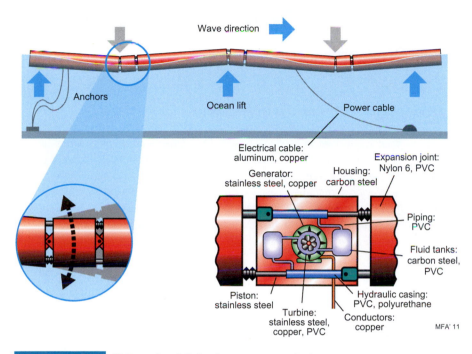

Wave direction

Anchors

Ocean lift

Power cable

Electrical cable:
aluminum, copper

Generator:
stainless steel, copper

Housing:
carbon steel

Expansion joint:
Nylon 6, PVC

Piping:
PVC

Fluid tanks:
carbon steel,
PVC

Piston:
stainless steel

Turbine:
stainless steel,
copper, PVC

Hydraulic casing:
PVC, polyurethane

Conductors:
copper

MFA' 11

FIGURE 12.24 *Schematic of Pelamis wave-power device*

none), but for those that do, wave power is an option. Once again the scale of the operation has to be vast to make a real contribution: to provide 1 kW per person to a country of 50 million inhabitants, the country would need something like 4,000 km of barrier. And any wave-driven device takes a battering, making maintenance a problem.

There are three schemes for harvesting wave power: *buoyancy devices, overtopping devices*, and *oscillating water-column devices*. Buoyancy devices use the motion of waves to pump a working fluid through a turbine. Overtopping devices uses a vertical axis turbine with a floating ramp; waves ride up the ramp and spill into the turbine below. Oscillating water-column devices trap water and air within the structure, and as the waves pass under it, the air is alternately compressed and allowed to expand through a turbine.

The Pelamis is an example of a buoyancy wave energy converter (Figure 12.24). It floats on the ocean surface with cables securing it to the seafloor. Links in the body allow it to flex with the motion of the waves, driving pistons that pump oil through a turbine inside the unit. To survive the harsh environment, the device requires a great deal of steel, making it material and energy intensive to construct.

The total world installed capacity of wave power is currently a mere 4 MW, but more devices are under development The demand for critical materials for harvesting wave power is relatively low with the exception of copper. The small power per unit and the long cables connecting them to shore means that building sufficient capacity to generate 2,000 GW over the next 10 years could consume up to 20% of current global copper production.

News-clip: The cruel sea

Power from the restless sea stirs the imagination.

For years, technological visionaries have painted a seductive vision of using ocean tides and waves to produce power. They foresee large installations off the coast and in tidal estuaries that could provide as much as 10 percent of the nation's electricity. But the technical difficulties of making such systems work are proving formidable. Last year, a $2.5 million wave-power machine sank off the Oregon coast. Blades have broken off experimental tidal turbines in New York's turbulent East River. Problems with offshore moorings have slowed the deployment of snakelike generating machines in the ocean off Portugal.

***The New York Times**, September 22, 2008*

The engineering challenges of harnessing wave and tidal power are considerable.

News-clip: Connecting wave power to shore

Wave Hub in uncertain waters.

After seven years Wave Hub was finally installed on the seabed off north Cornwall in September. Wave Hub provides the marine infrastructure to let wave energy projects hook up to the national grid. It is like a giant underwater socket that companies rent to avoid having to lay cables themselves.

***The Times**, November 2, 2010*

Distributed power systems like wave, wind, solar, and tide require an expensive infrastructure to get the power from the source to the consumer.

12.10 Tidal power

The moon orbits the earth and the earth orbits the sun. As they do so, gravitational and centrifugal forces act on seawater, pulling it into tidal bulges. As the earth rotates on its own axis, these bulges sweep across the planet's surface creating two tides each day. When sun, earth, and moon are aligned the forces pull together, creating high and low "spring" tides. When the moon is at right angles to the sun with respect to the earth, the gravitational fields work against each other, giving smaller "neap" tides.

The highest tidal ranges in the world are in the Bay of Fundy in Nova Scotia and in the Severn Estuary in Britain where tidal water funnels into a narrowing channel. Tidal power where tides are high can deliver 3 W/m² by making both the incoming and the outgoing tidal flow drive turbines. This is about the same as wind power, but few countries have the coastline to capture much of it. To those with tidal estuaries or those that lie at the mouths of land-locked seas (like the Mediterranean) harnessing tidal power is an option. There are two schemes for harvesting it: *tidal-stream* and *tidal-barrage* systems.

The Seagen tidal-stream power generator is an underwater wind turbine driven by the flow of the incoming and receding tides (Figure 12.25). One is in service. It has a power of 1.2 MW, a claimed capacity factor of 48%, and a design life of 20 years. A tidal barrage is a hydroelectric plant driven by a reservoir filled by tides rather than by rain. The largest tidal barrage is the 240 MW unit on the river La Rance in France, where the tidal range is 8 m. Tidal barrages have longer lifetimes than tidal-stream generators because the machinery is simpler. The attraction of tidal power is that it is completely predictable. The drawback, as Figures 12.4, 12.5 and 12.6 show, is that the systems are material, energy, and capital intensive.

Tidal power puts little pressure on critical materials. Using the same demand scenario as before, the consumption of copper might reach 5% of current production.

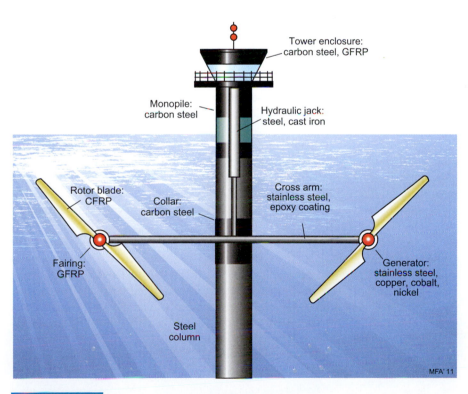

FIGURE 12.25 *Diagram of a Seagen tidal power generator*

12.11 Geothermal power

The core of the earth remains at a temperature above 5,000°C, heated by the
slow radioactive decay of elements at its center. Heat is carried to the earth's surface by conduction and by convection of the magma in the mantle, the layer
between the core and the crust. This heat leaks out at the surface, but little of it
is in a useful form. On average the near-surface temperature gradient is 20°C/km,
delivering 50 mW/m^2 at the surface. To generate electricity, it is necessary to heat
water to at least 200°C, and for most of the earth's crust that means drilling
down to about 10 km, making it expensive to harvest. Where magma wells up
close to the surface the gradient rises above 40°C/km. In such places (Iceland,
parts of the United States, New Zealand, and Italy) extracting geothermal heat is
a practical proposition.

To do so, water is injected into the hot rock from which it is recovered as hot
water or steam (Figure 12.26). High-temperature plants feed steam directly to a turbine and then pump the condensate back into the rock. Low-temperature plants
pass the hot water to a heat exchanger where it vaporizes a low-boiling-point working fluid—isopentane or isobutene—that drives the turbine. Typical power outputs
are 0.1 to 2 GW.

Geothermal power may have the largest potential of all renewable energy
sources; a USGS study in 2008 estimated that the US electrical power potential
from geothermal heat exceeded 500 GW. It is at present just potential, requiring
advances in deep drilling to make it a reality. Most of the cost of a geothermal
power plant is that of drilling, but aside from small maintenance costs, the electricity it generates is free. As with wave and tidal power, the only critical material used
in large quantities is copper. Using the same demand structure as before, the
deployment of 200 GW per year of geothermal power (could it be found) would
require about 2% of current copper production.

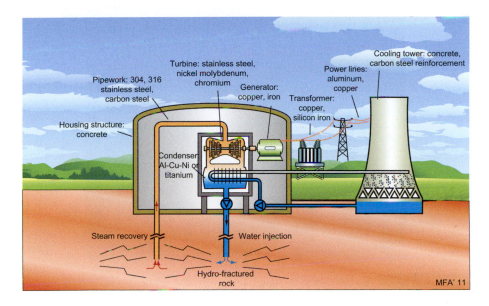

FIGURE 12.26 *Geothermal power plant*

News-clip: Sharing geothermal power

Iceland plans volcano power cable to Britain.
Landsvirkjun (Iceland's large state-owned electricity producer) is examining the feasibility of laying a 1,170 km cable between Iceland and Scotland.

The Sunday Times, March 6, 2011

The planned cable will transmit 0.68 GW of high-voltage electrical power from geothermal sources. The conductor of such a cable has a cross-section of about 2,000 mm^2, meaning that 1 kilometer of cable contains 2 m^3, or roughly 18 metric tons of copper. The entire cable will need 21,000 metric tons of it. That is a lot of copper—will sourcing it be a problem? Probably not. The world production of copper is 16 million metric tons per year. The cable will require 0.13% of annual production.

12.12 Biomass

Green plants capture the sun's energy and use it to photosynthesize carbohydrates, oils, and proteins from atmospheric CO_2 (Figure 12.27). The carbohydrates can be dried and burned to release the energy, or they can be fermented to give olefins (methane, ethane) and alcohols (methanol, ethanol) that can be used as fuels. Seed

FIGURE 12.27 *Soybean crop in Ohio, USA*

oil (soybean, sunflower, palm oil) can be burned or processed into biodiesel. Plant growth requires little fossil fuel energy, is relatively clean, and—until it is burned—sequesters carbon. But the efficiency of energy-capture by plants is low, typically 0.5%. The average annual flux of solar radiation in a temperate climate is about 100 W/m^2, so the area-intensity of biomass, before it has been converted to a useful fuel, is about 2000 m^2/kW, already greater than that of any other source of power. To make it useful it must be dried for combustion or fermented to make biodiesel, both with imperfect efficiency, driving the area intensity up to 5,000—10,000 m^2/kW$_{nom}$. Biomass is said to be a carbon-neutral fuel because the carbon dioxide it emits when burned was drawn from the atmosphere during its growth, but this is not quite true because farming and transportation generate some CO_2 that cannot be credited to biomass.

The use of biomass for liquid fuels is already generating competition for land to grow fuel, food, and materials. By 2007, almost a quarter of US coarse grain production and one half of EU vegetable oil was used for biofuels, yet together they provided only 0.36% of the global energy supply. According to Pimm (2001), the global production of natural and cultivated biomass is about 140 gigaton per year (Gt/year) and the global consumption of biomass as food, fodder for livestock, conversion of forest to pasture, firewood, construction, and fiber is currently 58 Gt/year, about 40% of the total. If we already use this fraction of biomass production, further consumption cannot grow much, even allowing for increased yields. It appears, then, that there is a limit to the supply of biomass. Only about 2% of the total, dominated by wood and natural fibers, is at present used to make engineering structures, so there is some scope for increasing this, but it will ultimately be at the expense of other uses of biomass.

12.13 Summary and conclusions

If the world is to have the electrical power that extrapolation suggests it will need by 2050, it will mean building an additional generating capacity of up to 6,000 GW. The global fluxes of solar radiation and of energy in the form of wind, wave, and tides all comfortably exceed this value, and the accessible coal and nuclear fuel reserves, too, could provide it. But each has its difficulties, and even a balanced combination of all of them presents problems.

Renewable power systems carry a carbon footprint that is 10 to 30 times less than that of the gas and oil-based systems on which we now largely depend. But almost all require greater surface area, more capital, more energy, and more materials to build them than to construct fossil fuel power stations with the same nominal generating capacity. Worse, most renewable power systems have lower capacity factors than fossil fuel plants, further increasing the differences.

A high material intensity is manageable if the materials in question are those with large reserves, low embodied energy, and low carbon footprint. Some materials meet this ideal—iron and carbon steel, concrete, wood, and commodity polymers are examples. But the sheer scale of construction, if just one third of the anticipated demand in 2050 is to be met over a 10-year time span, puts pressure on the supply of some critical materials and greatly exceeds the current production capacity of others.

The conclusions: no single renewable source can begin to supply energy on the scale we now use it. A combination of all of them might. But think of the difficulties. There is the low power density, meaning that a large fraction of the area of the country must be dedicated to capturing it. If you cover half the country with solar cells, you cannot also plant crops for biofuel on it, nor can you use it as we now do for agriculture and livestock for food. There is the cost of establishing the network needed to connect up such a dispersed system and—in the case of off-shore wave and wind farms—there is the cost of maintenance (even on land some 2% of wind turbines are disabled each year by lightning). And there is the opposition, much of it from environmentalists, to paving the country and framing the coast with machinery. MacKay's (2009) book examines all this in greater depth. For now we must accept that the dream of copious cheap, pollution-free and renewable energy from sun, wind, and wave is not realistic.

12.14 Further reading

The starting point—sources that help with the big picture

Andrews, J. and Jelley, N. (2007), *Energy science: principles, technologies and impacts*, Oxford University Press, Oxford, UK. ISBN 978-0-19-928112-1. *(An introduction to the science behind energy sources and energy storage systems)*

Boyle G. (editor) (2004), "Renewable energy," DE.158.

British Geological Survey (2011), "Risk list 2011," www.bgs.ac.uk/mineralsuk/statistics/riskList.html. Accessed December 2011. *(A supply risk index for critical elements or element groups that are of economic value)*

Cullen, J.M. (2010), "Engineering fundamentals of energy efficiency," PhD thesis, Engineering Department, University of Cambridge, Cambridge, UK. *(A revealing analysis of energy use and efficiency of energy conversion in modern society)*

Fay, J.A. and Golomb, D.S. (2002), *Energy and the environment*, Oxford University Press, Oxford, UK. ISBN 0-19-515092-9. *(The environmental background to energy production, with an exploration of the potential for replacing fossil fuels by lower carbon alternatives)*

Harvey, L.D.D. (2010), *Energy and the new reality 1: energy efficiency and the demand for energy services*, Earthscan Publishing, London, UK. ISBN 978-1-84971-072-5. *(An analysis of energy use in buildings, transportation, industry, agriculture, and services, backed up by comprehensive data)*

Harvey, L.D.D. (2010), *Energy and the new reality 2: carbon-free energy supply*, Earthscan Publishing, London, UK. ISBN 978-1-84971-073-2. *(A comprehensive analysis of low-carbon power generation systems)*

Hunt, W.H. (2010), "Linking transformational materials and processing for an energy efficient and low-carbon economy," The Minerals, Metals and Materials Society (TMS). www.energy.tms.org. *(A pair of reports assessing the materials challenges raised by a low-carbon economy)*

IAEA (1997), "Sustainable development and nuclear power," International Atomic Energy Agency, Vienna, Austria.

IEA (2008), *International Energy Agency Electricity information 2008*, OECD / IEA, Paris, France. ISBN 978-92-64-04252-0. *(An extraordinarily detailed, annual, compilation of historical and current statistics on electricity generation and use)*

Lund, H. (2010), *Renewable energy systems—the choice and modelling of 100% renewable solutions*, Elsevier, Amsterdam, the Netherlands. ISBN 978-0-12-375028-0. *(An analysis of the social and political challenges of deploying renewable energy systems on a large scale)*

MacKay, D.J.C. (2009), *Sustainable energy—without the hot air*, UIT Publishers, Cambridge, UK. SBN 978-0-9544529-3-3 and www.withouthotair.com. *(MacKay takes a critical look at the potential for replacing fossil-fuel-based energy by alternatives. A book noteworthy for clarity of argument and style.)*

McFarland, E.L., Hunt, J.L., Campbell, J.L.E. (2007), *Energy, physics and the environment*, Thompson Publishers, UK. ISBN 1-426-82433-6. *(The underpinning physics for alternative power systems and more)*

Meyer, P.J. (2002), "Life cycle assessment of electricity generation systems and applications for climate change policy analysis," PhD Thesis, Fusion Technology Institute, University of Wisconsin Madison, Wisconsin. (Report number UWFDM-1181). *(Case studies of alternative power generating systems, making use of triangle maps like that of Figure 12.2)*

Quaschning, V. (2010), *Renewable energy and climate change*, John Wiley, London, UK. ISBN 978-0 470-74707-0. *(A readable, well-illustrated introduction to renewable energy systems, with examples of deployment)*

San Martin, R.L. (1989), "Environmental emissions from energy technology systems: the total fuel cycle," US Department of Energy, Washington, DC, USA.

Sorensen, B. (2004), *Renewable energy—its physics, engineering, environmental impact, economics and planning*, 3rd edition, Elsevier, Amsterdam, the Netherlands. ISBN 978-0-12-656135-1. *(A densely written tome, but with much useful data)*

Tester, J.W., Drake, E.M., Driscole, M.J., Golay, M.W., and Peters, W.A. (2005), *Sustainable energy—choosing among the options*, MIT Press, Cambridge, MA, USA. ISBN 978-0-262-20153-7. *(A comprehensive [and very long] text exploring the economic and environmental issues raised by alternative sources of sustainable energy)*

Coal- and gas-fired power

CCSA (2010), "About CCS, carbon capture and storage association," www.ccsassociation.org.uk (Accessed August 2010).

EPRI (2009), "Program on technology innovation: integrated generation technology options," Energy Technology Assessment Centre, Palo Alto, CA. http://my.epri.com/. Accessed December 2011.

Hondo, H. (2005), "Life cycle GHG emission analysis of power generation systems: Japanese case," *Energy* 30, pp 2042−2056.

IEA (2002), "Environmental and health impacts of electricity generation," www .ieahydro.org/reports/ST3-020613b.pdf (Accessed December 2011).

Kaplan, S. (2008), "Power plants: characteristics and costs," CRS Report for Congress, www.fas.org/sgp/crs/misc/RL34746.pdf (Accessed December 2011).

Mayer-Spohn, O., (2009), "Parametrised life cycle assessment of electricity generation in hard-coal-fuelled power plants with carbon capture and storage," Universitat Stuttgart, Stuttgart, Germany. http://elib.uni-stuttgart.de/opus/volltexte/2010/5031/pdf/100114_Dissertation_Mayer_Spohn_FB105.pdf (Accessed December 2011).

Meier P.J. (2002), "Life cycle assessment of electricity generation systems and applications for climate change policy analysis," University of Wisconsin, Madison, Wisconsin. http://fti.neep.wisc.edu/pdf/fdm1181.pdf (Accessed August 2010).

National Grid (2010), "Calorific value description," www.nationalgrid.com/uk/Gas/Data/help/opdata (Accessed August 2010).

NaturalGas.org (2010), "Overview of natural gas," http://naturalgas.org/overview/overview.asp (Accessed December 2011).

White, S.W. and Kulcinski, G.L. (2000), "Birth to death analysis of the energy payback ratio and CO_2 gas emission rates from coal, fission, wind and DT-fusion electrical power plants," *Fusion Engineering and Design*, 48, pp 473−481.

Nuclear power

Andrews, J. and Jelley N. (2007), *Energy science*, Oxford University Press, Oxford, UK. ISBN 978-0-19-928112-1.

British Energy (2005), "Environmental Product Declaration of Electricity from Torness Nuclear Power Station," www.british-energy.com (Accessed August 2010).

Environmental Science and Technology (2008) Vol. 42, pp 2624–2630, http://ec.europa.eu/environment/integration/research/newsalert/pdf/109na4.pdf (Accessed August 2010).

Glasstone, S. and Sesonske, A. (1994), *Nuclear reactor engineering*, 4th edition, Chapman and Hall, New York, NY, USA. ISBN 0-412-98521-7.

Kaplan, S. (2008), "Power plants: characteristics and costs," CRS Report for Congress, www.fas.org (Accessed September 2010).

Rashad, S.M. and Hammad, F.H. (2000), "Nuclear power and the environment: comparative assessment of the environmental and health impacts of electricity-generating systems," *Applied Energy*, 65, pp 211–229.

Roberts, J.T.A. (1981), *Structural materials in nuclear power systems*, Plenum Press, New York, NY, USA. ISBN 0-306-40669-1.

Storm van Leeuwen, J.W. (2007), "Nuclear power—the energy balance," Ceedata Consultancy, www.stormsmith.nl/report20071013/partF.pdf (Accessed December 2011).

Voorspools, K.R., Brouwers, E.A., and D'Haeseleer, W.D. (2000), "Energy content and indirect greenhouse gas emissions embedded in 'emission-free' power plants: results for the Low Countries," *Applied Energy*, 67 pp 307–330.

White, S.W. and Kulcinski, G.L. (1998a), "Birth to death analysis of the energy payback ratio and CO_2 gas emission rates from coal, fission, wind and DT-fusion electrical power plants," *Fusion Engineering and Design*, 48 (248) pp 473–481.

White, S.W. and Kulcinski, G.L. (1998b), "Energy Payback Ratios and CO_2 Emissions Associated with the UWMAK-I and ARIES-RS DT-Fusion Power Plants, Fusion Technology Institute, Madison, Wisconsin, http://fti.neep.wisc.edu/pdf/fdm1085.pdf (Accessed August 2010).

WISE Uranium project (2009), "Nuclear fuel energy balance calculator," www.wise-uranium.org/nfce (Accessed August 2010).

World Nuclear Association (2009), "Nuclear power reactors," www.world-nuclear.org/info/inf32 (Accessed August 2010).

Solar power

AMP Blogs network (2009), "New solar panel materials studied," www.aboutmyplanet.com/alternative-energy/new-solar-panel-materials-studied/ (Accessed August 2010).

Ardente, F., Beccali, G., Cellura, M., and Lo Brano, V. (2005), "Life cycle assessment of a solar thermal collector," *Renewable Energy* 30 pp 1031–1054.

Bankier, C. and Gale, S. (2006), "Energy payback of roof mounted photovoltaic cells," *The environmental engineer*, www.rpc.com.au/pdf/ Environmental_Engineer_Summer_ 06_paper_2.pdf (Accessed August 2010).

Blakers, A. and Weber, K. (2000), "The energy intensity of photovoltaic systems," Centre for Sustainable Energy Systems, Australian National University, www.ecotopia.com (Accessed August 2010).

Carbon Free Energy Solutions (2010), "Philadelphia solar modules," www.carbonfreeenergy.co.uk (Accessed August 2010).

Denholm, P., Margolis, R.M., Ong, S., and Roberts, B. (2010), "Sun gets even," National Renewable Energy Laboratory, http://newenergynews.blogspot.com/ 2010/02/sun-gets-even.html (Accessed August 2010).

Energy Development Co-Operative Limited (2010), "Solar panels—solar PV modules," www.solar-wind.co.uk/solar_panels (Accessed August 2010).

Genersys Plc. (2007), "1000-10 Solar Panel Technical Datasheet," www.genersys-solar.com (Accessed August 2010).

Ginley, D., Green, M.A., and Collins, R. (2008), "Solar energy conversion towards 1 terawatt," *MRS Bulletin*, 33, 4.

Intelligent Energy Solutions (2010), "Solar panel cost," www .intelligentenergysolutions.com (Accessed August 2010).

Kannan, R., Leong, K.C., Osman, R., Ho, H.K., and Tso, C.P. (2005), *Life cycle assessment study of solar PV systems: An example of a 2.7 kWp distributed solar PV system in Singapore*, Elsevier, Amsterdam, the Netherlands.

Keoleian, G.A. and Lewis, G.McD. (1997), "Application of life cycle energy analysis to Photovoltaic Module Design," *Progress in Photovoltaics: Research and Applications*, 5. http://deepblue.lib.umich.edu (Accessed August 2010).

Knapp, K.E. and Jester, T.L. (2000), "An empirical perspective on the energy payback time for photovoltaic modules," Solar 2000 Conference, Madison, Wisconsin, USA.

Koroneos, C., Stylos, M., and Moussiopoulos N. (2005), "LCA of Multicrystalline silicon photovoltaic systems," Aristotle University of Thessaloniki, Thessalonika, Greece.

Lewis, G.M. and Keoleian, G.A. (1997), "Life cycle design of amorphous silicon photovoltaic modules," EPA. www.umich.edu/~nppcpub/research/pv.pdf (Accessed August 2010).

Meier, P.J. (2002), "Life cycle assessment of electricity generation systems and applications for climate change policy analysis," University of Wisconsin, Madison, Wisconsin, http://fti.neep.wisc.edu/pdf/fdm1181.pdf (Accessed August 2010).

Sanyo (2009), "HIT photovoltaic module," www.shop.solar-wind.co.uk (Accessed August 2010).

Sharp (2008), "*ND Series, 210W/200W,*" Photovoltaic solar panels, www.shop.solar-wind.co.uk/acatalog/Sharp_Solar_Panel_ND_210E1F_Brochure.pdf (Accessed August 2010).

Solar Systems (2010), "Projects," Mildura solar farm, www.solarsystems.com.au/ projects (Accessed August 2010).

Tritt, T.M., Bottner, H., and Chen, L. (2008), "Thermoelectrics: direct solar thermal energy conversion," *MRS Bulletin*, 33, 4.

US Department of Energy (2011), "Concentrating solar power," www1.eere.energy.gov/solar/csp_program.html. Accessed December 2011.

US Department of Energy (2011), "Dish/engine systems for concentrating solar power," www.eere.energy.gov/basics/renewable_energy/dish_engine.html. Accessed December 2011.

US Department of Energy (2011), "Power tower systems for concentrating solar power," www.eere.energy.gov/basics/renewable_energy/power_tower.html. Accessed December 2011.

Vidal, J. (2008), "World's biggest solar farm at centre of Portugal's ambitious energy plan," *The Guardian: Environment, Renewable Energy*, www.guardian.co.uk/environment/2008/jun/06/renewableenergy.alternativeenergy (Accessed August 2010).

Viebahn, P., Kronshage, S., Trieb, F., and Lechon Y. (2004), "Final report on technical data, costs, and life cycle inventories of solar thermal power plants," New Energy Externalities Developments for Sustainability (NEEDS) project, European Commission Sixth Framework Programme, www.needs-project.org/docs/results/RS1a/RS1a%20D12.2%20Final%20report%20concentrating%20solar%20thermal%20power%20plants.pdf (Accessed August 2010).

Woodall, B. (2006), "World's biggest solar farm planned for New Mexico," Planet Ark, www.planetark.com/dailynewsstory.cfm/newsid/36162/story.htm (Accessed August 2010).

Fuel cells

Karakoussis, V., Brandon, N.P., Leach, M., and van der Vorst, R. (2001), "The environmental impact of manufacturing planar and tubular solid oxide fuel cells," *J. Power Sources*, 101, pp 10–26.

US Department of Energy (2011), "Fuel cells," www1.eere.energy.gov/hydrogenandfuelcells/fuelcells/fc_types.html. Accessed December 2011.

Van Rooijen, J. (2006), "A life-cycle assessment of the PureCell stationary fuel cell system: providing a guide for environmental improvement," Report No. CSS06-09, Center for Sustainable Systems, University of Michigan, Ann Arbor, MI.

Wind power

Ardente, F., Beccali, M., Cellura, M., and Lo Brano, V. (2008), "Energy performances and life cycle assessment of an Italian wind farm." *Renewable and Sustainable Energy Reviews*, Vol. *12*(1), pp 200–217.

Hood, C.F. (2009), "The history of carbon fibre," www.carbon-fiber-hood.net/cf-history (Accessed August 2010).

Crawford, R.H. (2009), "Life cycle energy and greenhouse emissions analysis of wind turbines and the effect of size on energy yield," *Renewable and Sustainable Energy Reviews* 13: 9, pp 2653–2660.

Danish Wind Industry Association (2003), "What does a wind turbine cost?," http://guidedtour.windpower.org (Accessed August 2010).

EWEA (2003), "Costs and prices," *Wind Energy—The Facts*, Volume 2, www.ewea.org (Accessed August 2010).

Hau, E. (2006), *Wind turbines: fundamentals, technologies, application, economics*, 2nd Edition, Springer, New York, NY. ISBN 3-540-24240-6.

Hayman, B., Wedel-Heinen, J., and Brondsted, P. (2008), "Materials challenges in present and future wind energy," *MRS Bulletin*, 33, 4.

Martinez, E., Sanz, F., Pellegrini, S., Jimenez, E., and Blanco, J. (2007), "Life cycle assessment of a multi-megawatt wind turbine," University of La Rioja, Spain. www.assemblywales.org (Accessed August 2010).

McCulloch, M., Raynolds, M., Laurie, M. (2000), "Life cycle value assessment of a wind turbine," Pembina Institute, http://pubs.pembina.org/reports/windlcva.pdf (Accessed August 2010).

Musgrove, P. (2010), *Wind power*, Cambridge University Press, ISBN: 978-0-521-74763-9.

Nalukowe, B.B., Liu, J., Damien, W., and Lukawski, T. (2006), "Life cycle assessment of a wind turbine," www.infra.kth.se/fms/utbildning/lca/projects%202006/Group%2007%20(Wind%20turbine).pdf (Accessed August 2010).

Vestas (2008), "V82-1,65 MW," Wind turbine brochure, Vestas Wind Systems A/S, Aarhus, Denmark. www.Vestas.com. Accessed December 2011.

Vindmølleindustrien (1997), "The energy balance of modern wind turbines," *Wind Power Note*, 16, July. www.apere.org/manager/docnum/doc/doc1249_971216_wind.fiche37.pdf (Accessed August 2011).

Weinzettel, J., Reenaas, M., Solli, C., and Hertwich, E.G. (2009), "Life cycle assessment of a floating offshore wind turbine," *Elsevier Renewable Energy*, 34, pp 742–747.

Windustry (2010), "How much do wind turbines cost?" Windustry and Great Plains Windustry Project, Minneapolis, MN, USA. www.windustry.org/how-much-do-wind-turbines-cost (Accessed August 2010).

Hydropower

Davies, J., Wright, M., Monetti, K., and Van Den Berg, D. (2007), "Hydroelectric Section of the Energy Task Force," www.investfairbanks.com/documents/Hydronarrative12-07.pdf (Accessed September 2010).

Energy Saving Trust (2003), "Small scale hydroelectricity," www.savenergy.org (Accessed September 2010).

IEA (2000), "Hydropower and the world's energy future," www.ieahydro.org/reports/Hydrofut.pdf (Accessed September 2010).

Ribeiro, F., de M., and da Silva, G.A. (2010), "Life cycle inventory for hydroelectric generation: a Brazilian case study," *Journal of Cleaner Production*, 18 (2010) 44–54.

Sims, R.E.H. (2008), "Hydropower, geothermal, and ocean energy," *MRS Bulletin*, 33: 4, April 2008.

US Department of the Interior, Bureau of Reclamation (2008), "Hoover Dam, frequently asked questions and answers, The Dam," www.usbr.gov/lc/hooverdam/faqs/damfaqs.html (Accessed August 2010).

Vattenfall, A.B. (2008), "Certified environmental product declaration EPD of electricity from Vattenfall's Nordic hydropower," www.vattenfall.com (Accessed August 2010).

Wave power

Anderson, C. (2003), "Pelamis WEC—Main body structural design and materials selection," DTI, http://webarchive.nationalarchives.gov.uk/tna/ + /http:/www.dti .gov.uk/renewables/publications/pdfs/v0600197.pdf/ (Accessed September 2010).

Boronowski, S., Wild, P., Rowe, R., and van Kooten, G.C. (2010), "Integration of wave power in Haida Gwaii," *Renewable Energy*, 35 pp 2415–2421.

Carcas, M., "The Pelamis Wave Energy Converter—A phased array of heave + surge point absorbers," Ocean Power Delivery Ltd., http://hydropower.inel.gov/ hydrokinetic_wave/pdfs/day1/09_heavesurge_wave_devices.pdf (Accessed December 2010).

Energy Action Devon (2010), "Wave Power," Devon Association for Renewable Energy, www.devondare.org/wavepower (Accessed September 2010).

Henry, A., Doherty, K., Cameron, L., Whittaker, T., and Doherty, R. (2010), "Advances in the design of the Oyster wave energy converter," www .aquamarinepower.com/pub (Accessed September 2010).

Parker, R.P.M., Harrison, G.P., and Chick, J.P. (2007), "Energy and carbon audit of an offshore wave energy converter," School of Engineering and Electronics, University of Edinburgh, Edinburgh, UK.

Peak Energy (2010), "Wave power potential in Australia," http://peakenergy .blogspot.com/2010/09/wave-power-potential-in-australia.html (Accessed September 2010).

Pelamis Wave Power (2010), "Environmental characteristics," www.pelamiswave .com (Accessed September 2010).

Wave Dragon (2005), "Wave Dragon," www.wavedragon.net (Accessed September 2010).

Tidal power

Douglas, C.A., Harrison, G.P., and Chick, J.P. (2008), "Life cycle assessment of the Seagen marine current turbine," *Proc. IMechE* Part M: J. Engineering for the Maritime Environment 222, pp 1–12.

Harvey, L.D.D. (2010), *Energy and the New Reality 2, Carbon Free Energy Supply*, Earthscan Ltd., London, UK.

Miller, V.B., Landis, A.F., and Schaefer, L.A. (2010), "A benchmark for life cycle air emissions and life cycle impact assessment of hydrokinetic energy extraction using life cycle assessment," *Renewable Energy*, 35, pp 1–7.

Roberts, F. (1982), "Energy accounting of river severn tidal power schemes," *Applied Energy*, 11 pp 197–213.

Geothermal power

Frick, S., Kaltschmitt, M., and Schroder, G. (2010), "Life cycle assessment of geothermal binary power plants using enhanced low-temperature reservoirs," *Energy*, 35 pp 2281–2294.

Geothermal Energy Association (2010), "Geothermal basics—power plant costs," www.geo-energy.org/geo_basics_plant_cost.aspx (Accessed September 2010).

Huttrer, G.W. (2001), "The status of world geothermal power generation 1995–2000," *Geothermics*, 30 pp. 1–27.

Kaplan, S. (2008), "Power plants: characteristics and costs," CRS Report for Congress, www.fas.org/sgp/crs/misc/RL34746.pdf (Accessed September 2010).

Lawrence, S. (2008), "Geothermal energy," Leeds School of Business, University of Colorado, www.scribd.com/doc/6565045/Geothermal-Energy (Accessed September 2010).

MIT (2006), "The future of geothermal energy," Geothermal Program, Idaho National Laboratory, http://geothermal.inel.gov/publications/future_of_geothermal_energy.pdf (Accessed September 2010).

Saner, D., Juraske, R., Kubert, M., Blum, P., Hellweg, S., and Bayer, P. (2010), "Is it only CO_2 that matters? A life cycle perspective on shallow geothermal systems," *Renewable and Sustainable Energy Reviews*, 14, pp 1798–1815.

US Department of Energy (2005), "Buried treasure: the environmental, economic and employment benefits of geothermal energy," NREL, Energy Efficiency and Renewable Energy, www.nrel.gov/docs/fy05osti/35939.pdf (Accessed August 2010).

Usui, C. and Aikawa, K. (1970), "Engineering and design features of the otake geothermal power plant," Geothermics, O. N. Symposium on the Development and Utilization of Geothermal Resources, Pisa, Italy), Vol. 2, Part 2.

Biomass power

Asgard Biomass (2009), "Biomass fuels," www.asgard-biomass.co.uk/biomass_boiler_fuels.php (Accessed September 2010).

Bauer, C. (2008), "Life cycle assessment of fossil and biomass power generation chains," Paul Scherrer Institut, http://gabe.web.psi.ch/pdfs/PSI_Report/PSI-Bericht%2008-05.pdf (Accessed September 2010).

Ellington, R.T., Meo, M., and El-Sayed, D.A. (1993), "The net greenhouse warming forcing of methanol produced from biomass," *Biomass and Bioenergy*, Vol. 4: 6, pp 40–48.

HM Government (2010), "2050 pathways analysis," www.decc.gov.uk/assets/decc/What%20we%20do/A%20low%20carbon%20UK/2050/216-2050-pathways-analysis-report.pdf (Accessed September 2010).

Kim, S. and Dale, B.E. (2008), "Cumulative energy and global warming impact from the production of biomass for biobased products," Research and Analysis http://arrahman29.files.wordpress.com/2008/02/lca-use.pdf (Accessed September 2010).

Mann, M.K. and Spath, P.L. (1997), "Life cycle assessment of a biomass gasification combined-cycle power system," National Renewable Energy Laboratory, www.nrel.gov/docs/legosti/fy98/23076.pdf (Accessed September 2010).

Pollack, A. (2010), "His corporate strategy: the scientific method," *The New York Times*, www.nytimes.com/2010/09/05/business/05venter.html?_r = 1&src = me&ref = general (Accessed September 2010).

World Bank (2010), "Commodity prices," http://siteresources.worldbank.org/INTDAILYPROSPECTS/Resources/Pnk_0910.pdf (Accessed September 2010).

12.15 Appendix 1: Definitions of properties

Keeping track of units and the meanings of the various resource intensities can be a bit confusing. The definitions on this page will help keep them clear.

Energy and power

Energy   MJ or kW·hr (1 kW·hr = 3.6 MJ)

Energy appears in more than one form in this White Paper. It is important to choose one of these as the basis for comparison of conventional and renewable systems. By convention, the basic unit is MJ_{oe}, meaning megajoules, oil equivalent. Oil (like coal and gas) has a calorific value or heat content, 38 MJ_{oe}/liter or 44 MJ_{oe}/kg. The conversion efficiency when oil is used to generate electricity is about 36% (it depends on the age and type of generator), so one MJ of electrical energy (MJ_{elec}) requires the consumption of $1/0.36 = 2.8$ MJ_{oe}. Thus 1 kWh of electrical energy is equivalent to $3.6 \times 2.8 = 10$ MJ_{oe}.

Power   kW, MW, or GW

Power is energy per unit time, meaning J/sec (= Watt, W) or, in the context of power systems, kW, MW, GW, or even TW (terawatt = 10^{12} J/sec).

Rated or nominal power output kW_{nom}, etc.

The rated power output of a power system is the power it delivers under optimal conditions. Oil- and coal-fired power stations operate optimally for much of the time, but renewable power systems do not because they depend on a certain minimum level of solar radiation, wind velocity, wave height, or tidal flow, and for much of the time this minimum is not met.

Actual or real average power output kW_{actual}, etc.

The optimal conditions required for a power system to provide its rated power occur for only a fraction of the time, so the actual power output of the system, averaged over (say) a year, is much less than the rated value. The ratio of the real, averaged power output to the rated power output is called the "capacity factor," expressed as a %. Some systems have capacity factors as low as 10%, meaning that to generate an average of 1 kW of actual power it is necessary to install a system with a nominal capacity of 10 kW.

Resource intensities. Construction and commissioning of a power system require resources: capital, materials, energy, and space, meaning land or sea area. A resource intensity is the amount of the resource required to create one unit of power-generating capacity. Power systems have a rated power (kW_{nom}, see above), but none operates at full capacity all the time so it is convenient to define also the average delivered power (kW_{actual}). Because of this we need two intensities for each resource. The first is the intensity of the resource per unit of rated capacity—a well-defined quantity since the rated power is a fixed characteristic of the system (e.g., a 1 kW solar panel). The second is the intensity of the resource per unit of delivered power when averaged over a representative period such as a year. It is always larger, sometimes much larger, than the intensity per unit of rated capacity, and it is less well defined because it depends on how the system is operated, and, in the case of renewable systems, on the influence of the weather on sun, wind, wave, and tide.

Capital intensity, construction (rated power) GBP/kW_{nom}

The quantity of capital (money) used to construct the power system per unit of rated power (per kW_{nom}, e.g.) of the system. If you want the cost per kW of actual delivered power, as in Figure 12.5, you have to divide the capital intensity by the capacity factor (expressed as a fraction).

Capital intensity (fuel)   GBP/kW

The cost of the fuel used in the system per kW of power generated. This is based on input/output figures from plants and scaled to be a nominal value.

Area intensity m^2/kW_{nom}

The land area used by the power system per unit of rated power (kW) of the system. If you want the area per kW of actual delivered power capacity, as in Figure 12.4, you have to divide the area intensity by the capacity factor (expressed as a fraction).

Material intensities kg/kW_{nom}

The quantities of materials required to build a given power system per unit of nominal power. If you want the material per kW of actual delivered power, as in Figure 12.4, you have to divide the material intensity by the capacity factor (expressed as a fraction).

Energy intensity (construction) MJ/kW_{nom}

The energy is used to construct the power system, per unit of rated power (kW) of the system. To get the build-energy per kW of actual delivered power, you have to divide the energy intensity by the capacity factor (expressed as a fraction). This is not done in Figure 12.6 because the other axis, delivered energy per year, is also nominal. The point of Figure 12.6 is to illustrate the payback time.

Energy intensity (fuel)   MJ/kW · hr

The energy consumed as fuel by the power system to generate each kWh of delivered energy.

CO_2 intensity (construction)   kg/kW$_{nom}$

The quantity of carbon dioxide, in kg, released to the atmosphere during the construction of a given power system per unit of nominal power (kW$_{nom}$) of the system.

CO_2 intensity (fuel)   kg/kW · hr

The quantity of carbon dioxide, in kg, released to the atmosphere because of the burning of hydrocarbons by the power system per kWh of delivered energy.

Operational parameters

Capacity factor   %

The fraction of time, expressed as a percentage, that a power system operates at its rated power. This takes into account time when a system would be unavailable or generating less power than it potentially could due to maintenance or because the natural resource it uses is unavailable.

System efficiency   %

The efficiency with which the fuel or resource is converted into electricity.

Lifetime   yrs

The expected time that the power system will remain fully operational in years.

Status

Current installed capacity   GW

The total global rated capacity of a given power system.

Growth rate   %/year

The rate at which installed capacity currently grows each year expressed as a percentage.

Delivered cost   GBP/kWh

The cost of generating 1 kilowatt-hour of electrical energy for a given power system.

12.16 Appendix 2: Approximate material intensities for power systems

The bills of materials for the power systems assembled here are expressed as material intensities, I_m, meaning the mass (kg) of each material per unit of nominal generating capacity (kW_{nom}). As far as we know, no previous assembly of such data exists, so it has to be patched together from diverse sources. These differ in detail and scope. Some, for instance, are limited to the system alone; others include the iron, copper, and other materials needed to connect the system to the grid, and this is large for distributed systems like wind, wave, and solar power. Others give indirect information from which missing material content can be inferred. So be prepared for inconsistencies.

Despite these reservations there is enough information here to draw broad conclusions about the demand that large-scale deployment of any given low-carbon power system could put on material supply. The resource-demand plots in the text use data from these tables. They are based on an imagined scenario: that the global electric power demand will triple by 2050 and that, to meet it, the capacity of a chosen system is expanded by 200 GW (2×10^8 kW) per year. The metric for resource pressure, R_s, is the mass of each material required for the expansion of the system per year, $2 \times 10^8 I_m$, expressed in kg/year divided by the current (2009—the most recent available) global production per year, P_a, also expressed in kg/year.

$$R_s = \frac{2 \times 10^8 \times I_m}{P_a}$$

The resource-demand plots in the main text show this, expressed as the percent demand of current production.

The resource demand is not interesting when it is trivial. The plots show the demand on the materials that are deemed to be "critical," as explained in Section 12.2. They are starred (*) in the following tables.

Table 12.2 included the capital, energy, and carbon intensity of fuels. They are used to construct Figure 12.7. Some burn fuel for power. The others use much smaller quantities in operation and maintenance. The data used in the construction of Figure 12.7 are summarized here.

Fuel	Cost/kWh ($/kWh)	Energy/kWh (MJ/kWh)	CO_2/kWh (kg/kWh)
Coal	0.02–0.04	9.7–12	0.9–1.1
Gas	0.025–0.055	6–8	0.33–0.5
Fuel cell (Phosphoric acid)	0.025–0.055	9–10	0.49–0.55
Fuel cell (Solid oxide)	0.025–0.055	6–7.2	0.33–0.39
Nuclear	0.005–0.006	9.6–12	0.06–0.07
All other power systems	0.05–0.09	0.05–0.1	0.005–0.01

Coal-fired station

Material	Intensity (kg/kW$_{nom}$)
Aluminum	2.58–4.5
Bitumen	0.33–0.37
Brass	0.24–0.27
Carbon steel	30–614
Ceramic tiles	0.39–0.44
Chromium*	2.33–3.2
Concrete	460–1,200
Copper	1.47–5.17
Epoxy	0.21–0.23
GFRP	0.55–0.605
Glass	0.026–0.029
Glulam	0.004–0.005
HDPE	0.16–0.17
High alloy steel	0.5
Iron	50.2–809
Lead	0.04–0.23
Low alloy steel	13.6–15.1
Manganese*	0.084
Molybdenum*	0.032
Nickel	0.01
PP	0.08–0.09
PVC	1.82–2.02
Rock wool	3.9–4.3
Rubber	0.12–0.13
SAN	0.026–0.031

Silver*	0.001–0.007
Stainless steel	37–41
Vanadium*	0.003
Zinc	0.06–0.08
Total mass, all materials	**520–1,800**

Materials and quantities from White and Kulcinski (2007).

Pressurized water reactor

Material	Intensity (kg/kW$_{nom}$)
Aluminum	0.02–0.24
Boron	0.01
Brass/bronze	0.04
Cadmium	0.01
Carbon steel	10.0–65
Chromium*	0.15–0.55
Concrete	180–560
Copper*	0.69–2
Galvanized iron	1.26
Inconel	0.1–0.12
Indium*	0.01
Insulation	0.7–0.92
Lead	0.03–0.05
Manganese*	0.33–0.7
Nickel*	0.1–0.5
PVC	0.8–1.27
Silver*	0.01
Stainless steel	1.56–2.1
Uranium*	0.4–0.62
Wood	4.7–5.6
Zirconium*	0.2–0.4
Total mass, all materials	**170–625**

Materials and quantities from White and Kulcinski (2007).

Silicon-based PV

Material	(kg/kW)
Acids + Hydroxides	7.0–9
Aluminum	15–20
Ammonia	0.05–0.1
Argon	3.0–5.0
Carbon allotropes	10.0–20.0
Copper	0.2–0.3
Glass	60–70
Gold	0.05–0.1
Indium	0.02–0.08
Plastics	20–60
Silicon	25–40
Silicon carbide	6.0–10.0
Tin	0.1–0.2
Wood	10.0–20.0
Total mass, all materials	**150–250**

Materials and quantities from Phylipsen and Alsema (1995), Keoleion et al. (1997), and Tritt et al. (2008).

Solar: Concentrating thermal power

Material	Intensity (kg/kW$_{nom}$)
Aggregates	50–500
Aluminum	0.1–0.3
Borosilicate glass	3
Chromium (stainless steel)	2–10
Concrete	200–2,000
Copper	0.5–5
Glass	90–220
Magnesium	0.3–0.9
Manganese	0.008–0.2
Nickel	0.001
Paint	1–3
Silver	2.5–6.5
Steel and iron	300–800
Total mass, all materials	**650–3,500**

Materials and quantities from Viebahn et al. (2004).

CdTe thin film PV

Material	(kg/kW)
Aluminum	20
Cadmium	0.1–0.3
Tellurium	0.1–0.3
Copper	1
Glass	60
Indium	0.005–0.025
Lead	0.05
Plastics	30
Stainless steel	20
Tin	0.2
Total mass, all materials	**130**

Materials and quantities from Fthenakis and Kim (2005) and Pacca, Sivaraman, and Keoleian (2006).

Phosphoric acid fuel cell

Material	Intensity (kg/kW_{nom})
Aluminum	0.9–1.1
Carbon allotropes	5–9
Ceramics	1–5
Chromium	3–7
Concrete	10–20
Copper	3–8
Iron and steel	60–90
Molybdenum	0.02
Nickel	1.7
Palladium	0.0005
Phosphoric acid	0.5–2.5
Plastics	1.5–5
Platinum	0.005
Zinc	2.3
Total mass, all materials	**89–150**

Balance of plant for solar PV

Material	kg/kW
Aluminum	20–30
Concrete	500–550
Copper	1–2
Steel	1,000–1,200
Total mass, all materials	**1,500–1,800**

Materials and quantities from Pacca and Hovarth (2002) and Tahara et al. (1997).

Solid oxide fuel cell

Material	Intensity (kg/kW$_{nom}$)
Aluminum	0.5–2
Concrete	10–20
Chromium	0.5–3
Iron and steel	60–80
Lanthanum	0.01–3
Manganese	0.01–1
Nickel	1–6
Yttrium	0.1–0.4
Zirconium	0.1–3
Zinc	0.01–1
Total mass, all materials	**70–110**

Materials and quantities from Karakoussis et al. (2001) and Thijssen (2010).

Onshore wind

Material	kg/kW
Aluminum	0.8–3
CFRP	5.0–10
Concrete	380–600
Copper	1.0–2
GFRP	5.0–10
Steel	85–150
Neodymium	0.04
Plastics	0.2–10
Total mass, all materials	**500–750**

Materials and quantitites from Ardente et al. (2006), Crawford (2009), Vindmølleindustrien (2007), Vestas (2008), and Martinez et al. (2007).

Offshore wind

Material	kg/kW
Aluminum	0.5–3
Chromium (stainless steel)	4.5
Concrete and aggregates	400–600
Copper	10.0–20
GFRP	5.0–12
Steel	250–350
Neodymium	0.04
Plastics	1.0–10
Total mass, all materials	**650–1,000**

Materials and quantities from Ardente et al. (2006), Crawford (2009), Vindmølleindustrien (2007), and Weinzettel (2009).

Hydro power

Material	kg/kW
Aluminum	0.8–6
Brass	0.09
Cement and aggregates	7,500–30,000
Chromium	0.5–2.5
Copper	0.1–2
Iron	50–300
Lead	0.3
Magnesium	0.1
Manganese	0.2
Molybdenum	0.25
Plastics	1.0–6
Zinc	0.4
Wood	80
Total mass, all materials	**7,500–30,000**

Materials and quantities from Vattenfall (2008), Pacca and Hovath (2002), and Ribero and da Silva (2010).

Wave power (Pelamis)

Material	Intensity (kg/kW$_{nom}$)
Aluminum	25–30
Copper*	10–20

Nylon 6	8–12
Polyurethane	12–18
PVC	25–31
Sand (ballast)	640
Stainless steel	50–60
Steel	410
Total mass, all materials	**1,145–2,000**

*Materials and quantities from Anderson (2003).

Tidal current power: Seagen

Material	Intensity (kg/kW$_{nom}$)
CFRP (blades)	3.25
Copper*	3.88
Epoxy	0.25
GFRP (enclosure)	4.5
Iron	28.3
Neodymium or cobalt	0.9
Stainless steel	2.33
Steel	344
Total mass, all materials	**387**

*Materials and quantities from Douglas et al. (2008).

Tidal barrage power

Material	Intensity (kg/kW$_{nom}$)
ABS 30% glass fiber	0.019
Cement	1728
Copper*	0.004
Gravel	996
Pre-stressed concrete	3,416
Rock	28,686
Sand	20,488
Stainless steel	0.026
Steel	33
Total mass, all materials	**55,350**

*Materials and quantities from Roberts (1982) and Miller et al. (2010).

Geothermal power

Material	Intensity (kg/kW$_{nom}$)
Bentonite	20.9–45
Calcium carbonate	37.9
Carbon steel	10.8
Cement	3.3–41
Chalk	31
Concrete	21.9
Copper*	1.2–2.2
EVA	1
High alloy steel	342.4
LDPE	20.4
Low alloy steel	2–476
Portland limestone cement	133
PVC	0.1
Silica sand	39.6
Total mass, all materials	**61–1,200**

Materials and quantities from Saner et al. (2010).

Biomass

Material	Intensity (kg/kW$_{nom}$)
Aluminum	1.1–6.7
Bitumen	0.5
Brass*	0.37
Cast iron	1.47
Ceramic tiles	0.59–9.3
Chromium*	0.0024
Cobalt*	0.0018
Concrete	36–790
Copper*	1.04–3.5
Epoxy resin	0.31
GFRP	0.82
Glass	0.04
Glulam	0.006
HDPE	0.23

LDPE	3.25
Lead	0.104
Low alloy steel	20
Low carbon steel	33–112
Nickel*	0.02
PP	0.12
PVC	0.45–2.74
Rock wool	1.65–6
SAN	0.04
Stainless steel	4.5–5.5
Steel (electric)	0.82
Synthetic rubber	0.18
Titanium dioxide*	0.4
Zinc	0.16
Total mass, all materials	**69–922**

Materials and quantities from Bauer (2008) and Mann and Spath (1997).

12.17 Exercises

E12.1. What is the difference between energy and power? What are their units?

E12.2. What is the relationship between "oil equivalent" energy and "electrical" energy?

E12.3. What are the potential sources of renewable (non-hydrocarbon) power? What are the positive and negative aspects of converting to an economy based wholly on renewable sources?

E12.4. What is meant by the "nominal materials intensities" of a power system? What is meant by the "actual material intensities"?

E12.5. The land area of New York (the state) is 131,255 km^2. Its population is 19.5 million and the average power consumed per capita there is 10.5 kW. What fraction of the area of the state would be taken up by wind turbines in order to meet one half of its energy needs? How many metric tons of materials would be required? Use mean values of the data in Table 12.2 to find out.

E12.6. The land area of the state of New Mexico is 337,367 km^2. The population of the United States is 301 million and the average power consumed per capita there is 10.2 kW. Single crystal PV solar cells have an area intensity of about 80 m^2/kW$_{nom}$. If the capacity factor for solar power in New Mexico is 0.25, is the area of the state large enough to provide solar power that would meet one third of the current needs of the United States?

E12.7. The construction of PV solar cells uses a number of critical elements. One is indium, an ingredient of the transparent conducting coating on the surface of the cell. Find the material intensity of indium in solar cells from the table in Appendix 2 of this chapter and use it to calculate the quantity of indium that would be required for the project described in the previous question: that of providing PV solar power with capacity factor of 0.25, sufficient to meet one third of the present needs of the US population, 301 million, each using an average of 10.2 kW.

The current world production of Indium is 600 metric tons per year. If the solar project were to be implemented over a 5-year time span, would the supply of indium be adequate?

E12.8. A new generation of wind turbine is under development. The plan envisions land-based turbines with a rated power of 4 MW and a diameter of 125 m (about the same as the London Eye, one of the world's largest Ferris wheels). One turbine requires 410 metric tons of materials and construction consumes an energy of 2×10^7 MJ. The anticipated capacity factor, C, is 0.3.

   (a) What are the nominal material and energy intensities of the new turbines? How do these compare with the values for existing land-based turbines in Table 12.2?
   (b) What is the expected energy payback time?

E12.9. Use mean values of the material intensities of Table 12.2 to compare the number of metric tons of materials required to build a nominal 0.5 GW of new generating power by (a) nuclear, (b) single crystal solar PV power, and (c) land-based wind sources. How do these tonnages change if 0.5 GW of actual, not nominal, power is to be built?

E12.10. The Sahara Desert is a very sunny place. The capacity factor for solar PV power there could be as high as 40%, rather than the 10% of more temperate locations used in creating Figures 12.4 and 12.5.

   (a) Re-plot by hand the yellow ellipses for solar PV, single crystal, and polysilicon on copies of these two figures to show how this changes the relative material, area, energy, and capital intensities.
   (b) Do the same with Figure 12.6 (the changed capacity factor increases the delivered energy per year by a factor of 4 but leaves the construction energy per kW_{nom} unchanged). What is the energy payback time for the two power systems now?

E12.11. The energy of combustion of anthracite is 32 MJ/kg. If this energy is converted to electrical power at a conversion efficiency of 38%, how much anthracite is required per year to provide a steady 1 GW of electrical power?

E12.12. Indium is a strategic material. It plays a key role in solar PV technology as a transparent conductor. Use Google to explore alternative materials for transparent conductors, checking that they are not among the strategic elements plotted in Figure 12.8.

E12.13. Neodymium is a strategic material. Permanent magnets based on neodymium have the highest remnant induction (giving the most powerful magnetic field) and coercive force (the greatest resistance to demagnetization) per unit weight of any known material, making them the most attractive candidates for DC motors and generators like those used in hybrid and electric vehicles and in wind turbines. Currently over 90% of the global production of neodymium is in China, raising concerns about supply-chain concentration. Use Google to explore alternative permanent-magnet materials. Do any of them require the strategic elements plotted in Figure 12.8?

E12.14. The Energy Intensity (construction) (units: $MJ/kW_{nominal}$) is the energy needed to construct a power station divided by its rated generating capacity in kW. A power plant with a rating of $1~kW_{nominal}$ does not actually produce 1 kW continuously because its capacity factor C—the fraction of the time that it is actually running at its rated power—is less than 100%.

(a) Write an expression for the actual average power over 1 year, kW_{actual}, delivered by a system with a rated power of $1~kW_{nominal}$. If the capacity factor $C = 35\%$, how much energy will be generated by this power plant over 1 year in kWh?

(b) What is this energy, expressed in MJ?

(c) If the energy of construction is 20,000 MJ/kW_{nom} and the capacity factor is $C = 35\%$, how many years will it take before the energy generated by this power plant equals the energy it took to build it? This is the energy payback time.

Exercises using the CES Low Carbon Power database

E12.15. Use the CES Low Carbon Power database to make a bar chart with the energy payback time on the y-axis for power systems that do not burn fossil or nuclear fuels. How many years on average would it take a Solar Power—PV, single crystal power plant to payback its energy of construction?

E12.16. Use the CES Low Carbon Power database to plot a bar chart of Area Intensity. Which power system uses the most land area to generate 1 kW of actual delivered power? (Remember to divide the Area intensity by the Capacity factor (%/100).

E12.17. Use the CES Low Carbon Power database to create a chart with the same axes as Figure 12.4: actual material and area intensities. Remember that to evaluate these, the nominal values must be divided by the capacity factors, expressed as fractions.

E12.18. Use the CES Low Carbon Power database to create a chart with the same axes as Figure 12.5: actual construction-energy and capital intensities. Remember that to evaluate these, the nominal values must be divided by the capacity factors, expressed as fractions.

E12.19. If you want portable power, meaning a compact, light source of power, you want a system that has high power density. Find the power systems with the largest possible power per unit area, W/m^2, and power per unit weight, W/kg. These quantities are just the reciprocals of the Area and Material intensities, multiplied by 1,000 to convert kW to W. Make a chart with these axes and identify the most promising power system.

E12.20. Use the CES Edu Eco-audit tool with Level 2 data to estimate the construction-energy intensity of a Pelamis wave-power system. Data for the materials of Pelamis are listed in Appendix 2 of this chapter (use mean values of the ranges). Assume all the metals have their "typical" recycle content, and use Stone as a proxy for sand. Express the results as an energy per kW of nominal generating capacity. Does this value correspond to that listed in Table 12.2?

CHAPTER 13

Material efficiency

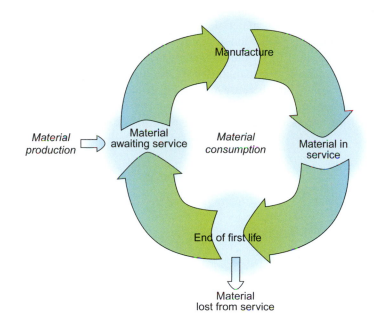

13.1 Introduction and synopsis

Nature uses materials efficiently. The tallest tree captures the most sunlight. The strongest predator captures the most prey. The most agile prey most often out run the predator. Plants, predators, and prey compete for resources with others in their forest, pack, or flock. Those that use these resources most efficiently are the ones that survive. Nature sets examples of material efficiency by favoring evolution that minimizes the resources consumed per unit of function.

In some parts of the world, humans still compete for resources in the harsh way familiar in nature; there, materials are not needlessly wasted. But in the developed world, resources have long been cheap and plentiful, allowing for unconstrained growth of material production. These societies treat materials in a prodigal

One aim of material efficiency is to retain as much material in the consumption loop with as little input from material production, and as little lost at end of life, as possible.

415

way, they consume natural resources to make them, use them once, and then discard them. Material efficiency in wealthy economies is low.

Let us be more specific. *Material efficiency* means providing more material services with less material production. Put another way, it means maintaining enough material in the Material Stock box of Figure 11.5 to provide the services we need while minimizing the new material production flowing in at the top. Remember that material *production* is not the same as material *consumption*. Material production measures the quantity of materials created per year from ores, feedstock, and energy. Material consumption measures the quantity of materials used in manufacturing of all types per year. Consumption is greater than production because of the contributions of material recycling and reuse. The aim of increased material efficiency is to provide the consumption required for necessary goods and services while producing as little material as possible.

Material efficiency

Example: Refrigerators come in many different sizes. One model—Fridge A—has a capacity of 0.35 m³ (12.5 cubic feet) and weighs 65 kg. Another—Fridge B—has a capacity of 0.28 m³ (10 cubic feet) and weighs 60 kg. Assuming that both fridges are made almost entirely from steel, which of the two is the most material efficient?

Answer: The function of a fridge is to provide cooled space. We therefore measure material efficiency as the mass of material per unit of cooled space. That of Fridge A is 186 kg/m³. That of Fridge B is 214 kg/m³. Fridge A is the more material efficient.

As we saw in Chapter 2, global stocks and reserves of most of the materials used to provide buildings, infrastructure, and products are still sufficient to meet anticipated demand, but the environmental impacts of material production and processing give cause for concern. These impacts and the risks associated with a dependence on material imports provide the motivation for improving material efficiency.

We first examine the case for increased material efficiency and then explore four strategies for achieving it:

- Improved material technologies
- Eco-oriented engineering design
- The use of economic instruments
- Social adaptation and change of lifestyle

The first of these, *material technology*—shown as the upper-left block of Figure 13.1—seeks ways to produce materials more efficiently, to improve their properties, to increase their functionality, and to develop new ways of reusing them. The second, *engineering design*—shown at the upper right—deploys the methods of engineering design to close the materials loop to retain materials in use in existing, modified, or new products. These two measures require little change in

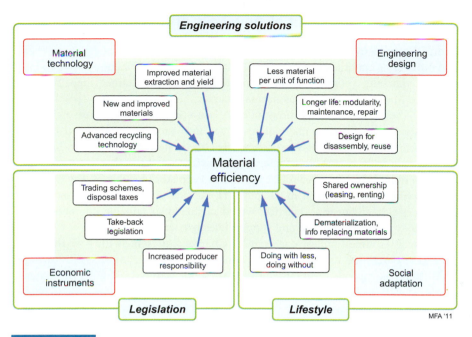

FIGURE 13.1 *Measures contributing to material efficiency*

consumer behavior beyond a willingness to retain and recycle, and an acceptance of recycled and reconditioned materials and products as being desirable. The two blocks are labeled "Engineering solutions" on Figure 13.1.

The lower part of the figure introduces strategies of a different sort. The lower-left block—*economic instruments*—lists some of the ways in which legislation can influence material use and reuse, most of which we have already met in Chapter 5. That box at the lower right—*social adaptation*—requires greater adaption. Increased material use-intensity means a changed view of possession, acknowledging that sole ownership carries responsibilities of product stewardship, accepting shared ownership and use of products, or abandoning ownership entirely and relying instead on service providers.

In the sections that follow we explore in more depth the case for increased material efficiency, examine the barriers that obstruct it, and list ways in which these might be overcome.[1]

---

[1]Allwood et al. (2011) analyzes the need for material efficiency, possible ways to achieve it, and the obstacles to implementing them. Much of the reasoning of this chapter follows arguments developed in that paper.

13.2 What is the point of materials efficiency?

Before the Industrial Revolution materials were expensive and labor was cheap. The number of materials in service was small and their high value relative to labor ensured that the products made from them were maintained, repaired, and upgraded. Material efficiency was normal practice.

Since the Industrial Revolution, industry has operated as an increasingly open-loop system, transforming material resources into products that are used once, then discarded. If materials were renewable, like water, we could live like that. But the only truly renewable engineering materials are those that grow as plants, and then only if they are grown fast enough to keep stocks stable. A few meet this criterion (they were described in Chapter 11), but any great reliance on renewables to replace, say, engineering plastics, would create unsustainable competition for land already used to grow food and support livestock.

Mineral resources are less constrained. Chapter 2 presented the case that most mineral resources are sufficient to meet current demand, but as ore grades dwindle, more energy is expended mining and refining them. Thus constraints on material supply may arise because of unmanageable energy requirements or because the environmental consequences of extraction and processing become unacceptable. And there is the geopolitical dimension: mineral scarcity gives power to resource-rich nations and is a concern to those that are resource-poor. It creates conditions for politically motivated barriers to free trade.

There are, then, at least four compelling motives for increasing material efficiency:

- To minimize the depletion of nonrenewable material resources
- To reduce dependence on imports, particularly of materials already in short supply
- To reduce the demand on other resources, particularly energy
- To reduce unwanted emissions generated by material production and use

News-clip: Material efficiency

Global lithium deposits enough to meet electric car demands.
The world has enough lithium resources to power electric vehicles for the rest of the century, according to a new report from the University of Michigan. Professor Greg Keoleian, co-director of the Center for Sustainable Systems at the university, said "Responsible use of the resource is a must even with abundant supplies. The key is to use it efficiently and not let it leak out of the economy after use." Still, the expected demand for lithium from electric vehicles has turned the element into a so-called critical material, with lawmakers on Capitol Hill working to secure US supply.

The New York Times, July 28, 2011

Visualizing material efficiency. The flow of material through the industrial system and the waste that is generated along the way can be documented in tables and text, but the complexity of the data make them almost impossible to grasp. Instead the flows can be plotted as a Sankey diagram, of which Figures 13.2 and 13.3, prepared by Allwood and Cullen (2010), are examples.

The metal we use most is iron, most of it in the form of steel. Its global consumption in products, 1.04 billion metric tons, starts as 928 million metric tons of new production and 568 million metric tons of scrap. Figure 13.2 shows these entering on the left. From there they are transformed by steelmaking, casting, rolling, and forming, and fabricating to create end products, shown on the right; the final yield is 74% of the incoming metal. The transformations appear as the five vertical slices of the figure. Threads run through these slices from left to right, each with a width proportional to the mass flow of metal. Values for the major flows are given in Mt (million metric tons). Useful metal has colored threads, scrap has gray threads, and metal loss is shown in black. Useful metal continues to flow to the next process step, while scrap loops back to the appropriate melting stage where it is recycled. Internal recycling loops, for example from the continuous casting processes for steel, are shown as small oval loops. Most of the metal losses are due to formation of dross and scale in hot metal processes.

Uses of steel

Example: Which end-use sector uses the most steel? Use the Sankey diagram for steel to find out.

Answer: The construction sector uses 583 million metric tons, accounting for roughly 56% of all steel consumption. Reinforcing bar for steel-reinforced concrete is the single largest consumer.

The diagram for aluminum, the second most-used metal, is simpler (Figure 13.3). The global consumption starts as 38 million metric tons of new production made by electrolysis and an equal quantity of scrap recovered by melting. From there it is transformed by casting, rolling, forming, and fabrication into end products. Useful metal is again shown as colored threads, scrap as grey. Loss as scrap is considerable: only 59% of the incoming metal makes it into a product.

The diagram has an upper and a lower loop. Virgin aluminum is almost all used in wrought products, which account for two thirds of all consumption. This stream occupies the upper part of the diagram. The remaining third is used for cast products; that stream occupies the lower part. The apparent separation of the two streams arises because most scrap cannot be cleanly separated by composition and is suitable only for recycling as casting alloys.

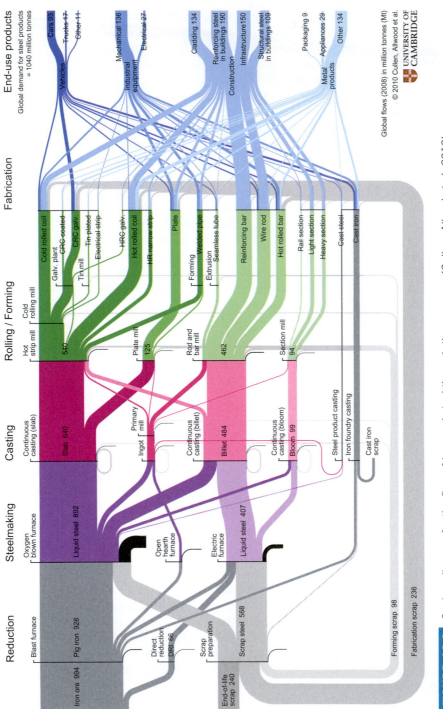

FIGURE 13.2 *Sankey diagram for the flow of iron and steel through the economy (Cullen, Allwod et al. 2010)*

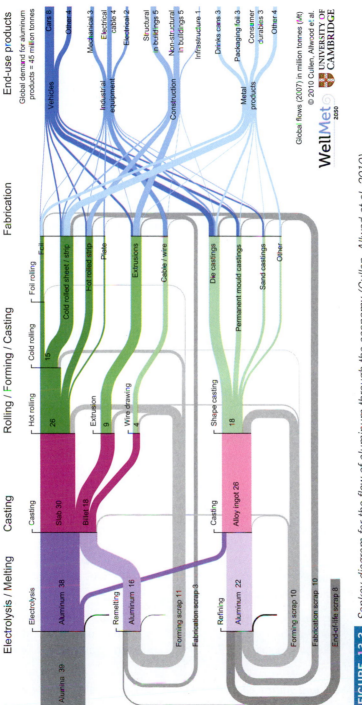

FIGURE 13.3   *Sankey diagram for the flow of aluminum through the economy (Cullen, Allwod et al. 2010)*

Scrap generation

Example: Which step in the production of aluminum products creates the greatest scrap fraction? Use the Sankey diagram for aluminum to find out.

Answer: Shape casting generates the greatest scrap fraction. Almost 40% of the metal entering this stage exits as scrap. (This arises from losses when aluminum ingots are scalped and trimmed.)

The diagrams suggest two strategies for reducing demand for metal: designing products that contain less metal and reducing the scrap created in metal manufacturing. If all products were redesigned to use less metal, the whole map would contract. However, if manufacturing scrap were reduced, the total demand for liquid metal would be reduced as well, but this would have little impact on total primary metal production because reducing scrap reduces the flow into recycling.

13.3 Increasing material efficiency (1): engineering solutions

The upper half of Figure 13.1 suggested engineering approaches to increased material efficiency. Here we examine them in more detail.

Material technology. Improved material extraction and yield. If 20% of all the energy we use is consumed in extracting and refining materials, then even an improvement of 1% in the efficiency of these processes would save much energy. New process routes such as the Fray process for producing titanium are both more efficient and give greater yield. The potential of such developments is great, and we need them: the ores we will mine in the future will be leaner than those we mine today.

Improved material extraction

Example: The embodied energy of titanium made by the present-day Kroll process is approximately 580 MJ/kg. There is a realistic hope that the new Fray process, at present under development, might reduce this by 20%. Use the annual production figure in the data sheet for Titanium in Chapter 15 to estimate the global energy saving this would permit.

Answer: The annual production of titanium is approximately 5.8×10^6 metric tons per year. A global change from the Kroll to the Fray process would save 116 MJ/kg, or a global annual total of 6.7×10^8 GJ, equivalent to about 20 million metric tons of oil.

New and improved materials. We saw in Chapter 9 how material properties can be "mapped" as material property charts. All the charts have one thing in common.

Parts of them are populated with materials but other parts are not: there are *holes*. Some parts of the holes are inaccessible for fundamental reasons that relate to the size of atoms and the nature of the forces that bind them together. But others are empty even though, in principle, they could be filled. Is anything to be gained by developing materials (or material combinations) that lie in these holes? To answer this we need criteria of excellence. These are provided by the material indices, described in Chapter 9. If a new material has better index values than any existing material, the newcomer has the potential to increase performance.

One approach to filling holes—the long-established one—is that of developing new metal alloys, new polymer chemistries, and new compositions of glass and ceramic that expand the populated areas of the property charts. But developing new materials can be expensive and uncertain, and the gains tend to be incremental rather than step-like. An alternative is to combine two or more existing materials so as to allow a superposition of their properties—in short, to create hybrids. The great success of carbon and glass-fiber, reinforced composites, of sandwich panels and shells, and of cellular structures (hybrids of material and space) in filling previously empty areas of the property charts is encouragement to explore ways in which further hybrids can be designed.

Carbon nanotube composites

Example: It is suggested that an epoxy-matrix composite reinforced with 30% of randomly dispersed, single-walled carbon nanotubes (SWNTs) could have a specific strength that exceeds that of any existing material. Explore this claim, assuming that one third of the nanotubes contribute to strength in any given direction and that the composite strength and density can be estimated by a rule of mixtures, using the data in the following table. Do so by plotting the composite properties onto a copy of the strength-density chart of Figure 9.13 of the text.

	Density (kg/m^3)	Strength (MPa)
SWNT	1,350	30,000
Epoxy	1,250	60

Answer: Using a rule of mixtures, the density and strength of the composite are

$$\rho = 0.3 \times 1,350 + 0.7 \times 1,250 = 1,280 \ \text{kg/m}^3$$
$$\sigma_y = 0.1 \times 30,000 + 0.7 \times 60 = 3,000 \ \text{MPa}$$

These values are plotted on the strength-density chart, of Figure 9.13 of the text. If the assumption that strength can be estimated by a rule of mixtures is valid, the nanotube composite lies in an empty space on the chart.

Process improvement. Both the material and energy efficiency of many manufacturing processes are low. If it was easy to improve either one, it would have happened—the economic gains are too obvious to ignore—so gains in efficiency here require radical innovation. The energy to shape metal by deformation is less than that to shape it by casting and the material lost during the process is less too. One approach, then, is to seek innovative deformation processes able to make shapes that are as complex as those possible by casting. Another is that of additive forming, creating a complex shape by electro- or vapor-forming through masks, or direct bonding of powder by selected laser sintering, adhesive bonding, or supersonic compaction.

Advanced recycling technologies. Current legislation aims at 90% recycling rates for materials. The key steps are separation and recognition. Airframe makers now stamp sheet metal with the alloy designation and polymer molders' imprint moldings with designated recycle marks (see the Appendix to Chapter 4). Recognition could be improved further by tagging components with bar codes or other identifiers that carry more detailed information, but none of these help if products are shredded before they are dismantled. To get further, we need sophisticated, if unglamorous, materials science. When shredding is practiced, separation is possible by using the optical, magnetic, electrical, dielectric, or inertial signatures of the materials, but this is imperfect; these techniques are able to distinguish material families but are unable to separate individual grades. What we need is an internal identifier, one that is preprogrammed by the manufacturer, can survive life, and can carry a designation tag to every shredded piece.

Tagging shreddings

Example: Many products are shredded at end of life. If the shredded material is to be recycled, the shredded fragments must be sorted. Techniques based on density, and magnetic and electrical signatures allow the primary sorting of metals, but polymers all have about the same density, are non-magnetic, and almost all are insulators. Thus there is need for an internal identifier for polymers that can survive both product life and subsequent shredding so that every piece can still be identified. How might this be achieved?

Answer: Tagging polymers to survive shredding requires that embedded tags are distributed within the material. A low concentration of sub-micron or nanoscale magnetic particles could provide the marker, and their concentration, externally measured, could code information about composition. Alternatively, distributed particles that fluoresce or absorb light in the UV range at low concentration could carry composition data via color spectrum.

The quantities of most metals currently in service are large, large enough to meet most needs. Recycling and reuse keep them in service; discarding them creates the need to produce more. Raising the levels of recycling and reuse toward 100% has to be a central aim of material efficiency.

Engineering design. Less material for the same service. Good design minimizes the material needed to do something by properly selecting the material and shape. The index-based methods described in Chapter 9 and their extension to include section shape aim to do just that. More sophisticated methods, known as *topological optimization* (e.g., Sigmund, 2000), work by redistributing material from regions where stresses are low to regions where they are high, incrementally refining the shape. This helps save material when shapes are cast in liquid metal to near-final form. The saving is less when complex parts are machined from the solid, as they often are in the aerospace industry. Weight is saved but the volume of scrap is greatly increased, swelling the gray threads in Figures 13.2 and 13.3.[2]

Longer life, more intense use. If the life of a product doubled, only half as many would need to be produced. That saves material, but it does not always save energy. Extending life saves energy when its manufacture is energy-intensive and its use is not. When the reverse is true and technology improvements reduce energy in use, making it last longer can actually increase energy consumption. The two schematics of Figure 13.4 show the contrast. The environmental impact depends on the balance between production energy (steps in the lines), use-phase energy (slope of the lines), lifespan (time between steps), and rate of technology improvement

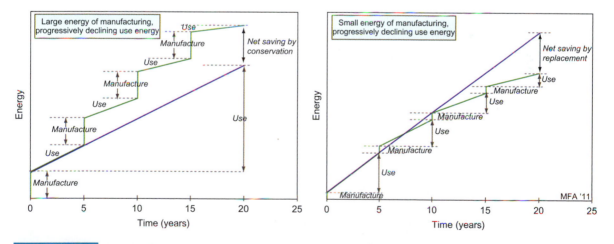

FIGURE 13.4   *Two scenarios illustrating the trade-off between conservation and replacement by more economical technology*

---

[2]In aerospace manufacturing, this problem is reflected in the "buy-to-fly" ratio (the ratio of the initial mass of material to that of the final finished component), which can be as poor as 20:1 for thin-walled shapes.

(reduction in the gradient of the slopes with time). As we saw in Chapter 8, most domestic appliances (refrigerators, washing machines, dishwashers, microwave and electric ovens) use much more energy during their life than it takes to make them. If the technology is advancing rapidly, replacement can be more energy efficient than extension of life.

It is not, however, more material efficient. Material efficiency means making products last longer and using them more intensively. There are many ways to do this:

- Transferring maintenance costs from purchaser to manufacturer (through extended warranties for instance)
- Leasing rather than purchasing
- Using products through a service provider (such as a launderette, car rental agency)
- Sharing ownership

Repair and resale. Extending product life by repair makes business sense where products are expensive and labor is cheap. But the business case for repair to extend product life is weak in developed economies where the reverse is true. The case for reuse is stronger, and for cars, boats, and houses, well-established systems are in place to enable it. For this to work there has to be a secure supply and consumers must trust it, needing clear quality assurance and price transparency. The success of eBay as an agent to enable reuse is—as we saw in Chapter 5—a model for how this can be achieved.

Design for upgrades, modularity, and remanufacturing. Remanufacturing is the upgrade or renewal of an existing product. There are many successful examples: the reconditioning of car and truck engines, the reuse of modules in photocopiers, the reuse of printer cartridges, and the remanufacturing of tires. Remanufacturing works for products that are at the mature end of their life cycle, in a market in which technology changes only slowly.

Remanufacturing needs appropriate design to allow

- Easy access and separation to allow fast, cheap disassembly
- Easy identification of components via the model and assembly numbers
- Easy verification of condition by using embedded monitoring to record use-history and predict residual life

That all sounds good, but it is not so simple. In practice the high costs of remanufacturing are dominated by those of storage, administration, and part replacement.

Component reuse. An alternative to remanufacturing is to separate the product into its subsystems or components and use them in new products—a kind of non-destructive recycling. It, too, sounds good. The problem lies with the lack of a developed supply chain. Without it, there are no large stocks on which to draw,

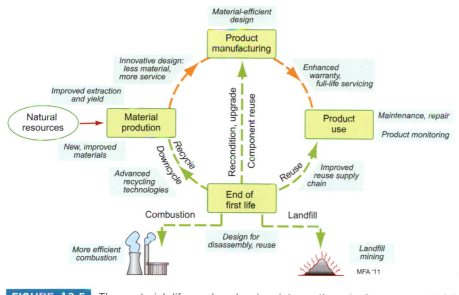

FIGURE 13.5 *The material life cycle, showing interventions to increase material efficiency*

making it difficult to source specific components or subsystems. By contrast, the supply chain for recycled *materials* rather than recycled *components* is easy to manage because materials re-enter the primary production cycle and thus can be used in a greater number of ways.

Figure 13.5 shows the material life cycle with the engineering solutions for increasing material efficiency superimposed. We turn now to solutions that lie in the hands of politicians and behavioral scientists.

13.4 Increasing material efficiency (2): legislation and social change

Economic instruments. Legislation. The lower-left block of Figure 13.1 lists some of the broad economic instruments—carrots and sticks—that governments have devised to make change happen. They are spread out in more detail in Figure 13.6, the most specific at the lower right, the most generic at the upper left. We have met all of these in Chapter 5; they are displayed here to illustrate the maze of legislation that has sprung up around environmental issues, most of it in the last 15 years. A maker of cars, or of household electrical products, or indeed of any material-intensive product must not only be aware of these, but must also demonstrate compliance with many of them. All are well-meant, but all increase the burden of tracking and reporting that manufacturers must accept.

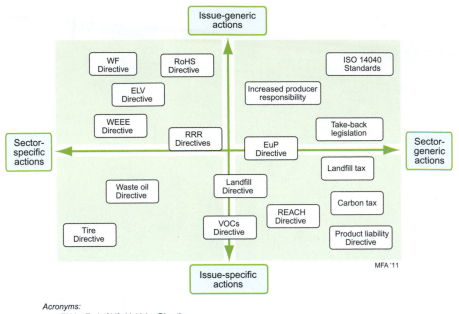

Acronyms:
- ELV = End of Life Vehicles Directive
- EuP = Energy Using Produces Directive
- REACH = Registration, Evaluation, Authorization and Restriction of Chemical Substances Directive
- RoHS = Restriction of Hazardous Substances Directive
- RRR = Reusability, Recyclability and Recoverability Directive
- VOC = Volatile Organic Compounds Directive
- WEEE = Waste Electrical and Electronic Equipment Directive
- WF = Waste Framework Directive

FIGURE 13.6 *Some of the legislative and economic restrictions currently in force in Europe that influence material efficiency (adapted from Loader (2008))*

Social adaptation. Shared ownership. Most families living in developed economies own a car. It is estimated that 98% of all licensed car seats are unoccupied at any time, a material efficiency of 2%. The use-statistics for yachts, motor cruisers, private aircraft, and second homes present a similar picture. Even commercial buildings are productively occupied for barely 25% of the time. If a product costs a lot and won't be used often, it makes sense to share it. Most of us share hospitals, schools, churches, parks, and public bathrooms, though with enough wealth you could own them all, as did the aristocracy of past ages.

Sharing makes material sense. A washing machine in a laundromat or a car in a taxi-rank is used more intensively than one that is individually owned. Carpools increase material usage without adding cost. Industrial equipment, aircraft, and trains can be leased rather than bought, allowing for their use elsewhere when local demand falls.

It is a small step from there to the idea of leasing the materials themselves, so that a manufacturer pays a rent for the time the materials are used but is obligated to return them at the end of the rental period. Some material are already treated in this way: uranium and thorium are examples.

Replacing materials by information. The applications on my computer cost more than the computer itself. The computer is made from materials but with it I can communicate, write, draw, do research, buy things, and be entertained without using any material at all. Tablet computers replace books and newspapers. At the commercial level, information technology allows customized products to be made when ordered, removing the need to hold a large stock. Similar dematerialization is spreading in retailing, manufacturing, banking, education, and entertainment.

Doing with less, doing without. Rich economies consume more resources than are necessary to achieve an adequate standard of living. In developed countries cars, television sets, hi-fis, and iPods are seen not as luxuries but as necessities. Wealth in many communities is linked to status; possessions are wealth-made-visible. There is every sign that as the poorer but more populous countries become richer, they will take the same view. Breaking this link means a drop in material consumption. Some communities and religious orders have made this choice, electing to live without amassing more possessions than are necessary for life, but they are a tiny minority.

Changing human behavior

Example: Is the idea of a social adaptation just a futile dream?

Answer: Not necessarily. Forty years ago no one wore a seat belt in a car or a crash helmet on a motorcycle, and it was permissible to smoke anywhere. Today the opposite is the norm, and failure to comply is seen as irresponsible.

13.5 What makes material efficiency difficult?

There are barriers to increased material efficiency, some technical, some economic, some regulatory or legal, and some social.

Technical barriers. Few products, even today, are designed to allow easy disassembly or are monitored over life with sufficient precision to know their condition or that of their subsystems. This and the difficulty of balancing supply and demand of used products or components discourage remanufacturing and component reuse or make it unnecessarily expensive.

Economic and business barriers. Why is it economically preferable to consume more material than is needed to deliver a service? There are two conflicting economic views of the need for material efficiency. One is that market forces will manage material scarcity by driving up prices, thereby reducing demand until it again matches supply. The other is that, globally, material resources are adequate; rather it is the energy demand and environmental impact of material production that is the real concern.

Materials are energy-intensive, so rising energy prices should reduce material demand. But this is not so for products. The final prices of products are less strongly linked to the cost of energy because so much (expensive) labor is required to make them. It is this that drives the outsourcing of manufacturing from developed economies to those with cheaper labor, despite the energy penalty that the additional transportation creates.

For much of the past century energy prices, when corrected for inflation, have been level. Over this period material prices have declined while GDP in richer countries has risen (the evidence is presented in the next chapter as Figure 14.1). Wages rise parallel with GDP, so production managers seek to reduce labor costs, even if this leads to an inefficient use of materials. Thus the economic motivation for material efficiency is weak because

- The cost of the environmental damage caused by material production (the "externalities" discussed in Chapters 3 and 14) is not fully reflected in its price.
- It is cheaper to use mass production in low-labor-cost countries than to implement material efficiency strategies.
- Businesses that invested in equipment and systems of mass production when materials were cheap may be locked into these legacy assets when the prices rise, leaving them unable to adapt to more material-efficient manufacturing.

Current economic systems are predicated on growth. One way to increase sales is to encourage product replacement. Material efficiency may not be in the interest of a manufacturing business unless it can reclaim value in some other way, such as an enhanced reputation for ethical business practice, for example.

Regulatory and legal barriers. Regulation can promote material efficiency when market forces, alone, don't work. We have already seen that the external costs of material production and use are not fully reflected in material prices. Carbon taxes, pollution taxes, and landfill taxes can correct this, but nobody likes them. Take-back legislation and recycling quotas retain material in service, reducing demand for new production. But prescriptive laws can hinder substitution and material reuse. Here are some examples:

- Building codes inhibit innovation in the use of more efficient materials in buildings.
- Standards are written with the assumption that all materials are new; there is a lack of government certification for recycled materials.
- Health and safety regulations promote use of virgin materials (for instance in medical equipment and in food packaging), inhibiting material reuse.
- Recycling legislation for vehicles inhibits the use of composites that can be both lighter and more durable than steel.

Social barriers. It is difficult to identify any period in peacetime in which a prosperous nation has chosen to reduce its habit of material acquisition. Consumerism started with the Industrial Revolution and continues today. Product ownership of goods can confer social status, with the result that change of fashion rather than loss of function determines the end of life of many products. Products made from recycled materials are seen as less desirable if they symbolize thrift. And today's lifestyle makes convenience a major driver of consumption, with the result that cultural attitudes to waste have moved from moral disapproval to complete acceptance. A "throw-away" society treats as normal the discarding of materials with reuse value.

13.6 Mechanisms to promote material efficiency

How might material efficiency be increased? There are a number of ways to do it.

Business opportunities. Material efficiency can be profitable.

- Environmental leadership can enhance brand image.
- New revenue streams can emerge when primary metals producers develop a "second hand" supply chain, much as car makers control their resale chains. Examples are the reconditioning, recertifying, and reselling of used I-beams, or the provision of recycled materials with a guarantee of composition.
- Leasehold as a business model—take, for example, Rolls Royce's leasing of aero-engines, or Xerox's leasing of copiers. The manufacturer retains ownership of materials and can control their life cycle and reuse in ways that maximize return.
- Increased emphasis on reducing embodied energy as use-phase energy efficiency improves. As buildings become more passive and vehicles more efficient, so their production energy becomes a larger fraction of their life energy and life cost.

Government interventions. Standards and directives aim to encourage change. Mandatory environmental labeling provides information that guides recycling. Preferential procurement programs by government agencies introduce economic incentives to meet environmental targets. Research agencies award grants to develop improved materials and material-efficient products. Take-back legislation makes manufacturers responsible for materials at end of life.

Consumer drivers. There is evidence that increased material possessions beyond a certain point does not increase well-being. Numerical measures of "happiness" or "quality of life" plotted against income show a plateau (Chapter 14, Figure 14.5), but it is not easy to persuade consumers that they have reached it. Contemporary social movements such as "ethical consumption," "downshifting," and "voluntary simplicity" urge living with less but are embraced by few. The consumer trend to larger houses, larger cars, and more possessions is not an easy one to reverse.

There are formidable societal, logistical, and even economic obstacles to increased material efficiency. Despite these, there is progress. Products ranging from cars to coffee makers are more durable, more efficient, and, often, more compact and lighter than in the past. The infrastructure for recycling and remanufacturing is more highly developed, and as energy prices rise, driving up material costs, the economic case for material efficiency grows steadily stronger.

13.7 Summary and conclusions

Making and using materials inefficiently depletes resources, generates emissions, and creates an uncomfortable dependence on imports. How much can we increase material efficiency before an enforced reduction in demand for services is the only option left? There are many ways to do so, some within the remit of the materials scientist and engineer, others involving government incentives and controls, or required changes to social perception and behavior. This chapter summarizes these and examines both the opportunities and the obstacles to improved material efficiency.

13.8 Further reading

Allwood, J.M., Cullen, J.M., Carruth, M.A., Cooper, D.R., McBrien, M., Milford, R.L., Moynihan, M., and Patel, A.C.H. (2012), *Sustainable materials: with both eyes open*, UIT Cambridge, Cambridge, UK. ISBN 978-1906860059.

Allwood, J.M., Ashby, M.F., Gutowski, T.G., and Worrell, E. (2011), "Material efficiency, a White Paper," *Resources, Conservation and Recycling*, 55, pp 362–381. *(An analysis of the need for material efficiency, possible ways of achieving it, and the obstacles to implementing them)*

Allwood, J.M., Cullen, J.M., Carruth, M.A., Milford, R.L., Patel, A.C.H., Moynihan, M., Cooper, D.R., and McBrien, M. (2011), *Going on a metal diet— using less liquid metal to deliver the same services in order to save energy and carbon*, University of Cambridge, Cambridge, UK. ISBN 978-0-903428-32-3. *(A report exploring ways of using less metal to deliver the same services in order to save energy and carbon, and detailing the Sankey diagrams shown in Figures 13.2 and 13.3)*

Allwood, J.M., Cullen, J.M., Cooper D.R., Milford, R.L., Patel, A.C.H., Moynihan, M., Carruth, M.A., and McBrien, M. (2011), *Conserving our metal energy*, University of Cambridge, Cambridge, UK. ISBN 978-0-903428-30. *(Ways of avoiding melting steel and aluminum scrap to save energy and carbon)*

Ashby, M.F. and Johnson, K. (2009), *Materials and design, the art and science of material selection in product design*, 2nd edition, Butterworth Heinemann, Oxford, UK. ISBN978-1-85617-497-8. *(Deals with the aesthetics, perceptions, and associations of materials and the bipolar influence of industrial design in manipulating consumer choice)*

Bendsoe, M.P. and Sigmund, O. (2003), *Topology optimization: theory, methods and applications*, 2nd edition, Springer Verlag. ISBN: 978-3540429920 *(The topology optimization method solves the basic engineering problem of distributing a limited amount of material in a design space, using iterative finite-element analysis)*

Eggert R.G., Carpenter, A.S., Freiman, S.W., Graedel, T.E., Meyer, D.A., and McNulty, T.P. (2008), *Minerals, critical minerals and the US economy*, National Academy Press, Washington, DC, USA. *(An examination of the criticality of 11 key elements, ranking them on scales of availability (supply risk) and importance of use (impact of supply restrictions))*

Geiser, K. (2001), *Materials matter: towards a sustainable materials policy*, MIT Press, Cambridge, MA, USA. ISBN 0-262-57148-X. *(A monograph examining the historical and present-day actions and attitudes relating to material conservation, with informative discussion of renewable materials, material efficiency, and dematerialization)*

Gutowski, T.G., Sahni, S., Boustani, A., and Graves, S.C. (2011), "Remanufacturing and energy savings," *Environmental Science and Technology*, Vol. 55, pp. 362–381, America Chemical Society.

Hertwich, E. G. (2005), "Consumption and the rebound effect: an industrial ecology perspective," *Journal of Industrial Ecology*, 9, pp 85–98. *(The rebound effect is the increase in consumption that occurs when improved technology reduces the energy consumption of products, cancelling the expected environmental gain.)*

IEA (International Energy Agency) (2007), "Tracking industrial energy efficiency and CO_2 emissions," OECD/IEA, Paris, France. *(The IEA provides the most reliable and complete statistics of energy consumption and consequent emissions, globally and nationally.)*

IEA (International Energy Agency) (2008), "Energy technology perspectives 2008: scenarios & strategies to 2050," OECD/IEA, Paris, France.

Loader, M. (2008), "Driving sustainability," *Materials World*, 16, pp 30–32. *(Loader presents figure with the same axes as Figure 13.3 of the text, detailing the complexity of automotive and materials legislation.)*

Lomberg, B. (2001), *The sceptical environmentalist—measuring the real state of the world*, Cambridge University Press, Cambridge, UK. ISBN 0-521-01068-3. *(A provocative and carefully researched challenge to the now widely held view of the origins and consequences of climate change, helpful in forming your own view of the state of the world)*

Lomberg, B., editor (2010), *Smart solutions to climate change: comparing costs and benefits*, Cambridge University Press, Cambridge, UK. ISBN 978-0-52113-856-7. *(A multiauthor text in the form of a debate—"The case for . . .," "The case against. . ." —covering climate engineering, carbon sequestration, methane mitigation, and market and policy-driven adaptation)*

MacKay, D.J.C. (2008), *Sustainable energy—without the hot air*, UIT Press, Cambridge, UK, and www.withouthotair.com/. ISBN 978-0-9544529-3-3. *(MacKay's analysis of the potential for renewable energy is particularly revealing)*

Pimm, S.L. (2001), *The world according to Pimm: a scientist audits the earth*, McGraw-Hill, New York, NY, USA. ISBN 0-07-137490-6. *(Pimm provides an enlightening discussion of biomass production and its limits.)*

Sigmund, O. (2000), "Topology optimization: a tool for the tailoring of structures and materials," *Philosophical Transactions of the Royal Society A*, 358, pp 211–227. *(An introduction to topological optimization, developed more fully in the book by Bendsoe and Sigmund, listed earlier)*

Szargut, J. (1989), "Chemical exergies of the elements," *Applied Energy* 32, pp 269–286. *(Szargut analyzes the thermodynamics of metal extraction)*

USGS (2010), "Mineral commodity summaries 2010," U.S. Geological Survey. minerals.usgs.gov/minerals/pubs/mcs/2010/mcs2010.pdf. Accessed December 2011. *(The ultimate source of data for mineral production and metal production)*

13.9 Exercises

E13.1. Origins of steel for castings. What is the primary source of steel for (a) continuously cast slab and (b) billet? Use the Sankey diagram for steel (Figure 13.2) to find out.

E13.2. Origins of scrap steel. Which stage in the processing of steel from pig iron and recycled scrap to end products generates the most scrap? Use the Sankey diagram for steel (Figure 13.2) to find out.

E13.3. Scrap generation. Which are the three highest volume semifinished aluminum products? Use the Sankey diagram for aluminum (Figure 13.3) to find out.

E13.4. Uses of aluminum, 1. What are the four main categories of end-use aluminum products? Use the Sankey diagram for aluminum (Figure 13.3) to find out.

E13.5. Uses of aluminum, 2. Which single product uses the most aluminum? Use the Sankey diagram for aluminum (Figure 13.3) to find out.

E13.6. The embodied energy of carbon steel is approximately 27 MJ/kg. Use the annual production figure in the data sheet for carbon steel in Chapter 14 to estimate the global energy saving if the efficiency of steel production could be improved by just 1%.

E13.7. The process by which primary aluminum is made consumes approximately 210 MJ/kg of energy (oil equivalent) and emits approximately 12 kg of CO_2 per kg of aluminum. Energy at present costs about $0.007/MJ, and the carbon tax stands at about $15/metric ton. If energy prices double and the carbon tax increases by a factor of 5, how much will it change the cost of producing aluminum? (The April 2011 spot price of aluminum is about $2.5/kg.)

E13.8. The embodied energy of primary aluminum is approximately 210 MJ/kg. If the global efficiency of primary aluminum use could be improved by just 1%, how much energy would be saved per year? Use the annual production figure in the data sheet for aluminum in Chapter 15 to find out.

E13.9. It is suggested that an epoxy-matrix composite reinforced with 24% of randomly dispersed, single-walled carbon nanotubes (SWNTs) could have a specific modulus that exceeds that of any existing material. Explore this claim, assuming that one third of the nanotubes contribute to stiffness in any given direction and that the composite modulus and density can be estimated by a rule of mixtures, using the data in the following table. Do so by plotting the composite properties onto a copy of the strength-density chart of Figure 9.11 of the text.

	Density (kg/m^3)	Modulus (GPa)
SWNT	1,350	940
Epoxy	1,250	2.4

E13.10. Redesign of an existing automobile allows a reduction of 3% in the total mass of steel, 490 kg, used for the body, and improvements in corrosion protection and warranty are expected to extend its average useful life from 250,000 km to 275,000 km. How would you estimate the gain in material efficiency of use of steel in the vehicle?

E13.11. The weight saving described in the previous example translates into improved fuel efficiency, expressed as MJ/1,000 km. The finished weight of the vehicle is 980 kg. Use data from Table 6.10 to calculate the increase in fuel efficiency if the vehicle is powered by a conventional gasoline engine.

E13.12. If the ore grade (G_m kg metal per kg ore) from which a metal is extracted falls, the energy required for mining, transportation, and subsequent concentration of the ore to a level at which it can be refined to metal rises at each stage, driving up its embodied energy. At its simplest we might expect the embodied energy to take the form

$$H_m = A + \frac{B}{G_m}$$

where A is the energy in MJ/kg metal for the refining process and B, also in MJ/kg, is the mining and transportation energy per unit ore grade. If, for copper, $A = 40$ MJ/kg and $B = 1.2$ MJ/kg, what is the embodied energy of copper mined from a deposit with a grade of 4%? How much does it increase if instead the available grade is 2%?

E13.13. Magnesium can be extracted by electrolysis of salts derived from sea water, or, better, from salt lakes such as the Dead Sea. The electrical energy necessary to electrolytically extract 1 kg of magnesium in this way is 40 kWh. The current price of magnesium is approximately $5/kg. What fraction of this price is the cost of the electricity used to make it? If electricity prices rise by a factor of 3 (responding to similar increases in the cost of oil or gas), how much will it increase the cost of making magnesium? Take the price of industrial electricity to be $0.06/kWh.

The bigger picture: future options

14.1 Introduction and synopsis

A free economy tends toward an economic optimum. If money can be gained or saved by doing something in another way, it tends to happen, provided the gain or saving is large enough to make the effort of the change seem worthwhile. Thus material production, manufacturing, and product design, use, and even disposal adjust themselves to the prevailing social and economic conditions—those of material and energy price, of concern for home, health, and happiness, and for the care of the environment.

There are times when these economic and social boundary conditions remain fixed or change only slowly. But when they do change, sometimes dramatically, as in times of armed conflict between nations, or more slowly, as in times of changing balance of economic power, what was optimal is no longer so. The altered conditions generate forces for change. Then, for survival (usually economic survival, but potentially, survival at a more basic level), two things are necessary: an ability to

Is this the future? Floods, drought, expanding desserts, and hurricanes.
(Images courtesy of Home.vicnet.net.au; Prisonplanet.com; Weathersavvy.com.)

perceive the change and its implications for the current optimum, and an ability to plan and implement a move toward a new optimum. In a word: *adaptability*.

This chapter is about forces for change. Some are perceived as threats. Others present opportunities. All require adaptability.

First, a justification of the bottom-up engineering approach of Chapters 2 to 12. There are good reasons for starting in this way. It builds on established, accepted methods; it avoids the controversy that plagues much discussion of environmental issues; and it advances understanding. The conclusions reached so far have a solid basis, giving perspective and replacing speculation and misinformation by fact. We have focused on methods to select materials to meet eco-objectives, taking energy and atmospheric carbon as the central actors. While the individual gains may be small, it is important to make them; we would fail in our obligations as engineers and scientists, *not* to try to do so. Much is learned in this process about where energy goes and where atmospheric carbon comes from. And it helps distinguish materials choices that contribute little to atmospheric carbon from those that contribute a great deal—it distinguishes the little fish from the big fish. If we are going to make a real difference, it is the big fish we need to catch.

But is this—will this be—enough? Can it provide a sustainable future? To answer this we must digress a little. If, through change of circumstance, your life is not going well, there are various steps you can take to fix it. The first and normal reaction is to examine the symptoms and try to remedy them with as little disruption to the rest of your life as possible. But if the problem persists, it becomes necessary to look more deeply into the *forces for change* that cause it. If these are real and unstoppable, a greater, more disruptive adjustment will be needed. There is a natural reluctance to do this until you are absolutely sure that the forces *are* real and the changes unavoidable. It is easier to do nothing, betting that things are not as bad as they seem and that the problem will go away. But if you lose your bet, you are caught unprepared. The adjustments you are then forced to make are not of your choosing. If you like to be in control, *anticipation* is better than *reaction*. It is called "the precautionary principle."

Using this little story as an analogy for global problems is an oversimplification but I'm going to make it anyway. The book, thus far, has dealt with minor inconveniences—a little local pollution, a spot of ozone depletion, a little global warming, occasional bits of restrictive legislation—and with corrective measures that disturb the rest of life very little. They can ameliorate the problems, but they will not fix them. Some of the forces for change are likely to prove too powerful to be dealt with in that way.

So let us look at the threats, the opportunities, and the options for the future. But first, three questions. Here is the first: why do we value materials so little?

14.2 Material value

How is it that materials are seen as of such little value? The waste stream is now so great that densely populated countries are running out of space to store it. We recycle some, but only under duress—it takes legislation, subsidies, penalties, and taxes to make us do it. Some materials don't get thrown away. Gold is one; its intrinsic value protects it. There was a time when the same was true of iron, aluminum, and glass, but not today. Why?

The three trends plotted in Figure 14.1 give clues. They show, in order, the aggregated price index of a spread of materials,[1] the price of energy,[2] and the purchasing power, expressed as GDP per capita,[3] of developed nations over the last 150 years, all normalized to the dollar value in 2000.[4] In real terms material prices have decreased steadily over time, making them cheaper than ever before. The price of energy (with some dramatic blips) remained pretty constant until 2008 when, in a very short period, it doubled, then doubled again. Purchasing power, measured by GDP/capita, increased enormously over the same period. Materials and the goods made from them are cheaper, in real terms, than they have ever been. But while the relative cost of materials has fallen, that of labor has risen, shifting the economic balance away from saving *material* toward that of saving *time*.

If the price of energy rises, what happens to the price of materials? Making them, as we have seen, requires energy; the embodied energy of some is much larger than that of others. The ratio of the cost of this energy to the price of the material is a measure of its sensitivity to energy-price rise. A value near 1 means that energy accounts for nearly all of the material price; if energy price doubles, so too will the price of the material. A value of 0.1 means that a doubling of energy price drives the material price up by 10%. Figure 14.2 shows the result. Aluminum, magnesium, and cement are particularly energy-intensive and thus vulnerable to a change in energy price. The same is true of the commodity polymers PE and PP. By contrast, materials such as CFRP and GFRP have a smaller energy component (here labor and equipment costs play a larger role) making them less sensitive to energy-price fluctuations.

Material price and product price. If energy price rises, the cost of making materials rises too. What happens to the price of products made from them? That depends on the material-intensity of the product. Figures 14.3 and 14.4 explain. The vertical axes are the price per unit weight ($/kg) of materials and of products: they give a common measure by which materials and products can be compared. The measure is a crude one but has the merit that it is unambiguous, easily determined, and bears some relationship to value added. A product with a price/kg that is only two or three times that of the materials of which it is made is material-intensive and is sensitive to change in material price; one with a price/kg that is 100 times that of its materials is insensitive to such change. On this scale the price

---

[1]Weighted average of six metals and six non-metallic commodities. Source: "Market indicators," *Economist*, January 15, 2000, cited by Lomberg (2001).

[2]Source: Annual energy reviews of the Energy Information Agency (EIA) of the US Department of Energy (www.eia.doe.gov).

[3]GDP per capita in constant $2000. Replotted from data assembled by Lomberg (2001).

[4]Correcting for inflation is not simple. For conversion factors, see www.measuringworth.com.

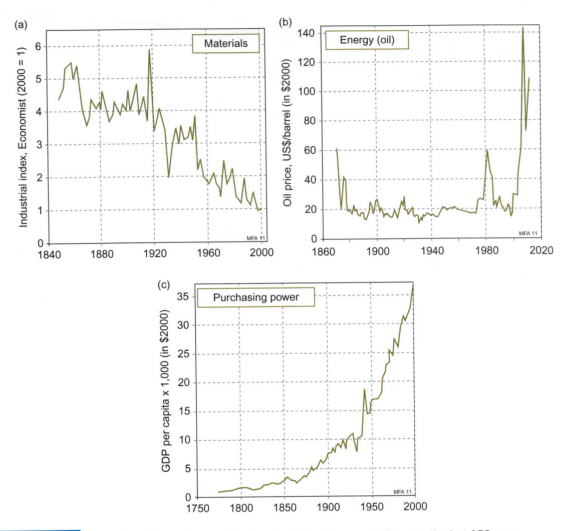

FIGURE 14.1 *The price of (a) materials, (b) oil, and (c) the GDP per capita, over the last 150 years*

per kg of a contact lens differs from that of a glass bottle by a factor of 10^5, even though both are made of almost the same glass. The cost per kg of a heart valve differs from that of a plastic bottle by a similar factor, even though both are made of polyethylene. There is obviously something to be learned here.

Look first at the price per unit weight of materials (Figure 14.3). The bulk "commodity" materials of construction and manufacturing lie in the shaded band; they all cost between $0.05 and $20/kg. Construction materials like concrete, brick, timber, and structural steel lie at the lower edge of the band; high-tech materials like titanium alloys lie at the upper. Polymers span a similar range: polyethylene at

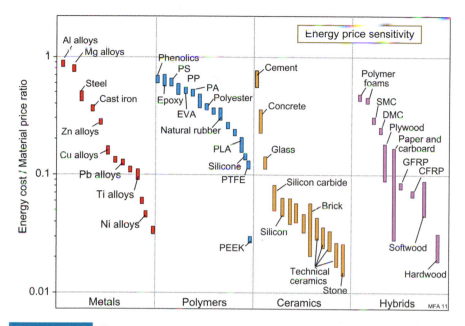

FIGURE 14.2 *The energy-price sensitivity of materials prices. Those with a value near 1 are the most vulnerable to rising energy price.*

the bottom, polytetrafluoroethylene (PTFE) near the top. Composites lie a little higher, with GFRP at the bottom and CFRP at the top of the range Engineering ceramics, at present, lie higher still, though this will change as production increases. Only low-volume "exotic" materials and precious metals lie much above the shaded band.

The price per kg of products (Figure 14.4) shows a different distribution. Eight market sectors are shown, covering much of the manufacturing industry. The shaded band on this figure spans the cost of commodity materials, exactly as on the previous figure. Sectors and their products within the shaded band have the characteristic that material cost is a major fraction of product price: up to 50% in civil construction, large marine structures, and some consumer packaging, and falling to perhaps 20% as the top of the band is approached (family car—around 20%). The value added in converting material to product in these sectors is relatively low, but the market volume is large. These constraints condition the choice of materials: they must meet modest performance requirements at the lowest possible cost. The associated market sectors generate a driving force for improved processing of conventional materials in order to reduce cost without loss of performance, or to increase reliability at no increase in cost, or (today) to reduce weight without serious cost penalty. For these sectors, incremental improvements in well-tried materials are more desirable than novel but less familiar high-performance materials. Slight improvements in steels, in manufacturing methods, or in lubrication technology are quickly assimilated and used.

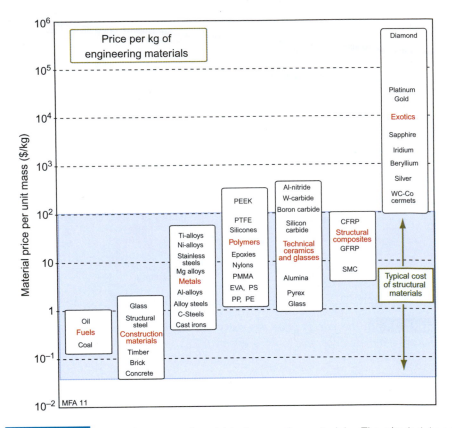

FIGURE 14.3 *The price per unit weight diagram for materials. The shaded band spans the range in which lies the most widely used commodity material of manufacture and construction.*

The products in the upper half of the diagram are technically more sophisticated. The materials of which they are made account for less than 10%—sometimes less than 1%—of the price of the product. The value added to the material during manufacturing is high. Product competitiveness is closely linked to material performance. Designers in these sectors have greater freedom in their choice of material and are more willing to adopt new ones if they have attractive properties. The objective here is *performance*, with cost as a secondary consideration. These smaller volume, higher value-added sectors drive the development of new or improved materials: materials that are lighter, or stiffer, or stronger, or tougher, or expand less, or conduct better—or all of these at once. They are often energy-intensive but are used in such small quantities that this is irrelevant.

The sectors have been ordered to form an ascending sequence, prompting the question: what does the horizontal axis measure? Many factors are involved here, one of which can be identified as "information content." The accumulated

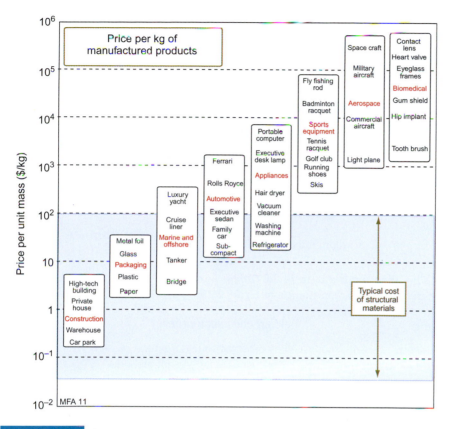

FIGURE 14.4 *The price per unit weight diagram for products. The shaded band spans the range in which lies most of the material of which they are made. Products in the shaded band are material-intensive; those above it are not.*

knowledge involved in the production of a contact lens or a heart valve is clearly greater than that in a beer glass or a plastic bottle. The sectors on the left make few demands on the materials they employ; those on the right push materials to their limits, and at the same time demand the highest quality and reliability. But there are also other factors: market size, competition (or lack of it), perceived value, fashion, and taste. For this reason the diagram should not be over-interpreted: it is a help in structuring information, but it is not a quantitative tool.

14.3 Carbon, energy, and GDP

The greater number of people that use a resource, the faster it is depleted. Global population—symbol P—is rising and so too is affluence, which we will for now equate to *GDP/P*, the gross domestic product per capita. Then material consumption grows as

$$\text{Material consumption} = P \times \frac{GDP}{P} \times \frac{Material}{GDP} \tag{14.1}$$

The last term is the *material intensity of GDP*: the amount of material consumed per unit of GDP. Energy consumption can be expanded in a similar way:

$$\text{Energy consumption} = P \times \frac{GDP}{P} \times \frac{Energy}{GDP} \tag{14.2}$$

Here the last term is the *energy intensity of GDP*. Finally, if we wish to focus on carbon emissions rather than energy, we have:

$$\text{Carbon emissions} = P \times \frac{GDP}{P} \times \frac{Energy}{GDP} \times \frac{Carbon}{Energy} \tag{14.3}$$

This time the last term is the *carbon intensity of energy*. We are back in the Material-Energy-Carbon triangle of Figure 2.19.

> ### Holding consumption level
>
> **Example:** The global population is expected to rise from the present 7.1 billion (mid 2011) to 8 billion by 2020. The global GDP, at present just under $1,000 per head of population, is expected to increase by 40% over the same period. By what factor must material intensity, measured by *Material/GDP*, decrease by 2020 to stabilize material consumption at the present-day level?
>
> **Answer:** The population is expected to grow by the factor $(8/7.1) = 1.13$. The GDP per head is expected to rise by the factor 1.4. If material consumption is to remain constant, the material required to generate one unit of GDP must fall by the factor $1/(1.13 \times 1.4) = 0.63$, a reduction of almost 40% from today's value.

Any move toward sustainability must stabilize and ultimately reduce the quantities on the left-hand side of these three equations. That means reducing at least one of the terms on the right. Population forecasts indicate continued growth at least in the near term, and to imagine that the world's population, *P*, will not continue to strive for increased affluence, *GDP/P*, is—well—unimaginable. You could ask: how much GDP is enough? Figure 14.5 is a plot of one measure of human fulfillment, the UN Human Development Index (HDI),[5] as a function of GDP per capita normalized to the value of the US dollar in 2008. There is clearly a plateau.

---

[5]The HDI is a compound index that combines provisions of education, level of health care, opportunity for individual development, and personal freedom. Other indices of "Happiness" show a similar pattern.

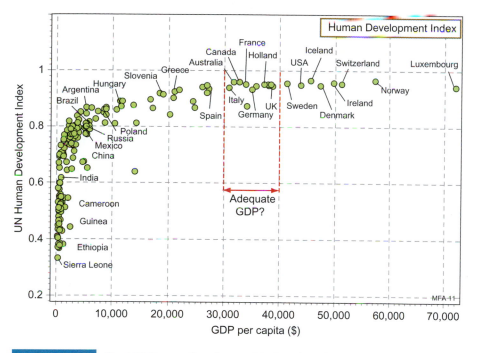

FIGURE 14.5   *The UN Human Development Index plotted against GDP/capita*

The plot suggests that a GDP per capita between $30,000 and $40,000 is more than enough to achieve it; any further increase buys no increase in HDI. But that, of course, is not how people see it. Attempting to limit GDP per capita is a non-starter.

So this leaves the last term in each equation as the target. Figure 14.6 is a plot of two of these—the energy/GDP and the CO_2/GDP—for 30 developed countries. The contours show the carbon emission per unit of energy, CO_2/energy. The three countries that do best are Sweden, Norway, and Iceland. Switzerland, Finland, and France are not far behind. Many other countries have values of these intensities that are less good, some by a factor of 5.

How do the best countries do it? Sweden, Norway, and Switzerland have terrain that allows for large scale hydroelectric generation, providing low CO_2 emissions, and they have efficient industry, which means they have a low energy/GDP. Iceland has a large geothermal capacity, meaning low CO_2, again, but it has an energy-intensive economy. Most countries do not have these energy resources. France, without much of either, does well with all three criteria through a combination of nuclear power and efficient manufacturing.

There is much more to this, of course. It depends on where the GDP comes from (agriculture, manufacturing, trade, finance, tourism …), on how hot or cold the country is (heating and air conditioning), on its natural resources (fertility of

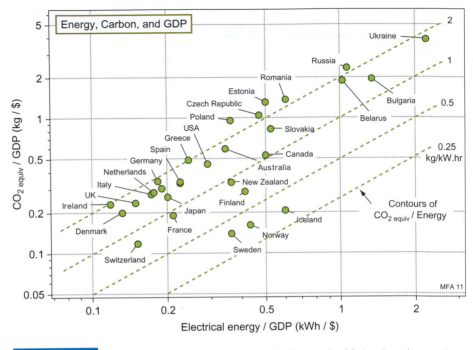

FIGURE 14.6   *The CO_2/GDP, Energy/GDP, and CO_2/Energy for 30 developed countries*

soil, mineral, oil and coal resources), and on the way it is governed. But Figure 14.6 provides some perspective: if all the countries plotted here could achieve the performance shown by France, global carbon emissions and energy consumption would fall by a factor of two straight away. But is GDP really the measure that we want?

14.4 Does GDP measure national wealth?[6]

"Economic growth," to an economist, means growth in Gross Domestic Product, GDP. Economic growth is seen as good, stagnation (no growth) or recession (negative growth) is seen as bad. But is GDP really the right measure? It measures the value of products and services delivered per year by an economy, but it does not include the diminution of what is called *natural capital*, meaning the resources we have inherited from nature: minerals and ores, clean air and water, productive land and sea, particularly with respect to forests and fisheries. As we have seen, there is evidence that the rate and way in which we now exploit natural capital is having a negative influence, and future growth in population and affluence will tend to make this worse.

[6]Dasgupta (2010) argues that measuring well-being by GDP, education, and health ignores the loss of natural capital through damage to the ecosystem.

If, through industrial activity, we are able to increase the accumulated *technical capital* (buildings, roads, industrial plants), *human capital* (health, education, scientific and engineering understanding), and *institutions* (schools, universities, hospitals, government), it might be argued that we have increased the nation's wealth. But we have left something out: the reduction or loss of natural capital that the industrial activity caused. This is a trade-off problem (Chapter 9): we want to maximize GDP while at the same time minimizing the loss of natural capital. We need a penalty function to determine whether, overall, we leave the nation richer or poorer. And to do that, we need an exchange constant that places a value on unit increase or decrease in natural capital. In some cases the diminution in natural capital and the activity causing it are obviously linked, allowing a value to be placed on it that the user of the natural capital must pay: a landfill tax (units: $ per metric ton of waste) is an example. But in other cases the diminution in natural capital is less easily quantified or attributed to a specific industrial site, and it has, until now, been externalized (Chapter 3), meaning that the consumer of the natural capital did not pay for it: emissions of carbon into the air or nitrates into water are examples. It then becomes necessary for governments to set a value for the exchange constant, as some have, in the form of a carbon tax (units: $ per metric ton of CO_2 to atmosphere) or a nitrate tax (units: $ per metric ton nitrate used as fertilizer).

When a value is assigned to a loss of natural capital, it is possible, at least in principle, to aggregate it with the growth in technical and human capital to assess whether the total wealth of the nation, per head, is really increasing or decreasing. Increasing GDP pushes it up, but loss of natural capital pulls it back, and if the population is increasing, there is a dilution as well because it is the wealth *per capita* that matters. In this picture one might see sustainable development by one generation as development that left as much total national wealth per person for the next generation as it had inherited from the previous one. In this view of sustainability the inescapable diminution of natural capital is offset by the increase in technical and human capability, allowing technical solutions that compensate for the loss of resources.

14.5 Forces for change: threats[7]

Now back to forces for change. Figure 14.7 is the roadmap for this section and the next. The central spine represents the design or redesign process, moving from market need through the steps of development (including choice of material and process) to the specification and ultimate production of products. The surrounding boxes summarize, on the left, some of the threats, those on the right, some of the opportunities that influence design decisions.

---

[7]For a full documentation and analysis of the facts listed in this section, see the book by Nielsen (2005) and the IPCC (2007) Report listed under Further reading.

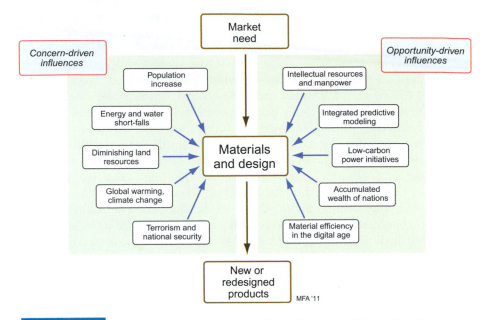

FIGURE 14.7 *Forces for change: threats on the left, opportunities on the right*

Population. For most of the history of man the population has been small and rising only very slowly (Figure 1.3) but in the last 150 years, the population has exploded, growing from 1.5 billion to over 7.1 billion. It is expected to rise above 9 billion before stabilizing, and there are already too many to live comfortably on the productive surface area of the planet (Chapter 1).

News-clip: Stemming population rise

Literacy and the population problem.
Researchers suggest that the single biggest factor influencing population growth rates is the educational status of women. If one wants to slow the rapid population growth in developing countries, the thinking goes, the highest priority should be putting more girls in school. If schools could be built and children educated at a rapid clip in all fast-growing countries, the global population in 2050 would hit 8.8 billion, the demographers projected. Under a far more pessimistic scenario in which education lags, the population in that year would likely approach 10 billion.

The New York Times, July 29, 2011

Implementing such an ambitious program is a huge engineering as well as educational challenge.

Energy. Fossil fuels provide most of the energy we now use (Chapter 2). The consumption has increased by a factor of 14 in the past hundred years and is still rising. Large reserves of easily accessible crude oil and gas are in the hands of a small group of nations: the energy-hungry OECD countries are heavily dependent on OPEC for supply. Such a one-sided market carries risk of supply shortages and volatile prices, with threat of economic disruption for the oil-importing nations (see the oil-price graph of Figure 14.1b).

News-clip: Consequences of imported energy dependence

Oil's roller coaster ride.

More than a century and a half after its discovery, oil continues to play an essential role in the global economy. Over the last decade oil price rose steadily from 2002 to 2007, soaring in 2008 to a peak of $147 a barrel before plummeting to $33 just five months later. In 2011, the increase in energy prices resembles the rise in 2008. Gasoline prices rose by nearly a third in 2010, and oil cost more than $100 a barrel for the first time in more than two years, driven by fears of extended Middle East supply disruptions and increased demand from an improving global economy.

The New York Times, August 23, 2011

Price fluctuations of this magnitude make economic planning very difficult.

Water. A second resource, water, is likely to exert an even greater constraint. Of the water on the earth's surface, only 2.5% is fresh water and much of this is inaccessible, locked up in ice and ground water. The growing population has created water shortages in many parts of the world: one third of the global population lacked adequate supplies at the end of the 20th century. Global warming is expected to make this worse. And without water it is not possible to grow crops or rear livestock.

News-clip: Emerging water crisis

Water security at extreme risk in Africa and Asia.

Clean, fresh water supply, which is fundamental to life and health—regardless of nationality, age, gender, profession or status—is at "extreme risk" in four African countries: Somalia, Mauritania, Sudan and Niger. The situation in Iraq, Uzbekistan, Pakistan, Egypt, Turkmenistan and Syria is equally precarious, says a new report emerging from the Water Security Risk Index of 165 countries around the world.

IDN (InDepthNews), June 25, 2010

Land. Industrialized countries use, on average, 6 hectares of productive land per person to support the way they live. The global average today is only 1.8 hectares per person and it is falling further as the population continues to grow. The developing nations aspire to a standard of living comparable with those of the developed nations but at least one resource—land—is insufficient to provide it unless there is a fundamental change in the way we live.

News-clip: Competition for land

African farmers losing land to investors.
Across Africa and the developing world, a new global land rush is gobbling up large expanses of arable land. Despite their ageless traditions, stunned villagers are discovering that African governments typically own their land and have been leasing it, often at bargain prices, to private investors and foreign governments for decades to come.

The New York Times, December 22, 2010

Land is a resource like any other.

Climate change. We are dumping more than 6 million tons of carbon into the atmosphere each year, pushing the atmospheric concentration from 270 ppm before the Industrial Revolution to above 400 ppm today (Figure 3.8). Global average temperature has risen by 0.75°C since the 19th century. Models for climate change suggest that, even if no further carbon entered the atmosphere (completely impossible scenario), the inertia of the ecosystem would still result in a further rise of 0.6°C before stabilizing. Increasingly precise meteorological modeling gives an idea of the consequences. They include melting of the Arctic ice cap (already evident), rising sea levels (already measureable), decreasing availability of fresh water, population migration, and loss of species (all already happening). A total rise of 4°C starts the melting of the Antarctic and Greenland ice caps, with more extreme rises in sea level and the flooding of low-lying countries. A total rise of 6°C has predicted consequences that one would rather not think about. All this, of course, is based on modeling, and there are those who dismiss its predictions.[8] The evidence that this progression has already started cannot be ignored. It is worth listening to the modelers—they are an early-warning system that did not exist in Malthus' day—they give us time to think out the best way to manage these changes.

National security. And finally there is a threat of a different nature. The increased availability and killing power of modern weapons together with the open nature of Western society makes it vulnerable to terrorism—the ability of small groups, even individuals, to inflict massive harm. This is a difficult devil to confront, when inspired, at least in part, by the large differences in wealth between rich and poor nations and by religious convictions.

---

[8]See, for example, the provocative but well-researched book by Lomberg (2001), or the less scientifically based treatise by Lawson (2008).

14.6 Opportunities

The concerns on the left of Figure 14.7 involve land, climate change, water, and food, but when you get down to it, it is *energy* that is the key to them all. The right-hand side of the figure shows some of the tools we have to deal with them.

Intellectual resources and manpower. We have enormous scientific and technological resources (the technical and human capital of Section 14.4). Recall (Chapter 2) that over three quarters of the scientists and engineers that have ever lived are alive and working today—it is one reason that the technologies of manufacturing, medicine, communication, surveillance, and defense have developed so rapidly in the last 50 years. They offer a resource that, if deployed to tackle the threats (as I am sure they will, ultimately, be), can do much to ease the pain. It is not clear, though, that technology alone can solve them all—some problems lie beyond its reach.[9]

Predictive modeling. As already said: it is better to *anticipate* than to *react*. Modeling is becoming one of the most valuable tools we have for mapping future strategy. To foresee a problem is the first step in tackling it. Fail to see it coming and you start with a self-inflicted handicap. Our ability to model material behavior from the scale of molecules to the scale of spacecraft, with intermediate scales of millimeters, microns, and meters, is very considerable. The challenge now is to connect the scales; linking them to enable the design of materials and structures that increase material efficiency and reduce energy consumption and emissions.

[9]See, for example, Hardin (1968).

the emerging industries that will address challenges in energy, national security, healthcare, and other areas. This initiative proposes a new national infrastructure for data sharing and analysis that will provide a greatly enhanced knowledgebase to scientists and engineers designing new materials. This effort will foster enhanced computational capabilities, data management, and an integrated engineering approach for materials deployment to better leverage and complement existing Federal investments.

US National Science and Technology Council, June 24, 2011

The central aim of the Initiative is to reduce the time it takes to develop new materials from 10 years to 2 years by integrated computational modeling, combinatorial experiments, and advanced data informatics.

Low-carbon power initiatives. As we saw in Chapter 10, it is possible but not easy to generate power on a global scale without burning carbon. Until now the cost of fossil fuel has been so low that there was little incentive to look elsewhere—all the alternatives were too expensive. Shifts to alternative sources, particularly nuclear power, become economically attractive when oil costs rises over $100 per barrel. Generating safe, low-carbon power on a large scale requires new technology and enormous capital investment, but it creates conditions under which materials innovation thrives, providing funds for research and development and subsidies for deployment of new systems.

Global wealth. How will it be paid for? The accumulated wealth of the developed and the oil-rich countries is huge. Estimates of the cost of a transition to low-carbon power, as an example, vary from 0.5% to 2% of global GDP, painful but not impossible. The banking crisis of 2008 and the loss of confidence in national economies in 2011 wiped more than that out of the global economy in those years.

News-clip: Global assets

Global wealth continues its strong recovery with $9 trillion gain.
Propelled by growth in nearly every region, global wealth continued a solid recovery in 2010, increasing by 8.0 percent, or $9 trillion, to a record of $121.8 trillion. That level was about $20 trillion above where it stood just two years prior during the depths of the financial crisis (of 2008), according to a new study by The Boston Consulting Group (BCG).

BCG, May 31, 2011

The digital economy. One way to reduce the man-made stress on the environment is to develop ways to use less material and energy per unit of GDP, issues we discussed earlier. The digital economy—one in which the trade of information is as central as that of goods—is one way of achieving this. It has already made a major contribution to increased material efficiency.

News-clip: Internet innovation

Can New York rival Silicon Valley?

The entire world is now a rival to Silicon Valley. No country, state, region, nor city has a lock on innovation in technology anymore. The Internet has made this so, and there's no going back. We will see Apples and Facebooks get built in China, India, Brazil, Eastern Europe, Western Europe, the Middle East, Africa, and plenty of other places. Until recently, "technology" was largely about moving electrons on wires. Now, "technology" is about building all kinds of interesting applications on top of the Internet. An increasing number of engineers and entrepreneurs are applying their ideas and energy to creating compelling services on the Internet.

***The New York Times**, August 3, 2011*

Adaptability. Perhaps the greatest unknown is the extent of human adaptability. Tools and resources exist to plan and implement strategies to deal with the concerns of Figure 14.7. The acceptability of the strategies, however, remains an unknown. But we should not forget those words of Thomas Malthus, uttered 214 years ago: *"The power of population is so superior to the power of the earth to produce subsistence for man, that premature death must in some shape or other visit the human race."* He didn't say when, just that it would. Others starting from quite different standpoints have converged on a similar view. The full weight of scientific evidence and of advanced climate modeling now points that way too. It begins to look as if Malthus might be right. Can we, in the 21st century, find ways to prove him wrong?

14.7 Summary and conclusions

For the past 150 years materials have become cheaper and labor more expensive. The ratio of the two has a profound effect on the way we develop, use, and value materials. In countries where this ratio is still high (India, China), materials are conserved and recovered. Where the ratio is low (most of the developed world), materials are used inefficiently and, after use, discarded. It is frequently cheaper to buy a new product than to pay the labor cost of having the old one repaired.

Things are now changing. One change is the steep rise in the price of energy shown in Figure 14.1b—at least part of it is here to stay. And there are others, all acting as drivers for change. The global population has now grown so large that there is insufficient productive land to support it adequately. Global warming caused by atmospheric carbon is causing climate change. Allowing it to ramp up further will have harmful effects on health, agriculture, water availability, and weather, all with economic penalties. The dependence on oil, much of it sourced from a few oil-rich countries, has bred a dependence that is increasingly trouble-some for the oil-hungry developed nations.

All have the effect of making materials more precious. They change the ways in which we design with and use materials. They create incentives to develop materi-als that better meet the constraints imposed by the design, to care for the health of materials in service, and to value them when they are retired so that they can be retrained, so to speak, to do a new job.

There are many challenges here. Many relate both to the materials and to the ways in which we design with them. Devising the best ways to tackle them is a task for you, the next generation of materials scientists and engineers.

14.8 Further reading

Dasgupta, P. (2010), "Nature's role in sustaining economic development," *Phil. Trans. Roy. Soc. B*, 365, pp 5–11. *(Dasgupta argues that measuring well-being by GDP, education, and health ignores the loss of natural capital through damage to the ecosystem.)*

Flannery, T. (2010), *Here on earth*, The Text Publishing Company, Victoria, Australia. ISBN 978-1-92165-666-8. *(The latest of a series of books by Flannery documenting man's impact on the environment)*

Hamilton, C. (2010), *Requiem for a species: why we resist the truth about climate change*, Allen and Unwin, NSW, Australia. ISBN 978-1-74237-210-5. *(A profoundly pessimistic view of the future for mankind)*

Hardin, G. (1968), "The tragedy of the commons," *Science*, 162, pp 1243–1248. *(Hardin argues, with convincing examples, that relying on technology alone to solve the problems listed on the left of Figure 14.7—particularly that of population growth— is a mistake. Some problems do not have technical solutions.)*

Lawson, N. (2008), *An appeal to reason—a cool look at global warming*, Duckworth Overlook, New York, NY, USA. ISBN 978-1-5902-0084-1. *(A politician's view of it all, starting "I readily admit that I am not a scientist." Worth examining—these are the views of the people who govern us.)*

Lomberg, B. (2001), *The skeptical environmentalist—measuring the real state of the world*, Cambridge University Press, Cambridge, UK. ISBN 0-521-01068-3. *(A provocative and carefully researched challenge to the now widely held view of the origins and consequences of climate change, helpful in forming your own view of the state of the world)*

Lomberg, B. editor (2010), *Smart solutions to climate change: comparing costs and benefits*, Cambridge University Press, Cambridge, UK. ISBN 978-0-52113-856-7. *(A multiauthor text in the form of a debate—"The case for …," "The case against …"—covering climate engineering, carbon sequestration, methane mitigation, market- and policy-driven adaptation)*

Lovelock, J. (2009) *The vanishing face of Gaia*, Penguin Books, Ltd. London, UK. ISBN 978-0-141-03925-1. *(Lovelock reminds us that humans are just another species, and that species have appeared and disappeared since the beginnings of life on earth.)*

MacKay, D.J.C.(2008), *Sustainable energy—without the hot air*, UIT Press, Cambridge, UK and www.withouthotair.com/. ISBN 978-0-9544529-3-3. *(MacKay's analysis of the potential for renewable energy is particularly revealing.)*

Malthus, T. R. (1798), "An essay on the principle of population," London, Printed for Johnson, St. Paul's Church-yard. www.ac.wwu.edu/~stephan/malthus/malthus. *(The originator of the proposition that population growth must ultimately be limited by resource availability)*

Meadows D.H., Meadows D.L., Randers J, and Behrens W.W. (1972), *The limits to growth*, Universe Books, New York, USA. *(The "Club of Rome" report that triggered the first of a sequence of debates in the 20th century on the ultimate limits imposed by resource depletion)*

Meadows, D.H., Meadows, D.L., and Randers, J. (1992), *Beyond the Limits*, Earthscan, London, UK. ISSN 0896-0615. *(The authors of "The Limits to Growth" use updated data and information to restate the case that continued population growth and consumption might outstrip the earth's natural capacities.)*

Nielsen, R. (2005), *The little green handbook*, Scribe Publications Pty Ltd, Carlton North, Victoria, Australia. ISBN 1-9207-6930-7. *(A cold-blooded presentation and analysis of hard facts about population, land, and water resources, energy, and social trends)*

Plimer, I. (2009), *Heaven and earth—global warming: the missing science*, Connor Publishing, Ballam, Victoria, Australia. ISBN 978-1-92142-114-3. *(Ian Plimer, Professor of Geology at the University of Adelaide, examines the history of climate change over a geological timescale, pointing out that everything that is happening now has happened many times in the past. A geo-historical perspective, very thoroughly documented.)*

Schmidt-Bleek, F. (1997), *How much environment does the human being need—factor 10—the measure for an ecological economy*, Deutscher Taschenbuchverlag, Munich, Germany. ISBN 3-936279-00-4. *(Both Schmidt-Bleek and von Weizsäcker, referenced below, argue that sustainable development will require a drastic reduction in material consumption.)*

von Weizsäcker, E., Lovins, A.B., and Lovins, L.H. (1997), *Factor four: doubling wealth, halving resource use*, Earthscan Publications, London, UK. ISBN 1-85383-406-8; ISBN-13: 978-1-85383406-6. *(Both von Weizsäcker and Schmidt-Bleek, referenced above, argue that sustainable development will require a drastic reduction in material consumption.)*

Walker, G. and King, D. (2008), *The hot topic—how to tackle global warming and still keep the lights on*, Bloomsbury Publishing, London, UK. ISBN 9780-7475-9395-9. *(A readable paraphrase of the IPCC (2007) Report on Climate Change, with discussion of the political obstacles to finding solutions—a topic on which Dr. King is an expert—he was, for some years, chief science advisor to the UK Government)*

14.9 Exercises

E14.1. **Energy price sensitivity 1.** An aluminum saucepan weighs 1.2 kg, and costs $10. The embodied energy of aluminum is 210 MJ/kg. If the cost of industrial electric power doubles from $0.0125/MJ to $0.025/MJ, how much will it change the cost of the saucepan?

E14.2. **Energy price sensitivity 2.** An MP3 player weighs 100 grams, and costs $120. The embodied energy of assembled integrated electronics of this sort is about 2,000 MJ/kg. If the cost of industrial electric power doubles from $0.0125/MJ to $0.025/MJ, how much will it change the cost of the MP3 player?

E14.3. **Achieving carbon neutrality.** The global population is expected to rise from the present 7.1 billion (mid 2011) to 8 billion by 2020. The global GDP, at present just under $1,000 per head of population, is expected to increase by 40% over the same period. If, over that period, improved manufacturing technology reduces the energy intensity of GDP by 20% by what factor must the carbon intensity of energy production be reduced to keep carbon emissions at today's level?

E14.4. Cars in Cuba are repaired and continue in use when they are 25 years old. The average life of a car in the US is 13 years. What is the underlying reason for this difference? Might this change?

E14.5. Use the Internet to research the meaning and history of "The precautionary principle." Select and report the definition that, in your view, best sums up the meaning.

E14.6. Use the Internet to research examples of problems that are approached by predictive modeling.

Exercises using the CES Edu and State of the World databases

E14.7. The energy-price sensitivity diagram, Figure 14.2, was made with data that were current at the time this book went to press. Material prices fluctuate, and data for embodied energies are continually being refined; annual updates for both appear in the CES Edu database. Use the current version of the software to reproduce the diagram, taking the price of energy to be $0.01/MJ. To do so, use the Advanced facility to plot the ratio of the cost of the embodied energy for a material divided by its price on the y-axis:

[Embodied energy, primary production] * 0.01/[Price]

To split the results by material class, choose **x-axis / Advanced / Trees**, and then successively click

Metals and alloys   Polymers and elastomers   Ceramics and glasses   Hybrids etc.

E14.8. A CES database called Wanner's State of the World Database, created from publiclly available data by Prof. Alexander Wanner of the University of Karlsruhe, can be downloaded free from the Granta Design web site. It contains economic and development data for 226 countries. Figures 14.5 and 14.6 of the text were constructed using it. Use it to explore other aspects of the economic development of nations.

As an example: make a bar chart of oil consumption/Oil production to find out which countries are self-sufficient in oil and which depend on imports.

Material profiles

CONTENTS

15.1 Introduction and synopsis

You can't calculate anything quantitative without numbers. This chapter provides them. It takes the form of single-page data sheets for 63 of the materials used in the greatest quantities. The sheets list the annual production and reserves, the embodied energy, process energies, and the carbon footprints associated with each. They list, too, the general, mechanical, thermal, and electrical properties, important because it is these that determine the environmental consequences of the use-phase of life. And they provide basic information about recycling at end of life.

459

Table 15.1   The material profiles

Metals and alloys	Polymers and elastomers	Ceramics and glasses	Hybrids: composites, foams, wood, and paper	Man-made and natural fibers
Low-carbon steel	ABS	Brick	CFRP	Aramid
Low-alloy steel	Polyamide PA	Stone	GFRP	Carbon
Stainless steel	Polypropylene, PP	Concrete	Sheet molding compound	Glass
Cast iron	Polyethylene, PE	Alumina	Bulk molding compound	Coir
Aluminum alloys	Polycarbonate, PC	Soda-lime glass	Furan-based composites	Cotton
Magnesium alloys	PET	Borosilicate glass	Rigid polymer foam	Flax
Titanium alloys	PVC		Flexible polymer foam	Hemp
Copper alloys	Polystyrene, PS		Softwood, along grain	Jute
Lead alloys	Polylactide, PLA		Softwood, across grain	Kenaf
Zinc alloys	PHB		Hardwood, along grain	Ramie
Nickel-chrome alloys	Epoxy		Hardwood, across grain	Sisal
Nickel-based super alloys	Polyester		Plywood	Straw
Silver	Phenolic		Paper and cardboard	
Gold	Natural rubber, NR			
	Butyl rubber, BR			
	EVA			
	Polychloroprene, CR			

Each section starts with a brief introduction to a material family: *metals and alloys, polymers and elastomers, ceramics and glasses, hybrids* (composites, wood, and paper), and *man-made and natural fibers*. Within a section the material profiles appear in the order shown in Table 15.1. Each data sheet has a description and an image of the material in use. The data that follow are listed as *ranges* spanning the typical spread of values of the property. When a single ("point") value is needed for exercises or projects, use the mean of the two values listed on the sheet.

A warning. The engineering properties of materials—their mechanical, thermal, and electrical attributes—are well characterized. They are measured with sophisticated equipment according to internationally accepted standards and are reported in widely accessible handbooks and databases. They are not *exact*, but their precision—when it matters—is reported; many are known to three-figure accuracy, some to more.

The eco-properties of materials are not like that. There are no sophisticated test machines to measure embodied energies or carbon footprints. International standards, detailed in ISO 14040 and discussed in Chapter 3, lay out procedures, but these are vague and not easily applied. The differences in the process routes by which materials are made in different production facilities, the differences in energy mix in electrical power in different countries, the difficulty in setting system boundaries, and the procedural problems in assessing energy, CO_2, and the other eco-attributes all contribute to the imprecision.

So just how far can values for eco-properties be trusted? An analysis, documented in Section 6.3, suggests a standard deviation of $\pm 10\%$ at best. To be significantly different, values of eco-properties must differ by *at least* 20%. The difference between materials with really large and really small values of embodied energy or carbon footprint is a factor of 1,000 or more, so the imprecision still allows firm distinctions to be drawn. But when the differences are small, other factors such as the recycled content of the material, its durability (and thus life time), and the ability to recycle it at end of life are more significant in making the selection.

15.2 Metals and alloys

Most of the elements in the periodic table are metals. Metals have "free" electrons, making them good electrical and thermal conductors and good reflectors of light. The metals used in product design are, almost without exception, alloys. Steels (iron with carbon and a host of other alloying elements to make them harder, tougher, or more corrosion resistant) account for more than 90% of all the metals consumed in the world; aluminum comes next, followed by copper, nickel, zinc, titanium, magnesium, and tungsten, in the order that was illustrated in Figure 2.3.

Compared to all other classes of materials, metals are stiff, strong, and tough, but they are heavy. They have relatively high melting points, allowing some metal alloys to be used at temperatures as high as 2,200°C. Only one metal—gold— is chemically stable as a metal; all the others will, given the chance, react with oxygen, sulfur, phosphorous, or carbon to form compounds that are more stable than the metal itself, making them vulnerable to corrosion. There are numerous ways of preventing or slowing this to an acceptable level, but they require maintenance. Metals are ductile, allowing them to be shaped by rolling, forging, drawing, and extruding; they are easy to machine with precision; and they can be joined in many different ways. This allows a flexibility of design with metals that is only now being challenged by polymers.

Primary production of metals is energy intensive. Many, among them aluminum, magnesium, and titanium, require at least twice as much energy per unit weight (or five times more per unit volume) than commodity polymers. But most metals can be recycled efficiently, and the energy required to do so is much less than that required for primary production. Some are toxic, particularly the heavy metals— lead, cadmium, and mercury. Some, however, are so inert that they can be implanted in the body: stainless steel, cobalt alloys, and certain alloys of titanium are examples. The following pages contain the data sheets for the 14 metals and alloys in the order in which they are listed Table 15.1.

Low carbon steel

The material. Think of steel and you think of railroads, oil rigs, tankers, and skyscrapers. And what you are thinking of is not just steel, it is carbon steel. That is the metal that made them possible; nothing else is at the same time so strong, so tough, so easily formed—and so cheap. Carbon steels are alloys of iron with carbon and, often a little manganese, nickel, and silicon. Low carbon or "mild" steels have the least carbon—less than 0.25%. They are relatively soft; easily rolled to plate, I-sections, or rod (for reinforcing concrete); and are the cheapest of all structural metals. It is these that are used on a huge scale for reinforcement, for steel-framed buildings and ship plate, and the like.

Composition
Fe/0.02−0.3C

General properties

Density	7,800	– 7,900	kg/m^3
Price	0.68	– 0.74	USD/kg

Mechanical properties

Young's modulus	200	– 215	GPa
Yield strength (elastic limit)	250	– 395	MPa
Tensile strength	345	– 580	MPa
Elongation	26	– 47	%
Hardness—Vickers	107	– 172	HV
Fatigue strength at 10^7 cycles	203	– 293	MPa
Fracture toughness	41	– 82	MPa·m$^{1/2}$

Thermal properties

Melting point	1480	– 1530	°C
Maximum service temperature	350	– 400	°C
Thermal conductor or insulator?	Good conductor		
Thermal conductivity	49	– 54	W/m·K
Specific heat capacity	460	– 505	J/kg·K
Thermal expansion coefficient	11.5	– 13	μstrain/°C

Electrical properties

Electrical conductor or insulator?	Good conductor		
Electrical resistivity	15	– 20	μohm·cm

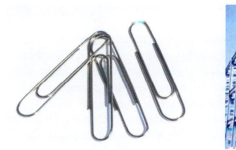 

Mild steel, the world's most versatile material

Eco properties: material

Global production, main component	2.3×10^9			metric ton/yr
Reserves	160×10^9			metric ton
Embodied energy, primary production	25	–	28	MJ/kg
CO_2 footprint, primary production	1.7	–	1.9	kg/kg
Water usage	23	–	69	L/kg
Eco-indicator	106			millipoints/kg

Eco properties: processing

Casting energy	11	–	12.2	MJ/kg
Casting CO_2 footprint	0.8	–	0.9	kg/kg
Deformation processing energy	3.0	–	6.0	MJ/kg
Deformation processing CO_2 footprint	0.22	–	0.46	kg/kg

End of life

Embodied energy, recycling	6.6	–	8.0	MJ/kg
CO_2 footprint, recycling	0.4	–	0.48	kg/kg
Recycle fraction in current supply	40	–	44	%

Typical uses. Low carbon steels are used so widely that no list would be complete without them. Reinforcement of concrete, steel sections for construction, sheet for roofing, car body panels, cans, and pressed sheet products give an idea of the scope.

Low alloy steel

The material. Addition of manganese (Mn), nickel (Ni), molybdenum (Mo), or chromium (Cr) to steel lowers the critical quench rate comes to create martensite, allowing thick sections to be hardened and then tempered. By adding some vanadium, V, as well, dispersion of carbides can be created that provides strength while retaining toughness and ductility. Chrome-molybdenum steels such as AIS 4140 are used for aircraft tubing and other high-strength parts. Chrome-vanadium steels are used for crank and propeller shafts and high-quality tools. Steels alloyed for this purpose are called *low-alloy steels*, and the property they have is called *hardenability*.

Composition
Fe/ < 1.0 C/ < 2.5 Cr/ < 2.5 Ni/ < 2.5 Mo/ < 2.5 V

General properties

Density	7,800	– 7,900	kg/m^3
Price	0.9	– 1.1	USD/kg

Mechanical properties

Young's modulus	205	– 217	GPa
Yield strength (elastic limit)	400	– 1,500	MPa
Tensile strength	550	– 1,760	MPa
Elongation	3	– 38	%
Hardness—Vickers	140	– 692	HV
Fatigue strength at 10^7 cycles	248	– 700	MPa
Fracture toughness	14	– 200	MPa $\cdot$ m$^{1/2}$

Thermal properties

Melting point	1,380	– 1,530	°C
Maximum service temperature	500	– 550	°C
Thermal conductor or insulator?	Good conductor		
Thermal conductivity	34	– 55	W/m $\cdot$ K
Specific heat capacity	410	– 530	J/kg $\cdot$ K
Thermal expansion coefficient	10.5	– 13.5	μstrain/°C

Electrical properties

Electrical conductor or insulator?	Good conductor		
Electrical resistivity	15	– 35	μohm $\cdot$ cm

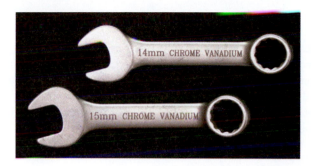

Low-alloy chrome-molybdenum and chrome-vanadium steels are used for high-quality tools, bike frames, and automobile engine and transmission components.

Eco properties: material

Global production, main component	2.3×10^9			metric ton/yr
Reserves	159×10^9			metric ton
Embodied energy, primary production	31	–	34	MJ/kg
CO_2 footprint, primary production	1.9	–	2.1	kg/kg
Water usage	37	–	111	L/kg
Eco-indicator	200			millipoints/kg

Eco properties: processing

Casting energy	10.9	–	12	MJ/kg
Casting CO_2 footprint	0.8	–	0.9	kg/kg
Deformation processing energy	7	–	14	MJ/kg
Deformation processing CO_2 footprint	0.5	–	1.1	kg/kg

End of life

Embodied energy, recycling	7.7	–	9.5	MJ/kg
CO_2 footprint, recycling	0.47	–	0.57	kg/kg
Recycle fraction in current supply	40	–	44	%

Typical uses. Springs, tools, ball bearings, rollers, crankshafts, gears, connecting rods, knives and scissors, pressure vessels.

Stainless steel

The material. Stainless steels are alloys of iron with chromium, nickel, and—often—four or five other elements. The alloying transmutes plain carbon steel that rusts and is prone to brittleness below room temperature into a material that does neither. Indeed, most stainless steels resist corrosion in most normal environments, and those that are "austenitic" (like AISI 302, 304, and 316) remain ductile to the lowest of temperatures.

Composition
Fe/ < 0.25 C/16−30 Cr/3.5−37 Ni/ <10 Mn+ Si,P,S (+ N for 200 series)

General properties

Density	7,600	– 8,100	kg/m^3
Price	8.2	– 9.1	USD/kg

Mechanical properties

Young's modulus	189	– 210	GPa
Yield strength (elastic limit)	170	– 1,000	MPa
Tensile strength	480	– 2,240	MPa
Elongation	5	– 70	%
Hardness—Vickers	130	– 570	HV
Fatigue strength at 10^7 cycles	175	– 753	MPa
Fracture toughness	62	– 150	MPa$\cdot$m$^{1/2}$

Thermal properties

Melting point	1,370	– 1,450	°C
Maximum service temperature	750	– 820	°C
Thermal conductor or insulator?	Poor conductor		
Thermal conductivity	12	– 24	W/m$\cdot$K
Specific heat capacity	450	– 530	J/kg$\cdot$K
Thermal expansion coefficient	13	– 20	μstrain/°C

Electrical properties

Electrical conductor or insulator?	Good conductor		
Electrical resistivity	64	– 107	μohm$\cdot$cm

On the left: Siemens toaster in brushed austenitic stainless steel (by Porsche Design)
On the right: scissors in ferritic stainless steel; it is magnetic, austenitic stainless is not

Eco properties: material

Global production, main component	30×10^6			metric ton/yr
Reserves	2.5×10^9			metric ton
Embodied energy, primary production	81	–	88	MJ/kg
CO_2 footprint, primary production	4.7	–	5.2	kg/kg
Water usage	112	–	336	L/kg
Eco-indicator	310			millipoints/kg

Eco properties: processing

Casting energy	10.0	–	12.0	MJ/kg
Casting CO_2 footprint	0.8	–	0.9	kg/kg
Deformation processing energy	5.0	–	11.4	MJ/kg
Deformation processing CO_2 footprint	0.4	–	0.8	kg/kg

End of life

Embodied energy, recycling	11	–	13	MJ/kg
CO_2 footprint, recycling	0.65	–	0.8	kg/kg
Recycle fraction in current supply	35	–	40	%

Typical uses. Railway cars, trucks, trailers, food-processing equipment, sinks, stoves, cooking utensils, cutlery, flatware, scissors and knives, architectural metal-work, laundry equipment, chemical-processing equipment, jet-engine parts, surgical tools, furnace and boiler components, oil-burner parts, petroleum-processing equipment, dairy equipment, heat-treating equipment, automotive trim. Structural uses in corrosive environments—for example, nuclear plants, ships, offshore oil installations, underwater cables, and pipes.

Cast iron, ductile (nodular)

The material. The foundations of modern industrial society are set, so to speak, in cast iron: it is the material that made the industrial revolution possible. Today it holds a second honor: that of being the cheapest of all engineering metals. Cast iron contains at least 2% carbon—most have 3% to 4%—and from 1% to 3% silicon. The carbon makes the iron very fluid when molten, allowing it to be cast to intricate shapes. There are five classes of cast iron: gray, white, ductile (or nodular), malleable, and alloy. The two that are most used are gray and ductile. This record is for ductile cast iron.

Composition
Fe/3.2−4.1% C/1.8−2.8% Si/<0.8% Mn/<0.1% P/<0.03% S

General properties

Density	7,050	–	7,250	kg/m^3
Price	0.5	–	0.65	USD/kg

Mechanical properties

Young's modulus	165	–	180	GPa
Yield strength (elastic limit)	250	–	680	MPa
Tensile strength	410	–	830	MPa
Elongation	3	–	18	%
Hardness—Vickers	115	–	320	HV
Fatigue strength at 10^7 cycles	180	–	330	MPa
Fracture toughness	22	–	54	MPa·m$^{1/2}$

Thermal properties

Melting point	1,130	–	1,250	°C
Maximum service temperature	350	–	745	°C
Thermal conductor or insulator?	Good conductor			
Thermal conductivity	29	–	44	W/m·K
Specific heat capacity	460	–	495	J/kg·K
Thermal expansion coefficient	10	–	12.5	μstrain/°C

Electrical properties

Electrical conductor or insulator?	Good conductor			
Electrical resistivity	49	–	56	μohm·cm

Ductile or malleable cast irons are used for heavily loaded parts such as gears and automotive suspension components.

Eco properties: material

Global production, main component	2.3×10^9			metric ton/yr
Reserves	159×10^9			metric ton
Embodied energy, primary production	16	–	20	MJ/kg
CO_2 footprint, primary production	1.4	–	1.6	kg/kg
Water usage	13	–	39	L/kg
Eco-indicator	80			millipoints/kg

Eco properties: processing

Casting energy	10	–	11	MJ/kg
Casting CO_2 footprint	0.75	–	0.83	kg/kg

End of life

Embodied energy, recycling	10	–	11	MJ/kg
CO_2 footprint, recycling	0.48	–	0.55	kg/kg
Recycle fraction in current supply	66	–	72	%

Typical uses. Brake discs and drums; bearings; camshafts; cylinder liners; piston rings; machine tool structural parts; engine blocks, gears, crankshafts; heavy-duty gear cases; pipe joints; pump casings; components in rock crushers.

Aluminum alloys

The material. Aluminum was once so rare and precious that the Emperor Napoleon III of France had a set of cutlery made from it that cost him more than silver. But that was 1860; today, nearly 150 years later, aluminum spoons are things you throw away—a testament to our ability to be both technically creative and wasteful. Aluminum, the first of the "light alloys" (with magnesium and titanium), is the third most abundant metal in the earth's crust (after iron and silicon), but extracting it costs much energy. It has grown to be the second most important metal in the economy (steel comes first), and the mainstay of the aerospace industry.

Composition
Al + alloying elements, e.g., Mg, Mn, Cr, Cu, Zn, Zr, Li

General properties

Density	2,500	–	2,900	kg/m^3
Price	2.4	–	2.7	USD/kg

Mechanical properties

Young's modulus	68	–	82	GPa
Yield strength (elastic limit)	30	–	550	MPa
Tensile strength	58	–	550	MPa
Elongation	1	–	44	%
Hardness—Vickers	12	–	150	HV
Fatigue strength at 10^7 cycles	22	–	160	MPa
Fracture toughness	22	–	35	$MPa \cdot m^{1/2}$

Thermal properties

Melting point	495	–	640	°C
Maximum service temperature	120	–	200	°C
Thermal conductor or insulator?	Good conductor			
Thermal conductivity	76	–	240	$W/m \cdot K$
Specific heat capacity	860	–	990	$J/kg \cdot K$
Thermal expansion coefficient	21	–	24	μstrain/°C

Electrical properties

Electrical conductor or insulator?	Good conductor			
Electrical resistivity	2.5	–	6	μohm · cm

Cast and wrought aluminum alloys, examples of the wide range of properties of this, the most widely used light alloy

Eco properties: material

Global production, main component	37×10^6			metric ton/yr
Reserves	2.0×10^9			metric ton
Embodied energy, primary production	200	–	220	MJ/kg
CO_2 footprint, primary production	11	–	13	kg/kg
Water usage	495	–	1,490	L/kg
Eco-indicator	710			millipoints/kg

Eco properties: processing

Casting energy	11	–	12.2	MJ/kg
Casting CO_2 footprint	0.82	–	0.91	kg/kg
Deformation processing energy	3.3	–	6.8	MJ/kg
Deformation processing CO_2 footprint	0.19	–	0.23	kg/kg

End of life

Embodied energy, recycling	22	–	30	MJ/kg
CO_2 footprint, recycling	1.9	–	2.3	kg/kg
Recycle fraction in current supply	41	–	45	%

Typical uses. Aerospace engineering; automotive engineering—pistons, clutch housings, exhaust manifolds; sports equipment such as golf clubs and bicycles; die cast chassis for household and electronic products; siding for buildings; reflecting coatings for mirrors, foil for containers and packaging; beverage cans; electrical and thermal conductors.

Magnesium alloys

The material. Magnesium is a metal almost indistinguishable from aluminum in color, but of lower density. It is the lightest of the light-metal trio (with partners aluminum and titanium) and light it is: a computer case made from magnesium is barely two thirds as heavy as one made from aluminum. Aluminum and magnesium are the mainstays of airframe engineering. Only beryllium is lighter, but its expense and potential toxicity limit its use to special applications only. Magnesium is flammable, but this is only a problem when it is in the form of powder or very thin sheet. It costs more than aluminum but nothing like the cost of titanium.

Composition
Mg + alloying elements, e.g., Al, Mn, Si, Zn, Cu, Li, rare-earth elements

General properties

Density	1,740	– 1,950	kg/m^3
Price	4.7	– 5.1	USD/kg

Mechanical properties

Young's modulus	42	– 47	GPa
Yield strength (elastic limit)	70	– 400	MPa
Tensile strength	185	– 475	MPa
Elongation	3.5	– 18	%
Hardness—Vickers	35	– 135	HV
Fatigue strength at 10^7 cycles	60	– 225	MPa
Fracture toughness	12	– 18	$MPa \cdot m^{1/2}$

Thermal properties

Melting point	447	– 649	°C
Maximum service temperature	120	– 200	°C
Thermal conductor or insulator?	Good conductor		
Thermal conductivity	50	– 156	$W/m \cdot K$
Specific heat capacity	955	– 1,060	$J/kg \cdot K$
Thermal expansion coefficient	24.6	– 28	µstrain/°C

Electrical properties

Electrical conductor or insulator?	Good conductor		
Electrical resistivity	4.15	– 15	$µohm \cdot cm$

Magnesium, the lightest of the light alloys, is increasingly used for components of cars and other vehicles.

Eco properties: material

Global production, main component	0.6×10^6			metric ton/yr
Reserves	1×10^9			metric ton
Embodied energy, primary production	300	–	330	MJ/kg
CO_2 footprint, primary production	34	–	38	kg/kg
Water usage	500	–	1,500	L/kg
Eco-indicator	1,500			millipoints/kg

Eco properties: processing

Casting energy	11.1	–	12.3	MJ/kg
Casting CO_2 footprint	0.83	–	0.92	kg/kg
Deformation processing energy	3.8	–	6.6	MJ/kg
Deformation processing CO_2 footprint	0.3	–	0.5	kg/kg

End of life

Embodied energy, recycling	23	–	26	MJ/kg
CO_2 footprint, recycling	2.6	–	3.2	kg/kg
Recycle fraction in current supply	36	–	41	%

Typical uses. Aerospace; automotive; sports goods such as bicycles; nuclear fuel cans; vibration damping and shielding of machine tools; engine case castings; crank cases; transmission housings; automotive wheels; ladders; housings for electronic equipment, particularly mobile phone and portable computer chassis; camera bodies; office equipment; marine hardware and lawnmowers.

Titanium alloys

The material. The alloys of titanium have the highest strength-to-weight ratio of any structural metal, about 25% greater than the best alloys of aluminum or steel. Titanium alloys can be used at temperatures up to 500°C—compressor blades of aircraft turbines are made of them. They have unusually poor thermal and electrical conductivity, and low expansion coefficients. The alloy Ti 6%/Al 4% V is used in quantities that exceed those of all other titanium alloys combined. The data in this record describe it and similar alloys.

Composition
Ti+ alloying elements, e.g., Al, Zr, Cr, Mo, Si, Sn, Ni, Fe, V

General properties
Density	4,400	– 4,800	kg/m^3
Price	57	– 63	USD/kg

Mechanical properties
Young's modulus	110	– 120	GPa
Yield strength (elastic limit)	750	– 1,200	MPa
Tensile strength	800	– 1,450	MPa
Elongation	5	– 10	%
Hardness—Vickers	267	– 380	HV
Fatigue strength at 10^7 cycles	589	– 617	MPa
Fracture toughness	55	– 70	MPa$\cdot$m$^{1/2}$

Thermal properties
Melting point	1,480	– 1,680	°C
Maximum service temperature	450	– 500	°C
Thermal conductor or insulator?	Poor conductor		
Thermal conductivity	7	– 14	W/m$\cdot$K
Specific heat capacity	645	– 655	J/kg$\cdot$K
Thermal expansion coefficient	8.9	– 9.6	µstrain/°C

Electrical properties
Electrical conductor or insulator?	Good conductor		
Electrical resistivity	100	– 170	µohm$\cdot$cm

Adiabatic heating heats the air in the compressor to about 500°C, requiring the use of titanium alloys for the blades (Reproduced with the permission of Rolls-Royce plc, copyright © Rolls-Royce plc 2004)

Eco properties: material

Global production, main component	2.0×10^5			metric ton/yr
Reserves	720×10^6			metric ton
Embodied energy, primary production	650	–	720	MJ/kg
CO_2 footprint, primary production	44	–	49	kg/kg
Water usage	470	–	1,410	L/kg
Eco-indicator	3,450			millipoints/kg

Eco properties: processing

Casting energy	12.6	–	13.9	MJ/kg
Casting CO_2	0.9	–	1.0	kg/kg
Forging, rolling energy	14	–	15	MJ/kg
Forging, rolling CO_2	1.1	–	1.2	kg/kg

End of life

Embodied energy, recycling	78	–	96	MJ/kg
CO_2 footprint, recycling	4.7	–	5.7	kg/kg
Recycle fraction in current supply	21	–	24	%

Typical uses. Aircraft turbine blades; general aerospace applications; chemical engineering; pressure vessels; high-performance automotive parts such as connecting rods; heat exchangers; bioengineering; medical; missile fuel tanks; compressors; valve bodies; light springs, surgical implants; marine hardware, paper-pulp equipment; sports equipment such as golf clubs and bicycles.

Copper alloys

The material. In Victorian times you washed your clothes in a "copper"—a vat or tank of beaten copper sheet, heated over a fire; the device exploited both the high ductility and the thermal conductivity of the material. Copper has a distinguished place in the history of civilization: it enabled the technology of the Bronze age (3,000 BC–1,000 BC). It is used in many forms: as pure copper, as copper-zinc alloys (brasses), as copper-tin alloys (bronzes), and as copper-nickel and copper-beryllium. The designation of "copper" is used when the percentage of copper is more than 99.3%.

Composition
Cu with up to 40% Zn or 30% Sn, Al, or Ni

General properties
Density	8,930	–	9,140	kg/m^3
Price	7.0	–	7.6	USD/kg

Mechanical properties
Young's modulus	112	–	148	GPa
Yield strength (elastic limit)	30	–	350	MPa
Tensile strength	100	–	400	MPa
Elongation	3	–	50	%
Hardness—Vickers	44	–	180	HV
Fatigue strength at 10^7 cycles	60	–	130	MPa
Fracture toughness	30	–	90	MPa$\cdot$m$^{1/2}$

Thermal properties
Melting point	982	–	1,080	°C
Maximum service temperature	180	–	300	°C
Thermal conductor or insulator?	Good conductor			
Thermal conductivity	160	–	390	W/m$\cdot$K
Specific heat capacity	372	–	388	J/kg$\cdot$K
Thermal expansion coefficient	16.9	–	18	µstrain/°C

Electrical properties
Electrical conductor or insulator?	Good conductor			
Electrical resistivity	1.74	–	5.01	µohm$\cdot$cm

Copper and brass are exceptionally ductile and can be worked to complex shapes.

Eco properties: material

Global production, main component	15×10^6		metric ton/yr
Reserves	540×10^6		metric ton
Embodied energy, primary production	56	– 62	MJ/kg
CO_2 footprint, primary production	3.5	– 3.9	kg/kg
Water usage	268	– 297	L/kg
Eco-indicator	2,170		millipoints/kg

Eco properties: processing

Casting energy	8.6	– 9.5	MJ/kg
Casting CO_2 footprint	0.65	– 0.72	kg/kg
Deformation processing energy	0.7	– 1.2	MJ/kg
Deformation processing CO_2 footprint	0.11	– 0.2	kg/kg

End of life

Embodied energy, recycling	12	– 15	MJ/kg
CO_2 footprint, recycling	0.75	– 0.9	kg/kg
Recycle fraction in current supply	40	– 45	%

Typical uses. Electrical wiring, cables, bus bars, high strength, high conductivity wires and sections, overheads lines, contact wires, resistance-welding electrodes, terminals, high conductivity items for use at raised temperatures, heat exchangers, coinage, pans, kettles and boilers, plates for etching and engraving, roofing and architecture, cast sculptures, pumps, valves, marine propellers.

Lead alloys

The material. When the Romans conquered Britain in 43 AD they discovered rich deposits of lead ore and started a mining and refining industry that was to continue for 1,000 years (the symbol for lead—Pb—derives from its Latin name: *Plumbum*). They used it for pipes, cisterns, and roofs—this last a use that continues to the present day. The biggest single use of lead (70% of the total) is as electrodes in lead acid batteries.

Composition
Pb + 0 to 25% Sb or 0 to 60% Sn, sometimes with some Ca

General properties
Density	11,300	— 11,400	kg/m^3
Price	2.35	— 2.5	USD/kg

Mechanical properties
Young's modulus	12.5	— 15	GPa
Yield strength (elastic limit)	8	— 14	MPa
Tensile strength	12	— 20	MPa
Elongation	30	— 60	%
Hardness—Vickers	3	— 6.5	HV
Fatigue strength at 10^7 cycles	2	— 9	MPa
Fracture toughness	5	— 15	MPa·m$^{1/2}$

Thermal properties
Melting point	183	— 31	°C
Maximum service temperature	70	— 120	°C
Thermal conductor or insulator?	Good conductor		
Thermal conductivity	22	— 36	W/m·K
Specific heat capacity	122	— 145	J/kg·K
Thermal expansion coefficient	18	— 32	μstrain/°C

Electrical properties
Electrical conductor or insulator?	Good conductor		
Electrical resistivity	20	— 22	μohm·cm

Lead weathers well and is exceptionally durable and corrosion-resistant.

Eco properties: material

Global production, main component	3.9×10^6		metric ton/yr
Reserves	79×10^6		metric ton
Embodied energy, primary production	26	– 28	MJ/kg
CO_2 footprint, primary production	1.8	– 2.2	kg/kg
Water usage	175	– 525	L/kg
Eco-indicator	290		millipoints/kg

Eco properties: processing

Casting energy	5.1	– 5.7	MJ/kg
Casting CO_2 footprint	0.38	– 0.43	kg/kg
Deformation processing energy	0.32	– 0.38	MJ/kg
Deformation processing CO_2 footprint	0.024	– 0.03	kg/kg

End of life

Embodied energy, recycling	6.7	– 8.2	MJ/kg
CO_2 footprint, recycling	0.406	– 0.49	kg/kg
Recycle fraction in current supply	69	– 76	%

Typical uses. Roofs, wall cladding, pipe work, window seals, and flooring in buildings; sculpture and table wear as pewter; solder for electrical circuits and for mechanical joining, bearings, printing type, ammunition, pigments, X-ray shielding, corrosion-resistant material in the chemical industry, and electrodes for lead acid batteries.

Zinc die-casting alloys

The material. Zinc is a bluish-white metal with a low melting point (420°C). The slang in French for a bar or pub is *le zinc*; bar counters in France used to be covered in zinc—many still are—to protect them from the ravages of wine and beer. Bar surfaces have complex shapes—a flat top, curved profiles, rounded or profiled edges. These two sentences say much about zinc: it is ductile; it is hygienic; it survives exposure to acids (wine), to alkalis (cleaning fluids), and to misuse (upset customers). These remain among the reasons it is still used today. Another is the "castability" of zinc alloys—their low melting point and fluidity gives them a leading place in die-casting.

Composition
Zn + 3-30% Al, typically, often with up to 3% Cu

General properties

Density	4,950	– 7,000	kg/m^3
Price	2.4	– 2.6	USD/kg

Mechanical properties

Young's modulus	68	– 100	GPa
Yield strength (elastic limit)	80	– 450	MPa
Tensile strength	135	– 510	MPa
Compressive strength	80	– 450	MPa
Elongation	1	– 30	%
Hardness—Vickers	55	– 160	HV
Fatigue strength at 10^7 cycles	20	– 160	MPa
Fracture toughness	10	– 70	MPa·m$^{1/2}$

Thermal properties

Melting point	375	– 492	°C
Maximum service temperature	80	– 110	°C
Thermal conductor or insulator?	Good conductor		
Thermal conductivity	100	– 130	W/m·K
Specific heat capacity	405	– 535	J/kg·K
Thermal expansion coefficient	23	– 28	μstrain/°C

Electrical properties

Electrical conductor or insulator?	Good conductor		
Electrical resistivity	5.4	– 7.2	μohm·cm

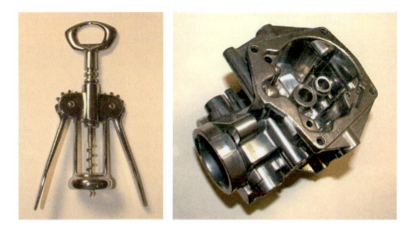

Zinc die-castings are cheap, have high surface finish, and can be complex in shape. On the left, a corkscrew; everything except the screw itself is die cast zinc alloy. On the right, a carburetor body.

Eco properties: material

Global production, main component	1.1×10^7			metric ton/yr
Reserves	1.8×10^9			metric ton
Embodied energy, primary production	57	–	63	MJ/kg
CO_2 footprint, primary production	3.9	–	4.3	kg/kg
Water usage	160	–	521	L/kg
Eco-indicator	472			millipoints/kg

Eco properties: processing

Casting energy	6.5	–	7.2	MJ/kg
Casting CO_2	0.48	–	0.53	kg/kg
Deformation processing energy	1.2	–	2.5	MJ/kg
Deformation processing CO_2 footprint	0.1	–	0.2	kg/kg

End of life

Embodied energy, recycling	10	–	12	MJ/kg
CO_2 footprint, recycling	0.62	–	0.72	kg/kg
Recycle fraction in current supply	21	–	25	%

Typical uses. Die castings; automotive parts and tools; gears; household goods; office equipment; building hardware; padlocks; toys; business machines; sound reproduction equipment; hydraulic valves; pneumatic valves; soldering; handles.

Nickel-chromium alloys

The material. Nickel forms a wide range of alloys, valued by the chemical engineering and food processing industries for their resistance to corrosion, and by the makers of furnaces and high temperature equipment for their ability to retain useful strength at temperatures up to 1,200°C. Typical of these are the nickel-chromium (Ni-Cr) alloys, often containing some iron (Fe) as well. The chromium increases the already good resistance to corrosion and oxidation by creating a surface film of Cr_2O_3, the same film that makes stainless steel stainless. The data given here are for nickel-chromium alloys. There are separate records for stainless steel and nickel-based super alloys.

Composition
Ni+ 10 to 30% Cr+ 0 to 10% Fe

General properties

Density	8,300	–	8,500	kg/m^3
Price	33	–	36	USD/kg

Mechanical properties

Young's modulus	200	–	220	GPa
Yield strength (elastic limit)	365	–	460	MPa
Tensile strength	615	–	760	MPa
Elongation	20	–	35	%
Hardness—Vickers	160	–	200	HV
Fatigue strength at 10^7 cycles	245	–	380	MPa
Fracture toughness	80	–	110	MPa·m$^{1/2}$

Thermal properties

Melting point	1,350	–	1,430	°C
Maximum service temperature	900	–	1,000	°C
Thermal conductor or insulator?	Poor conductor			
Thermal conductivity	9	–	15	W/m·K
Specific heat capacity	430	–	450	J/kg·K
Thermal expansion coefficient	12	–	14	μstrain/°C

Electrical properties

Electrical conductor or insulator?	Good conductor			
Electrical resistivity	102	–	114	μohm·cm

The heating elements of the dryer and toaster are Nichrome—an alloy of nickel and chromium that resists oxidation well.

Eco properties: material

Global production, main component	1.5×10^6		metric ton/yr
Reserves	63×10^6		metric ton
Embodied energy, primary production	173	– 190	MJ/kg
CO_2 footprint, primary production	11	– 12	kg/kg
Water usage	564	– 620	L/kg
Eco-indicator	2,800		millipoints/kg

Eco properties: processing

Casting energy	10.4	– 11.5	MJ/kg
Casting CO_2	0.78	– 0.85	kg/kg
Forging, rolling energy	3.3	– 6.5	MJ/kg
Forging, rolling CO_2	0.25	– 0.53	kg/kg

End of life

Embodied energy, recycling	30	– 36	MJ/kg
CO_2 footprint, recycling	1.8	– 2.2	kg/kg
Recycle fraction in current supply	29	– 32	%

Typical uses. Heating elements and furnace windings; bi-metallic strips; thermo-couples; springs; food processing equipment; chemical engineering equipment.

Nickel-based super alloys

The material. With a name like super alloy there has to be something special here. There is. Super alloy is a name applied to nickel-based, iron-based, and cobalt-based alloys that combine exceptional high-temperature strength with excellent corrosion and oxidation resistance. Without them, jet engines would not be practical: they can carry load continuously at temperatures up to 1,200°C. The nickel-based super alloys are the ultimate metallic cocktail: nickel with a good slug of chromium and lesser shots of cobalt, aluminum, titanium, molybdenum, zirconium, and iron. The data in this record span the range of high-performance nickel-based super alloys.

Composition
Ni+ 10 to 25% Cr+ Ti, Al, Co, Mo, Zr, B, and Fe in varying proportions

General properties
Density	7,750	– 8,650	kg/m$_3$
Price	31	– 33	USD/kg

Mechanical properties
Young's modulus	150	– 245	GPa
Yield strength (elastic limit)	300	– 1.9e3	MPa
Tensile strength	400	– 2.1e3	MPa
Compressive strength	300	– 1.9e3	MPa
Elongation	0.5	– 60	%
Hardness—Vickers	200	– 600	HV
Fatigue strength at 10^7 cycles	135	– 900	MPa
Fracture toughness	65	– 110	MPa$\cdot$m$^{1/2}$

Thermal properties
Melting point	1,280	– 1,410	°C
Maximum service temperature	900	– 1,200	°C
Thermal conductor or insulator?	Good conductor		
Thermal conductivity	8	– 17	W/m$\cdot$K
Specific heat capacity	380	– 490	J/kg$\cdot$K
Thermal expansion coefficient	9	– 16	µstrain/°C

Electrical properties
Electrical conductor or insulator?	Poor conductor		
Electrical resistivity	84	– 240	µohm$\cdot$cm

Nickel is the principal ingredient of super alloys used for high-temperature turbines and chemical engineering equipment. On the left, a gas turbine (image courtesy of Kawasaki Turbines). On the right, a single super alloy blade.

Eco properties: material

Global production, main component	1.5×10^6			metric ton/yr
Reserves	7.0×10^7			metric ton
Embodied energy, primary production	221	–	244	MJ/kg
CO_2 footprint, primary production	11	–	12.1	kg/kg
Water usage	134	–	484	L/kg
Eco-indicator	2,830			millipoints/kg

Eco properties: processing

Casting energy	10.0	–	11.0	MJ/kg
Casting CO_2	0.75	–	0.80	kg/kg
Forging, rolling energy	4.2	–	4.5	MJ/kg
Forging, rolling CO_2	0.31	–	0.34	kg/kg

End of life

Embodied energy, recycling	33.8	–	37.5	MJ/kg
CO_2 footprint, recycling	1.97	–	2.3	kg/kg
Recycle fraction in current supply	22	–	26	%

Typical uses. Blades, disks, and combustion chambers in turbines and jet engines; rocket engines; general structural aerospace applications; light springs; high-temperature chemical engineering equipment; bioengineering and medical.

Silver

The material. If gold is the king of metals, silver is the queen. Silver is a soft white metal with the highest electrical and thermal conductivities of any metal. It occurs as native silver but most is produced as a by-product of copper, lead, and zinc refining.

Silver is valued as a precious metal, used for jewelry, tableware, musical instruments, and currency. It has many industrial applications as electrical contacts and conductors, as a catalyst, in photographic film and photovoltaics, in batteries, in pharmaceuticals, in lead-free solders and in control rods of nuclear reactors. The important industrial uses of silver compete with its desirability as a hedge against inflation, leading to volatile pricing.

Silver is non-toxic and has useful antibacterial properties.

Composition
> 99.9 Ag

General properties
Density	10,500	–	10,600	kg/m^3
Price	1,850	–	2,000	USD/kg

Mechanical properties
Young's modulus	69	–	73	GPa
Yield strength (elastic limit)	190	–	300	MPa
Tensile strength	255	–	340	MPa
Elongation	1	–	2	% strain
Hardness—Vickers	90	–	110	HV
Fatigue strength at 10^7 cycles	100	–	170	MPa
Fracture toughness	40	–	60	$MPa \cdot m^{1/2}$

Thermal properties
Melting point	957	–	967	°C
Maximum service temperature	100	–	190	°C
Thermal conductor or insulator?	Good conductor			
Thermal conductivity	416	–	422	$W/m \cdot °C$
Specific heat capacity	230	–	240	$J/kg \cdot °C$
Thermal expansion coefficient	19.5	–	19.9	µstrain/°C

Electrical properties
Electrical conductor or insulator?	Good conductor			
Electrical resistivity	1.67	–	1.81	µohm · cm

Silver bullion and solid silver conductors

Eco properties: material

Global production, main component	21,800		metric ton/yr
Reserves	495,000		metric ton
Embodied energy, primary production	1,400	– 1,550	MJ/kg
CO_2 footprint, primary production	95	– 105	kg/kg
Water usage	1,150	– 3,460	L/kg

Eco properties: processing

Casting energy	6.8	– 7.3	MJ/kg
Casting CO_2 footprint	0.5	– 0.56	kg/kg
Deformation processing energy	1.7	– 3.5	MJ/kg
Deformation processing CO_2 footprint	0.13	– 0.26	kg/kg

End of life

Embodied energy, recycling	140	– 170	MJ/kg
CO_2 footprint, recycling	8.4	– 10.2	kg/kg
Recycle fraction in current supply	65	– 67	%

Typical uses. Electrical contacts, linings for chemical reactor vessels, linings for heavy-duty journal bearings, jewelry, tableware, photography, batteries, pharmaceuticals, lead-free solders, and control rods of nuclear reactors.

Gold

The material. Gold is the king of metals. Its name has entered the language as a symbol of perfection, achievement, good fortune, and security: golden ages, golden ratios, golden rules, gold standards, golden crowns, golden girl, go for gold, gilded lives. When times are bad and investments insecure, investors turn to gold. Its rich color, its malleability, its resistance to almost all corrosion, and its sheer scarcity combine to make it the most desirable of metals.

Some 90% of the global production of gold gets squirreled away as bullion and jewelry. The remaining 10% plays vital roles in the electronics industry as interconnectors and as surface layers on connectors. Gold's high electrical conductivity and resistance to corrosion allow extreme miniaturization. Its use in dentistry, once large, is diminishing.

Composition
$> 99.5\%$ Au

General properties

Density	19,300	— 19,350	kg/m^3
Price	53,100	— 53,500	USD/kg

Mechanical properties

Young's modulus	77	— 81	GPa
Yield strength (elastic limit)	165	— 205	MPa
Tensile strength	180	— 220	MPa
Elongation	2	— 6	% strain
Hardness—Vickers	50	— 70	HV
Fatigue strength at 10^7 cycles	70	— 110	MPa
Fracture toughness	40	— 70	MPa$\cdot$m$^{1/2}$

Thermal properties

Melting point	1,060	— 1,065	°C
Maximum service temperature	130	— 220	°C
Thermal conductor or insulator?	Good conductor		
Thermal conductivity	305	— 319	W/m$\cdot$°C
Specific heat capacity	125	— 135	J/kg$\cdot$°C
Thermal expansion coefficient	13.5	— 14.5	μstrain/°C

Electrical properties

Electrical conductor or insulator?	Good conductor		
Electrical resistivity	2	— 3	μohm$\cdot$cm

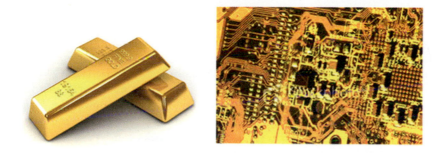

Gold bullion and gold in electronics, from which it can be recycled

Eco properties: material

Global production, main component	2,4000		metric ton/yr
Reserves	50,000		metric ton
Embodied energy, primary production	240,000	– 265,000	MJ/kg
CO_2 footprint, primary production	25,000	– 28,000	kg/kg
Water usage	126,000	– 378,000	L/kg

Eco properties: processing

Casting energy	6.0	– 6.6	MJ/kg
Casting CO_2 footprint	0.45	– 0.5	kg/kg
Deformation processing energy	1.0	– 1.7	MJ/kg
Deformation processing CO_2 footprint	0.11	– 0.14	kg/kg

End of life

Embodied energy, recycling	650	– 719	MJ/kg
CO_2 footprint, recycling	41	– 45	kg/kg
Recycle fraction in current supply	40	– 44	%

Typical uses. Jewelry; interconnects, printed circuit board edge connectors; electrical contacts; lining for chemical equipment; coinage; bullion; plating for space satellites; toning silver images in photography.

15.3 Polymers

Polymers are the chemist's contribution to the materials world. The facts that most are derived from oil (a nonrenewable resource) and that they are difficult to dispose of at the end of their lives (they don't easily degrade) have led to a view that polymers are environmental villains. There is some truth in this, but the present problems are solvable. Using oil to make polymers is a better primary use than just burning it for heat; the heat can still be recovered from the polymer at the end of its life. There are alternatives to oil: polymer feedstocks can be synthesized from agricultural products (notably starch and sugar, via methanol and ethanol). And thermoplastics can be (and, to some extent, are) recycled, provided they are not contaminated.

Thermoplastics soften when heated and harden again to their original state when cooled. This allows them to be molded to complex shapes. Some are crystalline, some amorphous, some a mixture of both. Most accept coloring agents and fillers, and many can be blended to give a wide range of physical, visual, and tactile effects. Their sensitivity to sunlight is decreased by adding UV filters, and their flammability is decreased by adding flame retardants. The properties of thermoplastics can be controlled by chain length (measured by molecular weight), by chain branching, by degree of crystallinity, and by blending and plasticizing. As the molecular weight increases, the resin becomes stiffer, tougher, and more resistant to chemicals, but it is more difficult to mold. Crystalline polymers tend to have better chemical resistance, greater stability at high temperature, and better creep resistance than those that are amorphous. For transparency, the polymer must be amorphous; partial crystallinity gives translucency.

Thermosets. If you are a do-it-yourself type, you have Araldite in your toolbox—two tubes, one a sticky resin, the other an even stickier hardener. Mix and warm them and they react to give a stiff, strong, durable polymer, stuck to whatever you put it on. Araldite is an epoxy resin. It is a typical *thermoset*: a resin that polymerizes when catalyzed and heated. When reheated thermosets do not melt—they degrade instead. Polyurethane thermosets are produced in the highest volume; polyesters come second; phenolics (Bakelite), epoxies, and silicones follow, and—not surprisingly—the cost rises in the same order. Once shaped, thermosets cannot be reshaped. They cannot easily be recycled.

Elastomers were originally called "rubbers" because they could rub out pencil marks, but that is the least of their many remarkable and useful properties. Unlike any other class of solid, elastomers remember their shape when they are stretched and return to it when released. This allows *conformability*—hence their use for seals and gaskets. High damping elastomers recover slowly; those with low damping snap back, returning the energy it took to stretch them—hence their use for springs, catapults, bungee cords, and other bouncy things. Conformability gives elastomers high friction on rough surfaces, part of the reason (along with comfort) that they are used for pneumatic tires and footwear, their two largest markets.

Elastomers are thermosets—once cured, you can't remold them or recycle them, a major problem with car tires. Tricks can be used to make them behave in some ways like thermoplastics (TPOs—thermoplastic elastomers, of which EVA is an example). Blending or copolymerizing elastomer molecules with a thermoplastic like polypropylene, PP, gives separated clumps of elastomer stuck together by a film of PP (Santoprene). The material behaves like an elastomer, but if heated so that the PP melts, it can be remolded and even recycled.

Profiles for 17 polymers follow, in the order in which they appear in Table 15.1.

Acrylonitrile butadiene styrene (ABS)

The material. ABS (acrylonitrile-butadiene-styrene) is tough, resilient, and easily molded. It is usually opaque, although some grades can now be transparent, and it can be given vivid colors. ABS-PVC alloys are tougher than standard ABS and, in self-extinguishing grades, are used for the casings of power tools.

Composition:
$(CH_2\text{-}CH\text{-}C_6H_4)_n$

General properties

Density	1,010	–	1,210	kg/m^3
Price	2.4	–	2.6	USD/kg

Mechanical properties

Young's modulus	1.1	–	2.9	GPa
Yield strength (elastic limit)	18.5	–	51	MPa
Tensile strength	27.6	–	55.2	MPa
Elongation	1.5	–	100	%
Hardness—Vickers	5.6	–	15.3	HV
Fatigue strength at 10^7 cycles	11	–	22.1	MPa
Fracture toughness	1.19	–	4.29	$MPa \cdot m^{1/2}$

Thermal properties

Glass temperature	88	–	128	°C
Maximum service temperature	62	–	77	°C
Thermal conductor or insulator?	Good insulator			
Thermal conductivity	0.188	–	0.335	$W/m \cdot K$
Specific heat capacity	1,390	–	1,920	$J/kg \cdot K$
Thermal expansion coefficient	84.6	–	234	µstrain/°C

Electrical properties

Electrical conductor or insulator?	Good insulator			
Electrical resistivity	3.3×10^{21}	–	3×10^{22}	µohm·cm
Dielectric constant	2.8	–	3.2	
Dissipation factor	0.003	–	0.007	
Dielectric strength	13.8	–	21.7	10^6 V/m

Eco properties: material

Global production, main component	5.7×10^6			metric ton/yr
Embodied energy, primary production	90	–	99	MJ/kg
CO_2 footprint, primary production	3.6	–	4.0	kg/kg
Water usage	250	–	277	L/kg
Eco-indicator	350			millipoints/kg

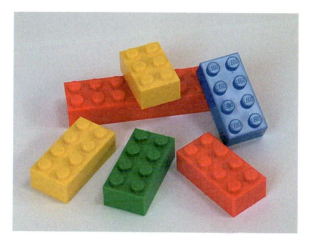

ABS allows detailed moldings, accepts color well, and is non-toxic and tough.

Eco properties: processing

Polymer molding energy	18	–	20	MJ/kg
Polymer molding CO_2 footprint	1.4	–	1.5	kg/kg
Polymer extrusion energy	5.8	–	6.4	MJ/kg
Polymer extrusion CO_2 footprint	0.44	–	0.48	kg/kg

End of life

Embodied energy, recycling	42	–	51	MJ/kg
CO_2 footprint, recycling	2.5	–	3.1	kg/kg
Recycle fraction in current supply	3.8	–	4.2	%
Heat of combustion	37.6	–	39	MJ/kg
Combustion CO_2	3.1	–	3.2	kg/kg
Recycle mark				

Other

Typical uses. Safety helmets; camper tops; automotive instrument panels and other interior components; pipe fittings; home-security devices and housings for small appliances; communications equipment; business machines; plumbing hardware; automobile grilles; wheel covers; mirror housings; refrigerator liners; luggage shells; tote trays; mower shrouds; boat hulls; large components for recreational vehicles; weather seals; glass beading; refrigerator breaker strips; conduit; pipe for drain-waste-vent (DWV) systems.

Polyamides (Nylons, PA)

The material. Back in 1945, the war in Europe had just ended, and the two most prized luxuries were cigarettes and nylons. Nylon (PA) can be drawn to fibers as fine as silk, and was widely used as a substitute for it. Today, newer fibers have eroded its dominance in garment design, but nylon-fiber ropes, and nylon as reinforcement for rubber (in car tires) and other polymers (PTFE, for roofs) remains important. It is used in product design for tough casings, frames, and handles, and—reinforced with glass—as bearings gears and other load-bearing parts. There are many grades (Nylon 6, Nylon 66, Nylon 11, ...) each with slightly different properties.

Composition
$(NH(CH_2)_5CO)_n$

General properties

Density	1,120	– 1,140	kg/m^3
Price	3.9	– 4.3	USD/kg

Mechanical properties

Young's modulus	2.62	– 3.2	GPa
Yield strength (elastic limit)	50	– 94.8	MPa
Tensile strength	90	– 165	MPa
Elongation	30	– 100	%
Hardness—Vickers	25.8	– 28.4	HV
Fatigue strength at 10^7 cycles	36	– 66	MPa
Fracture toughness	2.2	– 5.6	$MPa \cdot m^{1/2}$

Thermal properties

Melting point	210	– 220	°C
Maximum service temperature	110	– 140	°C
Thermal conductor or insulator?	Good insulator		
Thermal conductivity	0.23	– 0.25	$W/m \cdot K$
Specific heat capacity	1,600	– 1,660	$J/kg \cdot K$
Thermal expansion coefficient	144	– 149	$\mu strain/°C$

Electrical properties

Electrical conductor or insulator?	Good insulator		
Electrical resistivity	1.5×10^{19}	– 1.4×10^{20}	$\mu ohm \cdot cm$
Dielectric constant	3.7	– 3.9	
Dissipation factor	0.014	– 0.03	
Dielectric strength	15.1	– 16.4	10^6 V/m

Polyamides are tough, wear and corrosion resistant, and can be colored.

Eco properties: material

Global production, main component	3.7×10^6		metric ton/yr
Embodied energy, primary production	116	– 129	MJ/kg
CO_2 footprint, primary production	7.6	– 8.3	kg/kg
Water usage	250	– 280	L/kg
Eco-indicator	495		millipoints/kg

Eco properties: processing

Polymer molding energy	21	– 23	MJ/kg
Polymer molding CO_2 footprint	1.55	– 1.7	kg/kg
Polymer extrusion energy	5.9	– 6.5	MJ/kg
Polymer extrusion CO_2 footprint	0.44	– 0.49	kg/kg

End of life

Embodied energy, recycling	38	– 47	MJ/kg
CO_2 footprint, recycling	2.31	– 2.8	kg/kg
Recycle fraction in current supply	0.5	– 1	%
Heat of combustion	30	– 32	MJ/kg
Combustion CO_2	2.3	– 2.4	kg/kg
Recycle mark			

Other

Typical uses. Light duty gears, bushings, sprockets and bearings; electrical equipment housings, lenses, containers, tanks, tubing, furniture casters, plumbing connections, bicycle wheel covers, ketchup bottles, chairs, toothbrush bristles, handles, bearings, food packaging. Nylons are used as hot-melt adhesives for book bindings; as fibers—ropes, fishing line, carpeting, car upholstery, and stockings; as aramid fibers—cables, ropes, protective clothing, air filtration bags, and electrical insulation.

Polypropylene (PP)

The material. Polypropylene, PP, first produced commercially in 1958, is the younger brother of polyethylene—a very similar molecule with similar price, processing methods, and application. Like PE it is produced in very large quantities (more than 40 million tons per year in 2010), growing at nearly 10% per year, and like PE, its molecule lengths and side branches can be tailored by clever catalysis, giving precise control of impact strength, and of the properties that influence molding and drawing. In its pure form polypropylene is flammable and degrades in sunlight. Fire retardants make it slow to burn and stabilizers give it extreme stability, both to UV radiation and to fresh and salt water and most aqueous solutions.

Composition

$(CH_2-CH(CH_3))_n$

General properties

Density	890	–	910	kg/m^3
Price	1.85	–	2.05	USD/kg

Mechanical properties

Young's modulus	0.9	–	1.55	GPa
Yield strength (elastic limit)	21	–	37	MPa
Tensile strength	28	–	41	MPa
Elongation	100	–	600	%
Hardness—Vickers	6.2	–	11	HV
Fatigue strength at 10^7 cycles	11	–	17	MPa
Fracture toughness	3	–	4.5	$MPa \cdot m^{1/2}$

Thermal properties

Melting point	150	–	175	°C
Maximum service temperature	100	–	115	°C
Thermal conductor or insulator?	Good insulator			
Thermal conductivity	0.11	–	0.17	$W/m \cdot K$
Specific heat capacity	1,870	–	1,960	$J/kg \cdot K$
Thermal expansion coefficient	122	–	180	μstrain/°C

Electrical properties

Electrical conductor or insulator?	Good insulator			
Electrical resistivity	3.3×10^{22}	–	3×10^{23}	$\mu ohm \cdot cm$
Dielectric constant	2.1	–	2.3	
Dissipation factor	3×10^{-4}	–	7×10^{-4}	
Dielectric strength	22.7	–	24.6	10^6 V/m

Polypropylene is widely used in household products.

Eco properties: material

Global production, main component	44×10^6		metric ton/yr
Embodied energy, primary production	75	– 83	MJ/kg
CO_2 footprint, primary production	2.9	– 3.2	kg/kg
Water usage	189	– 209	L/kg
Eco-indicator	254		millipoints/kg

Eco properties: processing

Polymer molding energy	20.4	– 22.6	MJ/kg
Polymer molding CO_2 footprint	1.5	– 1.7	kg/kg
Polymer extrusion energy	5.9	– 6.5	MJ/kg
Polymer extrusion CO_2 footprint	0.44	– 0.49	kg/kg

End of life

Embodied energy, recycling	45	– 55	MJ/kg
CO_2 footprint, recycling	2.0	– 22	kg/kg
Recycle fraction in current supply	5	– 6	%
Heat of combustion	44	– 46	MJ/kg
Combustion CO_2	3.1	– 3.2	kg/kg
Recycle mark			

Typical uses. Ropes, general polymer engineering, automobile air ducting, parcel shelving and air-cleaners, garden furniture, washing machine tank, wet-cell battery cases, pipes and pipe fittings, beer bottle crates, chair shells, capacitor dielectrics, cable insulation, kitchen kettles, car bumpers, shatter proof glasses, crates, suitcases, artificial turf, thermal underwear.

Polyethylene (PE)

The material. Polyethylene, $(-CH_2-)_n$, first synthesized in 1933, looks like the simplest of molecules, but the number of ways in which the—CH_2—units can be linked is large. It is the first of the polyolefins, the bulk thermoplastic polymers that account for a dominant fraction of all polymer consumption. Polyethylene is inert, and extremely resistant to fresh and salt water, food, and most water-based solutions. Because of this it is widely used in household products, food containers, and chopping boards. Polyethylene is cheap, and particularly easy to mold and fabricate. It accepts a wide range of colors, can be transparent, translucent, or opaque, has a pleasant, slightly waxy feel, can be textured or metal coated, but is difficult to print on.

Composition

$(-CH_2-CH_2-)_n$

General properties

Density	939	– 960	kg/m^3
Price	1.7	– 1.9	USD/kg

Mechanical properties

Young's modulus	0.62	– 0.86	GPa
Yield strength (elastic limit)	18	– 29	MPa
Tensile strength	21	– 45	MPa
Elongation	200	– 800	%
Hardness—Vickers	5.4	– 8.7	HV
Fatigue strength at 10^7 cycles	21	– 23	MPa
Fracture toughness	1.4	– 1.7	$MPa \cdot m^{1/2}$

Thermal properties

Melting point	125	– 132	°C
Maximum service temperature	90	– 110	°C
Thermal conductor or insulator?	Good insulator		
Thermal conductivity	0.4	– 0.44	$W/m \cdot K$
Specific heat capacity	1,810	– 1,880	$J/kg \cdot K$
Thermal expansion coefficient	126	– 198	$\mu strain/°C$

Electrical properties

Electrical conductor or insulator?	Good insulator		
Electrical resistivity	3.3×10^{22}	– 3×10^{24}	$\mu ohm \cdot cm$
Dielectric constant	2.2	– 2.4	
Dissipation factor	3×10^{-4}	– 6×10^{-4}	
Dielectric strength	17.7	– 19.7	10^6 V/m

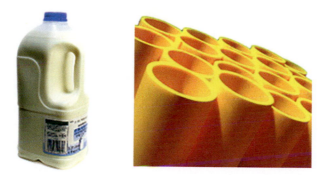

Low-density PE bottle and medium-density PE pipe

Eco properties: material

Global production, main component	69×10^6			metric ton/yr
Embodied energy, primary production	77	–	85	MJ/kg
CO_2 footprint, primary production	2.6	–	2.9	kg/kg
Water usage	38	–	114	L/kg
Eco-indicator	287			millipoints/kg

Eco properties: processing

Polymer molding energy	22.7	–	25.1	MJ/kg
Polymer molding CO_2 footprint	1.7	–	1.9	kg/kg
Polymer extrusion energy	6.0	–	6.6	MJ/kg
Polymer extrusion CO_2 footprint	0.45	–	0.49	kg/kg

End of life

Embodied energy, recycling	45	–	55	MJ/kg
CO_2 footprint, recycling	2.7	–	3.0	kg/kg
Recycle fraction in current supply	8	–	9.5	%
Heat of combustion	44	–	46	MJ/kg
Combustion CO_2	3.1	–	3.2	kg/kg
Recycle mark				

Typical uses. Oil container, milk bottles, toys, beer crates, food packaging, shrink wrap, squeeze tubes, disposable clothing, plastic bags, paper coatings, cable insulation, artificial joints, and as fibers—low cost ropes and packing tape reinforcement.

Polycarbonate (PC)

The material. PC is one of the "engineering" thermoplastics, meaning that they have better mechanical properties than the cheaper "commodity" polymers. The benzene ring and the −OCOO− carbonate group combine in pure PC to give it its unique characteristics of optical transparency and good toughness and rigidity, even at relatively high temperatures. These properties make PC a good choice for applications such as compact disks, safety hard hats, and housings for power tools.

Composition

$(O-(C_6H_4)-C(CH_3)_2-(C_6H_4)-CO)_n$

General properties

Density	1,140	–	1,210	kg/m^3
Price	3.8	–	4.2	USD/kg

Mechanical properties

Young's modulus	2	–	2.44	GPa
Yield strength (elastic limit)	59	–	70	MPa
Tensile strength	60	–	72.4	MPa
Compressive strength	69	–	86.9	MPa
Elongation	70	–	150	%
Hardness—Vickers	17.7	–	21.7	HV
Fatigue strength at 10^7 cycles	22.1	–	30.8	MPa
Fracture toughness	2.1	–	4.6	MPa$\cdot$m$^{1/2}$

Thermal properties

Glass temperature	142	–	205	°C
Maximum service temperature	101	–	144	°C
Thermal conductor or insulator?	Good insulator			
Thermal conductivity	0.189	–	0.218	W/m$\cdot$K
Specific heat capacity	1,530	–	1,630	J/kg$\cdot$K
Thermal expansion coefficient	120	–	137	μstrain/°C

Electrical properties

Electrical conductor or insulator?	Good insulator			
Electrical resistivity	1×10^{20}	–	1×10^{22}	μohm$\cdot$cm
Dielectric constant	3.1	–	3.3	
Dissipation factor	8×10^{-4}	–	0.0011	
Dielectric strength	15.7	–	19.2	10^6 V/m

Polycarbonate is tough and impact-resistant: hence its use in hard hats and helmets, transparent roofing, and riot shields.

Eco properties: material

Embodied energy, primary production	103	– 114	MJ/kg
CO_2 footprint, primary production	5.7	– 6.3	kg/kg
Water usage	142	– 425	L/kg
Eco-indicator	463		millipoints/kg

Eco properties: processing

Polymer molding energy	17.6	– 19.5	MJ/kg
Polymer molding CO_2	1.3	– 1.5	kg/kg
Polymer extrusion energy	5.8	– 6.4	MJ/kg
Polymer extrusion CO_2	0.43	– 0.48	kg/kg

End of life

Embodied energy, recycling	38	– 47	MJ/kg
CO_2 footprint, recycling	2.3	– 2.8	kg/kg
Recycle fraction in current supply	0.5	– 1	%
Heat of combustion	21	– 22	MJ/kg
Combustion CO_2	1.9	– 2.0	kg/kg
Recycle mark			

Other

Typical uses. Safety shields and goggles; lenses; glazing panels; business machine housing; instrument casings; lighting fittings; safety helmets; electrical switchgear; laminated sheet for bullet proof glazing; twin-walled sheets for glazing; kitchenware and tableware; microwave cookware; medical (sterilizable) components.

Polyethylene terephthalate (PET)

The material. The name *polyester* derives from a combination of "polymerization" and "esterification." Saturated polyesters are thermoplastic—examples are PET and PBT; they have good mechanical properties to temperatures as high as 175°C. PET is crystal clear and impervious to water and CO_2, but a little oxygen does get through. It is tough, strong, easy to shape, join, and sterilize—allowing reuse. Unsaturated polyesters are thermosets; they are used as the matrix material in glass fiber/polyester composites.

Composition
$(CO-(C_6H_4)-CO-O-(CH_2)_2-O)_n$

General properties

Density	1,290	– 1,400	kg/m^3
Price	1.65	– 1.8	USD/kg

Mechanical properties

Young's modulus	2.76	– 4.14	GPa
Yield strength (elastic limit)	56.5	– 62.3	MPa
Tensile strength	48.3	– 72.4	MPa
Compressive strength	62.2	– 68.5	MPa
Elongation	30	– 300	%
Hardness—Vickers	17	– 18.7	HV
Fatigue strength at 10^7 cycles	19.3	– 29	MPa
Fracture toughness	4.5	– 5.5	$MPa \cdot m^{1/2}$

Thermal properties

Melting point	255	– 265	°C
Glass temperature	67.9	– 79.9	°C
Maximum service temperature	66.9	– 86.9	°C
Thermal conductor or insulator?	Good insulator		
Thermal conductivity	0.138	– 0.151	$W/m \cdot K$
Specific heat capacity	1,420	– 1,470	$J/kg \cdot K$
Thermal expansion coefficient	115	– 119	$\mu strain/°C$

Electrical properties

Electrical conductor or insulator?	Good insulator		
Electrical resistivity	3.3×10^{20}	– 3.0×10^{21}	$\mu ohm \cdot cm$
Dielectric constant	3.5	– 3.7	
Dissipation factor	0.003	– 0.007	
Dielectric strength	16.5	– 21.7	10^6 V/m

*PET drinks containers, pressurized and unpressurized
(image courtesy of Tee design and printing Ltd.)*

Eco properties: material

Global production, main component	9.5×10^6			metric ton/yr
Embodied energy, primary production	81	–	89	MJ/kg
CO_2 footprint, primary production	3.7	–	4.1	kg/kg
Water usage	14.7	–	44.2	l/kg
Eco-indicator	276			millipoints/kg

Eco properties: processing

Polymer molding energy	18.7	–	20.6	MJ/kg
Polymer molding CO_2	1.4	–	1.55	kg/kg
Polymer extrusion energy	5.8	–	6.4	MJ/kg
Polymer extrusion CO_2	0.44	–	0.48	kg/kg

End of life

Embodied energy, recycling	35	–	43	MJ/kg
CO_2 footprint, recycling	2.1	–	2.6	kg/kg
Recycle fraction in current supply	20	–	22	%
Heat of combustion	23	–	24	MJ/kg
Combustion CO_2	2.3	–	2.4	kg/kg
Recycle mark				

Typical uses. Electrical fittings and connectors; blow molded bottles; packaging film; film; photographic and X-ray film; audio/visual tapes; industrial strapping; capacitor film; drawing office transparencies; fibers. Mylar film, metalized balloons, photography tape, videotape, carbonated drink containers, ovenproof cookware, windsurfing sails, credit cards.

Polyvinylchloride (tpPVC)

The material. PVC—vinyl—is one of the cheapest, most versatile and—with polyethylene—the most widely used of polymers and epitomizes their multifaceted character. In its pure form—as a thermoplastic, tpPVC—it is rigid, and not very tough; its low price makes it a cost-effective engineering plastic where extremes of service are not encountered. Incorporating plasticizers creates flexible PVC, elPVC, a material with leather-like or rubber-like properties, and used as a substitute for both. By contrast, reinforcement with glass fibers gives a material that is sufficiently stiff, strong, and tough to be used for roofs, flooring, and building panels.

Composition

$(CH_2CHCl)_n$

General properties

Density	1,300	– 1,580	kg/m^3
Price	1.36	– 1.5	USD/kg

Mechanical properties

Young's modulus	2.14	– 4.14	GPa
Yield strength (elastic limit)	35.4	– 52.1	MPa
Tensile strength	40.7	– 65.1	MPa
Compressive strength	42.5	– 89.6	MPa
Elongation	11.9	– 80	%
Hardness—Vickers	10.6	– 15.6	HV
Fatigue strength at 10^7 cycles	16.2	– 26.1	MPa
Fracture toughness	1.46	– 5.12	$MPa \cdot m^{1/2}$

Thermal properties

Glass temperature	74.9	– 105	°C
Maximum service temperature	60	– 70	°C
Thermal conductor or insulator?	Good insulator		
Thermal conductivity	0.147	– 0.293	$W/m \cdot K$
Specific heat capacity	1,360	– 1,440	$J/kg \cdot K$
Thermal expansion coefficient	100	– 150	μstrain/°C

Electrical properties

Electrical conductor or insulator?	Good insulator		
Electrical resistivity	1×10^{20}	– 1×10^{22}	μohm·cm
Dielectric constant	3.1	– 4.4	
Dissipation factor	0.03	– 0.1	
Dielectric strength	13.8	– 19.7	10^6 V/m

These boat fenders illustrate that PVC is tough, weather resistant, and easy to form and color.

Eco properties: material

Global production, main component	51×10^6			metric ton/yr
Embodied energy, primary production	56	–	62	MJ/kg
CO_2 footprint, primary production	2.4	–	2.6	kg/kg
Water usage	77	–	85	L/kg
Eco-indicator	170			millipoints/kg

Eco properties: processing

Polymer molding energy	13.9	–	15.4	MJ/kg
Polymer molding CO_2	1.05	–	1.16	kg/kg
Polymer extrusion energy	5.6	–	6.3	MJ/kg
Polymer extrusion CO_2	0.42	–	0.47	kg/kg

End of life

Embodied energy, recycling	32	–	40	MJ/kg
CO_2 footprint, recycling	1.9	–	2.4	kg/kg
Recycle fraction in current supply	1.5	–	2.0	%
Heat of combustion	17.5	–	18.5	MJ/kg
Combustion CO_2	1.37	–	1.44	kg/kg
Recycle mark				

PVC

Typical uses. tpPVC: pipes, fittings, profiles, road signs, cosmetic packaging, canoes, garden hoses, vinyl flooring, windows and cladding, vinyl records, dolls, medical tubes. elPVC: artificial leather, wire insulation, film, sheet, fabric, car upholstery.

Polystyrene (PS)

The material. Polystyrene is an optically clear, cheap, easily molded polymer, familiar as the standard "jewel" CD case. In its simplest form, PS is brittle. Its mechanical properties are dramatically improved by blending with polybutadiene, but with a loss of optical transparency. High-impact PS (10% polybutadiene) is much stronger even at low temperatures (meaning strength down to $-12°C$). The single largest use of PS is a foam packaging.

Composition
$(CH(C_6H_5)-CH_2)_n$

General properties

Density	1,040	– 1,050	kg/m^3
Price	2.1	– 2.3	USD/kg

Mechanical properties

Young's modulus	1.2	– 2.6	GPa
Yield strength (elastic limit)	28.7	– 56.2	MPa
Tensile strength	35.9	– 56.5	MPa
Compressive strength	31.6	– 61.8	MPa
Elongation	1.2	– 3.6	%
Hardness—Vickers	8.6	– 16.9	HV
Fatigue strength at 10^7 cycles	14.4	– 23	MPa
Fracture toughness	0.7	– 1.1	$MPa \cdot m^{1/2}$

Thermal properties

Glass temperature	73.9	– 110	°C
Maximum service temperature	76.9	– 103	°C
Thermal conductor or insulator?	Good insulator		
Thermal conductivity	0.121	– 0.131	$W/m \cdot K$
Specific heat capacity	1,690	– 1,760	$J/kg \cdot K$
Thermal expansion coefficient	90	– 153	$\mu strain/°C$

Electrical properties

Electrical conductor or insulator?	Good insulator		
Electrical resistivity	1×10^{25}	– 1×10^{27}	$\mu ohm \cdot cm$
Dielectric constant	3	– 3.2	
Dissipation factor	0.001	– 0.003	
Dielectric strength	19.7	– 22.6	10^6 V/m

Polystyrene is water-clear, easily formed and cheap.

Eco properties: material

Global production, main component	12.6×10^6			metric ton/yr
Embodied energy, primary production	92	–	102	MJ/kg
CO_2 footprint, primary production	3.6	–	4.0	kg/kg
Water usage	108	–	323	L/kg
Eco-indicator	320			millipoints/kg

Eco properties: processing

Polymer molding energy	16.5	–	18.3	MJ/kg
Polymer molding CO_2	1.24	–	1.37	kg/kg
Polymer extrusion energy	5.7	–	6.4	MJ/kg
Polymer extrusion CO_2	0.43	–	0.48	kg/kg

End of life

Embodied energy, recycling	43	–	52	MJ/kg
CO_2 footprint, recycling	2.6	–	3.1	kg/kg
Recycle fraction in current supply	5	–	6	%
Heat of combustion	40	–	42	MJ/kg
Combustion CO_2	3.3	–	3.5	kg/kg
Recycle mark				

PS

Typical uses. Toys; light diffusers; lenses and mirrors; beakers; cutlery; general household appliances; video/audio cassette cases; electronic housings; refrigerator liners.

Polyhydroxyalkanoates (PHA, PHB)

The material. PHAs, sold under the trade names Biopol and Biomer, are linear polyesters produced in nature by bacterial fermentation of sugar or lipids derived from soybean oil, corn oil, or palm oil. They are fully biodegradable. More than 100 different monomers can be combined within this family to give materials with a wide range of properties, from stiff and brittle thermoplastics to flexible elastomers. The most common type of PHA is PHB (poly-3-hydroxybutyrate) with properties similar to those of PP, though it is stiffer and more brittle. The data below are for PHB.

Composition

$(CH(CH_3)-CH_2-CO-O)_n$

General properties

Density	1,230	–	1,250	kg/m^3
Price	3.2	–	4	USD/kg

Mechanical properties

Young's modulus	0.8	–	4	GPa
Yield strength (elastic limit)	35	–	40	MPa
Tensile strength	35	–	40	MPa
Compressive strength	40	–	45	MPa
Elongation	6	–	25	%
Hardness—Vickers	11	–	13	HV
Fatigue strength at 10^7 cycles	12	–	17	MPa
Fracture toughness	0.7	–	1.2	MPa·m$^{1/2}$

Thermal properties

Melting point	115	–	175	°C
Glass temperature	4	–	15	°C
Maximum service temperature	60	–	80	°C
Thermal conductor or insulator?	Good insulator			
Thermal conductivity	0.13	–	0.23	W/m·K
Specific heat capacity	1,400	–	1,600	J/kg·K
Thermal expansion coefficient	180	–	240	µstrain/°C

Electrical properties

Electrical conductor or insulator?	Good insulator			
Electrical resistivity	1×10^{16}	–	1×10^{18}	µohm·cm
Dielectric constant	3	–	5	
Dissipation factor	0.05	–	0.15	
Dielectric strength	12	–	16	10^6 V/m

PHB containers (Kumar and Minocha, Trangenic Plant Research, Harwood Publishers)

Eco properties: material

Embodied energy, primary production	81	–	90	MJ/kg
CO_2 footprint, primary production	4.1	–	4.6	kg/kg
Water usage	100	–	300	L/kg

Eco properties: processing

Polymer molding energy	16.6	–	18.4	MJ/kg
Polymer molding CO_2	1.25	–	1.38	kg/kg
Polymer extrusion energy	5.8	–	6.4	MJ/kg
Polymer extrusion CO_2	0.43	–	0.48	kg/kg

End of life

Embodied energy, recycling	35	–	43	MJ/kg
CO_2 footprint, recycling	2.1	–	2.6	kg/kg
Recycle fraction in current supply	0.5	–	1	%
Heat of combustion	23	–	24	MJ/kg
Combustion CO_2	2.0	–	2.1	kg/kg
Recycle mark				

Other

Typical uses. Packaging, containers, bottles.

Polylactide (PLA)

The material. Polylactide, PLA, is a biodegradable thermoplastic derived from natural lactic acid from corn, maize, or milk. It resembles clear polystyrene, and provides good aesthetics (gloss and clarity), but it is stiff and brittle and needs modification using plasticizers for most practical applications. It can be processed like most thermoplastics into fibers, films, thermoformed, or injection molded.

General properties

Density	1,210	– 1,250	kg/m^3
Price	2.4	– 3	USD/kg

Mechanical properties

Young's modulus	3.45	– 3.83	GPa
Yield strength (elastic limit)	48	– 60	MPa
Tensile strength	48	– 60	MPa
Compressive strength	48	– 60	MPa
Elongation	5	– 7	%
Hardness—Vickers	14	– 18	HV
Fatigue strength at 10^7 cycles	14	– 18	MPa
Fracture toughness	0.7	– 1.1	MPa$\cdot$m$^{1/2}$

Thermal properties

Melting point	160	– 177	°C
Glass temperature	56	– 58	°C
Maximum service temperature	70	– 80	°C
Thermal conductor or insulator?	Good insulator		
Thermal conductivity	0.12	– 0.13	W/m$\cdot$K
Specific heat capacity	1,180	– 1,210	J/kg$\cdot$K
Thermal expansion coefficient	126	– 145	μstrain/°C

Electrical properties

Electrical conductor or insulator?	Good insulator		
Electrical resistivity	1×10^{17}	– 1×10^{19}	μohm$\cdot$cm
Dielectric constant	3.5	– 5	
Dissipation factor	0.02	– 0.07	
Dielectric strength	12	– 16	10^6 V/m

Cargill Dow polylactide food packaging

Eco properties: material

Embodied energy, primary production	49	– 54	MJ/kg
CO_2 footprint, primary production	3.4	– 3.8	kg/kg
Water usage	100	– 300	L/kg
Eco-indicator	278		millipoints/kg

Eco properties: processing

Polymer molding energy	15.4	– 17	MJ/kg
Polymer molding CO_2	1.15	– 1.27	kg/kg
Polymer extrusion energy	5.7	– 6.3	MJ/kg
Polymer extrusion CO_2	0.43	– 0.47	kg/kg

End of life

Embodied energy, recycling	33	– 40	MJ/kg
CO_2 footprint, recycling	2.0	– 2.4	kg/kg
Recycle fraction in current supply	0.5	– 1	%
Heat of combustion	18.8	– 20.1	MJ/kg
Combustion CO_2	1.8	– 1.9	kg/kg
Recycle mark			

Other

Typical uses. Food packaging, plastic bags, plant pots, diapers, bottles, cold drink cups, sheet and plastic wrap.

Epoxies

The material. Epoxies are thermosetting polymers with excellent mechanical, electrical, and adhesive properties and good resistance to heat and corrosion. They are used for adhesives (Araldite), surface coatings and, when filled with other materials such as glass or carbon fibers, as matrix resins in composite materials. Typically, as adhesives, epoxies are used for high-strength bonding of dissimilar materials; as coatings, they are used to encapsulate electrical coils and electronic components; when filled, they are used for tooling fixtures for low-volume molding of thermoplastics.

Composition
$(O-C_6H_4-CH_3-C-CH_3-C_6H_4)_n$

General properties

Density	1,110	–	1,400	kg/m^3
Price	8.0	–	10.0	USD/kg

Mechanical properties

Young's modulus	2.35	–	3.08	GPa
Yield strength (elastic limit)	36	–	71.7	MPa
Tensile strength	45	–	89.6	MPa
Compressive strength	39.6	–	78.8	MPa
Elongation	2	–	10	%
Hardness—Vickers	10.8	–	21.5	HV
Fatigue strength at 10^7 cycles	22.1	–	35	MPa
Fracture toughness	0.4	–	2.22	MPa·m$^{1/2}$

Thermal properties

Glass temperature	66.9	–	167	°C
Maximum service temperature	140	–	180	°C
Thermal conductor or insulator?	Good insulator			
Thermal conductivity	0.18	–	0.5	W/m·K
Specific heat capacity	1,490	–	2,000	J/kg·K
Thermal expansion coefficient	58	–	117	μstrain/°C

Electrical properties

Electrical conductor or insulator?	Good insulator			
Electrical resistivity	1×10^{20}	–	6×10^{21}	μohm·cm
Dielectric constant	3.4	–	5.7	
Dissipation factor	7×10^{-4}	–	0.015	
Dielectric strength	11.8	–	19.7	10^6 V/m

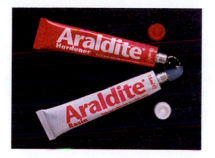

Epoxy paints are exceptionally stable and protective, and take color well. Epoxies are used at the matrix of high performance composite, and as high strength adhesives.

Eco properties: material

Global production, main component	0.14×10^6			metric ton/yr
Embodied energy, primary production	127	–	140	MJ/kg
CO_2 footprint, primary production	6.8	–	7.5	kg/kg
Water usage	107	–	322	L/kg
Eco-indicator	650			millipoints/kg

Eco properties: processing

Polymer molding energy	21	–	23	MJ/kg
Polymer molding CO_2	1.7	–	1.85	kg/kg

End of life

Recycle fraction in current supply	0			%
Heat of combustion	30	–	31	MJ/kg
Combustion CO_2	2.4	–	2.55	kg/kg

Typical uses. Pure epoxy molding compounds; the encapsulation of electrical coils and electronics components; epoxy resins in laminates; pultruded rods; girder stock; special tooling fixtures; mechanical components such as gears; adhesives, often for high-strength bonding of dissimilar materials; patterns and molds for shaping thermoplastics.

Polyester

The material. Polyesters can be a thermosets, thermoplastics, or elastomers. The unsaturated polyester resins are thermosets. Most polyester thermosets are used in glass fiber/polyester composites. They are less stiff and strong than epoxies, but they are considerably cheaper. This record is for thermosetting polyester. It cannot be recycled, but it can be used as a filler when ground.

Composition

$(OOC-C_6H_4-COO-C_6H_{10})_n$

General properties

Density	1,040	– 1,400	kg/m^3
Price	4.0	– 4.4	USD/kg

Mechanical properties

Young's modulus	2.07	– 4.41	GPa
Yield strength (elastic limit)	33	– 40	MPa
Tensile strength	41.4	– 89.6	MPa
Compressive strength	36.3	– 44	MPa
Elongation	2	– 2.6	%
Hardness—Vickers	9.9	– 21.5	HV
Fatigue strength at 10^7 cycles	16.6	– 35.8	MPa
Fracture toughness	1.09	– 1.69	$MPa \cdot m^{1/2}$

Thermal properties

Glass temperature	147	– 207	°C
Maximum service temperature	130	– 150	°C
Thermal conductor or insulator?	Good insulator		
Thermal conductivity	0.287	– 0.299	$W/m \cdot K$
Specific heat capacity	1,510	– 1,570	$J/kg \cdot K$
Thermal expansion coefficient	99	– 180	$\mu strain/°C$

Electrical properties

Electrical conductor or insulator?	Good insulator		
Electrical resistivity	3.3×10^{18}	– 3×10^{19}	$\mu ohm \cdot cm$
Dielectric constant	2.8	– 3.3	
Dissipation factor	0.001	– 0.03	
Dielectric strength	15	– 19.7	10^6 V/m

Thermosetting polyester is used as the matrix of fiber-reinforced boats and cars. The boat cover, too, is polyester.

Eco properties: material

Global production, main component	40×10^6			metric ton/yr
Embodied energy, primary production	68	–	75	MJ/kg
CO_2 footprint, primary production	2.8	–	3.2	kg/kg
Water usage	100	–	264	L/kg
Eco-indicator	437			millipoints/kg

Eco properties: processing

Polymer molding energy	26	–	28	MJ/kg
Polymer molding CO_2	2.1	–	2.3	kg/kg

End of life

Recycle fraction in current supply	0			%
Heat of combustion	28	–	29	MJ/kg
Combustion CO_2	2.5	–	2.6	kg/kg

Typical uses. Laminated structures; surface gel coatings; liquid castings; furniture products; bowling balls; simulated marble; sewer pipe gaskets; pistol grips; television tube implosion barriers; boats; truck cabs; concrete forms; lamp housings; skylights; fishing rods.

Phenolics

The material. Bakelite, commercialized in 1909, triggered a revolution in product design. It was stiff, fairly strong, could (to a muted degree) be colored, and—above all—was easy to mold. Products that, earlier, were handcrafted from woods, metals, or exotics such as ivory, could now be molded quickly and cheaply. At one time the production of phenolics exceeded that of PE, PS, and PVC combined. Now, although the ratio has changed, phenolics still have a unique value. They are stiff, chemically stable, have good electrical properties, are fire-resistant and easy to mold—and they are cheap. Thermosetting phenolics are recyclable, but by a different means than that for thermoplastics. Molded phenolic, ground into a fine powder, can be added to the raw material stream. Four to twelve percent ground phenolic does not degrade properties.

General properties

Density	1,240	–	1,320	kg/m^3
Price	1.65	–	1.87	USD/kg

Mechanical properties

Young's modulus	2.76	–	4.83	GPa
Yield strength (elastic limit)	27.6	–	49.7	MPa
Tensile strength	34.5	–	62.1	MPa
Compressive strength	30.4	–	54.6	MPa
Elongation	1.5	–	2	%
Hardness—Vickers	8.3	–	14.9	HV
Fatigue strength at 10^7 cycles	13.8	–	24.8	MPa
Fracture toughness	0.787	–	1.21	MPa·m$^{1/2}$

Thermal properties

Glass temperature	167	–	267	°C
Maximum service temperature	200	–	230	°C
Thermal conductor or insulator?	Good insulator			
Thermal conductivity	0.14	–	0.15	W/m·K
Specific heat capacity	1,470	–	1,530	J/kg·K
Thermal expansion coefficient	120	–	125	μstrain/°C

Electrical properties

Electrical conductor or insulator?	Good insulator			
Electrical resistivity	3.3×10^{18}	–	3×10^{19}	μohm·cm
Dielectric constant	4	–	6	
Dissipation factor	0.005	–	0.01	
Dielectric strength	9.84	–	15.7	10^6 V/m

Phenolics are good insulators, and resist heat and corrosion exceptionally well, making them a good choice for electrical switchgear like this telephone and distributor cap. (Telephone image courtesy of Eurocosm UK.)

Eco properties: material

Global production, main component	11×10^6		metric ton/yr
Embodied energy, primary production	75	– 83	MJ/kg
CO_2 footprint, primary production	3.4	– 3.8	kg/kg
Water usage	94	– 282	L/kg

Eco properties: processing

Polymer molding energy	26	– 29	MJ/kg
Polymer molding CO_2	2.1	– 2.3	kg/kg

End of life

Recycle fraction in current supply	0	%	
Heat of combustion	32	– 33	MJ/kg
Combustion CO_2	2.8	– 3.0	kg/kg

Typical uses. Electrical parts—sockets, switches, connectors, water-lubricated bearings, relays, pump impellers, brake pistons, brake pads, microwave cookware, handles, bottles tops, coatings, adhesives, bearings, foams, and sandwich structures.

Natural rubber (NR)

The material. Natural rubber was known to the natives of Peru many centuries ago and is now one of Malaysia's main exports. It made the fortune of Giles Macintosh who, in 1825, devised the rubber-coated waterproof coat that still bears his name. Latex, the sap of the rubber tree, is cross-linked (vulcanized) by heating it with sulfur; the amount of the cross-linking determines the properties. It is the most widely used of all elastomers—more than 50% of all produced.

Composition

$(CH_2-C(CH_3)-CH-CH_2)_n$

General properties

Density	920	— 930	kg/m^3
Price	3.6	— 4.9	USD/kg

Mechanical properties

Young's modulus	0.0015	— 0.0025	GPa
Yield strength (elastic limit)	20	— 30	MPa
Tensile strength	22	— 32	MPa
Compressive strength	22	— 33	MPa
Elongation	500	— 800	%
Fatigue strength at 10^7 cycles	4.2	— 4.5	MPa
Fracture toughness	0.15	— 0.25	$MPa \cdot m^{1/2}$

Thermal properties

Glass temperature	−78.2	— −63.2	°C
Maximum service temperature	68.9	— 107	°C
Thermal conductor or insulator?	Good insulator		
Thermal conductivity	0.1	— 0.14	$W/m \cdot K$
Specific heat capacity	1,800	— 2,500	$J/kg \cdot K$
Thermal expansion coefficient	150	— 450	$\mu strain/°C$

Electrical properties

Electrical conductor or insulator?	Good insulator		
Electrical resistivity	1×10^{15}	— 1×10^{16}	$\mu ohm \cdot cm$
Dielectric constant	3	— 4.5	
Dissipation factor	7×10^{-4}	— 0.003	
Dielectric strength	16	— 23	10^6 V/m

Natural rubber is used in medical equipment, fashion items, tubing, and tires.

Eco properties: material

Global production, main component	8.2×10^6		metric ton/yr
Embodied energy, primary production	64	– 71	MJ/kg
CO_2 footprint, primary production	2.0	– 2.2	kg/kg
Water usage	15,000	– 20,000	L/kg
Eco-indicator	24		millipoints/kg

Eco properties: processing

Polymer molding energy	15	– 17	MJ/kg
Polymer molding CO_2	1.2	– 1.4	kg/kg

End of life

Recycle fraction in current supply	0		%
Heat of combustion	42	– 45	MJ/kg
Combustion CO_2	3.2	– 3.3	kg/kg

Typical uses. Gloves, car tires, seals, belts, anti-vibration mounts, electrical insulation, tubing, rubber lining pipes, and pumps.

Butyl rubber

The material. Butyl rubbers (BR) are synthetics that resemble natural rubber in properties. They have good resistance to abrasion, tearing, and flexing, with exceptionally low gas permeability and useful properties up to 150°C. They have a low dielectric constant and loss, making them attractive for electrical applications.

Composition

$(CH_2-C(CH_3)-CH-(CH_2)_2-C(CH_3)_2)_n$

General properties

Density	900	– 920	kg/m^3
Price	3.8	– 4.1	USD/kg

Mechanical properties

Young's modulus	0.001	– 0.002	GPa
Yield strength (elastic limit)	2	– 3	MPa
Tensile strength	5	– 10	MPa
Compressive strength	2.2	– 3.3	MPa
Elongation	400	– 500	%
Fatigue strength at 10^7 cycles	0.9	– 1.35	MPa
Fracture toughness	0.07	– 0.1	MPa·m$^{1/2}$

Thermal properties

Glass temperature	−73.2	– −63.2	°C
Maximum service temperature	96.9	– 117	°C
Thermal conductor or insulator?	Good insulator		
Thermal conductivity	0.08	– 0.1	W/m·K
Specific heat capacity	1.8e3	– 2.5e3	J/kg·K
Thermal expansion coefficient	120	– 300	μstrain/°C

Electrical properties

Electrical conductor or insulator?	Good insulator		
Electrical resistivity	1×10^{15}	– 1×10^{16}	μohm·cm
Dielectric constant	2.8	– 3.2	
Dissipation factor	0.001	– 0.01	
Dielectric strength	16	– 23	10^6 V/m

Butyl rubber is one of the most important materials for inner tubes.

Eco properties: material

Global production, main component	11×10^6			metric ton/yr
Embodied energy, primary production	112	–	124	MJ/kg
CO_2 footprint, primary production	6.3	–	6.9	kg/kg
Water usage	63.8	–	191	L/kg
Eco-indicator	309			millipoints/ kg

Eco properties: processing

Polymer molding energy	14	–	16	MJ/kg
Polymer molding CO_2	1.2	–	1.4	kg/kg

End of life

Recycle fraction in current supply	2	–	4.1	%
Heat of combustion	42	–	44	MJ/kg
Combustion CO_2	3.2	–	3.3	kg/kg

Typical uses. Inner tubes, seals, belts, anti-vibration mounts, electrical insulation, tubing, brake pads, rubber lining pipes, and pumps.

EVA

The material. Ethylene-vinyl-acetate thermoplastic elastomers (EVA) are built around polyethylene. They are soft, flexible, and tough, and they retain these properties down to $-60°C$. They also have good barrier properties, which gains them FDA-approval for direct food contact. EVA can be processed by most normal thermoplastic processes: co-extrusion for films, blow molding, rotational molding, injection molding, and transfer molding.

Composition
$(CH_2)_n-(CH_2-CHR)_m$

General properties

Density	945	– 955	kg/m^3
Price	2.0	– 2.2	USD/kg

Mechanical properties

Young's modulus	0.01	– 0.04	GPa
Yield strength (elastic limit)	12	– 18	MPa
Tensile strength	16	– 20	MPa
Compressive strength	13.2	– 19.8	MPa
Elongation	730	– 770	%
Fatigue strength at 10^7 cycles	12	– 12.8	MPa
Fracture toughness	0.5	– 0.7	MPa·m$^{1/2}$

Thermal properties

Glass temperature	−73.2	– −23.2	°C
Maximum service temperature	46.9	– 51.9	°C
Thermal conductor or insulator?	Good insulator		
Thermal conductivity	0.3	– 0.4	W/m·K
Specific heat capacity	2,000	– 2,200	J/kg·K
Thermal expansion coefficient	160	– 190	μstrain/°C

Electrical properties

Electrical conductor or insulator?	Good insulator		
Electrical resistivity	3.1×10^{21}	– 1×10^{22}	μohm·cm
Dielectric constant	2.9	– 2.95	
Dissipation factor	0.005	– 0.022	
Dielectric strength	26.5	– 27	10^6 V/m

EVA is available in pastel or deep hues; it has good clarity and gloss. It has good barrier properties, little or no odor, and is UV resistant. (Running shoes courtesy of Adidas; slippers courtesy of Zhangzhou Yongxin Trade Co., Ltd.)

Eco properties: material

Embodied energy, primary production	75	– 83	MJ/kg
CO_2 footprint, primary production	2.0	– 2.2	kg/kg
Water usage	100	– 289	L/kg
Eco-indicator	268		millipoints/kg

Eco properties: processing

Polymer molding energy	13.8	– 15.2	MJ/kg
Polymer molding CO_2	1.1	– 1.2	kg/kg
Polymer extrusion energy	5.4	– 6.0	MJ/kg
Polymer extrusion CO_2	0.43	– 0.48	kg/kg

End of life

Embodied energy, recycling	42	– 52	MJ/kg
CO_2 footprint, recycling	2.5	– 3.1	kg/kg
Recycle fraction in current supply	6	– 10	%
Heat of combustion	40	– 42	MJ/kg
Combustion CO_2	2.8	– 3.0	kg/kg

Typical uses. Medical tubes, milk packaging, beer dispensing equipment, bags, shrink wrap, deep freeze bags, co-extruded and laminated film, closures, ice trays, gaskets, gloves, cable insulation, inflatable parts, running shoes.

Polychloroprene (Neoprene, CR)

The material. Polychloroprenes (Neoprene, CR)—the materials of wetsuits—are the leading non-tire synthetic rubbers. First synthesized in 1930, they are made by a condensation polymerization of the monomer 2-chloro−1,3 butadiene. The properties can by modified by copolymerization with sulfur, with other chlorobutadienes, and by blending with other polymers to give a wide range of properties. Polychloroprenes are characterized by high chemical stability and by resistance to water, oil, gasoline, and UV radiation.

Composition
$(CH_2-CCl-CH_2-CH_2)_n$

General properties
Density	1,230	–	1,250	kg/m^3
Price	5.2	–	5.7	USD/kg

Mechanical properties
Young's modulus	7e-4	–	0.002	GPa
Yield strength (elastic limit)	3.4	–	24	MPa
Tensile strength	3.4	–	24	MPa
Compressive strength	3.72	–	28.8	MPa
Elongation	100	–	800	%
Fatigue strength at 10^7 cycles	1.53	–	12	MPa
Fracture toughness	0.1	–	0.3	$MPa \cdot m^{1/2}$

Thermal properties
Glass temperature	−48.2	–	−43.2	°C
Maximum service temperature	102	–	112	°C
Thermal conductor or insulator?	Good insulator			
Thermal conductivity	0.1	–	0.12	$W/m \cdot K$
Specific heat capacity	2,000	–	2,200	$J/kg \cdot K$
Thermal expansion coefficient	575	–	610	µstrain/°C

Electrical properties
Electrical conductor or insulator?	Good insulator			
Electrical resistivity	1×10^{19}	–	1×10^{23}	µohm.cm
Dielectric constant	6.7	–	8	
Dissipation factor	1×10^{-4}	–	0.001	
Dielectric strength	15.8	–	23.6	10^6 V/m

Neoprene gives wetsuits flexibility and stretch.

Eco properties: material

Embodied energy, primary production	61	–	68	MJ/kg
CO_2 footprint, primary production	1.6	–	1.8	kg/kg
Water usage	126	–	378	l/kg

Eco properties: processing

Polymer molding energy	17.2	–	18.5	MJ/kg
Polymer molding CO_2	1.37	–	1.5	kg/kg

End of life

Recycle fraction in current supply	0			%
Heat of combustion	16.8	–	17.1	MJ/kg
Combustion CO_2	1.4	–	1.46	kg/kg

Typical uses. Brake seals, diaphragms, hoses and o-rings, tracked-vehicle pads, footwear, wetsuits.

15.4 Ceramics and glasses

Ceramics are materials both of the past and of the future. They are the most durable of all materials—ceramic pots and ornaments survive from 5,000 BC; Roman cement still bonds the walls of villas. It is their durability, particularly at high temperatures, that generates interest in them today. They are exceptionally hard (diamond—a ceramic—is the hardest of them all) and can tolerate higher temperatures than any metal. Ceramics are crystalline (or partly crystalline) inorganic compounds. They include high-performance technical ceramics like alumina (used for electronic substrates, nozzles, and cutting tools), traditional, pottery-based ceramics (including brick and porcelain for baths, sinks, and toilets), and hydrated ceramics—cements and concretes—used for construction. All are hard and brittle, have generally high melting points, and low thermal expansion coefficients, and most are good electrical insulators. When perfect they are exceedingly strong, but the tiny flaws they contain, almost impossible to eliminate, propagate as cracks when the material is loaded in tension or bending, drastically reducing the strength. The compressive strength, however, remains high (8 to 18 times the strength in tension). Impact resistance is low and stress caused by large temperature gradients or thermal shock can cause failure.

Glass. Discovered by the Egyptians and perfected by the Romans, glass is one of the oldest of man-made materials. For most of its long history it was a possession of the rich—as glass beads, ornaments, and vessels, and glass as a glaze on pottery. Its use in windows started in the 15th century, but it was not widespread until the 17th. Now, of course, it is so universal and cheap that—as bottles—we throw it away.

Glass is a mix of oxides, principally silica, SiO_2, that does not crystallize when cooled after melting. Pure glass is crystal clear. Adding metal oxides produces a wide range of colors. Nickel gives a purple hue, cobalt a blue, chromium a green, uranium a green-yellow, iron a green-blue. The addition of iron produces a material that can absorb wavelengths in the infrared range so that heat radiation can be absorbed. Colorless, non-metallic particles (fluorides or phosphates) are added to produce a translucent or an almost opaque white opalescence in glass and glass coatings. Photochromic glass changes color when exposed to UV. Filter glass protects from intense light and UV radiation—it is used in visors for welding.

Profiles for six ceramics and glasses follow, in the order in which they appear in Table 15.1.

Brick

The material. Brick is as old as Babylon (4,000 BC) and as durable. It is the most ancient of all man-made building materials. The regularity and proportions of bricks makes them easy to lay in a variety of patterns, and their durability makes them an ideal material for construction. Clay—the raw material from which bricks are made—is available almost everywhere; finding the energy to fire them can be more of a problem. Pure clay is gray-white in color; the red color of most bricks comes from impurities of iron oxide.

Composition

Bricks are fired clays—fine particulate alumino-silicates that derive from the weathering of rocks.

General properties

Density	1,600	–	2,100	kg/m^3
Price	0.62	–	1.7	USD/kg

Mechanical properties

Young's modulus	15	–	30	GPa
Yield strength (elastic limit)	5	–	14	MPa
Tensile strength	5	–	14	MPa
Elongation	0			%
Hardness—Vickers	20	–	35	HV
Fatigue strength at 10^7 cycles	6	–	9	MPa
Fracture toughness	1	–	2	MPa $\cdot$ m$^{1/2}$

Thermal properties

Melting point	927	–	1,230	°C
Maximum service temperature	600	–	1,000	°C
Thermal conductor or insulator?	Poor insulator			
Thermal conductivity	0.46	–	0.73	W/m $\cdot$ K
Specific heat capacity	750	–	850	J/kg $\cdot$ K
Thermal expansion coefficient	5	–	8	μstrain/°C

Electrical properties

Electrical conductor or insulator?	Good insulator			
Electrical resistivity	1×10^{14}	–	3×10^{16}	μohm $\cdot$ cm
Dielectric constant	7	–	10	
Dissipation factor	0.001	–	0.01	
Dielectric strength	9	–	15	10^6 V/m

The proportions and regularity of brick make it fast to assemble. Brick weathers well, and the texture and color make it visually attractive.

Eco properties: material

Global production, main component	51×10^6			metric ton/yr
Embodied energy, primary production	2.2	–	3.5	MJ/kg
CO_2 footprint, primary production	0.2	–	0.23	kg/kg
Water usage	2.8	–	8.4	L/kg
Eco-indicator value	11			millipoints/kg

Eco properties: processing

Construction energy	0.054	–	0.066	MJ/kg
Construction CO_2	0.009	–	0.011	kg/kg

End of life

Recycle fraction in current supply	15	–	20	%

Typical uses. Domestic and industrial building, walls, paths, and roads.

Stone

The material. Stone is the most durable of all building material. The pyramids (before 3,000 BC), the Parthenon (5th century BC), and the cathedrals of Europe (1,000–1,600 AD) testify to the resistance of stone to attack of every sort. It remained the principal material of construction for important buildings until the early 20th century—the railroads of the world could not have been built without stone for the viaducts and support structures. As the cost of stone increased and brick became cheaper, stone was increasingly used for the outer structure only; today it is largely used as a veneer on a concrete or breezeblock inner structure. Carefully selected samples of fully dense, defect-free stone can have very large compressive strengths—up to 1,000 MPa. But stone in bulk, as used in buildings, always contains defects. Then the average strength is much lower. The data given here are typical of bulk sandstone with a porosity of 5–30%. Bulk limestone is a little less strong, granites somewhat stronger.

Composition

There are many different compositions. The most common are made up of calcium carbonate, silicates, and aluminates.

General properties

Density	2,240	– 2,650	kg/m^3
Price	0.3	– 1	USD/kg

Mechanical properties

Young's modulus	20	– 60	GPa
Yield strength (elastic limit)	2	– 25	MPa
Tensile strength	2	– 25	MPa
Elongation	0		%
Hardness—Vickers	12	– 80	HV
Fatigue strength at 10^7 cycles	2	– 18	MPa
Fracture toughness	0.7	– 1.4	MPa$\cdot$m$^{1/2}$

Thermal properties

Melting point	1,230	– 1,430	°C
Maximum service temperature	350	– 900	°C
Thermal conductor or insulator?	Poor insulator		
Thermal conductivity	5.4	– 6	W/m$\cdot$K
Specific heat capacity	840	– 920	J/kg$\cdot$K
Thermal expansion coefficient	3.7	– 6.3	µstrain/°C

Stone, like wood, is one of man's oldest and most durable building materials.

Electrical properties

Electrical conductor or insulator?	Poor insulator		
Electrical resistivity	1×10^{10}	– 1×10^{14}	μohm·cm
Dielectric constant	6	– 9	
Dissipation factor	0.001	– 0.01	
Dielectric strength	5	– 12	10^6 V/m

Eco properties: material

Embodied energy, primary production	0.4	– 0.6	MJ/kg
CO_2 footprint, primary production	0.03	– 0.04	kg/kg
Water usage	1.7	– 5.1	L/kg
Eco-indicator value	3		millipoints/kg

Eco properties: processing

Construction energy	0.364	– 0.44	MJ/kg
Construction CO_2	0.054	– 0.066	kg/kg

End of life

Recycle fraction in current supply	1	– 2	%

Typical uses. Building and cladding, architecture, sculpture, optical benches for supports for high-performance or vibration-sensitive equipment such as microscopes.

Concrete

The material. Concrete is a composite, and a complex one. The matrix is cement; the reinforcement, a mixture of sand and gravel ("aggregate'") occupying 60–80% of the volume. The aggregate increases the stiffness and strength and reduces the cost (aggregate is cheap). Concrete is strong in compression but cracks easily in tension. This is countered by adding steel reinforcement in the form of wire, mesh, or bars ("rebar"), often with surface contours to lock it into the concrete; reinforced concrete can carry useful loads even when the concrete is cracked. Still higher performance is gained by using steel wire reinforcement that is pre-tensioned before the concrete sets. On relaxing the tension, the wires pull the concrete into compression.

Composition
6:1:2:4 Water:Portland cement:Fine aggregate:Coarse aggregate

General properties

Density	2,300	–	2,600	kg/m^3
Price	0.042	–	0.062	USD/kg

Mechanical properties

Young's modulus	15	–	25	GPa
Yield strength (elastic limit)	1	–	3	MPa
Tensile strength	1	–	1.5	MPa
Elongation	0			%
Hardness—Vickers	5.7	–	6.3	HV
Fatigue strength at 10^7 cycles	0.54	–	0.84	MPa
Fracture toughness	0.35	–	0.45	MPa·m$^{1/2}$

Thermal properties

Melting point	972	–	1,230	°C
Maximum service temperature	480	–	510	°C
Thermal conductor or insulator?	Poor insulator			
Thermal conductivity	0.8	–	2.4	W/m·K
Specific heat capacity	835	–	1,050	J/kg·K
Thermal expansion coefficient	6	–	13	µstrain/°C

Electrical properties

Electrical conductor or insulator?	Poor insulator			
Electrical resistivity	1.8×10^{12}	–	1.8×10^{13}	µohm·cm
Dielectric constant	8	–	12	
Dissipation factor	0.001	–	0.01	
Dielectric strength	0.8	–	1.8	10^6 V/m

Reinforced concrete enables large structures and complex shapes.

Eco properties: material

Global production, main component	16×10^9		metric ton/yr
Reserves	1×10^{12}		metric ton
Embodied energy, primary production	1.0	– 1.3	MJ/kg
CO_2 footprint, primary production	0.09	– 0.12	kg/kg
Water usage	1.7	– 5.1	L/kg
Eco-indicator	4		millipoints/kg

Eco properties: processing

Construction energy	0.0182	– 0.022	MJ/kg
Construction CO_2	0.0018	– 0.0022	kg/kg

End of life

Embodied energy, recycling	0.7	– 0.8	MJ/kg
CO_2 footprint, recycling	0.063	– 0.07	kg/kg
Recycle fraction in current supply	12.5	– 15	%

Typical uses. General civil engineering construction and building.

Alumina

The material. Alumina (Al_2O_3) is to technical ceramics what mild steel is to metals—cheap, easy to process, the workhorse of the industry. It is the material of spark plugs, electrical insulators, and ceramic substrates for microcircuits. In single crystal form it is sapphire, used for watch faces and cockpit windows of high-speed aircraft. More usually it is made by pressing and sintering powder, giving grades ranging from 80 to 99.9% alumina—the rest is porosity, glassy impurities, or deliberately added components. Pure aluminas are white; impurities make them pink or green. The maximum operating temperature increases with increasing alumina content. Alumina has a low cost and a useful and broad set of properties: electrical insulation, high mechanical strength, good abrasion, temperature resistance up to 1,650°C, excellent chemical stability, and moderately high thermal conductivity, but it has limited thermal shock and impact resistance. Chromium oxide is added to improve abrasion resistance; sodium silicate, to improve processability but with some loss of electrical resistance. Competing materials are magnesia, silica, and borosilicate glass.

Composition
Al_2O_3, often with some porosity and some glassy phase.

General properties
Density	3,800	– 3,980	kg/m^3
Price	18.2	– 27.4	USD/kg

Mechanical properties
Young's modulus	343	– 390	GPa
Yield strength (elastic limit)	350	– 588	MPa
Tensile strength	350	– 588	MPa
Compressive strength	690	– 5.5e3	MPa
Elongation	0		%
Hardness—Vickers	1.2e3	– 2.06e3	HV
Fatigue strength at 10^7 cycles	200	– 488	MPa
Fracture toughness	3.3	– 4.8	MPa·m$^{1/2}$

Thermal properties
Melting point	2,000	– 2,100	°C
Maximum service temperature	1,080	– 1,300	°C
Thermal conductor or insulator?	Good conductor		
Thermal conductivity	26	– 38.5	W/m·K
Specific heat capacity	790	– 820	J/kg·K
Thermal expansion coefficient	7	– 7.9	μstrain/°C

On the left: alumina components for wear resistance and for high temperature use (Kyocera Industrial Ceramics Corp.). On the right: an alumina spark plug insulator.

Electrical properties

Electrical conductor or insulator?	Good insulator			
Electrical resistivity	1×10^{20}	–	1×10^{22}	μohm·cm
Dielectric constant	6.5	–	6.8	
Dissipation factor	1×10^{-4}	–	4×10^{-4}	
Dielectric strength	10	–	20	10^6 V/m

Eco properties: material

Global production, main component	1.2×10^6			metric ton/yr
Embodied energy, primary production	49.5	–	54.7	MJ/kg
CO_2 footprint, primary production	2.67	–	2.95	kg/kg
Water usage	29.4	–	88.1	L/kg

Eco properties: processing

Ceramic powder forming energy	25.3	–	27.8	MJ/kg
Ceramic powder forming CO_2	2.02	–	2.23	kg/kg

End of life

Recycle fraction in current supply	0.5	–	1	%

Typical uses. Electrical insulators and connector bodies; substrates; high-temperature components; water faucet valves; mechanical seals; vacuum chambers and vessels; centrifuge linings; spur gears; fuse bodies; heating elements; plain bearings and other wear-resistant components; cutting tools; substrates for microcircuits; spark plug insulators; tubes for sodium vapor lamps; thermal barrier coatings.

Soda-lime glass

The material. Soda-lime glass is the glass of windows, bottles, and lightbulbs, used in vast quantities, the most common of them all. The name suggests its composition: 13−17% NaO (the "soda"), 5−10% CaO (the "lime"), and 70−75% SiO_2 (the "glass"). It has a low melting point, is easy to blow and mold, and is cheap. It is optically clear unless impure, when it is typically green or brown. Windows today have to be flat and that was not—until 1950—easy to do; now the float-glass process, solidifying glass on a bed of liquid tin, makes "plate" glass cheaply and quickly.

Composition
73% SiO_2/1% Al_2O_3/17% Na_2O/4% MgO/5% CaO

General properties

Density	2,440	–	2,490	kg/m^3
Price	0.8	–	1.7	USD/kg

Mechanical properties

Young's modulus	68	–	72	GPa
Yield strength (elastic limit)	30	–	35	MPa
Tensile strength	31	–	35	MPa
Elongation	0			%
Hardness—Vickers	439	–	484	HV
Fatigue strength at 10^7 cycles	29.4	–	32.5	MPa
Fracture toughness	0.55	–	0.7	$MPa \cdot m^{1/2}$

Thermal properties

Maximum service temperature	443	–	673	K
Thermal conductor or insulator?	Poor insulator			
Thermal conductivity	0.7	–	1.3	$W/m \cdot K$
Specific heat capacity	850	–	950	$J/kg \cdot K$
Thermal expansion coefficient	9.1	–	9.5	$\mu strain/°C$

Electrical properties

Electrical conductor or insulator?	Good insulator			
Electrical resistivity	7.94×10^{17}	–	7.94×10^{18}	$\mu ohm \cdot cm$
Dielectric constant	7	–	7.6	
Dissipation factor	0.007	–	0.01	
Dielectric strength	12	–	14	10^6 V/m

Glass is used in both practical and decorative ways.

Eco properties: material

Global production, main component	84×10^6			metric ton/yr
Reserves	1×10^{12}			metric ton
Embodied energy, primary production	10	–	11	MJ/kg
CO_2 footprint, primary production	0.7	–	0.8	kg/kg
Water usage	14	–	20.5	L/kg
Eco-indicator	75			millipoints/kg

Eco properties: processing

Glass molding energy	8.2	–	9.2	MJ/kg
Glass molding CO_2	0.66	–	0.73	kg/kg

End of life

Embodied energy, recycling	7.4	–	9.0	MJ/kg
CO_2 footprint, recycling	0.44	–	0.54	kg/kg
Recycle fraction in current supply	22	–	26	%

Typical uses. Windows, bottles, containers, tubing, lamp bulbs, lenses and mirrors, bells, glazes on pottery and tiles.

Borosilicate glass (Pyrex)

The material. Borosilicate glass is soda-lime glass with most of the lime replaced by borax, B_2O_3. It has a higher melting point than soda-lime glass and is harder to work, but it has a lower expansion coefficient and a high resistance to thermal shock, so it is used for glassware and laboratory equipment.

Composition
74% SiO_2/1% Al_2O_3/15% B_2O_3/4% Na_2O/6% PbO

General properties

Density	2,200	– 2,300	kg/m^3
Price	4.2	– 6.2	USD/kg

Mechanical properties

Young's modulus	61	– 64	GPa
Yield strength (elastic limit)	22	– 32	MPa
Tensile strength	22	– 32	MPa
Compressive strength	264	– 384	MPa
Elongation	0		%
Hardness—Vickers	83.7	– 92.5	HV
Fatigue strength at 10^7 cycles	26.5	– 29.3	MPa
Fracture toughness	0.5	– 0.7	$MPa \cdot m^{1/2}$

Thermal properties

Glass temperature	450	– 602	°C
Maximum service temperature	230	– 460	°C
Thermal conductor or insulator?	Poor insulator		
Thermal conductivity	1	– 1.3	$W/m \cdot K$
Specific heat capacity	760	– 800	$J/kg \cdot K$
Thermal expansion coefficient	3.2	– 4	μstrain/°C

Electrical properties

Electrical conductor or insulator?	Good insulator		
Electrical resistivity	3.16×10^{21}	– 3.16×10^{22}	$\mu ohm \cdot cm$
Dielectric constant	4.65	– 6	
Dissipation factor	0.01	– 0.017	
Dielectric strength	12	– 14	10^6 V/m

Borosilicate glass (Pyrex) is used for ovenware and chemical equipment.

Eco properties: material

Embodied energy, primary production	27	–	30	MJ/kg
CO_2 footprint, primary production	1.6	–	1.8	kg/kg
Water usage	26	–	37.5	L/kg
Eco-indicator	174			millipoints/kg

Eco properties: processing

Glass molding energy	8.8	–	8.98	MJ/kg
Glass molding CO_2	0.64	–	0.72	kg/kg

End of life

Embodied energy, recycling	20	–	23	MJ/kg
CO_2 footprint, recycling	1.2	–	1.4	kg/kg
Recycle fraction in current supply	18	–	23	%

Typical uses. Ovenware, laboratory ware, piping, lenses and mirrors, sealed beam headlights, tungsten sealing, bells.

15.5 Hybrids: composites, foams, wood, and paper

Composites are one of the great material developments of the 20th century. Those with the highest stiffness and strength are made with continuous fibers of glass, carbon, or Kevlar (an aramid) embedded in a thermosetting resin (polyester or epoxy). The fibers carry the mechanical loads, while the matrix material transmits loads to the fibers, provides ductility and toughness, and protects the fibers from damage caused by handling or the environment. It is the matrix material that limits the service temperature and processing conditions. Polyester-glass composites (GFRPs) are the cheapest; epoxy-carbon (CFRPs) and Kevlar-epoxy (KFRPs) are the most expensive. A recent innovation is the use of thermoplastics as the matrix material, using a co-weave of polypropylene and glass fibers that is thermoformed, melting the PP.

Continuous fiber CFRPs and GFRPs are the kings and queens of the composite world. The ordinary workers are polymers reinforced with chopped glass or carbon fibers (SMC and BMC), or with particulates (fillers) of silica sand, talc, or wood flour. They are used in far larger quantities, often in products so ordinary that most people would not guess that they were made of a composite: body panels of cars, household appliances, furniture, and fittings. It would, today, be hard to live without them.

So composites have remarkable potential. But the very thing that creates their properties—the hybridization of two very different materials—makes them near-impossible to recycle. In products with long lives, made in relatively small numbers (aircraft, for instance) this is not a concern; the fuel energy saved by the low weight of the composite far outweighs any penalty associated with the inability to recycle. But it is an obstacle to their use in high-volume, short-lived products (small cheap cars, for example). These environmental concerns can be offset a little by replacing the carbon, aramid, or glass with natural fibers, hemp, kenaf. You will find records for these in Section 15.6.

Foams are made by variants of the process used to make bread. Mix an un-polymerized resin (the dough) with a hardener and a foaming agent (the yeast), wait for a bit, and the agent releases tiny gas bubbles that cause the mixture to rise, then set. There are other ways to make foams: violent stirring, like frothing an egg white, or bubbling gas from below in the way you might make soap foam. All, suitably adapted, are used to make polymer foams. Those made from elastomers are soft and squashy, well adapted for cushions and packaging of delicate objects. Those made from thermoplastics or thermosets are rigid. They are used for more serious energy-absorbing and load-bearing applications: head protection in cycle helmets, cores for structural sandwich panels. And because they are mostly trapped gas, they are excellent thermal insulators.

Their eco-character, however, is mixed. The blowing agents used in the past—CFCs, chlorinated and fluoridated hydrocarbons—cause damage to the ozone layer; they have now been replaced. Some can be recycled, but only 1 to 10% of the foam that has to be collected, transported, and treated is real material (the rest is space), so you don't get much for your money.

Natural materials: wood, plywood, paper. Wood has been used for construction since the earliest recorded time. The ancient Egyptians used it for furniture, sculpture, and coffins before 2,500 BC. The Greeks at the peak of their empire (700 BC) and the Romans at the peak of theirs (around 0 AD) made elaborate buildings, bridges, boats, chariots, and weapons of wood, and established the craft of furniture-making that is still with us today. More diversity of use appeared in medieval times with the use of wood for large-scale building and mechanisms such as carriages, pumps, windmills, and even clocks. This meant that, right up to the end of the 17th century, wood was the principal material of engineering. Since then cast iron, steel, and concrete have displaced it in some of its uses, but timber continues to be used on a massive scale, particularly in housing and small commercial buildings.

Plywood is laminated wood, the layers glued together so that the grains in successive layers are at right angles, giving stiffness and strength in both directions. The number of layers varies, but is always odd (3, 5, 7, ...) to provide symmetry about the central ply—if it is asymmetric, it warps when wet or hot. Those with few plies (3, 5) are significantly stronger and stiffer in the direction of the outermost layers; with an increasing number of plies, the properties become more uniform. High-quality plywood is bonded with synthetic resin.

Papyrus, the forerunner of paper, was made from the flower stems of reeds native to Egypt; it has been known about and used for over 5,000 years. Paper, by contrast, is a Chinese invention (105 AD). It is made from pulped cellulose fibers derived from wood, cotton, or flax. Paper-making uses caustic soda (NaOH) and vast quantities of water—a bad combination if released back into the environment. Modern paper-making plants now release water that (they claim) is as clean as when it entered.

Profiles for 14 hybrid materials follow, in the order in which they appear in Table 15.1.

CFRP (Isotropic)

The material. Carbon fiber reinforced polymer (CFRP) composites offer greater stiffness and strength than any other type, but they are considerably more expensive than glass fiber reinforced polymer (GFRP). Continuous fibers in a polyester or epoxy matrix give the highest performance. The fibers carry the mechanical loads, while the matrix material transmits loads to the fibers and provides ductility and toughness as well as protecting the fibers from damage caused by handling or the environment. It is the matrix material that limits the service temperature and processing conditions.

Composition

Epoxy + continuous HS carbon fiber reinforcement (0, + −45, 90), quasi-isotropic lay-up

General properties

Density	1,500	–	1,600	kg/m^3
Price	40.0	–	44.0	USD/kg

Mechanical properties

Young's modulus	69	–	150	GPa
Yield strength (elastic limit)	550	–	1,050	MPa
Tensile strength	550	–	1,050	MPa
Elongation	0.32	–	0.35	%
Hardness—Vickers	10.8	–	21.5	HV
Fatigue strength at 10^7 cycles	150	–	300	MPa
Fracture toughness	6.12	–	20	MPa·m$^{1/2}$

Thermal properties

Maximum service temperature	140	–	220	°C
Thermal conductor or insulator?	Poor insulator			
Thermal conductivity	1.28	–	2.6	W/m·K
Specific heat capacity	902	–	1,037	J/kg·K
Thermal expansion coefficient	1	–	4	μstrain/°C

Electrical properties

Electrical conductor or insulator?	Poor conductor			
Electrical resistivity	1.65×10^5	–	9.46×10^5	μohm·cm

A CFRP bike frame, courtesy TREK

Eco properties: material

Global production, main component	2.8×10^4			metric ton/yr
Embodied energy, primary production	450	–	500	MJ/kg
CO_2 footprint, primary production	33	–	36	kg/kg
Water usage	360	–	1,367	L/kg

Eco properties: processing

Simple composite molding energy	9	–	12.9	MJ/kg
Simple composite molding CO_2	0.77	–	0.89	kg/kg
Advanced composite molding energy	21	–	23	MJ/kg
Advanced composite molding CO_2	1.7	–	1.8	kg/kg

End of life

Recycle fraction in current supply	0	–		%
Heat of combustion	31	–	33	MJ/kg
Combustion CO_2	3.1	–	3.3	kg/kg

Typical uses. Lightweight structural members in aerospace, ground transportation, and sports equipment such as bikes, golf clubs, oars, boats, and racquets; springs; pressure vessels.

GFRP (Isotropic)

The material. Composites are one of the great material developments of the 20th century. Those with the highest stiffness and strength are made of continuous fibers (glass, carbon, or Kevlar, an aramid) embedded in a thermosetting resin (polyester or epoxy). The fibers carry the mechanical loads, while the matrix material transmits loads to the fibers and provides ductility and toughness as well as protecting the fibers from damage caused by handling or the environment. It is the matrix material that limits the service temperature and processing conditions. Polyester-glass composites (GFRPs) are the cheapest and by far the most widely used. A recent innovation is the use of thermoplastics at the matrix material, either in the form of a co-weave of cheap polypropylene and glass fibers that is thermoformed, melting the PP, or as expensive high-temperature thermoplastic resins such as PEEK that allow composites with higher temperature and impact resistance. High-performance GFRP uses continuous fibers. Those with chopped glass fibers are cheaper and are used in far larger quantities. GFRP products range from tiny electronic circuit boards to large boat hulls, body and interior panels of cars, household appliances, furniture, and fittings.

Composition
Epoxy+ continuous E-glass fiber reinforcement (0, + − 45, 90), quasi-isotropic lay-up

General properties
Density	1,750	– 1,970	kg/m^3
Price	19	– 21	USD/kg

Mechanical properties
Young's modulus	15	– 28	GPa
Yield strength (elastic limit)	110	– 192	MPa
Tensile strength	138	– 241	MPa
Elongation	0.85	– 0.95	%
Hardness—Vickers	10.8	– 21.5	HV
Fatigue strength at 10^7 cycles	55	– 96	MPa
Fracture toughness	7	– 23	MPa·m$^{1/2}$

Thermal properties
Maximum service temperature	413	– 493	°C
Thermal conductor or insulator?	Poor insulator		
Thermal conductivity	0.4	– 0.55	W/m·K
Specific heat capacity	1,000	– 1,200	J/kg·K
Thermal expansion coefficient	8.6	– 32.9	μstrain/°C

GFRP body shell by MAS Design, Windsor, UK.

Electrical properties

Electrical conductor or insulator?	Good insulator			
Electrical resistivity	2.4×10^{21}	–	1.91×10^{22}	μohm·cm
Dielectric constant	4.86	–	5.17	
Dissipation factor	0.004	–	0.009	
Dielectric strength	11.8	–	19.7	10^6 V/m

Eco properties: material

Embodied energy, primary production	107	–	118	MJ/kg
CO_2 footprint, primary production	7.47	–	8.26	kg/kg
Water usage	105	–	309	L/kg

Eco properties: processing

Simple composite molding energy	9	–	12.9	MJ/kg
Simple composite molding CO_2	0.77	–	0.89	kg/kg
Advanced composite molding energy	21	–	23	MJ/kg
Advanced composite molding CO_2	1.7	–	1.8	kg/kg

End of life

Recycle fraction in current supply	0			%
Heat of combustion	12	–	13	MJ/kg
Combustion CO_2	0.9	–	1.0	kg/kg

Typical uses. Sports equipment such as skis, racquets, skateboards, and golf club shafts, ship and boat hulls; body shells; automobile components; cladding and fittings in construction; chemical plant.

Sheet molding compound, SMC

The material. Lay-up and filament winding methods of shaping composites are far too slow and labor-intensive to compete with steel pressings for car body panels and other enclosures. Sheet molding compounds (SMCs) overcome this by allowing molding in a single operation between heated dies. To make SMC, polyester resin containing thickening agents and cheap particulates like calcium carbonate or silica dust is mixed with chopped fibers—usually glass—to form a sheet. The fibers lie more or less parallel to the plane of the sheet, but are randomly oriented in-plane, with a volume fraction between 15 and 50%. This makes a "pre preg" with leather or dough-like consistency. When SMC sheet is pressed between hot dies it polymerizes, producing a strong, stiff sheet molding.

Composition

$(OOC-C_6H_4-COO-C_6H_{10})_n$ + $CaCO_3$ or SiO_2 filler + 15 to 50% chopped glass strand. This record is for 40% glass strand.

General properties

Density	1,800	–	2,000	kg/m^3
Price	5.0	–	5.5	USD/kg

Mechanical properties

Young's modulus	10.5	–	12.5	GPa
Yield strength (elastic limit)	89	–	108	MPa
Tensile strength	89	–	108	MPa
Compressive strength	223	–	270	MPa
Elongation	1.35	–	1.65	%
Hardness—Vickers	17	–	35	HV
Fatigue strength at 10^7 cycles	44	–	56	MPa
Fracture toughness	56	–	9	$MPa \cdot m^{1/2}$

Thermal properties

Glass temperature	150	–	197	°C
Maximum service temperature	140	–	170	°C
Thermal conductor or insulator?	Good insulator			
Thermal conductivity	0.71	–	1.1	$W/m \cdot K$
Specific heat capacity	1,050	–	1,210	$J/kg \cdot K$
Thermal expansion coefficient	16	–	20	$\mu strain/°C$

SMC cycle shed (image courtesy of the ACT program, McMaster University).

Electrical properties

Electrical conductor or insulator?	Good insulator		
Electrical resistivity	9×10^{18}	– 11×10^{18}	μohm·cm
Dielectric constant	4.2	– 4.7	
Dissipation factor	0.009	– 0.01	
Dielectric strength	9	– 11	10^6 V/m

Eco properties: material

Embodied energy, primary production	109	– 121	MJ/kg
CO_2 footprint, primary production	7.7	– 8.5	kg/kg
Water usage	68	– 280	L/kg

Eco properties: processing

Simple composite molding energy	3.5	– 4.0	MJ/kg
Simple composite molding CO_2	0.27	– 0.29	kg/kg

End of life

Recycle fraction in current supply	0		%
Heat of combustion	14	– 15	MJ/kg
Combustion CO_2	1.25	– 1.3	kg/kg

Typical uses. Sheet pressings of all types, competing with steel and aluminum sheet. Car body panels; enclosures; luggage and packing cases.

Bulk molding compound, BMC

The material. Lay-up and filament winding methods of shaping composites are far too slow and labor-intensive to compete with steel pressings for car body panels and other enclosures. Sheet molding compounds (SMCs) and bulk (or dough) molding compounds (BMCs or DMCs) overcome this by allowing molding in a single operation between heated dies. To make BMC, polyester resin containing thickening agents and cheap particulates like calcium carbonate or silica dust is mixed with chopped fibers—usually glass—to form a sheet. The fibers lie more or less parallel to the plane of the sheet, but are randomly oriented in three dimensions, with a volume fraction between 10 and 30%. This makes a "pre-preg" with leather or dough-like consistency. BMC is molded in closed, heated dies to make more complex shapes: door handles, shaped levers, parts for washing machines, and the like. This record is for 25% glass strand.

Composition

$(OOC-C_6H_4-COO-C_6H_{10})_n$ + $CaCO_3$ or SiO_2 filler + 10 to 30% chopped glass strand

General properties

Density	1,800	– 2,000	kg/m^3
Price	4.5	– 5.0	USD/kg

Mechanical properties

Young's modulus	11	– 12	GPa
Yield strength (elastic limit)	30	– 48	MPa
Tensile strength	30	– 48	MPa
Compressive strength	138	– 166	MPa
Elongation	0.5	– 1.0	%
Hardness—Vickers	7	– 16	HV
Fatigue strength at 10^7 cycles	132	– 15	MPa
Fracture toughness	2	– 4	MPa$\cdot$m$^{1/2}$

Thermal properties

Glass temperature	150	– 200	°C
Maximum service temperature	140	– 170	°C
Thermal conductor or insulator?	Good insulator		
Thermal conductivity	0.6	– 1.1	W/m$\cdot$K
Specific heat capacity	1,010	– 1,420	J/kg$\cdot$K
Thermal expansion coefficient	20	– 28	µstrain/°C

A BMC (or DMC) molding. BMC is used for door handles, casings for electrical and gas, and most small moldings in cars.

Electrical properties

Electrical conductor or insulator?	Good insulator			
Electrical resistivity	50×10^{18}	–	70×10^{18}	μohm·cm
Dielectric constant	6.8	–	7.2	
Dissipation factor	0.009	–	0.018	
Dielectric strength	10	–	16	10^6 V/m

Eco properties: material

Embodied energy, primary production	109	–	121	MJ/kg
CO_2 footprint, primary production	7.6	–	8.4	kg/kg
Water usage	89	–	280	L/kg

Eco properties: processing

Simple composite molding energy	3.3	–	0.3	MJ/kg
Simple composite molding CO_2	0.27	–	0.99	kg/kg

End of life

Recycle fraction in current supply	0			%
Heat of combustion	14	–	15	MJ/kg
Combustion CO_2	1.25	–	1.3	kg/kg

Typical uses. Car battery cases; door handles and window winders; washing machine parts such as lids; automotive vents, distributor caps, and other small moldings; casings for telephones; gas and electricity meters.

Furan-based composites

The material. Furan is a volatile organic compound with a five-member aromatic ring made up of four carbon atoms and one oxygen atom. It is made by the thermal decomposition of cellulose-containing materials, particularly corn cobs and pine wood. It polymerizes for give furan resin ("BioRes") that can be reinforced with flax, sisal, wood fibers, or glass to produce a composite ("Furolite") with properties comparable to natural fiber-reinforced epoxy. The composite has good fire and chemical resistance. It can be processed by hand lay-up, spray-up, prepreg, vacuum infusion, pultrusion, or filament winding.

Composition
Furan resin 70 wt% and flax or sisal 30 wt%

General properties

Density	1,200	–	1,350	kg/m^3
Price	7.0	–	9.0	USD/kg

Mechanical properties

Young's modulus	8	–	12	GPa
Yield strength (elastic limit)	30	–	35	MPa
Tensile strength	70	–	80	MPa
Elongation	3	–	3.5	% strain

Thermal properties

Glass temperature	135	–	145	°C
Maximum service temperature	127	–	147	°C
Thermal conductor or insulator?	Poor insulator			
Thermal conductivity	0.16	–	0.2	W/m·°C
Specific heat capacity	1,500	–	1,600	J/kg·°C
Thermal expansion coefficient	80	–	110	µstrain/°C

Electrical properties

Electrical conductor or insulator?	Good insulator

Flax and sisal reinforced furan interior door trim. (Image courtesy TransFurans Chemicals, Belgium)

Eco-properties: material
Unknown.

Typical uses. Interior panels of cars (Daimler, BMW), small boats, structural panels.

More information. TransFurans Chemicals bvba, Industriepark Leukaard 2, 2440 GEEL, Belgium

Rigid polymer foam

The material. Polymer foams are made by the controlled expansion and solidification of a liquid or melt through a blowing agent; physical, chemical, or mechanical blowing agents are possible. The resulting cellular material has a lower density, stiffness, and strength than the parent material, by an amount that depends on its relative density—the volume-fraction of solid in the foam. Rigid foams are made from polystyrene, phenolic, polyethylene, polypropylene, or derivatives of polymethylmethacrylate. They are light and stiff, and have mechanical properties that make them attractive for energy management and packaging, and for light-weight structural use. Open-cell foams can be used as filters, closed cell foams as flotation. Self-skinning foams, called "structural" or "syntactic," have a dense surface skin made by foaming in a cold mold. Rigid polymer foams are widely used as cores of sandwich panels.

Composition

Polystyrene, phenolic, polyethylene, polypropylene, or polymethylmethacrylate plus foaming agent. This record is for a polystyrene foam with a relative density of 0.05.

General properties

Density	47	–	53	kg/m^3
Price	3.0	–	10	USD/kg

Mechanical properties

Young's modulus	0.025	–	0.03	GPa
Yield strength (elastic limit)	0.8	–	1.0	MPa
Tensile strength	1.0	–	1.2	MPa
Compressive strength	0.8	–	1.0	MPa
Elongation	4	–	5	%
Hardness—Vickers	0.08	–	0.1	HV
Fatigue strength at 10^7 cycles	0.48	–	0.6	MPa
Fracture toughness	0.01	–	0.02	MPa$\cdot$m$^{1/2}$

Thermal properties

Glass temperature	82	–	92	°C
Maximum service temperature	90	–	110	°C
Thermal conductor or insulator?	Good insulator			
Thermal conductivity	0.033	–	0.034	W/m$\cdot$K
Specific heat capacity	1,200	–	1,250	J/kg$\cdot$K
Thermal expansion coefficient	60	–	80	μstrain/°C

Rigid polystyrene foam is used for packaging, thermal insulation, and sound absorption.

Electrical properties

Electrical conductor or insulator?	Good insulator			
Electrical resistivity	10×10^{18}	–	1×10^{21}	μohm·cm
Dielectric constant	1.05	–	1.1	
Dissipation factor	0.003	–	0.005	
Dielectric strength	1.9	–	2.1	10^6 V/m

Eco properties: material

Embodied energy, primary production	96	–	107	MJ/kg
CO_2 footprint, primary production	3.7	–	4.1	kg/kg
Water usage	299	–	865	L/kg
Eco-indicator	370			millipoints/kg

Eco properties: processing

Polymer molding energy	19	–	22	MJ/kg
Polymer molding CO_2	1.6	–	1.8	kg/kg
Polymer extrusion energy	7.7	–	8.5	MJ/kg
Polymer extrusion CO_2	0.6	–	0.7	kg/kg

End of life

Recycle fraction in current supply	0.5		1.0	%
Heat of combustion	40	–	42	MJ/kg
Combustion CO_2	3.3	–	3.5	kg/kg

Typical uses. Thermal insulation, cores for sandwich structures, panels, partitions, refrigeration, energy absorption, packaging, buoyancy, flotation.

Flexible polymer foam

The material. Polymer foams are made by the controlled expansion and solidification of a liquid or melt through a blowing agent; physical, chemical, or mechanical blowing agents are possible. The resulting cellular material has a lower density, stiffness, and strength than the parent material, by an amount that depends on its relative density—the volume-fraction of solid in the foam. Flexible foams can be soft and compliant, the material of cushions, mattresses, and padded clothing. Most are made from polyurethane, although latex (natural rubber) and most other elastomers can be foamed.

Composition
Polyurethane with a relative density of 0.16

General properties
Density	70	–	85	kg/m^3
Price	8.2	–	10.4	USD/kg

Mechanical properties
Young's modulus	0.0033	–	0.004	GPa
Yield strength (elastic limit)	0.025	–	0.03	MPa
Tensile strength	0.125	–	0.15	MPa
Compressive strength	0.025	–	0.03	MPa
Elongation	300	–	350	%
Hardness—Vickers	0.0025	–	0.003	HV
Fatigue strength at 10^7 cycles	0.09	–	0.1	MPa
Fracture toughness	0.006	–	0.007	MPa$\cdot$m$^{1/2}$

Thermal properties
Glass temperature	−33	–	−23	°C
Maximum service temperature	72	–	77	°C
Thermal conductor or insulator?	Good insulator			
Thermal conductivity	0.024	–	0.028	W/m$\cdot$K
Specific heat capacity	1,720	–	1,790	J/kg$\cdot$K
Thermal expansion coefficient	150	–	160	μstrain/°C

Electrical properties
Electrical conductor or insulator?	Good insulator			
Electrical resistivity	1×10^{18}	–	10×10^{18}	μohm$\cdot$cm
Dielectric constant	1.14	–	1.2	
Dissipation factor	0.0007	–	0.001	
Dielectric strength	5.5	–	6.5	10^6 V/m

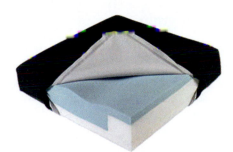

Flexible foams are used for cushions, mattresses, and packaging.
(Image courtesy of Sumed International UK Ltd.)

Eco properties: material

Embodied energy, primary production	104	–	114	MJ/kg
CO_2 footprint, primary production	4.3	–	4.7	kg/kg
Water usage	181	–	544	L/kg
Eco-indicator	385			millipoints/kg

Eco properties: processing

Polymer molding energy	17	–	18.7	MJ/kg
Polymer molding CO_2	1.36	–	1.5	kg/kg
Polymer extrusion energy	6.6	–	7.3	MJ/kg
Polymer extrusion CO_2	0.5	–	0.58	kg/kg

End of life

Recycle fraction in current supply	0			%
Heat of combustion	21	–	23	MJ/kg
Combustion CO_2	2.0	–	2.1	kg/kg

Typical uses. Packaging, buoyancy, cushioning, sleeping mats, soft furnishings, artificial skin, sponges, carriers for inks and dyes.

Softwood: pine, along grain

The material. Softwoods come from coniferous, mostly evergreen, trees such as spruce, pine, fir, and redwood. Wood must be seasoned before it is used. Seasoning is the process of drying to remove some of the natural moisture to make it dimensionally stable. In air-seasoning the wood is dried naturally in a covered but open-sided structure. In kiln-drying the wood is artificially dried in an oven. Wood has been used for construction and to make products since the earliest times. Timber continues to be used on a massive scale, particularly in housing and commercial buildings.

Composition

Cellulose/Hemicellulose/Lignin/12% H_2O

General properties

Density	440	–	600	kg/m^3
Price	0.7	–	1.4	USD/kg

Mechanical properties

Young's modulus	8.4	–	10.3	GPa
Yield strength (elastic limit)	35	–	45	MPa
Tensile strength	60	–	100	MPa
Compressive strength	35	–	43	MPa
Elongation	1.99	–	2.43	%
Hardness—Vickers	3	–	4	HV
Fatigue strength at 10^7 cycles	19	–	23	MPa
Fracture toughness	3.4	–	4.1	$MPa \cdot m^{1/2}$

Thermal properties

Glass temperature	77	–	102	°C
Maximum service temperature	120	–	140	°C
Thermal conductor or insulator?	Good insulator			
Thermal conductivity	0.22	–	0.3	$W/m \cdot K$
Specific heat capacity	1,660	–	1,710	$J/kg \cdot K$
Thermal expansion coefficient	2.5	–	9	$\mu strain/°C$

Electrical properties

Electrical conductor or insulator?	Poor insulator			
Electrical resistivity	6×10^{13}	–	2×10^{14}	$\mu ohm \cdot cm$
Dielectric constant	5	–	6.2	
Dissipation factor	0.05	–	0.1	
Dielectric strength	0.4	–	0.6	$10^6 V/m$

Wood remains one of the world's major structural materials, as well finding application in more delicate objects like furniture and musical instruments.

Eco properties: material

Global production, main component	9.7×10^8			metric ton/yr
Embodied energy, primary production	8.8	–	9.7	MJ/kg
CO_2 footprint, primary production	0.36	–	0.40	kg/kg
Water usage	500	–	750	L/kg
Eco-indicator	42			millipoints/kg

Eco properties: processing

Construction energy	0.455	–	0.55	MJ/kg
Construction CO_2	0.022	–	0.027	kg/kg

End of life

Recycle fraction in current supply	8	–	10	%
Heat of combustion	21	–	22	MJ/kg
Combustion CO_2	1.76	–	1.85	kg/kg

Typical uses. Flooring; furniture; containers; barrels; sleepers (when treated); building construction; boxes; crates and palettes; planing-mill products; subflooring; sheathing and as the feedstock for plywood, particleboard, and hardboard.

Softwood: pine, across grain

The material. Softwoods come from coniferous, mostly evergreen, trees such as spruce, pine, fir, and redwood. Wood must be seasoned before it is used. Seasoning is the process of drying to remove some of the natural moisture to make it dimensionally stable. In air-seasoning the wood is dried naturally in a covered but open-sided structure. In kiln-drying the wood is artificially dried in an oven. Wood has been used for construction and to make products since the earliest times. Timber continues to be used on a massive scale, particularly in housing and commercial buildings.

Composition
Cellulose/Hemicellulose/Lignin/12% H_2O

General properties

Density	440	– 600	kg/m^3
Price	0.7	– 1.4	USD/kg

Mechanical properties

Young's modulus	0.6	– 0.9	GPa
Yield strength (elastic limit)	1.7	– 2.6	MPa
Tensile strength	3.2	– 3.9	MPa
Compressive strength	3	– 9	MPa
Elongation	1	– 1.5	%
Hardness—Vickers	2.6	– 3.2	HV
Fatigue strength at 10^7 cycles	0.96	– 1.2	MPa
Fracture toughness	0.4	– 0.5	MPa·m$^{1/2}$

Thermal properties

Glass temperature	77	– 102	°C
Maximum service temperature	120	– 140	°C
Thermal conductor or insulator?	Good insulator		
Thermal conductivity	0.08	– 0.14	W/m·K
Specific heat capacity	1,660	– 1,710	J/kg·K
Thermal expansion coefficient	26	– 36	μstrain/°C

Electrical properties

Electrical conductor or insulator?	Poor insulator		
Electrical resistivity	2.1×10^{14}	– 7×10^{14}	μohm·cm
Dielectric constant	5	– 6.2	
Dissipation factor	0.03	– 0.07	
Dielectric strength	1	– 2	10^6 V/m

Wood remains one of the world's major structural materials, as well finding application in more delicate objects like furniture and musical instruments. The micrograph shows the section of softwood across the grain.

Eco properties: material

Global production, main component	9.7×10^8			metric ton/yr
Embodied energy, primary production	8.8	–	9.7	MJ/kg
CO_2 footprint, primary production	0.36	–	0.4	kg/kg
Water usage	500	–	750	L/kg
Eco-indicator	42			millipoints/kg

Eco properties: processing

Construction energy	0.455	–	0.55	MJ/kg
Construction CO_2	0.022	–	0.027	kg/kg

End of life

Recycle fraction in current supply	8	–	10	%
Heat of combustion	21	–	22	MJ/kg
Combustion CO_2	1.76	–	1.85	kg/kg

Typical uses. Flooring; furniture; containers; barrels; sleepers (when treated); building construction; boxes; crates and palettes; sub-flooring; sheathing and as the feedstock for plywood, particleboard, and hardboard.

Hardwood: oak, along grain

The material. Hardwoods come from broad leaf deciduous trees such as oak, ash, elm, sycamore, and mahogany. Although most hardwoods are harder than softwoods, there are exceptions: balsa, for instance, is a hardwood. Wood must be seasoned before it is used. Seasoning is the process of drying the natural moisture out of the raw timber to make it dimensionally stable, allowing its use without its shrinking or warping. In air-seasoning the wood is dried naturally in a covered but open-sided structure. In kiln-drying the wood is artificially dried in an oven or kiln.

Composition
Cellulose/Hemicellulose/Lignin/12% H_2O

General properties

Density	850	–	1,030	kg/m^3
Price	3.0	–	11	USD/kg

Mechanical properties

Young's modulus	20.6	–	25.2	GPa
Yield strength (elastic limit)	43	–	52	MPa
Tensile strength	132	–	162	MPa
Compressive strength	68	–	83	MPa
Elongation	1.7	–	2.1	%
Hardness—Vickers	13	–	15.8	HV
Fatigue strength at 10^7 cycles	42	–	52	MPa
Fracture toughness	9	–	10	MPa$\cdot$m$^{1/2}$

Thermal properties

Glass temperature	77	–	102	°C
Maximum service temperature	120	–	140	°C
Thermal conductor or insulator?	Good insulator			
Thermal conductivity	0.41	–	0.5	W/m$\cdot$K
Specific heat capacity	1,660	–	1,710	J/kg$\cdot$K
Thermal expansion coefficient	2.5	–	9	μstrain/°C

Electrical properties

Electrical conductor or insulator?	Poor insulator			
Electrical resistivity	6×10^{13}	–	2×10^{14}	μohm$\cdot$cm
Dielectric constant	5	–	6	
Dissipation factor	0.1	–	0.15	
Dielectric strength	0.4	–	0.6	10^6 V/m

Wood remains one of the world's major structural materials, as well finding application in more delicate objects like furniture. The micrograph shows the section of hardwood parallel to the grain.

Eco properties: material

Global production, main component	9.6×10^8			metric ton/yr
Embodied energy, primary production	9.8	–	10.9	MJ/kg
CO_2 footprint, primary production	0.8	–	0.94	kg/kg
Water usage	500	–	750	L/kg
Eco-indicator	19			millipoints/kg

Eco properties: processing

Construction energy	0.455	–	0.55	MJ/kg
Construction CO_2	0.022	–	0.027	kg/kg

End of life

Recycle fraction in current supply	8	–	10	%
Heat of combustion	21	–	22	MJ/kg
Combustion CO_2	1.76	–	1.85	kg/kg

Typical uses. Flooring; stairways; furniture; handles; veneer; sculpture; wooden ware; sash; doors; general millwork; framing; but these are just a few. Almost every load-bearing and decorative object has, at one time or another, been made from wood.

Hardwood: oak, across grain

The material. Hardwoods come from broad leaf deciduous trees such as oak, ash, elm, sycamore, and mahogany. Although most hardwoods are harder than softwoods, there are exceptions: balsa, for instance, is a hardwood. Wood must be seasoned before it is used. Seasoning is the process of drying the natural moisture out of the raw timber to make it dimensionally stable, allowing its use without shrinking or warping. In air-seasoning the wood is dried naturally in a covered but open-sided structure. In kiln-drying the wood is artificially dried in an oven or kiln.

Composition
Cellulose/Hemicellulose/Lignin/12% H_2O

General properties

Density	850	– 1,030	kg/m^3
Price	3.0	– 11	USD/kg

Mechanical properties

Young's modulus	4.5	– 5.8	GPa
Yield strength (elastic limit)	4	– 5.9	MPa
Tensile strength	7.1	– 8.7	MPa
Compressive strength	12.7	– 15.6	MPa
Elongation	1	– 1.5	%
Hardness—Vickers	10	– 12	HV
Fatigue strength at 10^7 cycles	2.1	– 2.6	MPa
Fracture toughness	0.8	– 1	$MPa \cdot m^{1/2}$

Thermal properties

Glass temperature	77	– 102	°C
Maximum service temperature	120	– 140	°C
Thermal conductor or insulator?	Good insulator		
Thermal conductivity	0.16	– 0.2	$W/m \cdot K$
Specific heat capacity	1,660	– 1,710	$J/kg \cdot K$
Thermal expansion coefficient	37	– 49	μstrain/°C

Electrical properties

Electrical conductor or insulator?	Poor insulator		
Electrical resistivity	2.1×10^{14}	– 7×10^{14}	μohm $\cdot$ cm
Dielectric constant	5	– 6	
Dissipation factor	0.1	– 0.15	
Dielectric strength	0.4	– 0.6	10^6 V/m

Wood remains one of the world's major structural materials, as well finding application in more delicate objects like furniture. The micrograph shows the section of hardwood across the grain. (Image of table courtesy Jia Design, UK.)

Eco properties: material

Global production, main component	9.6×10^8			metric ton/yr
Embodied energy, primary production	9.8	–	10.9	MJ/kg
CO_2 footprint, primary production	0.8	–	0.94	kg/kg
Water usage	500	–	750	L/kg
Eco-indicator	19			millipoints/kg

Eco properties: processing

Construction energy	0.455	–	0.55	MJ/kg
Construction CO_2	0.022	–	0.027	kg/kg

End of life

Recycle fraction in current supply	8	–	10	%
Heat of combustion	21	–	22	MJ/kg
Combustion CO_2	1.76	–	1.85	kg/kg

Typical uses. Flooring; stairways; furniture; handles; veneer; sculpture; wooden ware; sash; doors; general millwork; framing; but these are just a few. Almost every load-bearing and decorative object has, at one time or another, been made from wood.

Plywood

The material. Plywood is laminated wood; the layers are glued together so that the grain in successive layers is at right angles, giving it stiffness and strengthening it in both directions. The number of layers varies, but is always odd (3, 5, 7, …) to give it symmetry about the core ply—if it is asymmetric, it warps when wet or hot. Those with few plies (3, 5) are significantly stronger and stiffer in the direction of the outermost layers; with an increasing number of plies, the properties become more uniform. High-quality plywood is bonded with synthetic resin. The data listed below describe the in-plane properties of a typical 5-ply.

Composition
Cellulose/Hemicellulose/Lignin/12% H_2O/Adhesive

General properties

Density	700	–	800	kg/m^3
Price	0.5	–	1.1	USD/kg

Mechanical properties

Young's modulus	6.9	–	13	GPa
Yield strength (elastic limit)	9	–	30	MPa
Tensile strength	10	–	44	MPa
Compressive strength	8	–	25	MPa
Elongation	2.4	–	3	%
Hardness—Vickers	3	–	9	HV
Fatigue strength at 10^7 cycles	7	–	16	MPa
Fracture toughness	1	–	1.8	$MPa \cdot m^{1/2}$

Thermal properties

Glass temperature	120	–	140	°C
Maximum service temperature	100	–	130	°C
Thermal conductor or insulator?	Good insulator			
Thermal conductivity	0.3	–	0.5	$W/m \cdot K$
Specific heat capacity	1,660	–	1,710	$J/kg \cdot K$
Thermal expansion coefficient	6	–	8	µstrain/°C

Electrical properties

Electrical conductor or insulator?	Poor insulator			
Electrical resistivity	6×10^{13}	–	2×10^{14}	$\mu ohm \cdot cm$
Dielectric constant	6	–	8	
Dissipation factor	0.05	–	0.09	
Dielectric strength	0.4	–	0.6	10^6 V/m

Plywood dominates the market for both wood and steel stud construction. It is widely used, too, for furniture and fittings, boat building, and packaging.

Eco properties: material

Embodied energy, primary production	13	–	16	MJ/kg
CO_2 footprint, primary production	0.78	–	0.87	kg/kg
Water usage	500	–	1,000	l/kg
Eco-indicator	270			millipoints/kg

Eco properties: processing

Construction energy	0.455	–	0.55	MJ/kg
Construction CO_2	0.022	–	0.027	kg/kg

End of life

Recycle fraction in current supply	1	–	2	%
Heat of combustion	19	–	21	MJ/kg
Combustion CO_2	1.7	–	1.8	kg/kg

Typical uses. Furniture, building, and construction; marine and boat building; packaging; transportation and vehicles; musical instruments; aircraft; modeling.

Paper and cardboard

The material. Papyrus, the forerunner of paper, was made from the flower stem of the reed, native to Egypt; it has been known and used for over 5,000 years. Paper, by contrast, is a Chinese invention (105 AD). It is made from pulped cellulose fibers derived from wood, cotton, or flax. There are many different types of papers and paper boards—tissue paper, newsprint, craft paper for packaging, office paper, fine-glazed writing paper, cardboard—and a correspondingly wide range of properties. The data below span the range of newsprint and craft paper.

Composition

Cellulose fibers, usually with filler and colorant

General properties

Density	480	–	860	kg/m^3
Price	0.9	–	1.1	USD/kg

Mechanical properties

Young's modulus	3	–	8.9	GPa
Yield strength (elastic limit)	15	–	34	MPa
Tensile strength	23	–	51	MPa
Compressive strength	41	–	55	MPa
Elongation	0.75	–	2	%
Hardness—Vickers	4	–	9	HV
Fatigue strength at 10^7 cycles	13	–	24	MPa
Fracture toughness	6	–	10	MPa$\cdot$m$^{1/2}$

Thermal properties

Glass temperature	47	–	67	°C
Maximum service temperature	77	–	130	°C
Thermal conductor or insulator?	Good insulator			
Thermal conductivity	0.06	–	0.17	W/m$\cdot$K
Specific heat capacity	1,340	–	1,400	J/kg$\cdot$K
Thermal expansion coefficient	5	–	20	μstrain/°C

Electrical properties

Electrical conductor or insulator?	Good insulator			
Electrical resistivity	1×10^{13}	–	1×10^{14}	μohm$\cdot$cm
Dielectric constant	2.5	–	6	
Dissipation factor	0.015	–	0.04	
Dielectric strength	0.2	–	0.3	10^6 V/m

Cardboard, ready for recycling

Eco properties: material

Global production, main component	3.6×10^8			metric ton/yr
Embodied energy, primary production	49	–	54	MJ/kg
CO_2 footprint, primary production	1.1	–	1.2	kg/kg
Water usage	500	–	1,500	L/kg
Eco-indicator	110			millipoints/kg

Eco properties: processing

Construction energy	0.475	–	0.525	MJ/kg
Construction CO_2	0.023	–	0.026	kg/kg

End of life

Embodied energy, recycling	18	–	21	MJ/kg
CO_2 footprint, recycling	0.72	–	0.8	kg/kg
Recycle fraction in current supply	70	–	74	%
Heat of combustion	19	–	20	MJ/kg
Combustion CO_2	1.8	–	1.9	kg/kg

Typical uses. Packaging; filtering; writing; printing; currency; electrical and thermal insulation; gaskets.

15.6 Man-made and natural fibers

Man-made fiber. The rapid expansion of fiber-reinforced composites in the 20th century gives three man-made fibers—aramid, carbon, and glass—prominence as engineering materials. All three are stiff and strong, and two of them (aramid and carbon) are exceptionally light, making them ideal for reinforcement for aerospace structures, sports equipment, and, increasingly, ground transportation systems. Glass fiber is cheap, making composites containing it viable for cars and household products. Carbon and aramid are more expensive, limiting their use to higher-value applications.

Natural fibers. Natural fibers have been use to make fabrics, rope, and twine for thousands of years. They still are. Vegetable fibers are based on cellulose and are found in stem, leaf, seed, and fruit. Many derive from fast-growing plants, making them potentially renewable. Mammalian fibers—wool and hair—are based on the protein keratin.

Many natural fibers are remarkably strong and they have low densities, lower than those of carbon or glass fibers. Environmental concerns about the use of man-made fibers as reinforcement in composites has led to a renewed interest in using natural fibers instead. The idea is not new—Henry Ford tried it in 1910—but it has only now become economically viable. Difficulties remain: natural fibers have greater variability than those that are man-made, and fiber length, stiffness, and strength fluctuate from year to year and depend on the weather.

Profiles for 12 fiber types follow in the order in which they appear in Table 15.1.

Aramid (Kevlar 49)

The material. Originally produced by Dupont as Kevlar, aramid fibers are processed in such a way that the polymer chains are aligned parallel to the axis of the fiber. The chemical unit is an aromatic polyamide with a ring structure that produces high stiffness; the string covalent bonding produces high strength. They are available in low-density/high-strength form (Kevlar 29), as high-modulus fibers (Kevlar 49), and as ultra-high-modulus fibers (Kevlar 149). The first is used in ropes, cables, and armor; the second as reinforcement in polymers for aerospace, marine, and automotive components. Nomex fibers have excellent flame and abrasion resistance; they are made into a paper that is used to make honeycomb structures. These materials are exceptionally stable and have good strength, toughness, and stiffness up to 170°C. This record describes the properties of single Kevlar 49 aramid fibers.

Composition
Aramid—an aligned aromatic polyamide

General properties

Density	1,440	– 1,450	kg/m^3
Price	70.3	– 198	USD/kg

Mechanical properties

Young's modulus	125	– 135	GPa
Yield strength (elastic limit)	2,250	– 2,750	MPa
Tensile strength	2,500	– 3,000	MPa
Elongation	2.7	– 2.9	% strain

Thermal properties

Melting point	500	– 530	°C
Maximum service temperature	200	– 300	°C
Thermal conductor or insulator?	Poor insulator		
Thermal conductivity	0.2	– 0.3	W/m·°C
Specific heat capacity	1,350	– 1,450	J/kg·°C
Thermal expansion coefficient	−4	– −2	μstrain/°C

Electrical properties

Electrical conductor or insulator?	Good insulator

A weave of aramid fibers and sails woven from Kevlar (Image of sail courtesy of Ultimate Sails, Wynnum, Queensland, Australia)

Eco-properties: material

Global production, main component	41×10^3			metric ton/yr
Embodied energy, primary production	1,110	–	1,230	MJ/kg
CO_2 footprint, primary production	82.1	–	90.8	kg/kg
Water usage	890	–	980	L/kg

Eco-properties: processing

Fabric production energy	2.48	–	2.73	MJ/kg
Prepreg production energy	38.1	–	42	MJ/kg
Fabric production CO_2	0.198	–	0.218	kg/kg
Prepreg production CO_2	3.05	–	3.36	kg/kg

End of life

Recycle fraction in current supply	0			%
Heat of combustion (net)	27.4	–	28.8	MJ/kg
Combustion CO_2	2.52	–	2.65	kg/kg

Typical uses. As woven cloth: body armour protective clothing, bomb and projectile protection, and, in combination with boron carbide ceramic, bullet proof vests. As paper: honeycomb cores for sandwich panels. As fibers and weaves: reinforcement in polymer matrix composites. Kevlar has been used as a blast-resistant component of a building enclosure system. The US Pentagon building has been reinforced with such a system held in place with a welded hollow structural steel tube frame.

Carbon (HS carbon)

The material. Carbon fibers are made by pyrolizing organic fibers such as viscose, rayon, or polyacrylonitrile (PAN), or from petroleum pitch. The PAN type has the better mechanical properties, but those produced from pitch are cheaper. PAN fibers are first stretched for alignment, then oxidized in air at slightly elevated temperatures, then carbonized in an inert environment at very high temperatures, and finally heated under tension to convert the crystal structure to that of graphite. Carbon fibers have high strength and stiffness with low density, but they oxidize at high temperatures unless the atmosphere is reducing. They come in four grades: high modulus, high strength, ultra high modulus, and ultra high strength—with cost increasing in that order. The single fibers are very thin (<10 microns in diameter); they are generally spun into tows (threads and ropes) and woven into textiles. They are primarily used as reinforcement in polymer, metal, or carbon matrices. This record describes the properties of single high-strength (HS) carbon fibers.

Composition
Carbon, C

General properties
Density	1,800	–	1,840	kg/m^3
Price	124	–	166	USD/kg

Mechanical properties
Young's modulus	225	–	260	GPa
Yield strength (elastic limit)	3,750	–	4,000	MPa
Tensile strength	4,400	–	4,800	MPa
Elongation	0			% strain

Thermal properties
Melting point	3.69e3	–	3.83e3	°C
Maximum service temperature	530	–	580	°C
Thermal conductor or insulator?	Good conductor			
Thermal conductivity	80	–	200	W/m·°C
Specific heat capacity	705	–	715	J/kg·°C
Thermal expansion coefficient	0.2	–	0.4	μstrain/°C

Electrical properties
Electrical conductor or insulator?	Good conductor

Spools of carbon fiber will soon be woven into a strong and lightweight material.
(Image courtesy of Dusty Cline/Dreamstime.com)

Eco-properties: material

Global production, main component	48×10^3		metric ton/yr
Embodied energy, primary production	380	– 420	MJ/kg
CO_2 footprint, primary production	23.9	– 26.4	kg/kg
Water usage	399	– 441	L/kg

Eco-properties: processing

Fabric production energy	2.48	– 2.73	MJ/kg
Prepreg production energy	38.1	– 42	MJ/kg
Fabric production CO_2	0.198	– 0.218	kg/kg
Prepreg production CO_2	3.05	– 3.36	kg/kg

End of life

Recycle fraction in current supply	4.73	– 5.22	%
Heat of combustion (net)	32	– 33.6	MJ/kg
Combustion CO_2	3.58	– 3.76	kg/kg

Typical uses. Reinforcement of polymers to make CFRP, and of metals and ceramics to make metal matrix and ceramic matrix composites. Carbon fiber-reinforced carbon is used for brake pads in racing cars and aircraft.

Glass (E-glass)

The material. Glass fibers are made by drawing molten glass through a spinner, producing continuous fibers of diameter between 10 and 100 microns. Their perfection gives them exceptional strength in tension, yet they are flexible. They can be aggregated into loose felt that has very low heat conduction. Therefore they are used in an array of thermal insulation products for buildings; most commonly in a loose matte form. They can be woven into a fabric, and printed or colored to produce a fire-resistant substitute for curtains or covers (when the woven fabric is treated with silicone it can be used up to 250°C). As chopped strand or as continuous fibers or yarns (bundles of fibers), they form the reinforcement in glass fiber reinforced polymers, GFRPs. There are several grades of glass fiber, differing in composition and strength. E-glass is the standard reinforcement. C-glass has better corrosion resistance than E; R and S have better mechanical properties than E but are more expensive. AR-glass resists alkalis, allowing it to be used to reinforce cement. This record describes the properties of single E-glass fibers.

Composition

E-glass composition is complex: silica (SiO_2), 53–55%, alumina (Al_2O_3), 14–15.5%, calcium-magnesium oxide (CaO–MgO), 20–24%, boron oxide (B_2O_3), 5–9%.

General properties

Density	2,550	–	2,600	kg/m^3
Price	1.63	–	3.26	USD/kg

Mechanical properties

Young's modulus	72	–	85	GPa
Tensile strength	1,900	–	2,050	MPa
Elongation	0			% strain

Thermal properties

Maximum service temperature	347	–	377	°C
Thermal conductor or insulator?	Poor insulator			
Thermal conductivity	1.2	–	1.35	W/m·°C
Specific heat capacity	800	–	805	J/kg·°C
Thermal expansion coefficient	4.9	–	5.1	μstrain/°C

Electrical properties

Electrical conductor or insulator?	Good insulator

Glass fiber roving

Eco-properties: material

Global production, main component	4.9×10^6		metric ton/yr
Embodied energy, primary production	62.2	– 68.8	MJ/kg
CO_2 footprint, primary production	3.34	– 3.69	kg/kg
Water usage	89	– 99	L/kg

Eco-properties: processing

Fabric production energy	2.48	– 2.73	MJ/kg
Prepreg production energy	38.1	– 42	MJ/kg
Fabric production CO_2	0.198	– 0.218	kg/kg
Prepreg production CO_2	3.05	– 3.36	kg/kg

End of life

Recycle fraction in current supply	0.1	%

Typical uses. Glass fiber is used for thermal insulation, fire-resistant fabric, and reinforcement of polymers to make GFRP. E-glass is formulated specifically for reinforcing polymers.

Coir

The material. Coir (from Malayalam kayar cord) is a coarse fiber extracted from the fibrous outer shell of a coconut. The individual fiber cells are narrow and hollow, with thick walls made of cellulose. They are pale when immature but later become hardened and yellowed as a layer of lignin is deposited on their walls.

There are two varieties of coir. White coir is harvested from the coconuts before they are ripe. The fibers are white or light brown in color and are smooth and fine. They are generally spun to make yarn that is used in mats or rope. Brown coir is harvested from fully ripened coconuts. It is thicker and stronger and has greater abrasion resistance than white coir. It is typically used in mats, brushes, and sacking.

The coir fiber is relatively waterproof and is one of the few natural fibers resistant to damage by salt water. Both modulus and tensile strength depend strongly on the type of fiber (bundle or single filament) and moisture content.

Composition
Cellulose $(C_6-H_{10}-O_5)n$

General properties

Density	1,150	– 1,220	kg/m^3
Price	0.25	– 0.5	USD/kg

Mechanical properties

Young's modulus	4	– 6	GPa
Yield strength (elastic limit)	100	– 150	MPa
Tensile strength	135	– 240	MPa
Flexural strength (modulus of rupture)	135	– 240	MPa
Elongation	15	– 35	% strain

Thermal properties

Thermal conductor or insulator?	Poor insulator		
(Woven fabric has lower conductivity because of trapped air)			
Thermal expansion coefficient	37.4	– 49.3	$\mu strain/°C$

Electrical properties

Electrical conductor or insulator?	Good insulator

Coir. The individual fibers have a diameter of 150–400 microns and a length of 3.5–15 mm. (Image courtesy of Melbourne Museum, Melbourne, Victoria, Australia.)

Eco-properties: material

Global production, main component	0.25×10^6			metric ton/yr
Embodied energy, primary production	7.2	–	7.96	MJ/kg
CO_2 footprint, primary production	0.427	–	0.472	kg/kg
Water usage	2,320	–	3,100	L/kg
Eco-indicator	6.6			millipoints/kg

Eco-properties: processing

Fabric production energy	2.48	–	2.73	MJ/kg
Fabric production CO_2	0.198	–	0.218	kg/kg

End of life

Recycle fraction in current supply	8.55	–	9.45	%
Heat of combustion (net)	14.2	–	14.9	MJ/kg
Combustion CO_2	1.39	–	1.46	kg/kg

Typical uses. White coir is used in rope making and, when woven, for matting. White coir is also used to make fishing nets due to its excellent resistance to salt water. Brown coir is used in floor mats and doormats, brushes, mattresses, floor tiles, and sacking and twine. Brown coir pads are sprayed with rubber latex, which bonds the fibers together (rubberized coir); it is used as upholstery padding in the automobile industry.

Cotton

The material. Cotton is fiber from the blossom of the plant *Gossypium*, grown in hot climates. There are many species; all are roughly 95% cellulose, with a little wax. Fiber length and fineness determine quality. Cotton is used to make a vast range of different fabrics: sailcloth, canvas, voile, calico, gingham, muslin, crepe, cambric, organdy, crinoline, twill, and damask. They differ in the quality of the fiber and the nature of the weave. This record gives data for single cotton fibers. The properties of fabrics made from it depend strongly on the weave.

Composition
Cellulose $(C_6-H_{10}-O_5)_n$

General properties

Density	1,520	–	1,560	kg/m^3
Price	2.1	–	4.2	USD/kg

Mechanical properties

Young's modulus	7	–	12	GPa
Yield strength (elastic limit)	100	–	350	MPa
Tensile strength	350	–	800	MPa
Elongation	5	–	12	% strain

Thermal properties

Glass temperature	110	–	130	°C
Thermal conductor or insulator?	Poor insulator			
Thermal conductivity	0.57	–	0.61	$W/m \cdot °C$
(Woven fabric has lower conductivity because of trapped air)				
Specific heat capacity	1,200	–	1,230	$J/kg \cdot °C$
Thermal expansion coefficient	15	–	30	μstrain/°C

Electrical properties

Electrical conductor or insulator?	Good insulator

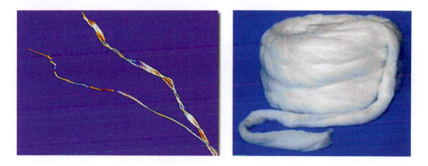

Cotton fiber and wool. (Image of fiber courtesy of Michael W. Davidson, Mortimer Abramowitz, Olympus America Inc., and The Florida State University; image of wool courtesy of American Fiber & Finishing, Inc., www.affinc.com)

Eco-properties: material

Global production, main component	27×10^{6}			metric ton/yr
Embodied energy, primary production	44	–	48	MJ/kg
CO_2 footprint, primary production	2.4	–	2.7	kg/kg
Water usage	7,400	–	8,200	L/kg

Eco-properties: processing

Fabric production energy	2.48	–	2.73	MJ/kg
Fabric production CO_2	0.198	–	0.218	kg/kg

End of life

Recycle fraction in current supply	0.1			%
Heat of combustion (net)	17	–	17.9	MJ/kg
Combustion CO_2	1.39	–	1.46	kg/kg

Typical uses. Fabric and ropes, bandages, textiles, reinforcing mesh.

Flax

The material. Flax is a bast-derived fiber from the stalk of the linseed plant (*Linum usitatissimum*). As with hemp, the fibers are freed from the stalk by an open-air retting process of an exceptionally smelly kind. (Many of those who have to put up with it live in Belgium and Ireland, two major producers.) It is used to make linen and rope, all highly valued for their strength and durability. Linen coverings and wall hangings were widely used in the 19th century for internal surfaces in grand houses.

Flax, like hemp, kenaf, ramie, and jute, belongs to the family of bast plants, known for their long, tough fibers. Bast plants have an outer stem layer that contains 10 to 40% of the stronger fibrous mass of the stem in fiber bundles surrounding a less strong inner fibrous material.

Composition
Cellulose $(C_6-H_{10}-O_5)_n$ with up to 12 wt% H_2O

General properties

Density	1,420	– 1,520	kg/m^3
Price	2.1	– 4.2	USD/kg

Mechanical properties

Young's modulus	75	– 90	GPa
Yield strength (elastic limit)	150	– 338	MPa
Tensile strength	750	– 940	MPa
Elongation	1.2	– 1.8	% strain

Thermal properties

Glass temperature	110	– 130	°C
Thermal conductor or insulator?	Poor insulator		
(Woven fabric has lower conductivity because of trapped air)			
Thermal conductivity	0.25	– 0.3	W/m·°C
Specific heat capacity	1,220	– 1,420	J/kg·°C
Thermal expansion coefficient	15	– 30	μstrain/°C

Electrical properties

Electrical conductor or insulator?	Good insulator

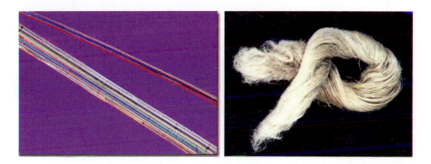

Flax fiber and spun twine. (Image of fiber courtesy of Michael W. Davidson, Mortimer Abramowitz, Olympus America Inc., and The Florida State University; image of spun fiber courtesy of Jeffrey Warren of Vestal Design, www.vestaldesign.com)

Eco-properties: material

Global production, main component	0.75×10^6			metric ton/yr
Embodied energy, primary production	10	–	12	MJ/kg
CO_2 footprint, primary production	0.37	–	0.41	kg/kg
Water usage	2,900	–	3,250	L/kg

Eco-properties: processing

Fabric production energy	2.48	–	2.73	MJ/kg
Fabric production CO_2	0.198	–	0.218	kg/kg

End of life

Recycle fraction in current supply	0.1			%
Heat of combustion (net)	17	–	17.9	MJ/kg
Combustion CO_2	1.39	–	1.46	kg/kg

Typical uses. Fabric and ropes, floor coverings, interior textiles, and furniture upholstery. Flax has been used as a reinforcement polymer matrix composite for natural fiber-reinforced concrete. It demonstrates good stability in the cured concrete while lending the composite a measure of strength. However, steel reinforcing is still necessary.

Hemp

The material. Hemp is a fiber derived from the stalk of the weed *Cannabis sativa*. It is chemistry is almost the same as cotton but its fibers are coarser and stronger. The fibers are freed from the stalk by a retting process remarkable for the unpleasant smells it creates. This record gives data for single hemp fibers. The properties of fabrics made from it depend strongly on the weave.

Composition

Cellulose $(C_6-H_{10}-O_5)_n$ with up to 12 wt% H_2O

General properties

Density	1,470	– 1,520	kg/m^3
Price	1.0	– 2.1	USD/kg

Mechanical properties

Young's modulus	55	– 70	GPa
Yield strength (elastic limit)	200	– 400	MPa
Tensile strength	550	– 920	MPa
Elongation	1.4	– 1.7	% strain

Thermal properties

Glass temperature	96.9	– 107	°C
Maximum service temperature	110	– 130	°C
Thermal conductor or insulator?	Poor insulator		
Thermal conductivity	0.25	– 0.3	W/m·°C
(Woven fabric has lower conductivity because of trapped air)			
Specific heat capacity	1,200	– 1,220	J/kg·°C
Thermal expansion coefficient	15	– 30	µstrain/°C

Electrical properties

Electrical conductor or insulator?	Good insulator

Hemp twine

Eco-properties: material

Global production, main component	83×10^3			metric ton/yr
Embodied energy, primary production	9.5	–	10.5	MJ/kg
CO_2 footprint, primary production	0.29	–	0.33	kg/kg
Water usage	2,500	–	2,780	L/kg

Eco-properties: processing

Fabric production energy	2.48	–	2.73	MJ/kg
Fabric production CO_2	0.198	–	0.218	kg/kg

End of life

Recycle fraction in current supply	0.1			%
Heat of combustion (net)	17.8	–	18.7	MJ/kg
Combustion CO_2	1.54	–	1.62	kg/kg

Typical uses. Rope; coarse fabric.

Jute

The material. Jute is a long, soft, shiny vegetable fiber made from plants in the genus *Corchorus*, family *Malvaceae*. Like kenaf, industrial hemp, flax (linen), and ramie, jute is a bast fiber plant, one in which the fibers are extracted from the stem or bast. Jute is one of the cheapest natural fibers and is second only to cotton in amount produced and variety of uses. It can be spun into coarse, strong threads. When woven it is called hessian or burlap. There is growing interest in using jute as a reinforcement in composites, replacing glass.

Composition
Jute fibers are primarily cellulose $(C_6-H_{10}-O_5)_n$ and lignin.

General properties
Density	1,440	–	1,520	kg/m^3
Price	0.35	–	1.5	USD/kg

Mechanical properties
Young's modulus	35	–	60	GPa
Yield strength (elastic limit)	145	–	530	MPa
Tensile strength	400	–	860	MPa
Elongation	1.7	–	2	% strain

Thermal properties
Glass temperature	107	–	117	°C
Maximum service temperature	127	–	147	°C
Thermal conductor or insulator?	Poor insulator			
Thermal conductivity	0.25	–	0.35	W/m·°C
(Woven fabric has lower conductivity because of trapped air)				
Specific heat capacity	1,200	–	1,220	J/kg·°C
Thermal expansion coefficient	15	–	30	μstrain/°C

Electrical properties
Electrical conductor or insulator?	Good insulator

Jute. The individual fiber diameter is about 200 microns. (Image courtesy of oriental-trading.com)

Eco-properties: material

Global production, main component	2.8×10^6		metric ton/yr
Embodied energy, primary production	30	– 33	MJ/kg
CO_2 footprint, primary production	1.2	– 1.4	kg/kg
Water usage	2,680	– 3,700	L/kg
Eco-indicator	6.6		millipoints/kg

Eco-properties: processing

Fabric production energy	2.48	– 2.73	MJ/kg
Fabric production CO_2	0.198	– 0.218	kg/kg

End of life

Recycle fraction in current supply	8.55	– 9.45	%
Heat of combustion (net)	16.9	– 17.7	MJ/kg
Combustion CO_2	1.39	– 1.46	kg/kg

Typical uses. Jute is used chiefly to make cloth for wrapping bales of raw cotton and to make sacks and coarse cloth. The fibers are also woven into curtains, chair coverings, carpets, area rugs, hessian cloth, and the backing for linoleum.

Kenaf

The material. Kenaf (*Hibiscus cannabinus*) is a fast growing fiber crop related to cotton, okra, and hibiscus. The plants, which reach heights of 8 to 20 feet, are harvested for their stalks from which the fiber is extracted. The fiber is used in the manufacture of industrial textiles, ropes, and twines. Kenaf is among the most widely utilized of the bast (stem) fibers.

Composition
Cellulose $(C_6-H_{10}-O_5)_n$ with up to 12 wt% H_2O

General properties

Density	1,435	–	1,500	kg/m^3
Price	0.26	–	0.52	USD/kg

Mechanical properties

Young's modulus	60	–	66	GPa
Yield strength (elastic limit)	195	–	666	MPa
Tensile strength	217	–	740	MPa
Elongation	1.3	–	5.5	% strain

Thermal properties

Glass temperature	107	–	117	°C
Maximum service temperature	127	–	147	°C
Thermal conductor or insulator?	Poor insulator			
(Woven fabric has lower conductivity because of trapped air)				
Thermal conductivity	0.25	–	0.35	W/m·°C
Specific heat capacity	1,200	–	1,220	J/kg·°C
Thermal expansion coefficient	15	–	30	μstrain/°C

Electrical properties

Electrical conductor or insulator?	Good insulator

Kenaf plant and fiber. (Images: plant courtesy of ATTRA—National Sustainable Agriculture Information Service attra.ncat.org, and fibers courtesy of N_FibreBase, www.n-fibrebase.net)

Eco-properties: material

Global production, main component	2.7×10^6			metric ton/yr
Embodied energy, primary production	31	–	34	MJ/kg
CO_2 footprint, primary production	1.3	–	1.4	kg/kg
Water usage	500	–	1,300	L/kg
Eco-indicator	6.6			millipoints/kg

Eco-properties: processing

Fabric production energy	2.48	–	2.73	MJ/kg
Fabric production CO_2	0.198	–	0.218	kg/kg

End of life

Recycle fraction in current supply	8.55	–	9.45	%
Heat of combustion (net)	17	–	17.9	MJ/kg
Combustion CO_2	1.39	–	1.46	kg/kg

Typical uses. The main uses of kenaf fiber have been rope, twine, coarse cloth (similar to that made from jute), and paper. Emerging uses of kenaf fiber include engineered wood, insulation, and clothing-grade cloth. It is also being used as a reinforcement in polymer-matrix composites.

Ramie

The material. Ramie (*Boehmeria nivea*) is a flowering plant in the nettle family. Ramie is one of the oldest fiber crops, used for at least 6,000 years; it is used for cordage and fabric production. Like jute, hemp, and flax (linen), ramie is a bast fiber plant, one in which the fibers are extracted from the stem or bast. It is one of the strongest natural fibers.

Composition
Cellulose $(C_6-H_{10}-O_5)_n$ with up to 12 wt% H_2O

General properties

Density	1,450	– 1,550	kg/m^3
Price	1.5	– 2.5	USD/kg

Mechanical properties

Young's modulus	38	– 44	GPa
Yield strength (elastic limit)	450	– 612	MPa
Tensile strength	500	– 680	MPa
Elongation	2	– 2.2	% strain

Thermal properties

Glass temperature	117	– 127	°C
Maximum service temperature	127	– 147	°C
Thermal conductor or insulator?	Poor insulator		
Thermal conductivity	0.25	– 0.35	W/m·°C
(Woven fabric has lower conductivity because of trapped air)			
Specific heat capacity	1,200	– 1,220	J/kg·°C
Thermal expansion coefficient	15	– 30	μstrain/°C

Electrical properties

Electrical conductor or insulator?	Good insulator

Ramie fiber. (Image of fiber courtesy of Michael W. Davidson, Mortimer Abramowitz, Olympus America Inc., and The Florida State University; image of cord courtesy the Victoria and Albert Museum, London)

Eco-properties: material

Global production, main component	0.28×10^6			metric ton/yr
Embodied energy, primary production	7.2	–	8.0	MJ/kg
CO_2 footprint, primary production	0.43	–	0.47	kg/kg
Water usage	3,750	–	4,250	L/kg
Eco-indicator	6.6			millipoints/kg

Eco-properties: processing

Fabric production energy	2.48	–	2.73	MJ/kg
Fabric production CO_2	0.198	–	0.218	kg/kg

End of life

Recycle fraction in current supply	8.55	–	9.45	%
Heat of combustion (net)	16.8	–	17.7	MJ/kg
Combustion CO_2	1.39	–	1.46	kg/kg

Typical uses. Ramie is used to make industrial sewing thread, packing materials, fishing nets, and filter cloths. It is also made into fabrics for household furnishings (upholstery, canvas) and clothing, frequently in blends with other textile fibers. Shorter fibers and waste are used in paper manufacturing.

Sisal

The material. Sisal fiber is derived from an agave, *Agave sisalana*. Sisal is valued for cordage use because of its strength, durability, ability to stretch, affinity for certain dyestuffs, and resistance to deterioration in saltwater.

Composition
Cellulose 70 wt% and lignin 12 wt%

General properties

Density	1,400	– 1,450	kg/m^3
Price	0.6	– 0.7	USD/kg

Mechanical properties

Young's modulus	10	– 25	GPa
Yield strength (elastic limit)	495	– 711	MPa
Tensile strength	550	– 790	MPa
Elongation	4	– 6	% strain

Thermal properties

Glass temperature	107	– 117	°C
Maximum service temperature	127	– 147	°C
Thermal conductor or insulator?	Poor insulator		
Thermal conductivity	0.25	– 0.35	W/m·°C
(Woven fabric has lower conductivity because of trapped air)			
Specific heat capacity	1,200	– 1,220	J/kg·°C
Thermal expansion coefficient	15	– 30	μstrain/°C

Electrical properties

Electrical conductor or insulator?	Good insulator

The agave and sisal cord derived from it. The individual fibers have a diameter of 50–300 microns and lengths of 10–30 mm.

Eco-properties: material

Global production, main component	0.24×10^6			metric ton/yr
Embodied energy, primary production	7.2	–	8.0	MJ/kg
CO_2 footprint, primary production	0.42	–	0.47	kg/kg
Water usage	7,400	–	8,300	L/kg
Eco-indicator 95	7			millipoints/kg

Eco-properties: processing

Fabric production energy	2.48	–	2.73	MJ/kg
Fabric production CO_2	0.198	–	0.218	kg/kg

End of life

Recycle fraction in current supply	8.55	–	9.45	%
Heat of combustion (net)	19.3	–	20.2	MJ/kg
Combustion CO_2	1.5	–	1.58	kg/kg

Typical uses. Sisal is used by industry in three grades, according to www.sisal.ws. The lower grade fiber is processed by the paper industry because of its high content of cellulose and hemicellulose. The medium-grade fiber is used in the cordage industry for making ropes and baling and binders twine. Ropes and twines are widely employed for marine, agricultural, and general industrial use. The higher-grade fiber, after treatment, is converted into yarns and used by the carpet industry. Sisal is now also used as a reinforcement in polymer-matrix composites.

Wool

The material. Wool is the hair or fleece of the sheep, alpaca, rabbit (angora), camel (camel hair), certain goats (mohair, cashmere), and llamas (vicuna). It remains, even today, an important commercial fiber because of its hard wearing qualities (as in wool carpets), the remarkable thermal insulation that it offers (as in woolen sweaters), and, at its best, its high quality (as in lamb's wool, mohair, and vicuna). It is widely used for carpets, seat covers, clothing, and blankets.

Wool is keratin. Its fibers have crimps or curls which create pockets that insulate and gives the wool a spongy feel. The outside surface of the fiber consists of a series of serrated scales that overlap each other much like the scales of a fish, making it possible for the fibers to cling together and produce felt. (The same serrations cling together when wool is improperly washed and shrinks; it will only return to its original position after being stretched. Its unique properties allow shaping and tailoring, making wool the most popular fabric for tailoring fine garments. It is woven in many different ways to produce cloth, tweed, plaid, flannel, or worsted. Wool fabrics are dirt resistant, flame resistant, and, in many weaves, resistant to wear and tear.

Composition
Keratin (protein)

General properties

Density	1,250	– 1,340	kg/m^3
Price	2.1	– 4.2	USD/kg

Mechanical properties

Young's modulus	3.8	– 4.2	GPa
Yield strength (elastic limit)	70	– 115	MPa
Tensile strength	190	– 230	MPa
Elongation	35	– 55	% strain

Thermal properties

Thermal conductor or insulator?	Good insulator		
Thermal conductivity	0.19	– 0.22	W/m·°C
Specific heat capacity	1,320	– 1,380	J/kg·°C
Thermal expansion coefficient	15	– 30	µstrain/°C

Electrical properties

Electrical conductor or insulator?	Poor insulator

Wool. (Image courtesy of the US Department of Agriculture)

Eco-properties: material

Global production, main component	1.6×10^6		metric ton/yr
Embodied energy, primary production	51	– 56	MJ/kg
CO_2 footprint, primary production	3.2	– 3.5	kg/kg
Water usage	160,000	– 180,000	L/kg

Eco-properties: processing

Fabric production energy	2.48	– 2.73	MJ/kg
Fabric production CO_2	0.198	– 0.218	kg/kg

End of life

Recycle fraction in current supply	0.1		%
Heat of combustion (net)	20	– 21	MJ/kg
Combustion CO_2	1.39	– 1.46	kg/kg

Typical uses. Clothing, carpets, upholstery fabrics, curtains, blankets, and insulation wool.

Straw bale

The material. Straw is what's left over when grains are harvested. About 128 million metric tons of straw are produced as a by-product of agriculture each year in North America. It is appealing as a building material for several reasons: in areas of grain production it is inexpensive, its embodied energy is low, and because it cannot be used to feed livestock, and there is an overabundance.

Straw has been used as a building material since the first settlements of ancient Egypt were constructed. Today, various construction methods use straw as both the structure and primary building enclosure material. Typically the straw is molded into bales and cut to dimension before being stacked to form thick insulating walls of small to medium-sized buildings. Straw bale arches and vaults have also been constructed to span significant distances.

Composition

Lignin and hemicellulose

General properties

Density	80	– 191	kg/m^3
Price	0.1	– 0.15	USD/kg

Mechanical properties

Young's modulus	5e-4	– 0.002	GPa
Yield strength (elastic limit)	0.16	– 0.48	MPa
Tensile strength	0.01	– 0.02	MPa
Compressive strength	0.16	– 0.48	MPa
Elongation	10	– 20	% strain
Hardness—Vickers	0.016	– 0.048	HV

Thermal properties

Glass temperature	90	– 102	°C
Maximum service temperature	90	– 110	°C
Thermal conductor or insulator?	Good insulator		
Thermal conductivity	0.05	– 0.06	W/m·K
Specific heat capacity	1.66e3	– 1.71e3	J/kg·°C
Thermal expansion coefficient	2	– 11	µstrain/°C

Electrical properties

Electrical resistivity	6e13	– 2e14	µohm.cm
Dielectric constant	7	– 8	
Dissipation factor	0.05	– 0.1	
Dielectric strength	0.2	– 0.4	MV/m

A straw building under construction (www.indymedia.org.uk) using in-fill methods

Eco-properties: material

Embodied energy, primary production	0.1	–	0.3	MJ/kg
CO_2 footprint, primary production	−1.1	–	−0.9	kg/kg

Eco-properties: processing

Assembly and construction energy	0.47	–	0.52	MJ/kg
Assembly and construction CO_2	0.038	–	0.042	kg/kg

End of life

Recycle fraction in current supply	0.1			%
Heat of combustion (net)	19.8	–	21.3	MJ/kg
Combustion CO_2	1.19	–	1.28	kg/kg

Typical uses. Walls, load-bearing and non-load bearing, in domestic and other small buildings.

Appendix—Useful numbers and conversions

A.1 Introduction

Quantitative analysis needs numbers. Many of those needed to understand and quantify eco-aspects of material production and use are presented in the chapters of the text.

- Table 2.1: approximate efficiency factors for energy conversion and the associated CO_2 emission per useful MJ.

- Table 6.7: the energy content of fossil fuels and the CO_2 they emit when burned.

- Table 6.8: the energy efficiency of electricity generation and the related CO_2 per useful kWh.

- Table 6.9: the energy and CO_2 costs of alternative modes of transportation.

- Table 10.11: the energy and CO_2 rating of cars as a function of mass.

- Chapter 15: data sheets listing the attributes of 63 of the most widely used materials.

This appendix assembles further useful numbers and conversion factors.

A.2 Physical constants in SI units

Physical constant	Value in SI units
Absolute zero temperature	$-273.2°C$
Acceleration due to gravity, g	9.807 m/s^2
Avogadro's number, N_A	6.022×10^{23}
Base of natural logarithms, e	2.718
Boltzmann's constant, k	$1.381 \times 10^{-23} \text{ J/K}$
Faraday's constant, k	$9.648 \times 10^4 \text{ C/mol}$

Physical constant	Value in SI units
Gas constant, $\overline{R}$	8.314 J/mol/K
Permeability of vacuum, μ_0	1.257×10^{-6} H/m
Permittivity of vacuum, ε_0	8.854×10^{-12} F/m
Planck's constant, h	6.626×10^{-34} J/s
Velocity of light in vacuum, c	2.998×10^{8} m/s
Volume of perfect gas at STP	22.41×10^{-3} m^3/mol

A.3 Conversion of units, general

Quantity	Imperial unit	SI unit
Angle, θ	1 rad	57.30°
Density, ρ	1 lb/ft^3	16.03 kg/m^3
Diffusion coefficient, D	1 cm^2/s	1.0×10^{-4} m^2/s
Energy, U	See Section A.5	
Force, F	1 kgf	9.807 N
	1 lbf	4.448 N
	1 dyne	1.0×10^{-5} N
Length, ℓ	1 ft	304.8 mm
	1 inch	25.40 mm
	1 Å	0.1 nm
Mass, M	1 metric ton	1,000 kg
	1 short ton	908 kg
	1 long ton	1,107 kg
	1 lb mass	0.454 kg
Power, P	See Section A.5	
Stress, σ	See Section A.4	
Specific heat, Cp	1 cal/gram $\cdot$ °C	4.186 kJ/kg $\cdot$ °C
	1 BTU/lb $\cdot$ °F	4.184 kJ/kg $\cdot$ °C
Stress intensity, K_{1c}	1 ksi $\sqrt{in}$	1.10 MN/m$^{3/2}$
Surface energy, γ	1 erg/cm^2	1 mJ/m^2

Quantity	Imperial unit	SI unit
Temperature, T	$1°F$	$0.556°K$
Thermal conductivity, λ	1 cal/s·cm·°C 1 Btu/h·ft·°F	418.8 W/m·°C 1.731 W/m·°C
Volume, V	1 Imperial gal 1 US gal	4.546×10^{-3} m^3 3.785×10^{-3} m^3
Viscosity, η	1 poise 1 lb ft·s	0.1 N·s/m^2 0.1517 N·s/m^2

A.4 Stress and pressure

The SI unit of stress and pressure is the N/m^2 or the Pascal (Pa), but from a materials point of view it is very small. The levels of stress large enough to distort or deform materials are measured in megaPascals (MPa). The following table lists the conversion factors relating MPa to measures of stress used in the older centimeter-gram-second (cgs) and metric systems (dyne/cm^2, kgf/mm^2) by the Imperial system (lb/in^2, ton/in^2) and by atmospheric science (bar).

Conversion of units—stress and pressure						
To →	MPa	dyne/cm^2	lb/in^2	kgf/mm^2	bar	long ton/in^2
From ↓			**Multiply by**			
MPa	1	10^7	1.45×10^2	0.102	10	6.48×10^{-2}
dyne/cm^2	10^{-7}	1	1.45×10^{-5}	1.02×10^{-8}	10^{-6}	6.48×10^{-9}
lb/in^2	6.89×10^{-3}	6.89×10^4	1	703×10^{-4}	6.89×10^{-2}	4.46×10^{-4}
kgf/mm^2	9.81	9.81×10^7	1.42×10^3	1	98.1	63.5×10^{-2}
bar	0.10	10^6	14.48	1.02×10^{-2}	1	6.48×10^{-3}
long ton/in^2	15.44	1.54×10^8	2.24×10^3	1.54	1.54×10^2	1

A.5 Energy and power

The SI units of energy is the joule (J), that of power is the watt (W = 1 J/sec), or multiples of them like MJ or kW. If energy and power were always listed in these units, life would be simple, but they are not. First there are Imperial units BTU, ft-lbf and ft-lbf/s. Then there are units of convenience: kWhr for electric power, hp for mechanical power. There are the units of the oil industry: barrels (7.33 barrels = 1 metric ton, 1000 kg) and toe (metric tons of oil equivalent—the weight of oil with the same energy content). Switching between these units is simply a case of multiplying by the conversion factors listed in the following tables.

Conversion of units—energy*						
To →	MJ	kWh	kcal	BTU	ft lbf	toe
From ↓				Multiply by		
MJ	1	0.278	239	948	0.738×10^6	23.8×10^{-6}
kWh	3.6	1	860	3.41×10^3	2.66×10^6	85.7×10^{-6}
kcal	4.18×10^{-3}	1.16×10^{-3}	1	3.97	3.09×10^3	99.5×10^{-9}
BTU	1.06×10^{-3}	0.293×10^{-3}	0.252	1	0.778×10^3	25.2×10^{-9}
ft lbf	1.36×10^{-6}	0.378×10^{-6}	0.324×10^{-3}	1.29×10^{-3}	1	32.4×10^{-12}
toe	41.9×10^3	11.6×10^3	10×10^6	39.7×10^6	30.8×10^9	1

*MJ = megajoules; kWh = kilowatt hour; kcal = kilocalorie; BTU = British thermal unit; ft lbf = foot-pound force; toe = metric tons oil equivalent.

Conversion of units—power*				
To →	kW (kJ/s)	kcal/s	hp	ft lbf/s
From ↓		Multiply by		
kW (kJ/s)	1	4.18	1.34	735
kcal/s	0.239	1	0.321	176
hp	0.746	3.12	1	545
ft lbf/s	1.36×10^{-3}	5.68×10^{-3}	1.82×10^{-3}	1

*kW = kilowatt; kcal/s = kilocalories per second; hp = horsepower; ft lb/s = foot-pounds/second.

A.6 Fuels

The energy content and CO_2 emission of fossil fuels are listed in Table 6.7 of Chapter 6. Here we list alternative measures of quantity for oil and gas. Coal is always quantified in metric tons or short tons.

Measures of quantity	
Crude oil	**Natural gas**
1 barrel = 35 Imperial gallons	1 billion m^3 (10^9 m^3) = 35.5×10^9 ft^3
= 42 US gallons	= 6.29×10^6 boe*
= 159 liters	= 0.9×10^6 toe**
≈135 kg	= 0.73×10^6 metric tons LNG***

*boe = barrel of oil equivalent.

**toe = metric ton of oil equivalent.

***LNG = liquid natural gas.

A.7 Energy prices (2011 data)

Energy prices fluctuate. All, ultimately, are tied to the price of oil, which in September 2011 was $112 per barrel.

Approximate energy prices, 2011 data*		
Energy form	**Price, usual units**	**Price per MJ**
Electricity, industrial	$0.09/kWh	$0.025/MJ
Electricity, domestic grid	$0.15/kWh	$0.041/MJ
Heavy fuel oil, industrial	$0.049/kWh	$0.013/MJ
Fuel oil, domestic heating	$0.093/kWh	$0.027/MJ
Natural gas, industrial	$0.024/kWh	$0.0067/MJ
Natural gas, domestic	$0.076/kWh	$0.021/MJ
Coal, industrial	$0.02/kWh	$0.0056

*Department of Energy and Climate Change (2011).

A.8 Further reading

Carbon Trust (2007), "Carbon footprint in the supply chain", www.carbontrust.co.uk. Department of Energy and Climate Change (2011), www.statistics.gov.uk/hub/business-energy/energy/energy-prices. Accessed December 2011.

Electricity Information (2008), IEA publications, ISBN 978-9264-04252-0. *(One of a series of IEA statistical publications, this one giving every statistic you ever wanted to know and plenty you don't about electricity)*

www.simetric.co.uk/sibtu.htm. Accessed December 2011. *(SI to Imperial conversion table)*

Jancovici, J-M. (2007), www.manicore.com. Accessed December 2011. *(A mine of useful information about energy)*

MacKay, D.J.C. (2008), *Sustainable energy—without the hot air*, Department of Physics, Cambridge University, Cambridge, UK. www.withouthotair.com/. Accessed December 2011. *(Helpful assembly of useful and quirky numbers relevant to energy generation and use)*

Nielsen, R. (2005), *The little green handbook*, Scribe Publications Pty Ltd, Carlton North, Victoria, Australia. ISBN 1-9207-6930-7. *(Well-researched tables of energy information)*

Shell Petroleum (2007), *How the energy industry works*, Silverstone Communications Ltd., Towchester, UK. ISBN 978-0-9555409-0-5. *(Both BP and Shell publish annual compilations of energy statisics)*

USGS (2007), Minerals Information: Mineral commodity summaries. http://minerals. usgs.gov/minerals/pubs/commodity/. Accessed December 2011. *(The bible of resource data)*

Index

A

ABS (acrylonitrile butadiene styrene), 492–493
acidification, 56–57
actual/real average power output
  defined, 398
adaptability, 437
  of human culture, 453
adobe, 5, 332
advanced recycling technologies (material efficiency), 424
Africa
  farmers losing land to investors in, 450
  water security at extreme risk in, 449
Agave sisalana, 590–591
Aggregain (2008) tool, 74
aggregated measures, 59–60, 59f, 60f
aggregate tax, 112, 116f
agreements, national, 106–112
alloying, 2
alloys, 461–489
  embodied energies of, 149–150
  material profiles, 461–489
alumina (Al$_2$O$_3$), material profile of, 534–535
aluminium
  consumption of, 419, 421f
aluminum and aluminum alloys, 2–4
  annual world consumption (production), 19f, 121t
  carbon footprint of, 150–152
  eco-attributes of, 121, 121t
  embodied energy, 149–150
  exponential growth of consumption, 30–31
  material profile, 470–471
  price trends, 439
animal-derived materials, 338–339

annual world production (consumption), 18–19, 19f, 121, 121t
anthracite, 363
aramid fibers, material profile for, 570–571
A-rated washing machine (eco-audit case study), 202–203
area intensity
  defined, 399
Argonne National Laboratory 212
asbestos, 333
Asia
  water security at extreme risk in, 449
asphalt, world consumption of, 18–19, 19f
auto bumpers
  crash barriers vs, 298–301
  eco-audit case study, 211–212
automobile selection strategies, 228–230
azurite, 2

B

bags
  paper. *See* Paper bags
  plastic. *See* Plastic grocery bags
Bakelite, 4–5
bamboos, 334–335
bar charts, 147–148, 237. *See also* Property charts
barriers (choice *vs.* purpose), 298–301
Battery Directive, 111
bill of materials
  for eco-audits, 176–177
bio-composites, 341–343
bio-derived materials, 339–344
  obstacles to use of, 343–344
biomass

as source of low-carbon power, 387–388
biomass energy, 328
Biomer, 508–509
Biopol, 508–509
biopolymers, 341
bituminous coal, 363
BMC. *See* Bulk molding compound (BMC)
Boehmeria nivea, 588–589
boiling water reactors (BWRs), 365
bonded board and linoleum, 340–341
bone, 339
borosilicate glass, material profile of, 538–539
bottled water
  carbonated-water bottles for, 278–281
  material selection case study, 276–278
boundaries of LCA systems, 55f
Boustead Model 5 (2007) tool, 72
brick
  annual world consumption (production), 19f
  embodied energy, 149–150
  fired and unfired, 332–333
  material profile, 528–529
bronze, 2
Bronze Age, 2, 3f
bubble charts, 147–148, 237. *See also* Property charts
building materials
  thermal properties, 330–331, 335
buildings
  initial and recurring embodied energy of, 283–285
  structural materials for, 281–283
buildings and construction
  energy mortgage, 122

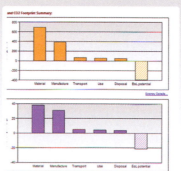
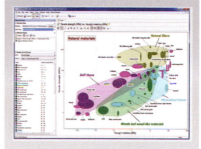